D0164549

BIOMETRIKA TABLES FOR STATISTICIANS

VOLUME I

To all those who have helped to compute numerical tables
which reduce the labour of statistical arithmetic

BIOMETRIKA TABLES FOR STATISTICIANS

VOLUME I

EDITED BY

E. S. PEARSON

AND

H. O. HARTLEY

WAGGONER LIBRARY
Trevecca Nazarene Univ
DISCARD

MACKEY LIBRARY
TREVECCA NAZARENE COLLEGE

CAMBRIDGE

Published for the Biometrika Trustees

AT THE UNIVERSITY PRESS

1970

PUBLISHED BY
THE SYNDICS OF THE CAMBRIDGE UNIVERSITY PRESS
Bentley House, 200 Euston Road, London, N.W. 1
American Branch: 32 East 57th Street, New York, N.Y. 10022
Standard Book Number 521 05920 8

First Edition 1954
Reprinted 1956
Second Edition 1958
Reprinted 1962
Third Edition 1966
Reprinted with additions 1970

First printed in Great Britain at the University Printing House, Cambridge
Reprinted in Great Britain by Lowe and Brydone (Printers) Ltd., London

LIST OF CONTENTS

References under each table heading to page numbers in the Introduction are given in three groups, (i) mathematical definition of function; (ii) method of interpolation in table; (iii) description of uses and relation to other tables.

I. TABLES OF THE NORMAL PROBABILITY FUNCTION

II. BASIC TABLES DERIVED FROM THE NORMAL FUNCTION

III. FURTHER TABLES OF PROBABILITY INTEGRALS, PERCENTAGE POINTS, ETC., OF DISTRIBUTIONS DERIVED FROM THE NORMAL FUNCTION

IV. TABLES RELATING TO CERTAIN DISCRETE DISTRIBUTIONS

V. MISCELLANEOUS TABLES

(Pearson type curves, rank correlation, orthogonal polynomials)

VI. AUXILIARY TABLES

VII. ADDENDUM TO SECOND IMPRESSION OF THIRD EDITION

PREFACE

The Founders of *Biometrika* in 1901 put forward as one of their objectives the provision of a series of numerical tables which would 'reduce the labour of statistical arithmetic'. During the next twelve years there appeared in the current volumes of the journal a considerable number of these tables, computed by many different workers associated in various ways with Karl Pearson's Biometric Laboratory. In 1914 these tables and a number of others hitherto unpublished were put together to form the first volume of *Tables for Statisticians and Biometricians.* This volume was reissued four times, between 1924 and 1948, but no alteration in its form was made.

In 1931 Karl Pearson issued a second volume of tables with the same title, again making use of many tables which had been published in *Biometrika* during the years 1914–30 and for which the type had been moulded. This volume, which ran out of print in 1945, was not reissued.

The rapid expansion of statistical theory in the decade 1925–35 with the development of new techniques made many of the older tables obsolete and called for the introduction of fresh ones. When, therefore, one of us became responsible for the management of *Biometrika* in 1936, it was soon clear that only a complete recasting of the two volumes would provide a series of tables meeting current needs. The prospect of a new issue of tables commonly inspires the table-maker to break fresh ground, and it was over this work, under the encouraging guidance of L. J. Comrie, that the present compilers were brought together.

The plan for computation of tables of the percentage points of the Beta distribution and of more extensive tables than then existed of percentage points of Student's t, of χ^2 and of the variance ratio, F, had been sketched out before *Statistical Tables for Biological, Agricultural and Medical Research* by Professor Fisher and Dr Yates appeared in 1938. But it is evident in these and other tables how much we have owed to the scheme of tabular presentation first used by Fisher in his *Statistical Methods for Research Workers.*

The onset of war in 1939 modified our plans. The issue of a new book of tables had to be postponed indefinitely, but it was decided to continue with the programme of computation as circumstances allowed and to publish each new table, when completed, in *Biometrika*, making it readily available in 'separate' form. This procedure, forced on us by war and post-war conditions, with its temptation to complete 'just one more table', has already continued too long, and we realize that the present publication of the tables in a collected series with a comprehensive Introduction is already overdue.

In compiling the tables, several considerations have determined the form adopted. In the first place it was clear, even after a fairly drastic process of weeding out, that more tables needing reissue had been published in *Biometrika* than could be contained in a single handy volume. Within the complete series, many of the tables fall into natural groupings, each dealing with a single function or set of allied functions. Within a group there may be one table in frequent use and another only required on rare occasions. One course would be to publish all tables of a group in the same volume; another, to put together in a first volume all the more commonly used tables and leave their more specialized companions to a second volume. We have adopted the latter plan, as more generally acceptable to the common user. It is for this reason that tables of percentage points of certain functions appear in the present volume while the issue of tables of the probability integral is deferred, e.g. for the mean deviation in samples from a normal population. In selecting the tables to be included, we have had valuable advice from our colleagues, in particular from Dr John Wishart and Professor M. G. Kendall. The final choice, however, has been ours, and the selection made will inevitably have been biased by our personal idiosyncrasies.

To keep the cost of production within bounds we have had to use a considerable number of stereos moulded from type of earlier tables. Thus the existing material from which we were able to draw has been largely conditioned by the scope of *Biometrika* and the work of those who have contributed to it in past years. Again, because of the need to use old material, we have felt unable in several places to introduce small improvements in lay-out and typographical detail which the specialist table-maker might expect of us. This requirement, for example, explains the retention of Sheppard's original format in Table 1 with its first and second difference columns. Partly for this reason also, we have reissued tables of several fundamental functions to the full number of decimal places originally published, although for most purposes a simpler form of table would be adequate. On the whole, we think that the majority of statisticians will be prepared to accept the slight inconvenience of additional page-turning for the sake of those few occasions when rather greater accuracy is required.

The compiler of a book of mathematical tables has always to face a somewhat difficult problem in deciding the character and scope of his Introduction. Broadly, we have tried to write our account so that within each of several connected sections the definitions and descriptions are developed in a rough logical sequence. The table-user will often only need to refer to a paged list of contents to locate the table which he knows he wants, but in so far as he does wish to see as a whole the range of tables we are providing, we have tried to write so that he can get a connected and orderly picture by consecutive reading of the Introduction.

In settling the further question of the amount of introductory matter to devote to each table, we have kept in mind a number of points: that a clear definition of the mathematical functions involved and of the links between them is essential; that, in presenting tables to be used in drawing conclusions from observational data in terms of probability, some comment on the processes of inference can hardly be avoided; that tables provided to assist in the several stages of a lengthy computational procedure, should be accompanied by rather detailed illustrative examples; that, while the uses of many of the fundamental tables are too well known to call for much comment, there are other tables concerned with methods of analysis recently introduced which require fuller explanation. On the whole, therefore, it will be found that the more special and possibly lesser known tables have been rather fully explained and illustrated, whilst essentials only are given for the well-known standard tables. Thus the allotment of introductory space is not closely related to the importance or expected usefulness of a table.

We have tried to use terminology which is generally familiar to statisticians. With notation, we have not allowed rigorous consistency to interfere with common usage, except perhaps in one point. The letter P has been used in statistical literature and tables indiscriminately to represent the integrals from the lower and upper terminals of a probability density function. We have used it solely as the integral from the lower terminal, i.e. to describe what is usually defined as the probability integral of the distribution; we have introduced $Q = 1 - P$ for the upper tail integral. Some account of the general principles we have adopted in our use of notation is given in a preliminary note (pp. xiii–xiv).

There is one general matter to which it is well to draw attention here. A large part of the techniques of modern statistical theory is based on the assumption that the underlying random variable is 'normally' distributed. The limitation may not hold rigorously in practice in the sense that the technique becomes invalid if the distribution law is different. Nevertheless, mathematically the assumption is present. Of the fifty-four tables in this volume, the main purpose of perhaps thirty lies in the analysis of data involving normally distributed variables. It would have been beyond the scope of our Introduction to attempt to indicate the extent to which different tests or methods of estimation are affected by departure from normality. Some of them are reasonably safe, others are very sensitive even in large samples. The price to be paid for the immense scope of 'normal theory' and the convenience of its derived tables is constant vigilance.

Many computers of all ranks have contributed to the construction of these tables. As workers in the early years we may mention W. F. Sheppard, W. P. Elderton, Alice Lee and H. E. Soper; in the later period, Catherine Thompson, Maxine Merrington and Joyce May. Recently some valuable help was obtained from the Mathematics Division of the National Physical Laboratory. In Appendix II, p. 101, we have indicated the origin of individual tables and the name or names of the computers. In the last stages of preparation we are indebted to Dr John Wishart, Mr P. G. Moore and Miss Joyce May for careful proof-reading which has enabled us to remove a number of obscurities and errors.

Special acknowledgement must be made for help received from other publications: to Sir Ronald Fisher, Dr Frank Yates and Messrs Oliver and Boyd Ltd.; to Dr D. van der Reyden and the Editors of the *Onderstepoort Journal of Veterinary Science and Animal Industry*; to Professor M. G. Kendall and Messrs Charles Griffin and Co. Ltd.; to Professor Truman Kelley and the Harvard University Press; to Messrs W. & R. Chambers Ltd. Details are given in Appendix II against the tables concerned.

Throughout our work we have been very conscious of how much the user of the statistical table owes to Karl Pearson. Many of the tables whose computation he planned have now been outdated or have required considerable remodelling; but much of his work remains, and we believe that this new series of revised tables is conceived in the tradition which he set. It should be recalled also that we have made use in many places of those two fundamental tables of the incomplete gamma- and beta-functions whose importance he had the foresight to realize and whose computation and publication he carried through against many odds.

Finally, we can repeat what Pearson wrote in 1914, in the Preface to his first volume of *Tables for Statisticians and Biometricians*: 'To those who have had experience of numerical tables prepared elsewhere, the excellence of the Cambridge first proof of columns of figures is a joy, which deserves the fullest acknowledgement.' We have indeed been fortunate in having still at hand the services of the Cambridge University Press.

Department of Statistics
University College, London
October 1953

E. S. PEARSON
H. O. HARTLEY

PREFACE TO SECOND EDITION

The opportunity of this re-issue has been taken to make a few minor alterations in the text of the Introduction and in the Tables. Any more extensive recasting must await the production of Volume II.

In one or two places the statements regarding accuracy of interpolation have been found to be too optimistic. Thus alterations have been made on pp. 14–15 where the $\frac{1}{2}\phi^2\delta^2$ term has now been included in equations (24)–(27). Again, the statement in paragraph (*a*) on p. 23 regarding interpolation in the tables of the *t*-distribution has needed modification. At the top of p. 27, there has been some rewording of § 6·1 which it is hoped will reduce the risk of misinterpretation of the meaning of Table 11. H. A. David's corrected values for the upper percentage points of the extreme studentized deviate from the sample mean have been inserted in Table 26, to which his columns for $n = 12$ have been added.

Finally, an index to the Tables (but not to the Introduction) has been added on pp. 239–40. We are indebted to Mrs Maxine Merrington for help in preparing this index. When the reader has identified the Table he wishes to use, the relevant pages in the Introduction can be found from the List of Contents on pp. v–vii.

E.S.P.
April 1958 H.O.H.

PREFACE TO THIRD EDITION

This new edition contains a number of alterations and additions which fall under the following main headings: (*a*) removal of errors detected in certain of the original tables; (*b*) extension of some of these tables; (*c*) inclusion of seven new tables. It is desirable to enlarge on these changes in a little detail.

(*a*) *Removal of errors.* The original volume contained a number of tables computed during the 1940s and first published in *Biometrika.* With the computation by hand-machine then used, we did not consider it justifiable to proceed to the large number of decimal places needed to ensure that there were no cut-off errors in the last figure tabled. It was therefore realized that our results would contain a number of entries out by one unit in the last place. With the aid of some extensive calculations on electronic computers recently carried out in the United States, we hope that we have been able to remove the majority of errors, particularly in Tables 8, 16, 20, 22 and 23. We are most grateful to Dr H. L. Harter and his colleagues, to Dr D. E. Amos and to Professor D. B. Owen for help in this connexion.

(*b*) *Extension of existing tables.* We have added to Table 11 two pages giving 2·5 % and 0·5 % critical levels of $v = (y-\eta)/\sqrt{(\lambda_1 s_1^2 + \lambda_2 s_2^2)}$ already published by W. H. Trickett, B. L. Welch & G. S. James. The table of the percentage points of the B-distribution (Table 16) has been extended to include the 0·25 % and 0·1 % points calculated by D. E. Amos. Table 26 for the percentage points of the extreme studentized deviate, $(x_n - \bar{x})/s_\nu$, has been recast; it now includes additional upper percentage points computed by H. A. David and a few values for degrees of freedom $\nu < 10$ computed by K. C. S. Pillai. The very slowly changing values of the lower percentage points have been omitted, but some remarks on how to determine these are included in §13 of the Introduction.

Table 29, giving percentage points of the studentized range $(x_n - x_1)/s_\nu$, has also been recast, now giving 10 % as well as more accurate 5 % and 1 % values; here again the less important lower percentage points have had to be omitted for reasons of space. The new table is based on tables published by J. Pachares and by H. L. Harter, D. S. Clemm & E. H. Guthrie.

Finally, it has been possible to add some new values for $50 \leqslant n < 200$ to Table 34C, giving percentage points of the distribution of $b_2 = m_4/m_2^2$ used in tests of normality.

(*c*) *New tables.* We have decided to take this opportunity of incorporating some additional tables in the present Volume I. All of these have already been published wholly or in part elsewhere. We hope that they will be of value to the working statistician in carrying out a variety of 'quick' tests which are often useful in a preliminary survey of data. As such, they seemed to us more appropriately included here than in the long promised Volume II of these *Tables.*

To avoid altering the present series of table numbers, the new tables have been given letters *a*, *b*, etc. attached to the number of the existing table which they seem to follow most appropriately.

Two of these new tables, *Tables* 26*a and b*, computed by C. P. Quesenberry & H. A. David, extend the use of the table of percentage points of the extreme studentized deviate $(x_n - \bar{x})/s_\nu$ by enabling the 'within sample of *n*' sum of squares to be added to the independent sum of squares νs_ν^2 in estimating the denominator standard deviation. As a special case, putting $\nu = 0$, the table provides percentage points of the ratio of the extreme deviate from the sample mean to the standard deviation in the same sample.

The next group of three tables are for use with some further tests involving the sample range. Table 29*a*, due to P. G. Moore, gives percentage points of the two-sample ratio $|\bar{x}_1 - \bar{x}_2|/(w_1 + w_2)$ and provides a test based on range analogous to Student's two-sample *t*-test. Table 29*b*, first published by H. L. Harter, contains percentage points of the ratio of two independent ranges, w_1/w_2, providing an analogue to the variance-ratio *F*-test. Table 29*c* gives percentage points of the ratio of range to standard deviation calculated from the same sample; this ratio provides a quick test which can be useful in detecting abnormality in sample pattern and, possibly, errors in computation.

Finally Tables 31*a* and *b* fit naturally after Table 31, which gives percentage points of the ratio of the maximum to the minimum variance in a set of k independent variance estimates s_t^2 $(t = 1, \ldots, k)$ each based on ν degrees of freedom. Table 31*a* gives corresponding points for the ratio $s_{\text{max.}}^2/\Sigma s_t^2$; the test involved was first put forward by W. G. Cochran and more fully tabled by C. Eisenhart, M. W. Hastay & W. A. Wallis. Table 31*b* is concerned with an analogous test using sample ranges, viz. involving the ratio $w_{\text{max.}}/\Sigma w_t$. The upper 5 % points tabled here are due to C. I. Bliss, W. G. Cochran & J. W. Tukey.

The incorporation of these new tables has inevitably involved some modification to our Introduction. This has been met by the enlargement of §§13 and 14 by the addition of six new pages.

We should like to thank most warmly the many statisticians who have made possible this expansion and we hope improvement in the present volume. If we have failed to mention any by name, we must ask their forgiveness. Full references to sources of the new material are included in Appendices I and II. Finally, we are grateful to Miss S. M. Burrough of the Mathematics Department, University College London for her careful work in computing and checking some of our figures.

E.S.P.
H.O.H.

April 1965

COMMENTS ON NOTATION

1. The following notes indicate certain general forms of notation we have used in connexion with measures of probability and the broad principles adopted in assigning letters and symbols to statistical quantities.

2. If x is a random variable distributed continuously in the interval (a_1, a_2), $f(x)$ has been written for its *probability density function*, often described as its *frequency function*. We have written* what is commonly known as the *probability integral* in the form

$$P(x) = \int_{a_1}^{x} f(u)\,du.$$

Further, we define
$$Q(x) = 1 - P(x) = \int_{x}^{a_2} f(u)\,du.$$

Where there is little chance of confusion and particularly as a column heading in a table, P or Q have been put in place of $P(x)$ or $Q(x)$.

3. When it is desirable to indicate that a frequency function or probability integral depends on the values of one or more *parameters*, say on θ_1, θ_2, whose values are changing throughout a table, we write $f(x \mid \theta_1, \theta_2)$, $P(x \mid \theta_1, \theta_2)$ or $Q(x \mid \theta_1, \theta_2)$. For example, in the case of a single parameter, the probability integral of Student's t based on ν degrees of freedom is written

$$P(t \mid \nu) = \int_{-\infty}^{t} f(u \mid \nu)\,du = \Gamma(\tfrac{1}{2}(\nu+1))\,(\Gamma(\tfrac{1}{2}\nu))^{-1}\,(\nu\pi)^{-\frac{1}{2}} \int_{-\infty}^{t} \left(1 + \frac{u^2}{\nu}\right)^{-\frac{1}{2}(\nu+1)} du.$$

4. Sometimes it has been convenient to express certain probability statements involving inequalities in a shortened form; for example, on p. 20 the expression

$$\Pr\{|\,t\,| > 1{\cdot}97 \mid \nu = 11\} = 0{\cdot}074$$

is written in place of the long sentence: the probability that the numerical value of t is greater than 1·97 (when the degrees of freedom of the variance estimate are 11) equals 0·074.

5. If a random variable can assume only *integral values*, e.g. $i = 0, 1, 2, \ldots$, $f(c)$ has been used to denote the probability that $i = c$. The dependence of this probability on the values of one or more parameters has been again indicated by writing $f(i \mid \theta)$, $f(i \mid \theta_1, \theta_2)$, etc. The *cumulative sum* of the terms of such a distribution has been indicated by a summation sign; for example, the sum of the first $c+1$ terms of the binomial series $(q+p)^n$, may be written

$$\sum_{i=0}^{c} f(i \mid n, p) = \sum_{i=0}^{c} \binom{n}{i} p^i (1-p)^{n-i}.$$

6. In dealing with tests of significance, according to the problem considered, the risk of rejecting the hypothesis tested when it is true† may be obtained from an integral calculated from the lower tail of a sampling distribution (P), from the upper tail (Q), or by summing integrals from both tails. To avoid confusion, a separate notation is required to denote this risk or *level of significance*; we have used the letter α, which in practice will assume values such as

* With the exception of the incomplete B-function integral $I_x(a, b)$; see p. 32.

† I.e. Neyman and Pearson's 'first kind of error'.

0·05, 0·01, etc. (or 100α, such values as 5%, 1%, etc.). Thus the term *significance level* is reserved for a probability whilst we denote by a *percentage point* the critical value of the statistic associated with this level. Moreover, α has also been used in connexion with the determination of *confidence intervals* in the problem of estimation.

7. The symbol $\mathscr{E}$ has been used to indicate the *expectation* of a random variable. Thus if x has a continuous frequency function $f(x)$ in the interval (a_1, a_2),

$$\mathscr{E}(x) = \int_{a_1}^{a_2} x f(x)\, dx.$$

8. Complete consistency in the use of statistical notation seems unattainable unless certain long-established usages are discarded. In general, Greek letters have been used for population parameters for which the corresponding Roman letters denote the sample estimators; thus we have ρ and r, σ^2 and s^2, β and b, etc. However, we have not hesitated to use χ^2 as a random variable nor $p = 1 - q$ for the probability of an event in binomial sampling. In this respect we have had to regard the Introduction and tables as a whole, which means that in some sections an obvious lettering cannot be used, since confusion would arise in following the same notation elsewhere. For example, having adopted p and q in connexion with the binomial, we cannot use the standard notation $I_x(p, q)$ for the incomplete B-function, as this provides the cumulative sum of binomial terms in a form in which the two p's would have different meanings. The function has therefore been taken as $I_x(a, b)$. On the other hand, commonly adopted notation has been retained where the same symbol is used to denote quantities unlikely to be confused. For example, w denotes the range in normal samples as well as the weight, $Z^2/(PQ)$, in probit analysis, while W is used both for the standardized range and the concordance coefficient.

9. The main details of notation are given most appropriately as they arise in the Introduction. It is sufficient to note here that with n, n_1, n_2, etc., reserved for sample sizes, group frequencies, etc., we have adopted the common practice of using ν, ν_1, ν_2, etc., for degrees of freedom; that s^2 is always used as an unbiased mean-square estimator of a variance σ^2, i.e. $\mathscr{E}(s^2) = \sigma^2$; and that we have retained Karl Pearson's β_1, β_2 notation for the well-known moment ratios, rather than introduce $\gamma_1 = \sqrt{\beta_1}$ and $\gamma_2 = \beta_2 - 3$, because we wish to make use of his classification of distributions into frequency curve types.

INTRODUCTION

I. TABLES OF THE NORMAL PROBABILITY FUNCTION

1. The standard normal tables (Tables 1–5)

1·1 *Definitions and notation*

The equation of the normal or Gaussian frequency function may be written in the standardized form

$$Z(X) = f(X) = \frac{1}{\sqrt{(2\pi)}} e^{-\frac{1}{2}X^2} \tag{1}$$

Regarded as a random variable, X has a mean or expectation of zero and a unit standard deviation. The probability integral is then

$$P(X) = \int_{-\infty}^{X} \frac{1}{\sqrt{(2\pi)}} e^{-\frac{1}{2}u^2} du \quad \text{while} \quad Q(X) = 1 - P(X). \tag{2}$$

If x is a random variable, normally distributed about a mean μ and with standard deviation σ (variance σ^2), the standard tables may be used by setting

$$X = (x - \mu)/\sigma \tag{3}$$

and remembering that the ordinate of the distribution curve for x is Z/σ. Fig. 1 presents the relationship between X, Z, $P(X)$ and $Q(X)$. The change in notation from W. F. Sheppard's (1902) original x, z and $\frac{1}{2}(1+\alpha)$, respectively, will be noted. From the symmetry we see that

$$Z(-X) = Z(X), \quad P(-X) = 1 - P(X) = Q(X).$$

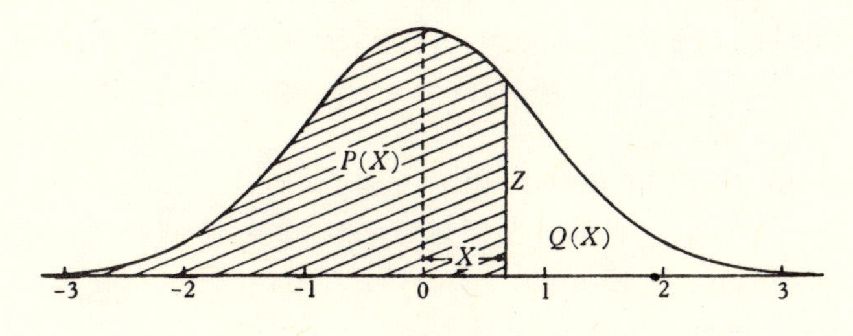

Fig. 1.

Table 1 is the fundamental table of the normal integral $P(X)$ and the ordinate $Z(X)$ due to W. F. Sheppard. Table 2 provides an extension for the tail area of the distribution in logarithmic form, i.e. it gives $-\log_{10}\{1 - P(X)\}$. Tables 3 and 4 give the normal deviates $X(P)$, and Table 5 the ordinates $Z(P)$ in terms of P and Q. Table 6, which introduces the additional notation, $Y = X + 5$, is primarily designed to facilitate probit analysis; its use is described in §§ 2·1 and 2·2 below.

1·2 *Applications of Tables* 1–5

The applications are so numerous and well known that we shall confine attention to illustration on a single series of data.

Example 1. The following results were given by Latter (1906) for the distribution of the length of 717 eggs of the cuckoo (*Cuculus canorus*) found in nests in the British Isles:

Length (central values in mm.)	19·25	20·25	21·25	22·25	23·25	24·25	25·25	26·25
Frequency	3	22	123	300	201	61	6	1

Graduate this distribution by a normal curve.

The values of the mean and standard deviation, using Sheppard's correction for grouping in calculating the latter, are found to be:

$$\text{mean} = 22{\cdot}48570\,\text{mm.}, \quad \text{standard deviation} = 0{\cdot}954631\,\text{mm.}^*$$

Taking these as estimates of μ and σ, the work required is set out in the following table:

Group boundary x-values (1)	$x-\mu$ (2)	$X=(x-\mu)/\sigma$ (3)	$P(X)$ (4)	$\Delta P(X)$ (5)	$717\times\Delta P(X)$ (6)	Observed frequencies (7)
18·75	−3·73570	−3·91324	0·00005			
				0·00203	1·5	3
19·75	−2·73570	−2·86571	·00208			
				·03244	23·3	22
20·75	−1·73570	−1·81819	·03452			
				·18593	133·3	123
21·75	−0·73570	−0·77066	·22045			
				·38861	278·6	300
22·75	0·26430	0·27686	·60906			
				·29825	213·8	201
23·75	1·26430	1·32439	·90731			
				·08384	60·1	61
24·75	2·26430	2·37191	·99115			
				·00854	6·1	6
25·75	3·26430	3·41944	·99969			
				·00031	0·2	1
26·75	4·26430	4·46696	1·00000			
				Total	716·9	717

The values of $P(X)$ in col. 4 are obtained from Table 1, using linear interpolation. Col. 6 shows the graduated frequencies which may be compared with the observed frequencies in col. 7. The agreement is satisfactory,† although there is a suggestion that the egg-length distribution has a rather more peaked top than the normal curve.

Two further questions might be asked which will illustrate the use of additional tables in this series.‡ Suppose that it was wished to classify cuckoos' eggs by length into three groups described as long, medium and short, in such a way that roughly $\frac{1}{6}$ of the eggs might be expected to fall into each of the extreme groups and $\frac{2}{3}$ into the medium group:

(*a*) Where should the divisions on the length scale be made?

(*b*) What would be the average lengths of the eggs in the extreme categories? (The expected value of the average of the medium group would clearly be μ.)

* In order to provide a detailed numerical illustration of the use of Table 1, rather more figures have been given in the values of mean and standard deviation than would ordinarily be retained. Thus to ensure that the figures in col. 6 of the table above were not in error by more than one unit in the last place, 4-decimal accuracy in mean and standard deviation would probably have been sufficient.

† Pooling together two groups at each tail and applying the usual test for goodness of fit (§§ 3·2 and 3·4 below), it is found that $\chi^2 = 3{\cdot}30$ with $6-3=3$ degrees of freedom, so that departure from normality cannot be established as significant by this test.

‡ We shall assume that egg-length in the population is normally distributed with the observed values of mean and standard deviation.

To answer (*a*) we require the values of $x=\mu+\sigma X$ for which $P(X)=0{\cdot}16\dot{6}$ and $0{\cdot}83\dot{3}$. Turning to Table 4 and using linear interpolation we find that these values of X are

$$\mp(0{\cdot}9701-\tfrac{2}{3}\,0{\cdot}0040)=\mp 0{\cdot}9674.$$

Hence, using the sample estimates of μ and σ given above, the divisions for x will lie at

$$22{\cdot}4857\mp 0{\cdot}9674\times 0{\cdot}9546=21{\cdot}56 \text{ and } 23{\cdot}41\text{ mm}.$$

To answer (*b*), we make use of the following result. If a truncated section of the standardized normal distribution is taken, bounded by ordinates at X_i and X_j, then the distance of the mean of this section from the origin, $X=0$, is

$$\{Z(X_i)-Z(X_j)\}/\{P(X_j)-P(X_i)\}.$$

For the category of 'short' eggs, $X_i=-\infty$, $Z(X_i)=0$, $P(X_i)=0$ and $P(X_j)=\frac{1}{6}$. The value of $Z(X_j)$ may be found by entering Table 5 with $P(X_j)=0{\cdot}1667$. Thus, using linear interpolation, we find $Z=(0{\cdot}24921+\frac{2}{3}\,0{\cdot}00096)=0{\cdot}24985$ and, from the expression given above, the average value of X for the lower $\frac{1}{6}$ tail area of the normal distribution is

$$-0{\cdot}24985\div\tfrac{1}{6}=-1{\cdot}4991.$$

Hence, in terms of x, the average lengths of eggs in the 'short' and 'long' categories may be expected to be $\quad 22{\cdot}4857\mp 1{\cdot}4991\times 0{\cdot}9546=21{\cdot}05$ and $23{\cdot}92$ mm.

1·3 *Note on the use of significance points*

Table 4 may be used to determine specified *significance points* for a normally distributed variable, i.e. to find the values of $X=(x-\mu)/\sigma$, the probability of exceeding which has certain conventional values, such as 0·10, 0·05, 0·01 or 0·001. It should be noted that according to the nature of his problem the statistician may be concerned with

(*a*) a *single-tail* test, e.g. he asks whether an observed value of x exceeds μ by an exceptional amount;

(*b*) a *double-tail* test, e.g. he asks whether the magnitude of the difference $|x-\mu|$ is exceptional.

This distinction is discussed more fully in connexion with Student's *t*-distribution in § 5·2 below.

1·4 *Interpolation in Tables* 1–5

To an accuracy of 5 decimals, which is usually adequate, linear interpolation can be used in Table 1 for both $P(X)$ and $Z(X)$. For example, to find the value of $P(X)$ for $X=0{\cdot}41667$ we have (to 5-decimal accuracy)

$$P(0{\cdot}41667)=0{\cdot}659097+0{\cdot}003660\ (0{\cdot}667)$$
$$=0{\cdot}66154.$$

In general, if X_0 is the tabular argument next to and below the given X, $\theta=100(X-X_0)$, the fraction of interpolation (since the tabular interval is 0·01), and $\delta_{\frac{1}{2}}=P(X_0+0{\cdot}01)-P(X_0)$, then

$$P(X)\simeq P(X_0)+\theta\times\delta_{\frac{1}{2}}.$$

Alternatively, we may use the Lagrangian formula

$$P(X)=(1-\theta)\,P(X_0)+\theta P(X_0+0{\cdot}01).$$

The first equation is usually preferable when working with a machine, as the settings of $P(X_0)$ and $\delta_{\frac{1}{2}}$ can first be checked by forming $P(X_0)+\delta_{\frac{1}{2}}$ and comparing with the tabled $P(X_0+0{\cdot}01)$.

In this connexion it should be noted that the values of δ and δ^2 do not always agree in the last figure with the differences of the corresponding tabulated values of P (or Z). These discrepancies have arisen because W. F. Sheppard's original calculations were made to two or three more places; the figures were then cut down to 7 decimals, the differences as well as the values of the function being rounded off separately so as to be as near their true values as possible (see Sheppard (1902), pp. 177, 180).

If 7-decimal accuracy is required, the further correction term $\frac{1}{4}\theta(\theta-1)(\delta_0^2+\delta_1^2)$ should be added, where δ_0^2 and δ_1^2 are the central second differences tabulated in line with $P(X_0)$ and $P(X_0+0{\cdot}01)$ and $\frac{1}{4}\theta(\theta-1)=B''(\theta)$ is the second Bessel coefficient for which tables are available (see, for example, Comrie, 1936). The signs of the δ^2 at the top of each column should be noted. Since $B''(\theta)<0$, the correction term will be of opposite sign to that of the δ^2. If $|\delta^2|<4$, no correction term is needed, and if $|\delta^2|<40$ no correction is needed to 6-decimal accuracy.

The rules for interpolating Z are identical.

Interpolation in Table 2 will not normally be required, since usually only the order of magnitude of the extreme tail area is needed, but interpolates accurate to about 4 decimals can be obtained up to $X=90$ with the 4-point Lagrangian formula

$$f_\theta = L_{-1}(\theta)\,f_{-1} + L_0(\theta)\,f_0 + L_1(\theta)\,f_1 + L_2(\theta)\,f_2,$$

where the Lagrangian multipliers $L(\theta)$ are tabulated, for example, by Comrie (1936).

The same type of formula may be used in Tables 3, 4 and 5, although here linear interpolation is adequate over most of the range; it is certainly sufficient for 3-decimal accuracy except in the top left-hand corner of each table where any interpolation will break down. In these latter cases, inverse interpolation in Tables 1 and 2 must be used.

Illustration. Find $X(0{\cdot}999356)$.

By linear interpolation in Table 3

$$0{\cdot}44(3{\cdot}1947)+0{\cdot}56(3{\cdot}2389)=3{\cdot}2195.$$

By four-point Lagrangian interpolation

$$-0{\cdot}0591(3{\cdot}1559)+0{\cdot}4942(3{\cdot}1947)+0{\cdot}6290(3{\cdot}2389)-0{\cdot}0641(3{\cdot}2905)=3{\cdot}2187.$$

The correct value is 3·21864.

2. Probit analysis (Table 6)*

2·1 *Theory and definitions*

The method of probit analysis is applied in situations where the individuals in a population require a varying 'stimulus' to produce a characteristic quantal 'response' and where our objective is to estimate the parameters of this distribution of critical stimuli. Such problems arise, for example, in dosage-mortality studies, in biological assay and in tests on the detonation of explosives. An ample description of the theory and its uses has been given in a number of publications (see, for example, Bliss, 1952; Finney, 1952), but the following summary may be helpful to users of the present table.

It is assumed that the stimulus, x, is measured on a scale for which individual tolerances are normally distributed; often, for example, the logarithm of the stimulating dose rather than

* This table was computed by D. J. Finney & W. L. Stevens (1948), and we are indebted to these authors for permission to reproduce. For convenience in use, abbreviated values of P have been added as a first column to their original table. These P-values are only given for $Y \geqslant 5{\cdot}00$. For the values of $Y<5{\cdot}00$ given in the last column, P must be obtained by subtracting from unity the corresponding figure in the first column. E.g. for $Y=4{\cdot}86$, $P=1-0{\cdot}556=0{\cdot}444$. For higher accuracy, Tables 1, 3 and 4 may be consulted.

the dose itself is used as a variable in order to satisfy this condition more closely. The relation between the stimulus x and the probability P that an individual selected at random will respond to it, is then given by equation (2), where $X=(x-\mu)/\sigma$. In order to avoid the necessity of computing with negative numbers it is customary to use the *probit*

$$Y=X+5 \tag{4}$$

in place of the so-called *normal equivalent deviate* X. Thus

$$P=\frac{1}{\sqrt{(2\pi)}}\int_{-\infty}^{Y-5} e^{-\frac{1}{2}u^2}\,du, \tag{5}$$

$$Q=1-P=\frac{1}{\sqrt{(2\pi)}}\int_{Y-5}^{\infty} e^{-\frac{1}{2}u^2}\,du. \tag{6}$$

If now out of n_t individuals who are given the stimulus x_t $(t=1,2,\ldots,k)$, r_t respond, we may write

$$p_t=1-q_t=r_t/n_t, \tag{7}$$

and the corresponding probits y_t will be the values of Y obtained by inserting p_t for P in the relation (5). The values y_t, when plotted as ordinates against the abscissae x_t, should then only differ through sampling fluctuations in the observed r_t from corresponding points on the population line

$$Y=5+(x-\mu)/\sigma. \tag{8}$$

The statistical problem is essentially that of estimating the parameters μ and σ in (8), where the observed y_t have unequal weights. This we do by fitting the regression line

$$y=a+bx.$$

Writing $n_t w_t$ for the weight of y_t and omitting the subscript t for convenience, standard regression theory provides the solution

$$\left.\begin{aligned} b &= \text{estimate of } \frac{1}{\sigma}=\frac{\Sigma nw(x-\bar{x})(y-\bar{y})}{\Sigma nw(x-\bar{x})^2}=\frac{\Sigma nw\,\Sigma nwxy-\Sigma nwx\,\Sigma nwy}{\Sigma nw\,\Sigma nwx^2-(\Sigma nwx)^2}, \\ a &= \text{estimate of } (5-\mu/\sigma)=\bar{y}-b\bar{x}, \end{aligned}\right\} \tag{9}$$

where $\bar{x}$ and $\bar{y}$ are the weighted means, or

$$\bar{x}=\Sigma nwx/\Sigma nw, \quad \bar{y}=\Sigma nwy/\Sigma nw. \tag{10}$$

The maximum-likelihood method of estimation leads to equations (9) and (10) in which the probits y_t are replaced by the *working probits* defined below, and these are used in conjunction with weights w_t, given by*

$$w_t=Z_t^2/(P_tQ_t), \tag{11}$$

P, Q and Z being as defined in equations (1) and (2) above. Since the working probits and weights can only be estimated from the data, the solution must follow an iterative process; this may be carried out as follows:

(*a*) From the observed p_t, defined in (7) above, find y_t, using the first two columns of Table 6, and make a rough plot of y_t against x_t.

(*b*) As a first approximation, a straight line fitted by eye to the points (x_t, y_t) may be used to obtain a set of provisional or *expected probits* Y'_t corresponding to x_t.

(*c*) The next step is to obtain a series of *working probits*, y'_t, and their weighting coefficients w'_t to use in relations (9) and so obtain a second approximation to the regression line. These working probits are obtained from either

$$y=Y+Q/Z-q/Z \tag{12}$$

or

$$y=Y-P/Z+p/Z, \tag{13}$$

* It may be noted that the sampling variance of y_t is approximately $P_tQ_t/(n_tZ_t^2)$.

as most convenient, where we find

$$\begin{aligned} &\text{the } \textit{range} && = 1/Z \\ \text{and} \quad &\text{the } \textit{maximum working probit}, Y_{\text{max.}} && = Y + Q/Z \\ \text{or} \quad &\text{the } \textit{minimum working probit}, Y_{\text{min.}} && = Y - P/Z, \end{aligned}$$

by entering Table 6 with the expected probits Y'_t, q or p being as before the observed proportion of individuals failing to respond (q) or responding (p). It will be noticed that under this procedure, if there is a zero or 100 % response ($p=0$ or 1), the solution allows us to use this result in the form of working probits of $Y-P/Z$ and $Y+Q/Z$.

(d) The appropriate weights $n_t w'_t$ based on (11) are also found, again entering Table 6 with Y'_t.

(e) We then use these weights and the working probits y'_t to compute the coefficients a and b of equations (9).

(f) The values of the probits Y''_t obtained from this second approximation to the regression line may be used to repeat the cycle of the iterative process, Table 6 being entered to obtain fresh working probits y''_t and weights $n_t w''_t$. The iteration is continued until the process has 'settled down'.

It will be noticed that the table gives

$$Y_{\text{max.}} \text{ for } Y = 3{\cdot}58\,(0{\cdot}01)\,9{\cdot}00, \qquad Y_{\text{min.}} \text{ for } Y = 1{\cdot}00\,(0{\cdot}01)\,6{\cdot}42,$$
$$1/Z \text{ and } w \text{ for } Y = 1{\cdot}00\,(0{\cdot}01)\,9{\cdot}00.$$

Below $Y=3{\cdot}58$, $Y_{\text{max.}}$ exceeds 10·00, and above $Y=6{\cdot}42$, $Y_{\text{min.}}$ is negative; it is then almost always more convenient to calculate the working probits from the other function, but the function not tabulated can easily be obtained from the relationship

$$Y_{\text{max.}} - Y_{\text{min.}} = 1/Z.$$

2·2 *Illustration of application of Table* 6

Example 2. As part of a large investigation into the nature of the fall-off with distance in the blast effect of explosive charges, an experiment was conducted as follows. A given weight of charge was detonated at a distance x from a standard disk of cardboard, and a record made of whether the disk was perforated or not. This was done 16 times at each of the five distances shown in the first column of the table on p. 8 below. This table shows in the second column the working-scale values, $x_t = \frac{1}{4}(53 - \text{distance})$, and in the third column the corresponding number of perforations, r_t. Throughout, $n_t = 16$.

Clearly there is no unique critical distance, such that no cards are perforated beyond this distance and all within it. If, however, we make the assumption that for a given weight of charge the critical distance in successive experiments is approximately normally distributed about a mean μ with standard deviation σ, we may use the technique described in the preceding section to estimate the values of these parameters. The investigation could then be extended to examine the effect on μ and σ of changes in the charge-weight or type of explosive.

The first section of the table on p. 8 gives, besides x_t and r_t, $p_t = \frac{1}{16} r_t$ and y_t, the last values obtained following the procedure described in para. (a) of p. 5. The three available values of y_t were then plotted against x_t and a rough line fitted to the points by eye as shown in Fig. 2;*

* It is known from experience that when the data include a 'no response' and/or an 'all response' record, one will usually do well to start with a line which is somewhat steeper than the 'eye estimate' based on the remaining points. This practice was not followed here so that the convergence is rather slow.

from this line the five values of the *expected probits* Y'_t corresponding to $x_t = 0, 1, 2, 3, 4$ were read off (para. (*b*) of p. 5).

We are now ready to obtain from Table 6 the quantities required for fitting the second approximation to the regression line (regarding the line fitted by eye as the first approximation). The number of decimals given in Table 6 makes provision for the rare occasions when high precision is required. In the present example, as in most practical situations, one decimal in Y_t and two in y_t would be adequate. The calculations, however, are here carried to a higher accuracy to illustrate both the interpolation in the table and the convergence of the process, which is taken beyond the accuracy warranted by the standard errors of the estimates.

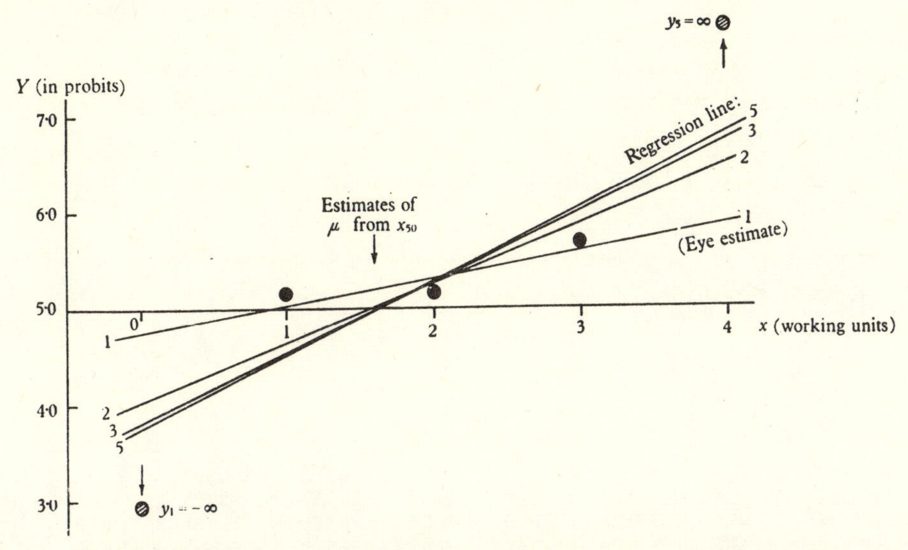

Fig. 2. Successive approximations to probit regression line (Example 2).

Entering Table 6 with Y'_t, we obtain the corresponding values of the minimum working probit ($Y'_{\text{min.}}$), the range ($1/Z'$) and then computing values of y'_t (see equation (13)) from

$$y'_t = Y'_{\text{min.}} + p_t \times 1/Z'$$

directly on a machine, we enter the answers in the next column of the table, as shown. Next, re-entering Table 6 at argument Y'_t, we obtain the weighting coefficients w'_t; these are copied down, as well as set on the machine to compute the products $w'_t x_t$ shown in the next column.*

We now calculate the constants in the second approximation to the regression line, noting that a constant multiplier, $n_t = 16$, may be omitted from all numerators and denominators. Equations (10) and (9) for purposes of computing become

$$\bar{x} = \Sigma w'_t x_t / \Sigma w'_t, \quad \bar{y} = \Sigma(w'_t)(y'_t)/\Sigma w'_t,$$

$$b = \frac{[\Sigma w'_t][\Sigma(w'_t x_t)(y'_t)] - [\Sigma w'_t x_t][\Sigma(w'_t)(y'_t)]}{[\Sigma w'_t][\Sigma(w'_t x_t)(x_t)] - [\Sigma w'_t x_t]^2}.$$

These formulae are arranged to show that the following operations must be performed on the appropriate columns of the table below:

Totals for	$w'_t x_t$,	w'_t
Sums of products for	w'_t by y'_t,	$w'_t x_t$ by y'_t and x_t

* If the n_t are not constant (as they are in the present example), the products $n_t w'_t$ and $n_t w'_t x_t$ must be entered instead.

The resulting values of the coefficients, $b=0{\cdot}619$ and $a=\bar{y}-b\bar{x}=4{\cdot}017$, are shown at the bottom of the table.

We have now to continue the iterative process. From the equation

$$y=4{\cdot}017+0{\cdot}619x,$$

we obtain the five probit values Y_t'' (at $x_t=0, 1, 2, 3, 4$) shown in the last column of the upper section of the table. These in turn, when used in Table 6, provide fresh values of working probits y_t and fresh weights, leading to the calculation of new values for a and b.

This process was carried out four times; we have recorded only the values of a and b, and of the final probits Y_t^{v} calculated from the fifth approximation to the probit line

$$y=3{\cdot}7747+0{\cdot}7568x.$$

The sixth approximation is shown merely to confirm the convergence.

To avoid accumulation of rounding-off errors more figures have sometimes been retained than are warranted by the accuracy of the quantities on which they are based. Y_t' and Y_t'' have been used to 2 decimals, Y_t''' to Y_t^{v} to 3 decimals, and the resulting values of y_t, $w_t x_t$, b and a have been taken to one more decimal.

x_t		r_t	p_t	y_t	Y_t'	y'	w_t'	$w_t' x_t$	Y_t''
ft.	Working scale								
53	0	0	0	$-\infty$	4·75	3·712	0·6223	0	4·02
49	1	9	0·5625	5·16	5·02	5·157	0·6365	0·636	4·64
45	2	9	0·5625	5·16	5·30	5·155	0·6161	1·232	5·26
41	3	12	0·7500	5·67	5·57	5·671	0·5652	1·696	5·87
37	4	16	1	∞	5·85	6·561	0·4873	1·949	6·49

x_t	Y_t^{v}	y_t^{v}	w_t^{v}	$w_t^{\mathrm{v}} x_t$	P_t	$16P_t$	r_t
0	3·775	3·1896	0·3617	0	0·1103	1·76	0
1	4·532	5·2105	0·5876	0·5876	0·3199	5·12	9
2	5·288	5·1552	0·6177	1·2354	0·6133	9·81	9
3	6·045	5·6036	0·4235	1·2705	0·8520	13·63	12
4	6·802	7·2567	0·1794	0·7176	0·9642	15·43	16

Successive approximations:

	a	b	$x_{50}=(5-a)/b$	
			Working scale	ft.
Second	4·017	0·619	1·59	46·6
Third	3·820	0·731	1·61	46·6
Fourth	3·7791	0·7546	1·618	46·53
Fifth	3·7747	0·7568	1·619	46·52
[Sixth	3·7744	0·7569	1·619	46·52]

Est. S.E. $=0{\cdot}90$.

We may also use the solution to obtain expected frequencies of perforations at the five distances. Either from the first two columns of Table 6 or through equation (5) and Table 1, the final approximations Y_t^{v} provide estimates of P_t and hence of the expectations $16P_t$. These are given in the second section of the table, where they are compared with the observed r_t.

If none of the expected frequencies n_tP_t or n_tQ_t is very small, we may calculate

$$\chi^2 = \sum_t (r_t - n_tP_t)^2/(n_tP_tQ_t),$$

and so obtain a test of the significance of discrepancies between observation and theory. The degrees of freedom will be two less than the number of levels of x used in the experiment, since the two parameters, μ and σ, have been estimated from the data.* Finney (1952, pp. 52–4) describes an alternative method of computing χ^2 as the weighted sum of squares of deviations of the working probits from the regression line. The two values of χ^2 are not identical until the maximum-likelihood solution is reached.

In the present example, both methods give a value, $\chi^2 = 8{\cdot}4$, with degrees of freedom $5-2=3$, which is just significant at the 5% level (Table 8). Since, however, some of the expectations are very small ($16Q_5 = 0{\cdot}57$ and $16P_1 = 1{\cdot}76$), it is a little doubtful whether reference to the χ^2 distribution is appropriate.

It will be noted that the estimate of μ is $(5-a)/b$. This is the estimate of the value of x at which there is a 50:50 chance of a 'response'; in dosage-mortality work it is often described as the *median lethal dose* (L.D. 50) or *median effective dose* (E.D. 50). In the table we have given the successive estimates of E.D. 50 under the heading x_{50}. It is seen that to 2-decimal accuracy (in ft.) these have settled to a constant value at the fourth approximation. Using the large-sample formula for the standard error of x_{50} (see, for example, Finney, 1952, equation (3·6)), namely,

$$\sigma(x_{50}) = \{1/\Sigma nw + (x_{50}-\bar{x})^2/\Sigma nw(x-\bar{x})^2\}^{\frac{1}{2}}/b, \tag{14}$$

we obtain, for the final estimate, a standard error of 0·225 in working units or 0·90 in ft. For a description of a small-sample technique for obtaining an interval estimate for E.D. 50, Finney (1952, § 19) may be consulted. The final estimate of σ is the reciprocal of b, i.e. 1·321 (4) or 5·29 ft.

II. BASIC TABLES DERIVED FROM THE NORMAL FUNCTION

3. The χ^2-integral (Tables 7 and 8)

3·1 *Definition*

As is well known, the χ^2-integral, the incomplete Γ-function, the integral of Karl Pearson's Type III distribution and the cumulative sum of terms of the Poisson distribution are all different forms of the same mathematical function. We shall first define this function in terms of the χ^2-integral. If $X_1, X_2, \ldots, X_\nu$ are ν independent normal variates each with expectation zero and unit standard deviation, then

$$\chi^2 = \sum_{i=1}^{\nu} X_i^2$$

has for its probability integral

$$P(\chi^2 \mid \nu) = 2^{-\frac{1}{2}\nu}\{\Gamma(\tfrac{1}{2}\nu)\}^{-1}\int_0^{\chi^2} e^{-\frac{1}{2}x}x^{\frac{1}{2}\nu-1}\,dx. \tag{15}$$

The quantity ν is generally called the *degrees of freedom* of χ^2.

* This is a particular application of the χ^2 test for goodness of fit described in § 3·2 below.

Because of its use in tests of significance, Table 7 gives the complement of the probability integral, i.e.

$$Q(\chi^2 \mid \nu) = 1 - P(\chi^2 \mid \nu),$$

that is to say, the chance that ΣX_i^2 exceeds a given value, χ^2. Table 8 gives certain percentage points of the distribution, i.e. the values of χ^2 for which $Q(\chi^2 \mid \nu) = 0{\cdot}995, 0{\cdot}99, \ldots, 0{\cdot}01, 0{\cdot}005$ and $0{\cdot}001$.

It will be useful to put on record here, for later reference, the values of certain of the moment coefficients of the frequency function $f(\chi^2)$ of χ^2, for which equation (15) gives the probability integral. Writing κ_r for the rth cumulant and μ_r for the rth moment about the mean, then

$$\kappa_r = \nu 2^{r-1}(r-1)! \tag{16}$$

Hence

$$\kappa_1 = \mathscr{E}(\chi^2) = \nu,$$

$$\mu_2 = 2\nu, \quad \mu_3 = 8\nu, \quad \mu_4 = 48\nu + 12\nu^2,$$

$$\beta_1 = \mu_3^2/\mu_2^3 = 8/\nu, \quad \beta_2 = \mu_4/\mu_2^2 = 3 + 12/\nu.$$

3·2 *Applications; analysis of variance and goodness of fit*

The integral and its associated frequency function occur in numerous ways in the theory of statistical tests. In the analysis of variance, for example, the fundamental sum of squares of standardized deviates, which may be broken up into component parts each following a χ^2-distribution, is obtained by an appropriate orthogonal transformation of the observable variables.

In other problems the χ^2-distribution may be used as a mathematical approximation. An illustration of this type of application is provided in Example 4, but the best known case in point concerns the test of goodness of fit where the observations are the frequencies of occurrence of events in a number of alternative categories. Here, if an observation must fall into one or other of k alternative groups, if n_t are found in the tth group ($t = 1, 2, \ldots, k$), if $N = \sum_t n_t$, and if the probability, on the hypothesis tested, of an observation falling in this group is p_t, then

$$\chi^2 = \sum_{t=1}^{k} \frac{(n_t - Np_t)^2}{Np_t} = \sum_{t=1}^{k} \frac{n_t^2}{Np_t} - N \tag{17}$$

will have a sampling distribution whose probability integral is represented approximately by equation (15), if the hypothesis tested is true. The degrees of freedom, ν, will equal the number of categories, k, less the number of linear or approximately linear independent relations holding among the k differences $n_t - Np_t$ (see Example 3, p. 11).* To avoid stretching the approximation too far, none of the expected frequencies Np_t used in the test should be less than about 4 or 5.

3·3 *Relation to the Poisson distribution*

For even values of ν (for which the entries are given in bold type in Table 7) the integral (15) can be written as

$$Q(\chi^2 \mid \nu) = 1 - P(\chi^2 \mid \nu) = \sum_{j=0}^{c-1} \frac{e^{-m} m^j}{j!}, \tag{18}$$

with $m = \frac{1}{2}\chi^2$ and $c = \frac{1}{2}\nu$. Thus Table 7 gives the cumulative sum of the terms of the Poisson series with mean $m = \frac{1}{2}\chi^2$, up to but excluding the term for which $c = \frac{1}{2}\nu$. The table, therefore, supplements the later table of individual Poisson terms (Table 39) which does not adequately cover values of m less than unity.

* If the hypothetical p_t are functions of certain parameters whose values are determined by fitting a theoretical law to the observations, certain conditions must, of course, hold concerning the method of fitting employed.

3·4 *Applications. Determination of Poisson frequencies and the test of goodness of fit*

Example 3. Snedecor (1946, p. 441) quotes data of Leggatt (1935) on the pollution of seeds of *Phleum pratense* by the presence of a few noxious weed seeds. The table below shows the frequency with which 0, 1, 2, ... of these weed seeds were found in 98 quarter-ounce subsamples of *Ph. pratense*. We shall now use Table 7* to graduate the observed frequencies n_t by a Poisson distribution and afterwards to test for goodness of fit.

No. of noxious seeds x_t	Observed frequency of occurrence n_t		Poisson frequency Np_t		$n_t - Np_t$	$\frac{(n_t - Np_t)^2}{Np_t}$
0	3		4·8		−1·8	0·67
1	17		14·4		2·6	0·47
2	26		21·8		4·2	0·81
3	16		22·0		−6·0	1·64
4	18		16·6		1·4	0·12
5	9		10·0		−1·0	0·10
6	3	9	5·0	8·4	0·6	0·04
7	5		2·2			
8	0		0·8			
9	1		0·3			
10	—		0·1			
Total	98		98·0		0·0	3·85

The mean of the observed frequencies in col. 2, $\bar{x} = 3{\cdot}020$, provides the estimate of m with which to enter Table 7. The nearest tabular column is $m = 3{\cdot}0$, from which our value differs by $\phi = 3{\cdot}020 - 3{\cdot}0 = 0{\cdot}020$. The cumulative Poisson distribution for $m = 3{\cdot}020$ is now obtained† by multiplying, for each c, the entry in the column $m = 3{\cdot}0$ by $1 - \phi = 0{\cdot}98$ and adding the bold-type entry above it (i.e. the entry for $c - 1$) multiplied by $\phi = 0{\cdot}020$. Thus we find:

Cumulative Poisson sum for $c = 1$: $(0{\cdot}04979)(0{\cdot}98) + (0)(0{\cdot}02) = 0{\cdot}0488$.

Cumulative Poisson sum for $c = 2$: $(0{\cdot}19915)(0{\cdot}98) + (0{\cdot}04979)(0{\cdot}02) = 0{\cdot}1962$.

The differences between these cumulative entries multiplied by $N = 98$ give the expected values Np_t shown in the third column of the table.

We may now test the goodness of fit of the Poisson distribution. In order to compute the χ^2 of equation (17) we note that the last five expected frequencies do not exceed 5 and hence club these together as shown, so that the total number of differences, $n_t - Np_t$, is $k = 7$. The seven values of $(n_t - Np_t)^2/(Np_t)$ are shown in the last column together with their total, $\chi^2 = 3{\cdot}85$. Since the seven differences $n_t - Np_t$ must satisfy the two conditions

$$\sum_t (n_t - Np_t) = 0 \quad \text{and} \quad \sum_t (n_t - Np_t)\, x_t = 0,$$

the degrees of freedom, ν, for χ^2 are $\nu = 7 - 2 = 5$. Entering Table 7 for $\nu = 5$ and $\chi^2 = 3{\cdot}85$ we find that $Q(3{\cdot}8 \mid 5) = 0{\cdot}58$, so that a value of χ^2, as large or larger than this, might have arisen through chance sampling fluctuations on nearly six occasions out of ten; on the evidence available, therefore, we have no reason to question the appropriateness of the Poisson law, i.e. the random distribution of weed seeds among the others.

* For conditions under which the use of Table 39 may be preferable, see §21.

† This method of interpolation is described in §3·8*b*, except that in this example the ϕ^2 terms may be omitted.

Instead of obtaining the exact value of $Q(\chi^2 \mid \nu)$ from Table 7 we may be content to know whether the observed value of χ^2 does or does not exceed some specified percentage point, e.g. the 5 % point. Table 8 shows that for $\nu = 5$ the 5 % point for χ^2 is 11·070. Since the observed value of 3·85 is well within this limit we can say that the departure from the Poisson law is not significant at the 5 % level, thereby reaching a similar conclusion to that found above.

3·5 *Relation to the incomplete Γ-function, and to the Pearson Type III curve; other applications*

If in the integral (15) we regard $\frac{1}{2}x$ as the variable of integration, we recognize the expression as the incomplete Γ-integral or incomplete Γ-function for Γ-argument $\frac{1}{2}\nu$, normalized by division by $\Gamma(\frac{1}{2}\nu)$ to make the complete integral equal unity. The most extensive table of this integral is Karl Pearson's (1922) 7-decimal table; his notation is related to the present notation as follows:

Pearson	$I(u, p)$	p	u
Present table	$P(\chi^2 \mid \nu)$	$\frac{1}{2}\nu - 1$	$\frac{1}{2}\chi^2/\sqrt{(p+1)} = \chi^2/\sqrt{(2\nu)}$

Pearson's argument u was chosen to standardize the variable by division by its standard deviation which, for χ^2, is $\sqrt{(2\nu)}$. Whilst this device has the advantage of providing a uniform tabular range and interval for all ν, the transformation to argument u necessitates extra labour and often additional interpolation which is here avoided.

By changing the scale and origin of the Γ-integral we obtain the general form of the probability integral of the Pearson Type III distribution (see § 23·1 below), namely,

$$P(x \mid \gamma, a) = y_0 \int_{-a}^{x} \left(1 + \frac{v}{a}\right)^{\gamma a} e^{-\gamma v}\, dv.$$

This can clearly be related to the χ^2-integral by setting

$$\nu = 2\gamma a + 2, \quad \chi^2 = 2\gamma(x + a).$$

Hence Table 7 could be used to compute values of the probability integral of Type III frequency curves, the interpolation to fractional degrees of freedom, ν, being discussed in § 3·8 (*c*). In so far as these curves have been found useful in approximating to a number of the less tractable distributions of statistical theory, Tables 7 and 8 provide approximate solutions to tests involving such statistics. The χ^2-approximation used in testing for heterogeneity of variances is a case in point (see § 16·1); so are numerous likelihood criteria in multivariate analysis (see, for example, Box, 1949).

As an illustration of this type of application we shall take the case of obtaining an approximation to the distribution of a special form of *non-central* χ^2, namely,

$$\chi'^2 = \sum_{i=1}^{n} \frac{(x_i + \alpha_i - \bar{x} - \bar{\alpha})^2}{\sigma^2},$$

where the x_i are n independent normal deviates (i.e. have expectation zero) with a common variance σ^2 and the α_i are n constant parameters with a mean of $\bar{\alpha}$. It can be shown (Patnaik, 1949) that χ'^2/C has approximately the probability integral (15) of χ^2 with ν' degrees of freedom, where

$$C = (\nu + 2\lambda)/(\nu + \lambda), \quad \nu' = (\nu + \lambda)^2/(\nu + 2\lambda), \quad \nu = n - 1, \tag{19}$$

these relations being expressed in terms of a single 'non-centrality' parameter defined by

$$\lambda = \sum_{i=1}^{n} \frac{(\alpha_i - \bar{\alpha})^2}{\sigma^2}. \tag{20}$$

It will be noted that ν' will in general be fractional.

3·6 *Applications. Approximation to the non-central χ^2-distribution*

Example 4. A sample of twenty observations x_i ($i = 1, 2, \ldots, 20$) was to have been drawn from a normal population with a known variance of 4. It is, however, suspected that accidentally an unspecified subsample of ten of the x_i may all have been increased by the same amount of $\Delta = 4$. The sum of squares of deviations $\Sigma(x_i - \bar{x})^2$ is found to be 148. Does this result indicate that the suspected increase has taken place?

If no increase has occurred, the sum $\frac{1}{4}\Sigma(x_i - \bar{x})^2 = 37$ would be a random sample value of χ^2 for $\nu = 19$ degrees of freedom; Table 8 shows that it exceeds the 1 % value of χ^2 (36·2), throwing doubt on the hypothesis of no increase.

On the other hand, if the increase of Δ *has* in fact occurred, ten of the x_i would have come from a population with mean $\mu + \Delta$ and ten from a population with mean μ. Hence we are dealing with a non-central χ^2 for which from (20)

$$\lambda = 20(\tfrac{1}{2}\Delta)^2/4 = 20.$$

Further, from the relations (19),

$$C = (19 + 40)/(19 + 20) = 1{\cdot}51, \quad \nu' = (19 + 20)^2/(19 + 40) = 25{\cdot}8.$$

Accordingly
$$\chi'^2/C = \tfrac{1}{4}\sum_i (x_i - \bar{x})^2/1{\cdot}51$$

would be distributed approximately as χ^2 with 26 degrees of freedom. The observed numerical value is $\frac{1}{4}\,148/1{\cdot}51 = 24{\cdot}5$ and is now slightly less than the expected value, $\nu' = 25{\cdot}8$. The observations are therefore clearly consistent with the hypothesis that for ten of the observations an increase in mean of $\Delta = 4$ has occurred.

3·7 *Applications to expansions in series*

The tables also can be used as a basis for the expansion of distribution functions in Gram-Charlier or Laguerre series, which may be shown to be equivalent to series of χ^2 distributions progressing in integral steps of degrees of freedom.

3·8 *Interpolation in Table* 7

(*a*) *Single-entry interpolation χ^2-wise*

With no differences tabulated, Lagrangian interpolation or Aitken's iterative method may be used. We describe here a somewhat simpler method based on a Taylor expansion which utilizes the relations

$$\left.\begin{aligned} \frac{\partial Q}{\partial \chi^2} &= \tfrac{1}{2}\{Q(\chi^2 \mid \nu - 2) - Q(\chi^2 \mid \nu)\}, \\ \frac{\partial^2 Q}{\partial \chi^{2\,2}} &= \tfrac{1}{4}\{Q(\chi^2 \mid \nu - 4) - 2Q(\chi^2 \mid \nu - 2) + Q(\chi^2 \mid \nu)\}. \end{aligned}\right\} \qquad (21)$$

Let χ_0^2 denote the tabular argument nearest to the given value of χ^2 for which the answer is required and let $\theta = \chi^2 - \chi_0^2$.* Then from the second-order Taylor expansion we obtain

$$\begin{aligned} 2Q(\chi^2 \mid \nu) = Q(\chi_0^2 \mid \nu - 4)\,(\tfrac{1}{2}\theta)^2 &+ \\ Q(\chi_0^2 \mid \nu - 2)\,\{\theta - 2(\tfrac{1}{2}\theta)^2\} &+ \\ Q(\chi_0^2 \mid \nu)\,\{2 - \theta + (\tfrac{1}{2}\theta)^2\}&. \end{aligned} \qquad (22)$$

* Note that θ is here the argument increment and not the fraction of interpolation as used elsewhere, e.g. in § 1·4.

Illustration. To find $Q(3{\cdot}64132 \mid 12)$ the following method is suggested:

Copy down $\theta = 0{\cdot}04132$, form and copy $(\frac{1}{2}\theta)^2 = 0{\cdot}00043$,

and finally compute the sum of products

$$\begin{aligned} Q(3{\cdot}64132 \mid 12) = \tfrac{1}{2}\{&0{\cdot}89129\,(0{\cdot}00043) + \\ &0{\cdot}96359\,(0{\cdot}04132 - 0{\cdot}00086) + \\ &0{\cdot}98962\,(2 - 0{\cdot}04132 + 0{\cdot}00043)\} \\ = 0{\cdot}98907.& \end{aligned}$$

On a calculating machine the Q entries would be set as multiplicands and the terms inside the parentheses ( ) applied in turn as multipliers; the sum of products so formed is then immediately divided by 2.

The fifth decimal computed from (22) may be a few units of error. Where 3-decimal accuracy is adequate, linear interpolation is usually sufficient and the $(\frac{1}{2}\theta)^2$ terms of (22) may be omitted.

(*b*) *Single-entry interpolation m-wise*

The method is almost identical with that described above. Let m_0 denote the tabular argument nearest to m and $\phi = m - m_0$; then

$$\begin{aligned} Q(m \mid c) = Q(m_0 \mid c-2)\,&\tfrac{1}{2}\phi^2 + \\ Q(m_0 \mid c-1)\,&(\phi - \phi^2) + \\ Q(m_0 \mid c)\,&(1 - \phi + \tfrac{1}{2}\phi^2). \end{aligned} \qquad (23)$$

For details of computation see the illustration given in § 3·4 above.

(*c*) *Double-entry interpolation*

When it is intended to use the present table for the evaluation of the χ^2-integral for fractional degrees of freedom (or odd degrees of freedom, $\nu > 30$), double-entry interpolation for both intermediate ν and χ^2 is necessary. Similarly, when using the present tables for computing the general Type III probability integral quoted on p. 12 above, fractional ν and χ^2 will usually arise. In this case the relations (21) are particularly useful, as they allow the differential with regard to χ^2 to be expressed in terms of the very entries $Q(\chi_0^2 \mid \nu)$ which are required for the Lagrangian interpolation ν-wise.

Let χ_0^2 denote the tabular argument nearest to the given χ^2 and ν_0 the tabular line heading nearest to, but smaller than the given ν; then the following formulae and methods are suggested in which we must distinguish two cases:

(i) $\nu > 30$

Let $\omega = \frac{1}{2}(\nu - \nu_0)$, $\phi = \frac{1}{2}(\chi^2 - \chi_0^2)$. Compute and copy $\omega\phi$, ϕ^2 and ω^2 and finally obtain the interpolate as the sum of products:

$$\begin{aligned} Q(\chi^2 \mid \nu) = Q(\chi^2 \mid \nu_0 - 4)\,(\tfrac{1}{2}\phi^2) + Q(\chi_0^2 \mid \nu_0 - 2)\,&(\tfrac{1}{2}\omega^2 - \tfrac{1}{2}\omega + \phi - \phi^2 - \omega\phi) + \\ Q(\chi_0^2 \mid \nu_0)\,&(1 - \omega^2 - \phi + \tfrac{1}{2}\phi^2 + 2\omega\phi) + \\ Q(\chi_0^2 \mid \nu_0 + 2)\,&(\tfrac{1}{2}\omega^2 + \tfrac{1}{2}\omega - \omega\phi). \end{aligned} \qquad (24)$$

On a calculating machine, the multiplicands Q are set and the terms within the parentheses applied in turn as multipliers.

Illustration. Find $Q(25{\cdot}2154 \mid 39)$.

We have $\omega=0{\cdot}5$, $\phi=0{\cdot}1077$, whence $\omega^2=0{\cdot}25$, $\phi^2=0{\cdot}0116$, $\phi\omega=0{\cdot}0538$. Hence we obtain

$$\begin{aligned} Q(25{\cdot}2154 \mid 39) = {} & 0{\cdot}86931\ (0{\cdot}0058) + 0{\cdot}91584\ (0{\cdot}125 - 0{\cdot}25 + 0{\cdot}1077 - 0{\cdot}0116 - 0{\cdot}0538) + \\ & 0{\cdot}94815\ (1 - 0{\cdot}25 - 0{\cdot}1077 + 0{\cdot}0058 + 0{\cdot}1077) + \\ & 0{\cdot}96941\ (0{\cdot}125 + 0{\cdot}25 - 0{\cdot}0538) \\ = {} & 0{\cdot}95729. \end{aligned}$$

The exact interpolate is 0·95706.

The main terms neglected in equation (24) are $B''\delta^3$ and $-2B''\phi\delta^3$, where $B''=\frac{1}{4}\omega(\omega-1)$ and all differences δ are taken with regard to ν. Of these terms, the first is usually the largest and may result in an error of a few units in the fourth decimal. If higher, accuracy is needed, a table of Lagrangian interpolation coefficients L_{-1}, L_0, L_1 and L_2 may be used in conjunction with the following formula:

$$\begin{aligned} Q(\chi^2 \mid \nu) = {} & Q(\chi_0^2 \mid \nu_0 - 4)\ (\tfrac{1}{2}\phi^2) + \\ & Q(\chi_0^2 \mid \nu_0 - 2)\ (L_{-1}(\omega) + \phi - \phi^2 - \omega\phi) + \\ & Q(\chi_0^2 \mid \nu_0)\ (L_0(\omega) - \phi + \tfrac{1}{2}\phi^2 + 2\omega\phi) + \\ & Q(\chi_0^2 \mid \nu_0 + 2)\ (L_1(\omega) - \omega\phi) + \\ & Q(\chi_0^2 \mid \nu_0 + 4)\ L_2(\omega). \end{aligned} \tag{25}$$

With this formula, the first of the above error terms is eliminated and the remaining error may sometimes amount to a few units in the fifth decimal, but is often smaller than 5×10^{-6}. In the present example equation (25) yields 0·95710.

(ii) $\nu<30$

Let $\omega=\nu-\nu_0$ and again $\phi=\frac{1}{2}(\chi^2-\chi_0^2)$. Compute and copy $\omega\phi$ and ω^2, and hence find the interpolate as the sum of products

$$\begin{aligned} Q(\chi^2 \mid \nu) = {} & Q(\chi_0^2 \mid \nu_0 - 4)\ \tfrac{1}{2}\phi^2 + \\ & Q(\chi_0^2 \mid \nu_0 - 2)\ (\phi - \phi^2 - \omega\phi) + \\ & Q(\chi_0^2 \mid \nu_0 - 1)\ (\tfrac{1}{2}\omega^2 - \tfrac{1}{2}\omega + \omega\phi) + \\ & Q(\chi_0^2 \mid \nu_0)\ (1 - \omega^2 - \phi + \tfrac{1}{2}\phi^2 + \omega\phi) + \\ & Q(\chi_0^2 \mid \nu_0 + 1)\ (\tfrac{1}{2}\omega^2 + \tfrac{1}{2}\omega - \omega\phi). \end{aligned} \tag{26}$$

For operation on a calculating machine proceed as when using equation (24).

Illustration. Find $Q(5{\cdot}8764 \mid 10{\cdot}3148)$.

We have $\omega=0{\cdot}3148$, $\phi=0{\cdot}0382$, whence $\omega^2=0{\cdot}09910$, $\omega\phi=0{\cdot}01203$, $\phi^2=0{\cdot}00146$, so that

$$\begin{aligned} Q(5{\cdot}8764 \mid 10{\cdot}3148) = {} & 0{\cdot}44596\ (0{\cdot}00073) + 0{\cdot}66962\ (0{\cdot}0382 - 0{\cdot}00146 - 0{\cdot}01203) + \\ & 0{\cdot}75976\ (0{\cdot}04955 - 0{\cdot}1574 + 0{\cdot}01203) + \\ & 0{\cdot}83178\ (1 - 0{\cdot}09910 - 0{\cdot}0382 + 0{\cdot}00073 + 0{\cdot}01203) + \\ & 0{\cdot}88637\ (0{\cdot}04955 + 0{\cdot}1574 - 0{\cdot}01203) \\ = {} & 0{\cdot}84503. \end{aligned}$$

The main terms neglected in equation (26) are $B''\delta^3$ and $-4B''\phi\delta^3$, where $B''=\frac{1}{4}\omega(\omega-1)$ and all differences δ are taken with regard to ν. The error may be a few units in the fourth decimal except near the singularity $\nu=0$, $\chi^2=0$, where interpolation ν-wise by this method is not possible.* If higher accuracy is required, a table of the four-point Lagrangian coefficients may be used in conjunction with the following formula:

$$Q(\chi^2 \mid \nu) = Q(\chi_0^2 \mid \nu_0 - 4)\,\tfrac{1}{2}\phi^2 + Q(\chi_0^2 \mid \nu_0 - 2)\,(L_{-1}(1+\omega) + \phi - \phi^2 - \omega\phi) +$$
$$Q(\chi_0^2 \mid \nu - 1)\,(L_0(1+\omega) + \omega\phi) + Q(\chi_0^2 \mid \nu_0)\,(L_1(1+\omega) - \phi + \tfrac{1}{2}\phi^2 + \omega\phi) +$$
$$Q(\chi_0^2 \mid \nu_0 + 1)\,(L_2(1+\omega) - \omega\phi). \qquad (27)$$

(*d*) *Single-entry interpolation ν-wise*

The necessity for this type of interpolation does not arise very often, as interpolation ν-wise usually occurs in conjunction with interpolation χ^2-wise, thereby leading to double-entry interpolation. The formulae (25), (26) and (27) may be used with $\phi = 0$. For higher accuracy, a six-point Lagrangian formula may be employed.

3·9 *Interpolation in Table* 8

Although values of χ^2 for intermediate probability levels are obtainable by Lagrangian formulae or by special methods† (Simaika, 1942), we recommend in such cases inverse interpolation in Table 7, and shall confine ourselves here to interpolation to intermediate values of ν, whether integral or fractional. For $\nu < 30$, Lagrangian interpolation at unit interval is convenient. Except for the top left-hand corner of the table, the four-point formula will be found to yield sufficient accuracy, and often simple linear interpolation will be adequate. For small ν and large $Q = 1 - P$ (top left-hand corner) approximate values χ_0^2 can be directly computed from

$$\tfrac{1}{2}\chi_0^2 = \{P\,\Gamma(\tfrac{1}{2}\nu + 1)\}^{2/\nu},$$

and their accuracy improved by the next iterate

$$\tfrac{1}{2}\chi_1^2 = \tfrac{1}{2}\chi_0^2\{1 + \tfrac{1}{2}\chi_0^2/(\tfrac{1}{2}\nu + 1)\},$$

These formulae are readily obtained from an approximate expansion of (15) for small χ^2.

For $30 < \nu < 100$ Lagrangian interpolation at interval 10 is needed, and we give below the four-point Lagrangian coefficients required for integer ν. The accuracy, to 4 or 5 figures, increases with ν.

Unit figure of ν	L_{-10}	L_0	L_{10}	L_{20}	Unit figure of ν
	−	+	+	−	
0	0·0000	1·0000	0·0000	0·0000	
1	0·0285	0·9405	0·1045	0·0165	9
2	0·0480	0·8640	0·2160	0·0320	8
3	0·0595	0·7735	0·3315	0·0455	7
4	0·0640	0·6720	0·4480	0·0560	6
5	0·0625	0·5625	0·5625	0·0625	5
	−	+	+	−	
	L_{20}	L_{10}	L_0	L_{-10}	

* For $\chi^2 \leq 1$, instead of interpolating ν-wise in the table we may compute any required value of $Q(\chi^2 \mid \nu)$ directly from the series

$$Q(\chi^2 \mid \nu) = 1 - e^{-m} \sum_{j=0}^{\infty} \frac{m^{c+j}}{\Gamma(c+j+1)},$$

where $c = \tfrac{1}{2}\nu$; this converges rapidly. A few values of the Γ-function for fractional arguments are required for its evaluation.

† As a rough approximation, we may use linear interpolation for χ^2 with argument $\log Q$ or $\log(1-Q) = \log P$, whichever is the smaller.

For $\nu > 100$ the percentage points of χ^2 can be computed from either the Fisher or, for higher accuracy, from the cubic Wilson-Hilferty formula shown at the bottom of the table, the required normal percentage points X being shown in the last tabular line.

4. Moments of the distribution of $s/\sigma = \chi/\sqrt{\nu}$ and confidence limits for σ (Table 35)*

4·1 *Definitions: the distribution of s^2 and s*

Since the sum of squares of ν independent unit normal deviates is distributed as χ^2 (see § 3·1), it follows that if the sum of squares of ν independent normal deviates, each having a variance σ^2, is divided by ν, a statistic is obtained which is distributed as $\chi^2\sigma^2/\nu$. Further, since the expectation of χ^2 is ν, the expectation of this statistic will be σ^2. This result, taken when necessary in conjunction with an orthogonal transformation of the variates, is used again and again in statistical theory to obtain estimates of unknown variances and to use these estimates in a variety of statistical tests. We shall here denote a statistic of this character by s^2 and shall speak of it as a mean square estimate† of σ^2, based on ν degrees of freedom.

As $s^2 = \chi^2\sigma^2/\nu$ and in view of the equation (15) giving the probability integral of χ^2, it follows that the frequency function of s^2 has the form

$$f(s^2) = (\tfrac{1}{2}\nu)^{\frac{1}{2}\nu}\{\Gamma(\tfrac{1}{2}\nu)\}^{-1}\sigma^{-\nu}(s^2)^{\frac{1}{2}\nu-1}e^{-\frac{1}{2}\nu s^2/\sigma^2}. \tag{28}$$

While it is generally simpler, both from the mathematical and computational points of view, to work in terms of s^2, and therefore of χ^2, there are certain problems, e.g. in connexion with Student's ratio, t, discussed below, in which it is more convenient to work with s (or χ). The frequency function of s may be derived at once from (28), namely,

$$f(s) = 2(\tfrac{1}{2}\nu)^{\frac{1}{2}\nu}\{\Gamma(\tfrac{1}{2}\nu)\}^{-1}\sigma^{-\nu}s^{\nu-1}e^{-\frac{1}{2}\nu s^2/\sigma^2}. \tag{29}$$

Table 35 contains values of the expectation, the standard deviation and of the moment ratios $\beta_1 = \mu_3^2/\mu_2^3$ and $\beta_2 = \mu_4/\mu_2^2$ of s/σ. The moments about the mean cannot be expressed in any simple form, but have been calculated from the moments about zero, which are given by

$$\mu_r' = (2/\nu)^{\frac{1}{2}r}\frac{\Gamma\{\frac{1}{2}(\nu+r)\}}{\Gamma(\frac{1}{2}\nu)}. \tag{30}$$

The figures in cols. 7–10 of the table, used in the determination of confidence limits for σ, are defined in § 4·3.

4·2 *Applications of moments*

The expectation of s is not, of course, equal to σ; col. 2 of Table 35 indicates the rate of approach of $\mathscr{E}(s/\sigma)$ to unity. If s is divided by the appropriate factor taken from this column, then we have an unbiased estimator of σ, as distinct from s^2 which is an unbiased estimator of σ^2. Col. 4 of the table gives the ratio of col. 3, the standard deviation of s/σ, to $1/\sqrt{(2\nu)}$, and therefore shows the approach of the true standard error of s to its large sample limit $\sigma/\sqrt{(2\nu)}$.

It should be noted that since S.D. of $\chi = \sqrt{\nu} \times$ S.D. of (s/σ), the standard deviation of χ (with ν degrees of freedom) is obtained by multiplying the figures in col. 4 by $1/\sqrt{2} = 0{\cdot}7071$. The figures in cols. 5 and 6 give an indication of the approach of the distributions of s and χ to the normal.

A practical use of these results is illustrated below in §§ 12·6 and 12·7.

* In planning the arrangement of pages, it was found convenient to place this table as no. 35, rather late in the sequence of tables.

† To make its character quite clear it might alternatively be termed an unbiased χ^2-estimate of σ^2.

4·3 *Confidence limits for a standard deviation*

The distribution of χ^2 may be used to obtain what is termed an *interval estimate* for the unknown value of a population variance (or standard deviation). A rather fuller description of the principles underlying the theory of confidence intervals is given in § 22 below; a brief description of the derivation and use of the factors given in cols. 7–10 of Table 35 will be here sufficient.

Suppose that a random set of observations provides a mean square estimate, s^2, having ν degrees of freedom, of a population variance σ^2. Then s^2 is distributed as $\chi^2\sigma^2/\nu$. It follows that there is a probability, α, that $s^2 > \chi^2_\alpha \sigma^2/\nu$ and a probability, α, that $s^2 < \chi^2_{1-\alpha}\sigma^2/\nu$, where χ^2_α and $\chi^2_{1-\alpha}$ are, respectively, the upper and lower 100α % points of the χ^2 distribution for ν degrees of freedom, i.e. (using the notation of § 3·1*) where

$$1 - P(\chi^2_\alpha \mid \nu) = \alpha = P(\chi^2_{1-\alpha} \mid \nu).$$

These two probability statements are clearly equivalent to the following: there is a probability of α that $s\sqrt{\nu}/\chi_\alpha > \sigma$ and a probability of α that $s\sqrt{\nu}/\chi_{1-\alpha} < \sigma$. Since, for $\alpha < 0{\cdot}5$, $\sqrt{\nu}/\chi_\alpha < \sqrt{\nu}/\chi_{1-\alpha}$, it follows that the probability is $1-2\alpha$ that the unknown σ is included between the limits $s\sqrt{\nu}/\chi_\alpha$ and $s\sqrt{\nu}/\chi_{1-\alpha}$, or

$$\Pr\left\{\frac{s\sqrt{\nu}}{\chi_\alpha} \leqslant \sigma \leqslant \frac{s\sqrt{\nu}}{\chi_{1-\alpha}}\right\} = 1 - 2\alpha. \qquad (31)$$

It will be seen that the two limits are random variables whose values depend only on the observations; their position and distance apart will vary from one sampling to another. The probability statement must be interpreted in the sense that if we determine the limits for σ from the observations by the formulae given above, we are following a procedure which will give in the long run an interval estimate including the true σ in a proportion of about $1-2\alpha$ cases. Further, this statement is true even if the successive samples are drawn from populations having different standard deviations. The limits are termed *confidence limits* and $1-2\alpha$, often expressed as a percentage, is termed the *confidence coefficient.*

Table 35 gives in cols. 7–10 the factors $\sqrt{\nu}/\chi_\alpha$ and $\sqrt{\nu}/\chi_{1-\alpha}$, by which s must be multiplied to obtain 'central' intervals, with $\alpha = 0{\cdot}025$ and $0{\cdot}005$ or $1-2\alpha = 0{\cdot}95$ and $0{\cdot}99$. These factors were computed from the appropriate percentage points of the χ^2-distribution given in Table 8. Clearly considerable variety in the form of interval is possible; for example, we might use Table 8 to derive 'non-central' intervals based on $\chi^2_{\alpha_1}$, and $\chi^2_{1-\alpha_2}$, where $\alpha_1 \neq \alpha_2$, giving a confidence coefficient $1-\alpha_1-\alpha_2$.

In making use of these results it is essential to bear in mind two underlying assumptions; the probability statement associated with the interval is only valid if (*a*) the observed variables are approximately normally distributed and (*b*) the sampling is random, so that the frequency law for s^2 of equation (28) holds good.

4·4 *Application of confidence interval theory*

Example 5. The following (taken from E. S. Pearson, 1935*b*, p. 22) are breaking strengths in lb. of twelve strips of a certain cotton fabric:

163	175	161	154	166	174
166	173	164	181	170	158

* Clearly in the notation of Table 8, $\chi^2_\alpha = \chi^2(Q \mid \nu)$ for $Q = \alpha$.

Assuming that the figures represent random and independent observations of a normally distributed variable, determine the 95 and 99 % central confidence intervals for the population standard deviation.

It is found that $\Sigma x = 2005$ $\Sigma x^2 = 335669$ $\bar{x} = 167\cdot08$ lb.

$\Sigma(x-\bar{x})^2 = 666\cdot92$ $s^2 = 60\cdot63$ $s = 7\cdot786$ lb.

Hence entering Table 35 with $\nu = 11$, we obtain

95 % confidence limits: $0\cdot708 \times 7\cdot79 = 5\cdot52$ lb. and $1\cdot70 \times 7\cdot79 = 13\cdot2$ lb.

99 % confidence limits: $0\cdot641 \times 7\cdot79 = 4\cdot99$ lb. and $2\cdot06 \times 7\cdot79 = 16\cdot0$ lb.

Thus in adopting this procedure, which in the present instance leads to the assertion that $5\cdot5 \leqslant \sigma \leqslant 13\cdot2$, we shall be correct in the long run in 95 % of cases. If we wish to reduce the risk of error, we must at the same time widen the interval; thus the statement $5\cdot0 \leqslant \sigma \leqslant 16\cdot0$ can be associated with a 99 % confidence coefficient.

5. The t-distribution (Tables 9, 10 and 12)

5·1 *Definition*

If (a) y is a normally distributed variable with expectation zero and standard deviation σ_y, and (b) s_y^2 is a mean square estimate of σ_y^2 based on ν degrees of freedom* and independent of y, then the probability integral of the ratio

$$t = \frac{y}{s_y}$$

assumes the Student form

$$P(t \mid \nu) = \Gamma\{\tfrac{1}{2}(\nu+1)\}\{\Gamma(\tfrac{1}{2}\nu)\}^{-1}(\nu\pi)^{-\frac{1}{2}} \int_{-\infty}^{t} \left(1+\frac{u^2}{\nu}\right)^{-\frac{1}{2}(\nu+1)} du. \tag{32}$$

Clearly the t-distribution is symmetrical, so that if

$$Q(t \mid \nu) = 1 - P(t \mid \nu), \quad \text{then} \quad P(-t \mid \nu) = Q(t \mid \nu).$$

It will be noted that this integral depends only on ν, the number of *degrees of freedom* of the estimate of σ_y^2.

Table 9 gives values of $P(t \mid \nu)$ for argument t and ν, while Table 12 is the corresponding table of percentage points, i.e. it gives the values of t for certain conventional 'levels' of the integral.

5·2 *Applications. The significance of a mean of a single sample*

The t-distribution, with slightly different notation, was first derived by W. S. Gosset (Student) in 1908 to deal with the simple problem in which y is the deviation of the mean, $\bar{x}$, of a random sample of n normally distributed observations $x_1, x_2, \ldots, x_n$ from a population value μ, and s_y^2 is the mean-square estimate of the variance of $\bar{x}$ obtained from the same sample. Thus

$$\left.\begin{aligned} \bar{x} = \sum_i x_i/n, \quad s^2 = \sum_i (x_i - \bar{x})^2/(n-1), \quad s_y^2 = s^2/n, \\ t = (\bar{x}-\mu)\sqrt{n}/s, \quad \nu = n-1. \end{aligned}\right\} \tag{33}$$

* For a definition of such an estimate, see §4·1; s_y^2 will have the probability distribution of equation (28).

Example 6. Consider the twelve observations of breaking strength of cotton fabric used in Example 5, p. 18. We may ask whether these results are consistent with the hypothesis that the fabric sampled is one for which the average strength of strips tested in this way is $\mu = 171{\cdot}5$ lb. We have

$$\bar{x} = 167{\cdot}08\,\text{lb.}, \quad s = 7{\cdot}786\,\text{lb.}, \quad s/\sqrt{n} = 2{\cdot}248,$$

$$t = (167{\cdot}08 - 171{\cdot}5)/2{\cdot}248 = -1{\cdot}97, \quad \nu = 11.$$

Turning to Table 9, visual inspection shows that to 3-decimal accuracy $P(1{\cdot}97 \mid 11) = 0{\cdot}963$. The interpretation of this result depends to some extent on the nature of the problem investigated.

In carrying out the test of a given statistical hypothesis* we have in general two considerations to bear in mind. In the first place we do not want to run an undue risk of rejecting the hypothesis tested when it is true; in the second place we wish to use a form of test which provides as large a chance as possible of establishing the existence of that type of effect or change we think most likely to occur or regard it as most important not to overlook, if it is present. If, then, in the present illustration, we were on the look-out for a change in mean μ, in either direction from 171·5 lb., we should regard large positive and negative values of t as significant, and it would be appropriate to use the t-test in its symmetrical or double-tail form. Thus we should note that

$$\Pr\{|t| > 1{\cdot}97 \mid \nu = 11\} = 2 \times 0{\cdot}037 = 0{\cdot}074.\dagger$$

Thus as large a departure (+ or −) in $\bar{x}$ from $\mu = 171{\cdot}5$ (expressed in terms of the estimated standard error derived from the data) would occur through chance about once in every thirteen or fourteen samples. The discrepancy is therefore hardly significant.

If, however, as seems likely in the present illustration, we are only concerned with a change in mean in one direction, i.e. here, in a fall in average strength, we should only want to know whether a large negative value of t was statistically significant. We should therefore use the asymmetrical or single-tail form of the test and note that

$$\Pr\{t < -1{\cdot}97 \mid \nu = 11\} = 0{\cdot}037.$$

Thus we should regard the difference between $\bar{x} = 167{\cdot}08$ and $\mu = 171{\cdot}5$ as of greater significance than in the first situation.

It is common practice not to determine a precise value for $P(t \mid \nu)$, but to base conclusions on whether $|t|$ or t falls beyond a pre-selected *significance point*, chosen according to the risk the statistician is prepared to accept of rejecting the hypothesis tested when it is in fact true. Thus if his practice is to note for further investigation any result for which the appropriate test statistic falls beyond its 5 % significance point, he will refer the value $t = 1{\cdot}97$ to Table 12. Here he will find that, when $\nu = 11$, for the double-tail test the 5 % point for t is 2·201 and for the single-tail test it is 1·796. Thus if he is looking for a change in mean in either direction, he will find that t is not significant at the 5 % level; but if he is interested in a decrease in mean only, the result is significant at this level.

* In this case where we ask whether $\mu - 171{\cdot}5 = 0$, it may appropriately be termed a *null hypothesis*.

† It is convenient to use this form of summary statement in place of the longer sentence: the probability that the numerical value of t is greater than 1·97 (when the degrees of freedom of the variance estimate is 11) equals 0·074.

5·3 *The difference in the means of two samples*

Probably the most common application of the t-test occurs when two independent samples are available and, while we believe that the observations are subject to a common variance σ^2 and are approximately normally distributed, we are on the look-out for a possible difference in the means. Thus if we have:

	No. of observations	Mean	Sum of squares
1st sample, $x_{11}, x_{12}, \ldots, x_{1n_1}$.	n_1	$\bar{x}_1$	$\sum_{i=1}^{n_1} (x_{1i}-\bar{x}_1)^2$
2nd sample, $x_{21}, x_{22}, \ldots, x_{2n_2}$.	n_2	$\bar{x}_2$	$\sum_{i=1}^{n_2} (x_{2i}-\bar{x}_2)^2$

the parts of y and σ_y^2 of § 5·1 will be played by $\bar{x}_1-\bar{x}_2$ and $\sigma^2(1/n_1+1/n_2)$, where σ^2 is the assumed common variance of x in the two populations. σ^2 will now be estimated from the pooled sum of squares, i.e. by

$$s^2 = \left\{\sum_{i=1}^{n_1} (x_{1i}-\bar{x}_1)^2 + \sum_{i=1}^{n_2} (x_{2i}-\bar{x}_2)^2\right\}\Big/ (n_1+n_2-2), \tag{34}$$

which is a mean-square estimate of the common variance based on n_1+n_2-2 degrees of freedom. It follows that

$$t = (\bar{x}_1-\bar{x}_2)\Big/\left\{s\sqrt{\left(\frac{1}{n_1}+\frac{1}{n_2}\right)}\right\} \tag{35}$$

will be distributed as Student's t, i.e. will have $P(t \mid \nu)$, with $\nu = n_1+n_2-2$, for its probability integral, if the populations sampled have a common mean. The test may again be used in double- or single-tail form according to the problem investigated.

5·4 *Application of the t-distribution in regression problems*

The distribution is also applied in a number of regression problems. Suppose, for example, that we have a sample of n paired observations (x_i, y_i) and may assume that y_i is distributed normally about a mean of $\alpha+\beta x_i$ with variance σ^2, independent of x_i. Then an estimate of β is obtained from the sample regression coefficient

$$b = \sum_{i=1}^{n} (y_i-\bar{y})(x_i-\bar{x}) \Big/ \sum_{i=1}^{n} (x_i-\bar{x})^2.$$

If the x_i are regarded as fixed, the y_i alone varying from sample to sample, b is then a weighted linear function of the normal y_i and is itself normally distributed about a mean of β with variance $\sigma^2/\sum_i (x_i-\bar{x})^2$. σ^2 may be independently estimated from the residual sum of squares about the sample regression line, i.e. by

$$s^2 = \frac{\sum_i (y_i-\bar{y})^2 - b^2 \sum_i (x_i-\bar{x})^2}{n-2},$$

where s^2 has the frequency function $f(s^2)$ of equation (28) with $\nu = n-2$. It follows that

$$t = (b-\beta)\sqrt{\{\sum_i (x_i-\bar{x})^2\}}/s$$

is distributed as the Student ratio with $\nu = n-2$. The application to a test of significance or to an interval estimate for β follows at once.

Since the distribution of t is a function of ν only, being independent of the particular values of x_i, the restriction of fixed x-values can be removed.

A natural extension of the theory is made in the case of two independent samples of n_1 and n_2 paired observations, which provide estimates of regression, b_1 and b_2. The estimate of the assumed common array variance of the y's is now obtained from the pooled sums of squares of the residuals from the two fitted regression lines, the degrees of freedom of this estimate, s^2, being now $\nu = n_1 + n_2 - 4$. If the regression coefficient β is the same in both populations sampled, then

$$t = \frac{b_1 - b_2}{s\sqrt{\left(1\Big/\sum_{i=1}^{n_1}(x_{1i}-\bar{x}_1)^2 + 1\Big/\sum_{i=1}^{n_2}(x_{2i}-\bar{x}_2)^2\right)}}$$

will follow Student's distribution.

Numerous other applications of the t-test follow from the simple basis of the fundamental distribution, namely, that it is the distribution of the ratio of a normal deviate to an independent root-mean-square estimate of its standard deviation.

5·5 *Confidence limits for a mean*

Apart from its use in tests of significance, the t-distribution may be used to obtain an *interval estimate* for an unknown population mean μ, for the difference between two means $\mu_1 - \mu_2$, for the difference between two regression coefficients, etc.

The argument is similar to that used in deriving confidence limits for σ (see § 4·3), this time inverting the probability statement regarding Student's ratio t. Thus if a random sample of n observations $x_1, x_2, \ldots, x_n$ is drawn from a normal population with mean μ, if $\bar{x}$ is the sample mean and if $s^2 = \sum_i (x_i - \bar{x})^2/(n-1)$, then the probability is α that

$$\mu_A = \bar{x} - t_\alpha s/\sqrt{n} \geqslant \mu.$$

Similarly, it is α that

$$\mu_B = \bar{x} + t_\alpha s/\sqrt{n} \leqslant \mu,$$

where t_α is the value of t satisfying

$$Q(t_\alpha \mid n-1) = 1 - P(t_\alpha \mid n-1) = \alpha,$$

i.e. t_α is the 100α % point of t for $\nu = n-1$ degrees of freedom. Further, it follows that there is a probability of $1 - 2\alpha$ that the population mean μ is included between the limits μ_A and μ_B which are random variables, depending on the sample x-values above. μ_A and μ_B are the *confidence limits* and $1 - 2\alpha$ the *confidence coefficient.*

Example 7. Let us suppose we have the sample of twelve observations used in Examples 5 and 6, pp. 18, 20. Take $\alpha = 0·025$; Table 12 shows that if $Q(t_\alpha \mid 11) = 0·025$, $t_\alpha = 2·201$. Then

$$\mu_A = 167·08 - 2·201 \times 2·248 = 162·1,$$

$$\mu_B = 167·08 + 2·201 \times 2·248 = 172·0.$$

Thus we say that there is a probability of $1 - 2\alpha = 0·95$ that the limits 162·1 and 172·0 lb. include the unknown population mean. The interval will of course vary in length and position from sample to sample, and the probability statement must be interpreted as meaning that if we determine μ_A and μ_B from a sample as indicated, we are following a procedure which will give an interval including the true mean in the long run in about nineteen cases out of twenty.

5·6 *Interpolation in Table* 9*

(a) *Single-entry interpolation for t*

Since no differences are provided in the table, a brief discussion of Lagrangian interpolation formulae is appropriate. Let t_{-1}, t_0, t_1 denote three consecutive tabular arguments such that $t_0 \leqslant t \leqslant t_1$, and let

$$P_i = P(t_i \mid \nu), \quad \theta = (t - t_0)/(t_1 - t_0).$$

Then we have the following formulae:

$$\left.\begin{aligned} &\text{(linear)} && P(t \mid \nu) \sim P_0(1-\theta) + P_1\theta, \\ &\text{(three-point)} && P(t \mid \nu) = P_{-1}(\tfrac{1}{2}\theta^2 - \tfrac{1}{2}\theta) + P_0(1-\theta^2) + P_1(\tfrac{1}{2}\theta^2 + \tfrac{1}{2}\theta). \end{aligned}\right\} \qquad (36)$$

The three-point formula (36) gives 5-decimal accuracy for $\nu \geqslant 20$. For $\nu \leqslant 20$ the maximum error is about (*a*) 10^{-5} for $t \geqslant 0{\cdot}7$ and (*b*) 4×10^{-5} for $0 \leqslant t \leqslant 0{\cdot}7$. To obtain 5-decimal accuracy throughout, use (36) and repeat the computation with θ replaced by $(1-\theta)$ and P_2, P_1 and P_0 substituted for P_{-1}, P_0 and P_1, respectively. Compare the two answers as a check on the calculations: if they agree to 4 decimals, their average should yield an interpolate accurate to 5 and checked to 4 decimals.

The accuracy of the linear formula depends on the argument range as follows: For $1 \leqslant \nu \leqslant 20$, $0 < t < 4$, the maximum error is three units in the fourth decimal; for $1 \leqslant \nu \leqslant 20$, $4 < t$, it is four units in the fifth decimal; for $20 \leqslant \nu < \infty$, $0 < t < 2$, it is eight units in the fifth decimal and for $20 \leqslant \nu \leqslant \infty$, $2 < t$, it is twelve units in the fifth decimal.

(b) *Single-entry interpolation for integer ν*

For intermediate integers, $\nu > 20$, linear harmonic interpolation with argument $120/\nu$ will yield interpolates accurate to one unit in the fifth decimal.

(c) *Double-entry interpolation*

Interpolation to fractional ν may sometimes be required, as, for instance, when using the table to compute the probability integral of the general Pearson Type VII frequency curve

$$f(x) = \frac{\Gamma(m)}{a\sqrt{\pi}\,\Gamma(m-\frac{1}{2})}\left(1 + \frac{x^2}{a^2}\right)^{-m}$$

Here we can obtain $\int_{-\infty}^{x} f(x)\,dx$ by entering the present table with $\nu = 2m-1$, $t = x\sqrt{\nu}/a$, and double-entry interpolation becomes necessary. For most practical purposes double-linear interpolation will be adequate.

Let $4 \leqslant \nu = 2m - 1 \leqslant 24$, and let ν_0 and ν_1 denote two consecutive tabular ν such that $\nu_0 \leqslant \nu \leqslant \nu_1$. Then compute $\phi = \nu - \nu_0$, θ as defined above and $\theta\phi$. The interpolate is

$$P(t \mid \nu) = P_{00}(1 - \theta - \phi + \theta\phi) + P_{10}(\theta - \theta\phi) + P_{01}(\phi - \theta\phi) + P_{11}\theta\phi, \qquad (37)$$

where $P_{ij} = P(t_i \mid \nu_j)$. If $\nu > 24$ the same formula may be used, defining ϕ harmonically by $\phi = 120(\nu_1 - \nu)/(\nu_1\nu)$ and interchanging ν_0 and ν_1 in (37). If $\nu < 4$ it may be preferable to obtain $P(t \mid \nu)$ directly by numerical quadrature.

(d) *Values outside range of Table* 9

In Table 9 the slowly convergent tail of the t-distribution for small ν has been sacrificed in order to keep the tabular t-interval uniform for all ν below 20. To cover the tail region, a few

* Linear interpolation in Table 12, using log Q as argument, will give approximate values for t at intermediate probability levels (see Simaika, 1942, p. 271).

landmarks are provided at the bottom of the second page of Table 9 in the form of percentage points. Other values may be readily obtained from the approximate formula

$$1-P(t\mid\nu)=c_\nu t^{-\nu}, \tag{38}$$

with the c_ν given below:

ν	1	2	3	4	5	6	7
c_ν	0·317	0·488	1·04	2·7	8·2	26	84

For $\nu=1$, only four accurate decimals are given by (38), but since

$$P(t\mid 1)=\tfrac{1}{2}+(\tan^{-1}t)/\pi,$$

more accurate values can be obtained from tables of the inverse tangent if required. Where the table terminates before the bottom of a column is reached, the entry in the first blank space is 1·00000 to 5-decimal accuracy.

5·7 *The power of the t-test; definition of the concept* (*Table* 10)

The chart of Table 10 is arranged to answer questions of the following type. If the t-test is applied at a specified level of significance, say at the 5% level,

(*a*) for a fixed size of sample, what is the probability that the test will establish a given effect as significant, or

(*b*) how large a sample is required in order that the odds may be, say, 99 to 1 on establishing the significance of this effect?

The solution depends on the distribution of a *non-central* Student ratio,

$$t'=(y+\Delta)/s_y=y'/s_y,$$

where y and s_y are defined as before and Δ is a parameter representing the degree of non-centrality. For example, in the application discussed in §5·3, Δ would be the difference between the true means of the populations from which the two samples have been drawn. It may be shown that the distribution of t' depends only on ν, the degrees of freedom of the variance estimate s_y^2 and on the ratio Δ/σ_y. For discussion of the distribution of t' and for references to other available tables see Johnson & Welch (1940), Greenwood & Hartley (1962).

If we write $f(t'\mid\nu,\Delta/\sigma_y)$ as the frequency function of t', then answers to the questions raised above require a knowledge of the following integrals:

(i) For the double-tail t-test,

$$\int_{-\infty}^{-t_{\frac{1}{2}\alpha}} f(t'\mid\nu,\Delta/\sigma_y)\,dt'+\int_{t_{\frac{1}{2}\alpha}}^{\infty} f(t'\mid\nu,\Delta/\sigma_y)\,dt'. \tag{39}$$

(ii) For the single-tail t-test (with $\Delta\geqslant 0$)

$$\int_{t_\alpha}^{\infty} f(t'\mid\nu,\Delta/\sigma_y)\,dt', \tag{40}$$

where $Q(t_\alpha\mid\nu)=\alpha$ defines the significance point for t used in the test. The value of the integral, (39) or (40), considered as a function of Δ/σ_y provides the *power function* of the corresponding test.

The chart shows on a logarithmic scale the integral (39) for $\alpha=0{\cdot}05$ and $0{\cdot}01$. The horizontal scale is in terms of $\phi=\Delta/\sigma_y\times 1/\sqrt{2}$.* The degrees of freedom associated with s_y are denoted by ν, and separate curves are shown for $\nu=6\,(1)\,10$ and afterwards for values at the harmonic

* This chart forms one of a series of 8 charts, already published, constructed for the more general problem of analysis of variance tests, where the ϕ-notation was more appropriate (see Pearson & Hartley, 1951).

intervals 12, 15, 20, 30, 60 and ∞. The limiting case with $\nu = \infty$ corresponds to the situation where the value of σ_y is known and the distributions of both t and of t' are normal curves.

Since the first of the two integrals in (39) becomes rapidly negligible as Δ/σ_y increases above zero (and similarly for the second term as Δ/σ_y decreases below zero), it follows that the chart can also be used to give the integral (40) associated with the single-tail t-test for significance levels $\alpha = 0{\cdot}025$ and $0{\cdot}005$.

5·8 *Illustration of the use of the power chart*

Example 8. Consideration is being given to an experiment in which the effect of two treatments A and B on a variable characteristic x are to be compared. It is believed that x is roughly normally distributed and that the variance σ^2 among the individuals will be the same for either treatment, but it is possible that the mean values of the character, μ_A and μ_B, will differ in the two cases. It is further held that the experiment should be planned on a sufficient scale to detect any real difference $|\mu_A - \mu_B|$ which exceeds 2σ. Supposing each treatment is applied to an equal number of individuals, n, how large should n be? In putting the question in this way, it is realized that the precise value of σ is not known, so that the variance must be estimated from the experimental results and the t-test used. However, it will generally be the case that experience has given a rough lead on the value of σ, so that the experimenter can specify broadly at what multiple of σ the difference $|\mu_A - \mu_B|$ would become of practical significance.

To come to grips with the problem, the statistician must be prepared to assign to the risks involved some numerical values which, though somewhat arbitrary, will help him to explore the position. Suppose, then, that he asks how large the sample sizes n must be so that, using the 5 % level to judge significance of t (double-tail test, $\alpha = 0{\cdot}05$), the chance of establishing significance if $|\mu_A - \mu_B|/\sigma \geqslant 2$ is at least $0{\cdot}98$. The test will be based on the expression for t of equations (34) and (35) with $n_1 = n_2 = n$ and $\nu = 2(n-1)$.

We must first consider the form assumed by Δ/σ_y. In this case σ_y is the standard error of the difference in sample means $\bar{x}_1 - \bar{x}_2$, i.e. $\sigma_y = \sigma\sqrt{(2/n)}$, so that

$$\phi = \frac{\Delta}{\sigma_y}\frac{1}{\sqrt{2}} = \tfrac{1}{2}\sqrt{n}\,\frac{\mu_A - \mu_B}{\sigma}.$$

We have therefore to find the value of n just large enough to make $\phi = \sqrt{n}$ exceed the value needed to give a $0{\cdot}98$ chance of establishing significance when $\nu = 2(n-1)$. The table below gives the relevant information, where the critical values of ϕ are the abscissae of the points on the chart where the members of the family of curves with $\nu = 10$, 12, 15 and 20 and $\alpha = 0{\cdot}05$ cut the horizontal grid drawn at $\beta = 0{\cdot}98$:

$\nu = 2(n-1)$	10	12	15	20
$n = \frac{1}{2}\nu + 1$	6	7	8·5	11
$\sqrt{n}$	2·45	2·65	2·92	3·32
Critical ϕ	3·16	3·10	3·04	2·98

From a rough plot, it will be seen that the cross-over occurs very nearly at $n = 9$, so that we may conclude that with two samples of nine observations the chance would be at least $0{\cdot}98$ of establishing a difference of $|\mu_A - \mu_B| \geqslant 2\sigma$ as significant at the 5 % level.

6. Test for comparisons involving two variances, which must be separately estimated (Table 11)

6·1 *Definition*

In order that s_y^2, the estimate of the variance of y used in the Student ratio, should have the frequency function of equation (28), it is essential that it should be based on the sum of squares of ν independent normal deviates, all having a common variance. Thus in the two sample comparisons discussed in § 5·3, it was necessary that the variance of x should be the same, or approximately so, in both populations sampled.

It will sometimes happen, however, that the variance of y cannot be simply estimated in this way. Thus two samples may have been drawn from normal populations with standard deviations σ_1 and σ_2, where the ratio σ_1/σ_2 is unknown and thought likely to differ considerably from unity. In this case the sampling variance of $\bar{x}_1-\bar{x}_2$ is $\sigma_1^2/n_1+\sigma_2^2/n_2$ and not $\sigma^2(1/n_1+1/n_2)$. This variance may be estimated from $s_1^2/n_1+s_2^2/n_2$, where

$$s_j^2=\sum_{i=1}^{n_j}\frac{(x_{ji}-\bar{x}_j)^2}{n_j-1}\quad (j=1,2),$$

but this estimator no longer has the sampling distribution of equation (28), although s_1^2 and s_2^2, separately, will still follow this law.

R. A. Fisher (following Behrens) has proposed a test for the equality of the population means for use in this situation and tables have been computed by P. V. Sukhatme (see Fisher & Yates, 1963, Tables VI and VI 1). This work, however, involves the use of an argument which is not universally accepted by statisticians. One of the consequences of the use of the Fisher-Behrens test is that, if the null hypothesis is true, the probability (in the usual direct sense) of rejecting it will not equal the figure specified as the level of significance. Fisher, indeed, does not consider this to be a drawback. It should be noted, however, that the table given in the present volume is not related to the Fisher-Behrens test in any way. It is the result of an attempt to produce a test which *does* satisfy the condition that the probability of rejection of the hypothesis tested (or of the truth of a confidence statement) will be equal to a specified figure. The theory of this test is due to B. L. Welch (1947) and the table is computed by A. A. Aspin, W. H. Trickett, B. L. Welch and G. S. James.

The test is applicable to a more general situation than the comparison of two sample means. Let y be a normally distributed variable with expectation η, and suppose that the sampling variance of y is of the form $\lambda_1\sigma_1^2+\lambda_2\sigma_2^2$, where

(*a*) λ_1 and λ_2 are known positive constants,

(*b*) σ_1^2 and σ_2^2 can be estimated independently by statistics s_1^2 and s_2^2, with degrees of freedom ν_1 and ν_2, which have frequency functions given by equation (28), and are independent of each other and of y.

Then the table gives, for four probability levels, critical values of the ratio

$$v=(y-\eta)/(\lambda_1 s_1^2+\lambda_2 s_2^2)^{\frac{1}{2}}. \tag{41}$$

The nature of the problem and its solution is such that the critical levels for v cannot be fixed independently of all sample statistics. They are in fact functions of (i) ν_1 and ν_2, (ii) the significance level α and (iii) the ratio of the observed sample variances, which may be conveniently expressed in the form

$$c=\lambda_1 s_1^2/(\lambda_1 s_1^2+\lambda_2 s_2^2). \tag{42}$$

The quantity tabled may therefore be written as $V(c;\, \nu_1, \nu_2, \alpha)$ and has the property that

$$\Pr\{v \geqslant V(c;\, \nu_1, \nu_2, \alpha)\} = \alpha,$$

whatever may be the value of the unknown population variances σ_1^2 and σ_2^2. The reference set in which probabilities are calculated is that of the totality of possible samples with y, s_1^2 and s_2^2 all free to vary. The numerical values in the table are not therefore comparable with any which may be derived from other probability interpretations.

Four probability levels are given for the single-tail test, namely for $\alpha = 0{\cdot}05$, $0{\cdot}025$, $0{\cdot}01$ and $0{\cdot}005$. For the double-tail test, $\Pr\{|v| \geqslant V(c;\, \nu_1, \nu_2, \alpha)\} = 2\alpha$. Although the values of V have been rounded off to 2 decimal places, the probability of v exceeding the tabled value will practically always equal α to 3 decimal places for $\alpha = 0{\cdot}05$ and $0{\cdot}025$ and to 4 decimal places for $\alpha = 0{\cdot}01$ and $0{\cdot}005$.

Linear interpolation is adequate everywhere except, for each ν, in the panel which includes $\nu = \infty$; here harmonic linear interpolation with $20/\nu$ or $30/\nu$ as argument will be required.

The method of computation described by Welch does not, in general, provide two decimal place accuracy in $V(c;\, \nu_1, \nu_2, \alpha)$ for degrees of freedom smaller than those shown in the table. As is seen, this limitation is more severe the smaller α.

6·2 *Application of Table* 11

Example 9. A useful application occurs in the case already discussed of the test of significance of the difference between two means, when the population variances cannot be assumed equal. If the population means are μ_1 and μ_2, set $\eta = \mu_1 - \mu_2$ while $y = \bar{x}_1 - \bar{x}_2$. Then

$$\lambda = 1/n_1, \quad \lambda_2 = 1/n_2,$$

$$v = \frac{(\bar{x}_1 - \bar{x}_2) - (\mu_1 - \mu_2)}{(s_1^2/n_1 + s_2^2/n_2)^{\frac{1}{2}}}, \qquad c = \frac{s_1^2/n_1}{s_1^2/n_1 + s_2^2/n_2}, \quad \nu_1 = n_1 - 1, \quad \nu_2 = n_2 - 1.$$

The tables may then be used to make inferences regarding $\mu_1 - \mu_2$. Thus if $n_1 = 10$, $n_2 = 15$, $\bar{x}_1 = 73{\cdot}4$, $\bar{x}_2 = 47{\cdot}1$, $s_1^2 = 51$, $s_2^2 = 141$, we shall have

$$v = \frac{26{\cdot}3 - (\mu_1 - \mu_2)}{3{\cdot}81}, \qquad c = 0{\cdot}352, \quad \nu_1 = 9, \quad \nu_2 = 14.$$

From the tables $\qquad V(c;\, \nu_1, \nu_2, \alpha) = V(0{\cdot}352;\, 9, 14, 0{\cdot}05) = 1{\cdot}71.$

If it were a matter of testing the consistency of the data with the hypothesis that $\mu_1 = \mu_2$, we should have $v = 26{\cdot}3/3{\cdot}81 = 6{\cdot}90$, which is clearly far beyond the tabled 5% point.

In obtaining an *interval estimate* for $\mu_1 - \mu_2$ we should note that in the above description of the tables we have been dealing explicitly with single-tail probabilities. The chance that v exceeds the tabled value *numerically*, either in the positive or negative direction, is 2α. Thus in the sense of confidence interval theory, (see §§ 4·3, 5·5)

$$\Pr\{-1{\cdot}71 \leqslant [26{\cdot}3 - (\mu_1 - \mu_2)]/3{\cdot}81 \leqslant 1{\cdot}71\} = 1 - 2\alpha = 0{\cdot}90,$$

i.e.

$$\Pr\{19{\cdot}8 \leqslant \mu_1 - \mu_2 \leqslant 32{\cdot}8) = 0{\cdot}90,$$

so that the 90% confidence limits for $\mu_1 - \mu_2$ are 19·8 and 32·8.

The tables may also be used for the comparison of two regression coefficients, where the array distributions in the two populations are normal and homoscedastic, but the two within-array variances are thought likely to be unequal.

7. The correlation coefficient in normal samples (Tables 13, 14 and 15)

7·1 *Definitions and tables*

The bivariate normal distribution law has the form

$$f(x,y) = [2\pi\sigma_x\sigma_y\sqrt{(1-\rho^2)}]^{-1}\exp\left\{-\frac{1}{2(1-\rho^2)}\left[\left(\frac{x-\mu_x}{\sigma_x}\right)^2 - 2\rho\left(\frac{x-\mu_x}{\sigma_x}\right)\left(\frac{y-\mu_y}{\sigma_y}\right) + \left(\frac{y-\mu_y}{\sigma_y}\right)^2\right]\right\}. \tag{43}$$

It depends on five parameters: the means of x and y, μ_x and μ_y; the standard deviations of x and y, σ_x and σ_y; the coefficient of correlation, ρ. Given that this law holds, the intensity of the relationship between x and y is thus completely specified by the single parameter ρ.

If x_i, y_i $(i = 1, 2, \ldots, n)$ denote a random sample of n paired observations from (43), then ρ may be estimated from the sample product-moment correlation coefficient r, defined by

$$r = \frac{\sum\limits_i (x_i - \bar{x})(y_i - \bar{y})}{\{\sum\limits_i (x_i - \bar{x})^2 \cdot \sum\limits_i (y_i - \bar{y})^2\}^{\frac{1}{2}}},$$

where $\bar{x}$ and $\bar{y}$ are the mean values of x and y in the sample.

The frequency function of r, which clearly depends on ρ and n only, was first derived by R. A. Fisher (1915) and is given by

$$f(r \mid n, \rho) = \frac{(1-\rho^2)^{\frac{1}{2}(n-1)}}{\pi(n-3)!}(1-r^2)^{\frac{1}{2}(n-4)}\frac{d^{n-2}}{d(r\rho)^{n-2}}\left(\frac{\cos^{-1}(-r\rho)}{\sqrt{(1-r^2\rho^2)}}\right). \tag{44}$$

Tables of this function and of its probability integral were computed by F. N. David (1938), for $\rho = 0{\cdot}0\,(0{\cdot}1)\,0{\cdot}9$ and $n = 3\,(1)\,25, 50, 100, 200, 400$. The present volume contains three tables which will be found useful in handling some, if not all, aspects of the correlation problem.

(*a*) *Charts of Table* 15

If we write
$$P(r \mid n, \rho) = \int_{-1}^{r} f(u \mid n, \rho)\, du$$

for the probability integral of the distribution (44), Table 15 shows in graphical form the roots ρ_A and ρ_B $(>\rho_A)$ of

$$\alpha = P(r \mid n, \rho_B) \quad \text{and} \quad 1 - \alpha = P(r \mid n, \rho_A),$$

plotted against r for a number of selected sample sizes n. These graphs are assembled in two charts, taken from David's 1938 Tables, corresponding to $\alpha = 0{\cdot}025$ and $0{\cdot}005$.

(*b*) *Percentage points of the distribution* $f(r \mid n, \rho = 0)$ (*Table* 13)

In the special case, $\rho = 0$, we have from (44)

$$f(r \mid n, 0) = \frac{\Gamma(\frac{1}{2}n - \frac{1}{2})}{\sqrt{\pi}\,\Gamma(\frac{1}{2}n - 1)}(1-r^2)^{\frac{1}{2}(n-4)}, \tag{45}$$

which can be transformed into the distribution of t (§ 5·1) for $\nu = n - 2$ degrees of freedom, through the relation
$$r^2 = t^2/(\nu + t^2). \tag{46}$$

Table 13 gives the percentage points of the distribution (45); alternatively, these could be calculated from the percentage points of t (Table 12) by means of (46).

The distribution (45) has a variance of $1/(n-1)$, and it tends fairly rapidly to the normal form as n increases.

(c) *Table to assist the application of Fisher's z-transformation* (*Table* 14)

In order to deal with the general distribution (44), Fisher (1921*a*) introduced the transformation

$$z = \tfrac{1}{2}\log_e \frac{1+r}{1-r} = \tanh^{-1} r, \quad \zeta = \tanh^{-1}\rho, \tag{47}$$

and showed that the moments of z could be expanded in series of inverse powers of $n-1$. These expressions are as follows:*

$$\left.\begin{aligned} \text{Mean } z &= \zeta + \frac{\rho}{2(n-1)}\left\{1 + \frac{5+\rho^2}{4(n-1)} + \dots\right\}, \\ \mu_2(z) &= \frac{1}{n-1}\left\{1 + \frac{4-\rho^2}{2(n-1)} + \frac{22-6\rho^2-3\rho^4}{6(n-1)^2} + \dots\right\}, \\ \beta_1(z) &= \frac{\mu_3^2}{\mu_2^3} = \frac{\rho^6}{(n-1)^3} + \dots, \\ \beta_2(z) &= \frac{\mu_4}{\mu_2^2} = 3 + \frac{2}{n-1} + \frac{4+2\rho^2-3\rho^4}{(n-1)^2} + \dots. \end{aligned}\right\} \tag{48}$$

From these results he concluded that z could be regarded as approximately normally distributed about a mean of $\zeta = \tanh^{-1}\rho$, with variance $1/(n-3)$. This approximation has been shown to be remarkably effective, the greatest error, which can in some situations be allowed for, arising from the neglect of the term $\tfrac{1}{2}\rho/(n-1)$ in the expression for mean z.

The function $z = \tanh^{-1} r$ is given in Table 14; with its help certain tests regarding correlation coefficients can be reduced to familiar tests based on the normal distribution. Methods for interpolation in the table are given at the foot. These may result in errors of a unit in the last figure.

7·2 *Applications. Tests regarding the correlation in a single sample*

Example 10*a*. In a sample of twenty-five pairs of normally correlated observations, (x, y), a correlation of $r = 0{\cdot}65$ was found. Is this result consistent with a population correlation coefficient of $\rho = 0{\cdot}5$?

We find from Table 14 that $z = 0{\cdot}775$, $\zeta = 0{\cdot}549$, and therefore the question asked is equivalent to the following: can

$$X = (z-\zeta)\sqrt{(n-3)} = (0{\cdot}775 - 0{\cdot}549)\sqrt{22} = 1{\cdot}06$$

be reasonably regarded as a random normal deviate, having unit standard deviation? Clearly this is so, and the observed result is consistent with the hypothesis. It will be noted that if we introduced the second term in the series for mean z (see equations (48)), namely,

$$\tfrac{1}{2}\rho/(n-1) = \tfrac{1}{2}\,0{\cdot}50/24 = 0{\cdot}010,$$

the conclusion would not be altered.

We may also deal with this question by reference to the charts of Table 15. The first of these shows that for $n = 25$ and $\rho = 0{\cdot}5$ the upper 2·5 % level for r lies at about 0·75, well beyond the observed $r = 0{\cdot}65$.

The significance points for the special case $\rho = 0$, given in Table 13, are of particular importance, as they enable us to judge whether an observed sample correlation is or is not indicative of any real correlation in the population. For medium and large samples we may test this result

* The equations (48) are those given by Gayen (1951), who found that an error in one of Fisher's original formulae (1921 *a*, first equation on p. 13) made certain corrections to the moment expansions necessary.

by dividing the observed r by $1/\sqrt{(n-1)}$, its standard error on the assumption that $\rho=0$ and referring to the normal probability scale, i.e. by asking whether $r\sqrt{(n-1)}$ can be regarded as a unit normal deviate. For small samples, Table 13 provides an immediate and more precise answer.

Example 10*b*. We may ask whether the correlation of $r=0{\cdot}65$ referred to above is clearly indicative of a real correlation in the population sampled, when the sample consists of only twenty-five observations. We see at once from Table 13, using the double-tail test for $\nu=n-2=23$, that the observed r falls beyond even the 0·1 % point which lies at 0·62. The correlation is, therefore, clearly significant.

7·3 *Confidence limits for ρ*

The charts of Table 15 may be used to derive an *interval estimate* for ρ from a sample value of r. The probability argument will follow the lines already used in §§ 4·3 and 5·5. The *confidence coefficients*, using the central interval, are $1-2\alpha=0{\cdot}95$ for the first chart and $1-2\alpha=0{\cdot}99$ for the second.

Example 10*c*. Take the case of a sample of twenty-five pairs of observations, giving a correlation $r=0{\cdot}65$. Assuming that this sample has been randomly drawn from a bivariate normal population, what limits may be assigned to ρ? Turning to the first chart, we note that the ordinate at $r=0{\cdot}65$ cuts the curves for $n=25$ at $\rho_A=0{\cdot}34$ and $\rho_B=0{\cdot}82$. Hence we may say that

$$0{\cdot}34 \leqslant \rho \leqslant 0{\cdot}82,$$

with a probability 0·95 (in the sense defined) of being correct. If we do not wish to accept so large a risk of error, we may turn to the second chart and, again, reading off the ρ-values of the points where the ordinate at $r=0{\cdot}65$ cuts the pair of curves for $n=25$, obtain the inequality

$$0{\cdot}21 \leqslant \rho \leqslant 0{\cdot}87$$

for a confidence coefficient 0·99 (see §§ 4·3, 5·5)

7·4 *The significance of the difference between two sample correlations*

If two independent samples of n_1 and n_2 paired observations give correlation coefficients r_1 and r_2, respectively, we may ask whether these values are consistent with the hypothesis that the population correlations ρ_1 and ρ_2 are the same. Using the z-transformation, it is only necessary to refer the ratio

$$X=(z_1-z_2)/\{1/(n_1-3)+1/(n_2-3)\}^{\frac{1}{2}}$$

to the normal probability scale.*

Example 11. Suppose that

$$n_1=30, \quad r_1=0{\cdot}612; \qquad n_2=40, \quad r_2=0{\cdot}423.$$

Is the difference between r_1 and r_2 significant? We find from Table 14 that

$$z_1=0{\cdot}712, \quad z_2=0{\cdot}451.$$

Hence $$X=(0{\cdot}712-0{\cdot}451)/\{\tfrac{1}{27}+\tfrac{1}{37}\}^{\frac{1}{2}}=1{\cdot}03.$$

Clearly the result is not significant.

* The error involved in omitting the corrective terms $\frac{1}{2}\rho/(n-1)$ and assuming that if $\rho_1=\rho_2$ then $\zeta_1=\zeta_2$, can always be assessed roughly, e.g. by inserting $\frac{1}{2}(r_1+r_2)$ as an assumed common value of ρ.

7·5 *Tests involving more than two correlation coefficients*

The property that $(z-\zeta)\sqrt{(n-3)}$ is approximately a unit normal deviate makes available a variety of tests based on the distribution of χ^2.

Example 12. The following results from a paper by Tschepourkowsky (1905) have been used by Tippett (1952, § 9·23) and F. N. David (1938, p. xxiv) to illustrate tests for heterogeneity among correlation coefficients. Col. 3 of the table gives the values of the correlation, r_t, between cephalic index and upper face form, for samples of n_t (col. 2) skulls from each of the $k=13$ different races shown in col. 1. The form of analysis will depend on what are the alternatives to the hypothesis which it is wished to test. Thus, assuming that the two characters are normally correlated:

(*a*) We may ask whether the data are consistent with the hypothesis that there is a common, but unknown, value of ρ for all thirteen races. The alternative is, here, that there are different correlations ρ_t for the races.

(*b*) Or, again, we may ask whether the assumed common correlation ρ has a specified value ρ_0, the alternative being that $\rho \neq \rho_0$.

(*c*) In certain situations we may wish to test directly whether a set of correlation coefficients r_t are consistent with a specified population correlation ρ_0, the alternative being different correlations ρ_t.

Race (1)	No. of skulls n_t (2)	Correlation r_t (3)	z_t (4)	n_t-3 (5)
Australians	66	+0·089	+0·089	63
Negroes	77	+0·182	+0·184	74
Duke of York Islanders	53	−0·093	−0·093	50
Malays	60	−0·185	−0·187	57
Fijians	32	+0·217	+0·221	29
Papuans	39	−0·255	−0·261	36
Polynesians	44	+0·002	+0·002	41
Alfourous	19	−0·302	−0·312	16
Micronesians	32	−0·251	−0·256	29
Copts	34	−0·147	−0·148	31
Etruscans	47	−0·021	−0·021	44
Europeans	80	−0·198	−0·201	77
Ancient Thebans	152	−0·067	−0·067	149
Total	735			696

Case (*a*). Col. 4 of the table shows the value of z_t obtained from Table 14, while col. 5 shows n_t-3, the reciprocal of the approximate sampling variance of z_t. We now calculate

$$\chi^2 = \sum_t (n_t-3)(z_t-\bar{z})^2, \tag{49}$$

where

$$\bar{z} = \sum_t (n_t-3)z_t \Big/ \sum_t (n_t-3), \quad k=13. \tag{50}$$

We find $\bar{z} = -0{\cdot}0609$, $\chi^2 = 14{\cdot}65$. The appropriate degrees of freedom will be $\nu = k-1 = 12$. Referring to Table 8, we find that the observed χ^2 falls well below the 5 % point (21·03), so that there appears no reason to reject the assumption of a common correlation coefficient. A good estimate of this common value is given by $\hat{\rho} = \tanh \bar{z} = -0{\cdot}061$.

Case (*b*). In view of (50), the sampling variance of $\bar{z}$ is $1/\{\sum_t (n_t - 3)\}$. We therefore deal with this case by calculating the ratio

$$X = (\bar{z} - \zeta_0)\sqrt{(\sum_t (n_t - 3))}, \tag{51}$$

where $\zeta_0 = \tanh^{-1}\rho_0$, and referring to the normal probability scale. If, for the data of the example, we wish to test whether $\rho_0 = 0$, i.e. whether $\zeta_0 = 0$, we find

$$X = -0{\cdot}0609\sqrt{696} = -1{\cdot}607,$$

a figure which is not significant at the 5 % level (Table 4).

Case (*c*). This may be regarded as a fusion of tests (*a*) and (*b*). Indeed, on adding X^2 from (51) to the χ^2 of (49) we obtain $\sum_t (n_t - 3)(z_t - \zeta_0)^2$, a χ^2 for $k = 13$ degrees of freedom which, in our example, is 17·23, confirming the insignificant results of (*a*) and (*b*). Normally, however, we would compute the separate components (49) and (51) in order not to confuse the issues, (*a*) and (*b*).

Other cases. It is clear that certain more complex problems, with samples falling into different groups, will lead to tests on the transformed variables *z* following the appropriate analysis of variance patterns.

8. The incomplete B-function (Tables 16 and 17)

8·1 *Definitions*

The incomplete B-function ratio is defined as

$$I_x(a,b) = \frac{B_x(a,b)}{B(a,b)}$$

$$= \frac{\Gamma(a+b)}{\Gamma(a)\,\Gamma(b)}\int_0^x u^{a-1}(1-u)^{b-1}\,du. \tag{52}$$

In the notation which we have used hitherto for a probability integral, we should write

$$P(x \mid a,b) = I_x(a,b).$$

It appears, however, preferable to retain the traditional notation for this integral.*

When the fundamental 7-decimal-place tables of this function were published by Karl Pearson (1934) it was realized that they might form a basis for shorter tables suited for use in special problems. The present Table 16 satisfies a requirement of this kind; it gives lower percentage points of the distribution of *x*, i.e. the roots $x(I \mid a, b)$ of $I_x(a,b) = P$ for $P = 0{\cdot}001$, 0·0025, 0·005, 0·01, 0·025, 0·05, 0·10, 0·25 and 0·50. The corresponding upper percentage points can be obtained from the relation

$$I_x(a,b) = 1 - I_{1-x}(b,a), \tag{53}$$

i.e. by interchanging *a* and *b* and taking $1-x$ as the percentage point.

The chart of Table 17, whose use is explained in § 8·2 below, gives $I_x(a,b)$ directly to graphical accuracy, but for ranges of argument slightly exceeding those of Pearson's tables. It is believed that it will be of value in a number of applications where the percentage point Tables 16 and 18 are not appropriate, and where Karl Pearson's (1934) tables are not available.

* Because of the importance of the application to the binomial, where *x* corresponds to the *p* in the expansion of $(q+p)^n$, we have altered the traditional notation from $I_x(p,q)$ to $I_x(a,b)$ to avoid confusion.

Probably the most important applications of the incomplete B-function follow from:

(*a*) *Its relation to the distribution of the variance ratio, F* (see §§ 9·1, 9·2 below)

If χ_1^2 and χ_2^2 are two independent statistics, each distributed as χ^2 (i.e. with the probability integral of equation (15)) and having ν_1 and ν_2 degrees of freedom, respectively, then the frequency function of

$$x=\chi_1^2/(\chi_1^2+\chi_2^2)$$

is

$$f(x\,|\,\nu_1,\nu_2)=\frac{\Gamma(\tfrac{1}{2}\nu_1+\tfrac{1}{2}\nu_2)}{\Gamma(\tfrac{1}{2}\nu_1)\,\Gamma(\tfrac{1}{2}\nu_2)}\,x^{\frac{1}{2}\nu_1-1}(1-x)^{\frac{1}{2}\nu_2-1}. \tag{54}$$

If we now divide each χ^2 by its degrees of freedom and take the ratio of the quotients

$$F=\nu_2\chi_1^2/(\nu_1\chi_2^2),$$

it will follow that the probability that F exceeds a specified value, F_0, is

$$\int_{F_0}^{\infty} f(F\,|\,\nu_1,\nu_2)\,dF=I_{x_0}(\tfrac{1}{2}\nu_2,\tfrac{1}{2}\nu_1), \tag{55}$$

where

$$x_0=\nu_2/(\nu_2+\nu_1F_0). \tag{56}$$

Table 18 was, in fact, derived in this way from Table 16.

(*b*) *Its relation to the cumulative sum of the terms of the binomial expansion* $(q+p)^n$

Here

$$\sum_{i=a}^{n}\binom{n}{i}p^i(1-p)^{n-i}=I_p(a,n-a+1). \tag{57}$$

For a few selected values of n, individual terms of the binomial distribution $\binom{n}{i}p^i(1-p)^{n-i}$ are given in Table 37 which is mainly intended for illustrative purposes. This table is discussed in § 18 below. More comprehensive tables of the binomial distribution are now available elsewhere (see, for example, National Bureau of Standards, Applied Mathematics Series, no. 6, 1950).

The charts of Table 41, discussed in § 22·3 give a method of obtaining confidence intervals for the binomial probability p, based on an observed proportion of successes i/n, for selected values of sample sizes n.

(*c*) *Its relation to the Pearson Type I curve*

The probability integral of this curve (see § 23·1 below) namely, of

$$y=y_0\left(1+\frac{x'}{a_1}\right)^{m_1}\left(1-\frac{x'}{a_2}\right)^{m_2},\quad(-a_1\leqslant x'\leqslant a_2), \tag{58}$$

can be obtained from the integral (52) by setting

$$x=(a_1+x')/(a_1+a_2),\quad a=m_1+1,\quad b=m_2+1.$$

The Type I curve has been found to play a part similar to that of the Type III curve in approximating to some of the less tractable distributions of statistical theory (see §§ 3·5 and 8·4).

Numerous other applications of the integral are given by Karl Pearson (1934), and some of these are discussed below.

8·2 *Description and use of the chart in Table* 17

On the top edge of the chart a scale of $b\ (=\tfrac{1}{2}\nu_1)$ is provided which is split into three sections:

A, $1\leqslant b\leqslant 4$; B, $4\leqslant b\leqslant 15$; C, $15\leqslant b\leqslant 60$.

These scales form the upper margins of the three corresponding chart sections A, B and C.

Each section is further divided into upper and lower parts. The lower part contains a family of curves corresponding to the values of the parameter a ($=\frac{1}{2}\nu_2$) shown at the right-hand end of the curves, while the top part contains a family of curves corresponding to selected values of the probability $I_x(a, b)$. Both families are referred to the same abscissa scale b at the top edge. The two families are linked by the aid of x-scales provided on the ruler enclosed in the pocket inside the back cover of the volume. Each x-scale is marked with an arrow ($\rightarrow$) at the 'pivot point', whose significance is explained below.

To find $P = I_x(a, b)$ for given a, b and x, with $a \geqslant b$, we proceed as follows. Choosing the appropriate scale (A, B or C), place the ruler vertically on the chart so that its x-scale intersects the abscissa scale (at top of chart) at the given value of b. Slide the rule up or down this vertical until its pivot mark ($\rightarrow$) is at the intersection of the vertical with the curve corresponding to the given value of a in the bottom part. With the x-scale in this position the required value of $P = I_x(a, b)$ can be read off in the top part opposite the given x.

When $a < b$ find $I_{1-x}(b, a)$ and use relation (53).

8·3 *Illustration of applications of the chart*

Example 13*a*. Given $a = 30$, $b = 5$, it is required to find $I_x(a, b)$ for $x = 0{\cdot}70$, $0{\cdot}94$ and $0{\cdot}96$.

Following the above procedure we find from section B of the chart:

$$I_{0\cdot 70}(30, 5) = 0{\cdot}012, \quad I_{0\cdot 94}(30, 5) = 0{\cdot}95\,(2), \quad I_{0\cdot 96}(30, 5) = 0{\cdot}99\,(0).$$

Example 13*b*. In binomial theory, given that $p = 0{\cdot}2$ and $n = 25$, it is required to find the chance of obtaining at least $a = 10$ successes. Reference to equation (57) shows that this chance is given by $I_{0\cdot 2}(10, 16)$ or $1 - I_{0\cdot 8}(16, 10)$, so that, using section B, we find it to be equal to $1 - 0{\cdot}985 = 0{\cdot}015$.

Example 13*c*. Given $\nu_1 = 40$ and $\nu_2 = 120$ degrees of freedom, find the chance that $F \geqslant 1{\cdot}5$. We have from (56),

$$x_0 = 120/(120 + 1{\cdot}5 \times 40) = 120/180 = 0{\cdot}667,$$

and using section C we find this chance to be a little over 0·05, a result confirmed by the 5% value of F, given in Table 18 as 1·50.

Example 14.* A choice between two manufactured products, I and II, is to be made. The relative variability, σ_1^2/σ_2^2, of a characteristic of the two products is relevant, but of secondary importance in the sense that: (*a*) it is immaterial which is chosen if $\sigma_1^2/\sigma_2^2 = 2$, (*b*) product I will be preferred if $\sigma_1^2/\sigma_2^2 < 2$, and (*c*) product II if $\sigma_1^2/\sigma_2^2 > 2$. The characteristic is to be measured for samples of 41 items from each of the products, the two sample mean-square estimates of variance s_1^2 and s_2^2 are to be computed, and it is decided to accept product I if $s_1^2/s_2^2 < 2$ and product II if $s_1^2/s_2^2 > 2$. What is the chance that product I will be accepted if, in fact, $\sigma_1^2/\sigma_2^2 = 1{\cdot}2$?

If $\sigma_1^2/\sigma_2^2 = 1{\cdot}2$, the ratio s_1^2/s_2^2 is distributed as $1{\cdot}2F$ for $\nu_1 = 40$, $\nu_2 = 40$ degrees of freedom, so that we must find

$$\Pr\{1{\cdot}2F < 2\} = \Pr\{F < 1{\cdot}67\}.$$

From (56) we compute the equivalent x_0 value as

$$x_0 = 40/(40 + 1{\cdot}67 \times 40) = 1/2{\cdot}67 = 0{\cdot}375.$$

Placing the ruler x-scale for section C on the vertical marked $b = \frac{1}{2}\nu_1 = 20$ and pivoting it to $a = \frac{1}{2}\nu_2 = 20$, we read opposite $x = 0{\cdot}375$ that $I_{0\cdot 375}(20, 20) = 0{\cdot}05$ so that our chance is 0·95 or 95%.

* A similar example is discussed by Eisenhart *et al.* (1947).

8·4 *Applications of Table* 16

In a number of problems there is a direct requirement for the percentage levels of x rather than those of the transformed variables F or Fisher's $z=\frac{1}{2}\log_e F$. A case in point concerns the multiple correlation coefficient R in samples from uncorrelated normally distributed material. Thus if for each of n 'subjects', $k+1$ variables have been measured and it is desired to assess the dependence of one of these (say y) on the k remaining variables $(x_1, \ldots, x_k)$, then R^2 is given by

$$R^2=S_1/(S_1+S_2),$$

where S_1 is the sums of squares component of y due to its multilinear regression on the x_i and S_2 is the residual sum of squares about this regression. A comparison with (54) shows that, in the case of independence, R^2 is directly distributed as a B-variable with $a=\frac{1}{2}k$, $b=\frac{1}{2}(n-k-1)$.

Again, if a random variable can assume only values between 0 and 1, if it has a mean value of μ_1' and a second moment about 0 of μ_2', then $I_x(a,b)$ with

$$\left.\begin{aligned} a&=\mu_1'(\mu_1'-\mu_2')/(\mu_2'-\mu_1'^2),\\ b&=(1-\mu_1')\,(\mu_1'-\mu_2')/(\mu_2'-\mu_1'^2),\end{aligned}\right\} \qquad (59)$$

will often give a very close approximation to the true probability integral. The assumption here made is that the distribution may be represented by a Type I curve having the correct terminal points and first two moments. Use has been made of this approximation by Neyman & Pearson (1931), Bishop (1939) and others in determining probability levels for test statistics derived by application of the likelihood-ratio principle. Earlier, Karl Pearson had used a four-moment fit of a Type I curve to graduate the hypergeometric distribution (see Karl Pearson, 1934, p. xxxvi).

8·5 *Interpolation in Table* 16

The methods of Lagrangian interpolation which may be used have been discussed in detail by Hartley (1941) and are partly based on a special table of Lagrangian coefficients for harmonic interpolation (Comrie & Hartley, 1941). Here we confine ourselves to interpolation in the neighbourhood of the singular point $a=b=\infty$. If x is required for values $2a=\nu_2>60$, $2b=\nu_1>40$, interpolation between the tabular values is not possible and x has to be computed *ab initio*.

Numerous approximate formulae for the incomplete B-function are valid in this range (Soper, 1921; Wishart, 1927); others give the percentage points $x(I\mid a,b)$ directly (Halton Thomson, 1947; Carter, 1947; Wise, 1950; and also Cochran, 1940, who deals with Fisher's z-transformation of $x(I\mid a,b)$). When a and b are both moderately large, Carter's formula appears to be convenient and may be used as follows:

If X denotes the standardized normal deviate corresponding to $P=1-I$, and if $\lambda=\frac{1}{6}(X^2-3)$, $\tau=\frac{1}{2}\lambda+\frac{5}{12}$ (all tabled below), we compute in turn

$$A=\frac{1}{12}\left(\frac{1}{a-\frac{1}{2}}+\frac{1}{b-\frac{1}{2}}\right),\quad h=\frac{1}{3A},$$

$$z=X\sqrt{(h+\lambda)/h}-\left(\frac{1}{b-\frac{1}{2}}-\frac{1}{a-\frac{1}{2}}\right)(\tau-A),$$

$$x(I\mid a,b)=a/(a+be^{2z}).$$

P	0·50	0·25	0·10	0·05	0·025	0·01	0·005
X	0	0·6745	1·2816	1·6449	1·9600	2·3263	2·5758
λ	−0·5000	−0·4242	−0·2263	−0·0491	0·1402	0·4020	0·6058
τ	0·1667	0·2046	0·3035	0·3921	0·4868	0·6177	0·7196

As an illustration we may recompute the tabular value $x(0{\cdot}01 \mid 60, 20)$, i.e. the entry for $2a = 120$ and $2b = 40$ in the section of Table 16 for $I = 0{\cdot}01$.

We obtain

$$1/(a-\tfrac{1}{2}) = 0{\cdot}0168067, \quad 1/(b-\tfrac{1}{2}) = 0{\cdot}0512821,$$

$$A = 0{\cdot}00567407, \quad h = 58{\cdot}747,$$

$$z = \frac{2{\cdot}3263\sqrt{59{\cdot}149}}{58{\cdot}747} - 0{\cdot}0344754 \times 0{\cdot}6120 = 0{\cdot}28345.$$

Hence $x = 0{\cdot}62988$, which agrees with the tabular entry.

9. The F-distribution (Tables 18 and 19)

9·1 *Definition and relation to other tables*

Consider two independent χ^2-values, χ_1^2 and χ_2^2, based respectively on ν_1 and ν_2 degrees of freedom; the variance ratio F is defined by

$$F = \frac{\chi_1^2}{\nu_1} \Big/ \frac{\chi_2^2}{\nu_2}, \tag{60}$$

and has for its frequency function

$$f(F) = \frac{\Gamma(\tfrac{1}{2}\nu_1 + \tfrac{1}{2}\nu_2)}{\Gamma(\tfrac{1}{2}\nu_1)\,\Gamma(\tfrac{1}{2}\nu_2)} \nu_1^{\frac{1}{2}\nu_1} \nu_2^{\frac{1}{2}\nu_2} F^{\frac{1}{2}\nu_1 - 1} (\nu_2 + \nu_1 F)^{-\frac{1}{2}(\nu_1+\nu_2)}. \tag{61}$$

It is useful to note that

$$\text{Mean } F = \nu_2/(\nu_2 - 2) \quad \text{for} \quad \nu_2 > 2,$$

$$\sigma_F = \frac{\nu_2}{\nu_2 - 2}\left\{\frac{2(\nu_1 + \nu_2 - 2)}{\nu_1(\nu_2 - 4)}\right\}^{\frac{1}{2}} \quad \text{for} \quad \nu_2 > 4.$$

For large ν_2, F tends to be distributed as χ^2/ν_1 with a mean of unity and standard deviation $\sqrt{(2/\nu_1)}$. Table 18 gives seven upper percentage points of F, that is to say, it gives the roots, $F(Q \mid \nu_1, \nu_2)$, of the equation

$$Q = \int_{F(Q \mid \nu_1, \nu_2)}^{\infty} f(F)\, dF$$

for $Q = 1 - P = 0{\cdot}25$, $0{\cdot}10$, $0{\cdot}05$, $0{\cdot}025$, $0{\cdot}01$, $0{\cdot}005$ and $0{\cdot}001$. Except in the case of the 0·1 % points, all the results were obtained by rounding off the values in Merrington & Thompson's (1943) table of percentage points of F. Where the rounding-off was uncertain, values were recomputed from Karl Pearson's *Tables of the Incomplete* B*-function* or *ab initio*. A table of 0·1 % points of $z = \frac{1}{2}\log_e F$ was originally computed by Colcord & Deming (1935); the 0·1 % tables of F and z given by Fisher & Yates (1938) were based on this table. Norton (1952) has recently checked the Fisher & Yates tables and given a list of corrections. The present 0·1 % table for F utilizes all this information. In addition, a recalculation has been made of all entries given by Colcord & Deming but not by Fisher & Yates, i.e. entries with $\nu_1 = 7, 9, 10, 11.$ 15, 20, 30, 40, 60 and 120.

For consistency throughout the volume the symbol P is retained to denote the conventional probability integral; it follows that, as in the case of the χ^2-test, $Q = 1 - P$ is used to denote the upper-tail area of the distribution usually required in tests of significance. To obtain lower percentage points it is only necessary to interchange the values of ν_1 and ν_2 in entering the table and to take for F the reciprocal of the value so obtained. For example, if $\nu_1 = 12$, $\nu_2 = 40$ the upper 0·5 % point is seen to be $F(0{\cdot}005 \mid 12, 40) = 2{\cdot}95$ whilst the lower 0·5 % point is

$$F(0{\cdot}995 \mid 12, 40) = 1/F(0{\cdot}005 \mid 40, 12) = 1/(4{\cdot}23) = 0{\cdot}236.$$

The marginal row of each table under the heading $\nu_2 = \infty$ provides the corresponding upper percentage point of the distribution of χ^2/ν with $\nu = \nu_1$ degrees of freedom. The marginal column of each table for $\nu_1 = \infty$ gives the upper percentage points of ν/χ^2 or its reciprocal gives the lower percentage point of χ^2/ν with $\nu = \nu_2$ degrees of freedom. The percentage points of the t-distribution for $\nu = \nu_2$ degrees of freedom may be obtained from the columns of the tables headed $\nu_1 = 1$, since in this case $t = \sqrt{F}$.

The link between the distribution of F and the B-variable, x, has already been referred to (§ 8·1). This well-known relation enables us:

(*a*) To compute percentage points of F from those of x, utilizing the greater accuracy of x in Table 16.

(*b*) To obtain values of the probability integral of F, either from Karl Pearson's (1934) tables or, more approximately, from the chart of Table 17 for $I_x(a, b)$.

In both these cases we use the relation

$$\left.\begin{aligned} &\int_{F_0}^{\infty} f(F \mid \nu_1, \nu_2)\, dF = I_{x_0}(a, b), \\ \text{where} \quad &F_0 = a(1-x_0)/(bx_0), \quad a = \tfrac{1}{2}\nu_2, \quad b = \tfrac{1}{2}\nu_1. \end{aligned}\right\} \tag{62}$$

9·2 *Application to the analysis of variance*

In the terminology of the analysis of variance, let S_1 and S_2 be two component sums of squares having, respectively, ν_1 and ν_2 degrees of freedom, which have arisen in the analysis of a system of normal variates. If all the variates have a common standard deviation σ and if S_1 and S_2 are independent, then S_1/σ^2 and S_2/σ^2 are distributed independently as χ^2 with ν_1 and ν_2 degrees of freedom respectively. Introducing the mean squares, $s_1^2 = S_1/\nu_1$ and $s_2^2 = S_2/\nu_2$, as two independent estimates of σ^2, the variance ratio

$$F = \frac{s_1^2}{s_2^2} = \frac{S_1}{\nu_1\sigma^2} \Big/ \frac{S_2}{\nu_2\sigma^2}$$

is clearly of the form (60) and is accordingly distributed in the standard F-distribution (61), which is independent of the unknown value of σ^2. This ratio may therefore be used for the comparison of any two independent variance estimates obtained in an analysis of variance decomposition. We confine ourselves here to illustrating the use of Table 18 in terms of one simple example.

Example 15. The table on p. 38 shows measurements of tensile strength x_{ti} $(i = 1, \ldots, n;\ t = 1, \ldots, k)$ made on $n = 6$ specimens of rubber randomly selected from each of $k = 5$ different batches.

If all the x_{ti} were random observations from the same normal population, having a variance σ^2, we could obtain two different estimates of σ^2 by computing the mean squares:

(i) $$s_1^2 = \frac{S_1}{\nu_1} = n \sum_{t=1}^{k} \frac{(\bar{x}_{t.} - \bar{x}_{..})^2}{k-1},$$

described as the *between-batch* mean square, based on $\nu_1 = k - 1$ degrees of freedom, and

(ii) $$s_2^2 = \frac{S_2}{\nu_2} = \sum_{t=1}^{k} \sum_{i=1}^{n} \frac{(x_{ti} - \bar{x}_{t.})^2}{nk - k},$$

described as the *within-batch* mean square, based on $\nu_2 = k(n-1)$ degrees of freedom.

Measurements of tensile strength (in kg./cm.²) of specimens of rubber

	Batch no.				
Specimen no.	$t=1$	$t=2$	$t=3$	$t=4$	$t=5$
$i=1$	177	116	170	181	177
2	172	179	156	190	186
3	137	182	188	210	199
4	196	143	212	173	202
5	145	156	164	172	204
6	168	174	184	187	198
Mean, $\bar{x}_{t.}$	165·8	158·3	179·0	185·5	194·3
Mean square	468·6	653·1	406·0	196·3	111·5
Range	59	66	56	38	27
Grand mean, $\bar{x}_{..}=176·6$					

It can be shown that under these conditions S_1/σ^2 and S_2/σ^2 are independent χ^2-values, so that the ratio $F=s_1^2/s_2^2$ would follow the distribution (61). Suppose, however, that while the batch variances are the same, their means differ. Then while s_2^2 will still remain an unbiased estimator of σ^2, the expectation of s_1^2 will now be greater than σ^2. We investigate this possibility of departure from the null hypothesis of equal means by finding whether the observed value of F is displaced exceptionally towards the upper tail of the F-distribution.

In our example

$$s_1^2=1273·5, \quad \nu_1=4; \qquad s_2^2=367·1, \quad \nu_2=25.$$

Hence $F=3·47$, a value considerably larger than 1. Turning to Table 18, we find that the observed F falls between the 2·5 % point at 3·35 and the 1 % point at 4·18. It is thus significant at the 2·5 % level (and *a fortiori* at the 5 % level), though not at the 1 % level. On the hypothesis tested it is, therefore, unusually large, and we should suspect the presence of real differences in batch means.

9·3 *Tests using both upper and lower tails of the F-distribution*

In the majority of applications in the analysis of variance we are concerned with comparing (i) a 'treatment' mean square which possible systematic effects or other disturbing causes may have enhanced, with (ii) an 'error' mean square, which will be an unbiased estimate of the error variance whether the treatment effects are present or not. Hence, as in the preceding example, we set out to detect effects causing large values of F and are therefore concerned with a single-tail test based on the *upper* percentage points of the distribution. Sometimes, however, we may need to use a double-tail test, because large excesses of s_2^2 over s_1^2 *or* of s_1^2 over s_2^2 are of equal importance.

This is the case, for example, if we have two samples of size n_1 and n_2, one from each of two normal populations, and wish to use the variance estimates, s_1^2 and s_2^2, to judge whether the population variances, σ_1^2 and σ_2^2, differ. Suppose that we wish to arrange the test so that the probability is 0·05 of rejecting the hypothesis that $\sigma_1^2=\sigma_2^2$ when, in fact, it is true. Then, if the labels '1' and '2' were randomly assigned to the two samples it would seem reasonable to regard the difference as significant if $F = s_1^2/s_2^2$ is either (*a*) below the lower 2·5 % point or (*b*) above the upper 2·5 % point of the appropriate F-distribution, i.e. of

$$f(F \mid \nu_1 = n_1 - 1, \nu_2 = n_2 - 1).$$

Since, if $\sigma_1^2 = \sigma_2^2$, the probability of (*a*) equals the probability that $F = s_2^2/s_1^2$ is above the upper 2·5 % point of $f(F \mid \nu_1 = n_2 - 1, \nu_2 = n_1 - 1)$, it follows that a double-tail test at the 5 % level will be achieved by taking *whichever of the two variance estimates is the larger as the numerator*, and comparing the ratio with the 2·5 % F-value.

Example 16. Let us examine whether the variance estimates from two (randomly selected) batches from the data of Example 15, say batches 2 and 4, differ significantly. We calculate $F = 653/196 = 3{\cdot}33$, and note that it does not exceed the 2·5 % point for F with $\nu_1 = \nu_2 = 5$ degrees of freedom, which Table 18 shows to fall at 7·15. Clearly no significant difference between the variances in batches 2 and 4 is indicated.

A test of whether there is evidence of heterogeneity among the $k = 5$ variances, taken as a whole, is discussed and applied in § 16 below.

9·4 *Relation to the Pearson Type VI curve*

The F-distribution is related to the Pearson Type VI curve (see § 23·1 below)

$$y = y_0(x' - a)^{q_2} x'^{-q_1} \quad (a \leqslant x' \leqslant \infty), \tag{63}$$

through the transformation $x' = a(1 + \nu_1 F/\nu_2)$, where

$$\nu_1 = 2(q_2 + 1), \quad \nu_2 = 2(q_1 - q_2 - 1).$$

Table 18, therefore, provides percentage points for these distributions, but if the reciprocal transformation (62) is employed as well, the percentage points can be obtained to a higher accuracy from those for the incomplete B-variable, x, given in Table 16. In so far as the Type VI distribution has been found useful as an approximation to less tractable distributions in statistical theory (see, for example, Box, 1949), the present tables provide approximate percentage points for such statistics.

9·5 *Tests of significance for the largest F-ratio* (*Table* 19)

In the analysis of variance it frequently occurs that a number, k, of independent 'treatment' mean squares $s_1^2, \ldots, s_k^2$, each based on m degrees of freedom and measuring different 'effects', must all be gauged against an independent 'error' mean square s_0^2, based on ν degrees of freedom. It is clear that if we were to form for each of the k mean squares, s_i^2, the F-ratio s_i^2/s_0^2 and refer these to the ordinary F-tables, the largest of these F-ratios, say $s_{\text{max.}}^2/s_0^2$, would have a greater chance of being declared significant than one *randomly* selected from the set (to which the F-tables would be applicable). This source of bias can be avoided if the percentage points (in random sampling) of the largest F-ratio, $s_{\text{max.}}^2/s_0^2$, in a set of k ratios are used, as shown below. Such tables would depend on k, the number of independent random mean squares s_i^2, on m, their degrees of freedom and on ν, the degrees of freedom of the error mean square s_0^2.

Table 19 (taken from K. R. Nair, 1948*a*) gives such upper 5 and 1 % points for the special case $m = 1$, which arises in the important class of factorial experiments of the 2^l type (such as that analysed in Example 17 below).

In other cases, an approximation suggested by Hartley (1938) and examined in detail by Finney (1941) for $m=2$, regards the F-ratios as independent, so that the probability that $s^2_{\max.}/s^2_0$ exceeds F^* (say) is approximately evaluated from

$$1-\left\{\int_0^{F^*} f(F\,|\,m,\nu)\,dF\right\}^k \sim k\int_{F^*}^{\infty} f(F\,|\,m,\nu)\,dF.$$

From the form of this expression we note that the upper $100\alpha/k\,\%$ point of F for m and ν degrees of freedom is approximately equal to the $100\alpha\,\%$ point of the distribution of $s^2_{\max.}/s^2_0$. Thus we may be able to utilize the extreme percentage points of F for this test, provided k is not too great.

Example 17. Wishart (1938) gives an example of a uniformity trial on asparagus with dummy treatments (N, P, K) superimposed. The analysis of variance table (with NPK confounded), is shown below:

Source of variation	Degrees of freedom	Mean square	F-ratio
Blocks	7	6543·2	—
N	1	488·3	0·35
P	1	69·0	0·05
K	1	34·0	0·02
NP	1	830·3	0·60
NK	1	57·8	0·04
PK	1	9765·0	7·02
Error	18	1391·2	—

The largest mean square is that for PK, and the variance ratio $s^2_{\max.}/s^2_0=7\cdot02$ would be returned as significant if compared with the 5 % point of F for $\nu_1=1$, $\nu_2=18$, namely, 4·41. However, using the appropriate test and entering Table 19 for $k=6$, $\nu=18$, we find by interpolation a 5 % point of 8·59 which is not exceeded; thus the dummy PK effect should be returned as insignificant. If we were to use the approximate test mentioned above at the 5 % level, we would compare an observed value of 7·02 with the $\frac{5}{6}$ % point of F for $\nu_1=1$, $\nu_2=18$. Interpolating for a 0·833 % point between the 2·5, 1 and 0·5 % points for $\nu_1=1$, $\nu_2=18$, using a 3-point Lagrangian, we obtain a value of 8·87, which is slightly in excess of the exact value (8·59) and again returns the observed ratio of 7·02 as insignificant.

No exact tests are available

(*a*) for the second largest, third largest, etc., mean square,

(*b*) for the ordered mean squares based on different degrees of freedom.

III. FURTHER TABLES OF PROBABILITY INTEGRALS, PERCENTAGE POINTS, ETC., OF DISTRIBUTIONS DERIVED FROM THE NORMAL FUNCTION (TABLES 20–35)

The tables collected together in this group, while of less importance than those of the preceding group, will be found of considerable use in a number of directions. Contained in the section are tables dealing with the distribution of statistics based on ordered observations, e.g. on the extreme values in a sample. These distributions are of importance both in quick assessments of significance in connexion with outlying observations or groups of observations and, in the case of range, in providing alternative means of estimating variance. The section also contains tables which can be used to investigate the validity of the assumptions, in general necessary in the application of analysis of variance technique, regarding (*a*) homogeneity of variance and (*b*) normality of distribution. In so far as is possible the tables will be described and illustrated in a logical sequence. For obvious reasons of economy in printing, the arrangement of the tables in the book does not always follow in quite this order.

10. The mean deviation: moments and percentage points (Tables 20 and 21)

10·1 *Definitions*

If $x_1, x_2, \ldots, x_n$ is a random sample of n observations, the mean deviation (from the mean) is defined by

$$m = \sum_{i=1}^{n} | x_i - \bar{x} | / n,$$

where $\bar{x}$ is the arithmetic mean of the observations. If the population sampled is normal with standard deviation σ, no simple expression exists for the frequency function $f(m)$ of the distribution of m, except in the cases $n = 2, 3$. Computations have, however, been made of the probability integral of m/σ for $n = 2\,(1)\,10$, from the formula for the distribution of m derived by H. J. Godwin (1945), who showed that (for $\sigma = 1$)

$$f(m) = n^{\frac{3}{2}}\, 2^{-\frac{1}{2}(n+1)}\, \pi^{-\frac{1}{2}(n-1)} \sum_{i=1}^{n-1} \binom{n}{i} \exp\left\{ -\frac{m^2 n^3}{8i(n-i)} \right\} G_{i-1}(\tfrac{1}{2}mn)\, G_{n-i-1}(\tfrac{1}{2}mn),$$

where $G_r(x)$ is defined by the recurrence formula (70), p. 49. For a description of the method of computation, see Hartley (1945). This table has not been reproduced in the present volume, but Table 21 gives certain lower and upper percentage points of the standardized mean deviation m/σ calculated from the probability integral table.

The expectation and variance of m for a normal population, derived first by Helmert (1876), and later by Fisher (1920), are as follows:

$$\mathscr{E}(m) = \sigma \left\{ \frac{2(n-1)}{n\pi} \right\}^{\frac{1}{2}}, \tag{64}$$

$$\operatorname{var}(m) = \sigma^2 \frac{2(n-1)}{n^2 \pi} \left\{ \tfrac{1}{2}\pi + [n(n-2)]^{\frac{1}{2}} - n + \sin^{-1} \frac{1}{n-1} \right\}. \tag{65}$$

Geary (1936)* has provided expansions in inverse powers of $n-1$ from which the third and fourth moments, μ_3 and μ_4 of m can be calculated. Table 20 gives the expectation, variance and standard deviation calculated from equations (64) and (65), as well as the moment ratios $\beta_1 = \mu_3^2/\mu_2^3$, $\beta_2 = \mu_4/\mu_2^2$ calculated from Geary's formulae. For $n \leqslant 6$, exact values for β_1 were obtained from Geary's work. For $n = 3$, Mr A. R. Kamat has provided us with an exact value for β_2 while he and Mr J. H. Cadwell (1953) have pointed out a correction needed to Fisher's (1920) value for $n = 4$. All values of β_1 and β_2 for $n \leqslant 6$ were checked by applying numerical integration to Godwin and Hartley's table of the probability integral of m.† When $n = 2$, m is distributed as the modulus of a normal deviate. The extent of error involved when $n = 10$ by assuming that m is normally distributed with an expectation and variance given by equations (64) and (65) is shown by the values of the percentage points, under the heading 'normal approximation' in Table 21. For the more extreme points, the

* See also E. S. Pearson (1945, 1948).

† *Biometrika*, **33**, 259.

error is still appreciable. A considerably more accurate approximation, by which the lower and upper 5, 2·5, 1 and 0·5 % points may be obtained using Table 42, is described below in Example 49, p. 83.

10·2 *Applications*

Example 18. The probability integral and percentage points of m/σ were originally calculated in connexion with quality control, because in certain fields of production it was customary to estimate within-lot variation from the mean deviation rather than the standard deviation or range. For example, if (*a*) it is desired to control the variability in running time of certain fuses so that the standard deviation for the fuses in a large lot does not exceed 0·5 sec. under specified firing conditions, and (*b*) standard proof routine involves firing eight fuses per lot under these conditions, then inner and outer *control limits* for the mean deviation may be obtained as follows from the 2·5 and 0·1 % points of the table:

Inner control limits at

$$0{\cdot}5 \times 0{\cdot}372 = 0{\cdot}19 \text{ sec.} \quad \text{and} \quad 0{\cdot}5 \times 1{\cdot}196 = 0{\cdot}60 \text{ sec.}$$

Outer control limits at

$$0{\cdot}5 \times 0{\cdot}220 = 0{\cdot}11 \text{ sec.} \quad \text{and} \quad 0{\cdot}5 \times 1{\cdot}499 = 0{\cdot}75 \text{ sec.}$$

While the upper control limits will naturally be of most value in detecting first signs of trouble in the shape of increasing variability, the lower limits may also be useful, since factors leading to greater consistency in production may be worth while following up.

The figures for the expected value and the standard deviation of m/σ given in Table 20 may also be used to obtain an estimate of σ, and of the standard error of this estimate, based on the mean deviation found from one or a number of small samples. This procedure is illustrated in Example 22, p. 47 below.

11. Percentage points of the extreme standardized deviate, $X_n = (x_n - \mu)/\sigma$ (Table 24)

11·1 *Definitions*

If the observations in a sample of n are arranged in ascending order of magnitude $x_1, x_2, \ldots, x_n$, the extreme values will be denoted by x_1 and x_n, while their values standardized with regard to the population mean and standard deviation may be written

$$X_1 = (x_1 - \mu)/\sigma, \quad X_n = (x_n - \mu)/\sigma.$$

If the sample is drawn from a normal population the probability integral of X_n is given by

$$P(X_n) = \left\{\int_{-\infty}^{X_n} \frac{1}{\sqrt{(2\pi)}} e^{-\frac{1}{2}u^2} du\right\}^n. \tag{66}$$

Tables giving values of this probability integral for certain sample sizes were first computed by Tippett (1925); these were reproduced in *Tables for Statisticians and Biometricians*, vol. II (1931), together with a table giving certain percentage points of X_n (Tables XXI and XXI *bis*). The first table is not reproduced here; the second appears in a modified form as Table 24.

As given, the table relates to X_n; the percentage points for X_1 will be obtained by changing the sign of the figures and reversing the headings 'lower' and 'upper'. Thus while the probability is 0·05 that the *highest* value of x in a sample of $n = 4$ will have a value *below* $\mu - 0{\cdot}068\sigma$, it is also clearly 0·05 that the *lowest* value of x will exceed $\mu + 0{\cdot}068\sigma$.

11·2 *Applications*

The upper percentage points of X_n are helpful in determining whether outlying observations belong to a specified normal distribution. The lower percentage points are perhaps less useful, but they may be of value as rough exploratory yardsticks, and they are instructive in showing the vagaries to which random samples are sometimes subject. For example, since the lower 2·5 % point of X_n is just negative, and hence the upper 2·5 % point of X_1 is just positive, it is seen that in 1 out of 20 samples of $n=5$, all the observations will fall either above or below the population mean. The table may also be useful in certain quality control problems, of which the following is an illustration.

Example 19. In the mass production of a certain article a firm aims at an average breaking strength of 176 lb., and a variability in strength which should not exceed a value measured by a standard deviation of 12 lb. In order to ensure that the production is kept under control, tests of breaking strength are applied at intervals to samples of 20 articles, and a simple check rule is required to be given to the foreman in charge of these tests. There is reason to believe that the distribution of breaking strength of this product when the manufacture is properly controlled is approximately normal.

The following rule is suggested:

The foreman should report when

(*a*) the sum of the breaking strength of the 20 articles in the sample is less than 3395 lb.; or when

(*b*) the lowest breaking strength is less than 136 lb.

This rule has the following basis:

The standard error of the mean of samples of 20 is $12/\sqrt{20}=2\cdot683$ lb. Table 24 shows that the deviation corresponding to the lower 1 % point of X_1 is $-2\cdot326$ in samples of 1 and $-3\cdot289$ in samples of 20. Hence the mean in a random sample of 20 should only once in a hundred times be less than $176-2\cdot326\times2\cdot683=169\cdot76$ lb., and the sum of the 20 breaking strengths should not be less than $20\times169\cdot76=3395\cdot2$ lb. Again, the lowest value in the sample should only be less than $176-3\cdot289\times12=136\cdot53$ lb. in 1 % of samples. Of course the strengths of the mean and of the weakest individuals are correlated, and a more exhaustive test might be applied, based on the mean and standard deviation of the sample. But if one of the main purposes in controlling variability is to prevent articles appearing below a certain level of strength, the use of the lower limit seems to be suitable. The test is also much simpler in application than one involving the calculation of the standard deviation of the sample.

12. The distribution of the range (Tables 20, 22, 23 and 27)

12·1 *Definitions*

If x_1 and x_n are the lowest and highest values of the variate in a sample of n observations,

$$w=x_n-x_1$$

is the *range* of the sample.

In standardized form $\quad W=X_n-X_1=(x_n-x_1)/\sigma.$

Where it is necessary to make clear the size of the sample involved, we shall write w_n or W_n.

In general, if $f(X)$ is the frequency function of a standardized variable X, then the probability integral of W for a sample of n may be expressed in the form

$$P(W\mid n)=n\int_{-\infty}^{\infty}f(X)\left\{\int_X^{X+W}f(u)\,du\right\}^{n-1}dX. \tag{67}$$

Table 23 gives values of $P(W \mid n)$ for the normal frequency function $f(X) = (2\pi)^{-\frac{1}{2}} e^{-\frac{1}{2}X^2}$, and for $n = 2$ (1) 20, $W = 0{\cdot}00$ (0·05) 7·25. The method of computation has been described by Hartley (1942). The table is perhaps mainly of value as a fundamental table which has formed the starting point for certain theoretical investigations into the properties of range distribution and for several derived tables of considerable practical utility described below. Small errors in the original table have been corrected by reference to Harter *et al.* (1959).

12·2 *Application*

We give below a rather special example of the direct application of Table 23 to a problem in machine-part assembly.

Example 20. Electromagnetic relays are manufactured to a specified-setting-up time. This is the time elapsing between the primary impulse in the coil and the complete contact in the secondary circuit of the relay. For each individual relay the actual setting-up time may differ from specification but will stay practically constant in time. Sets of, say, 15 relays are now assembled in a machine. To prevent 'arcing', a cam must keep the secondary circuit broken from before the first relay has set up until after the last relay has set up. The length of the break interval is a fixed characteristic of the cam and samples of 15 relays whose range in setting-up time exceeds this interval cannot, therefore, be fitted. Thus, on testing, the slowest or the fastest relay will have to be replaced, and it is necessary to keep the frequency of such replacements below a reasonably low percentage. Table 23 gives this frequency if we have adequate information on the standard deviation σ of the setting up times. Suppose, for example, that $\sigma = \frac{1}{25}$ sec. and that the cam has a break interval of $\frac{1}{5}$ sec., then $W = 0{\cdot}20/0{\cdot}04 = 5$, so that entering Table 23 for $n = 15$, $W = 5$ we find $P(5 \mid 15) = 0{\cdot}9688$; thus the expected percentage frequency of replacements is $100\,(1 - P(5 \mid 15)) = 3{\cdot}12$.

In general, the formulae

$$W = \frac{\text{Cam break interval}}{\text{Standard deviation of setting-up times}},$$

$$100\,(1 - P(W \mid n)) = \text{percentage frequency of necessary replacements},$$

where n is the number of relays assembled in the machine, relate the variation of the relays to the cam-break interval. The table may therefore be used as a guide when deciding on tolerance limits in the manufacture of relays or in designing a cam to make it fit the relays.

12·3 *Interpolation in Table* 23

If $W = W_0 + 0{\cdot}05\theta$ (where W_0 is the tabular argument nearest to and smaller than W), interpolates from the linear formula

$$P(W \mid n) = (1-\theta)\, P(W_0 \mid n) + \theta P(W_0 + 0{\cdot}05 \mid n)$$

should be (a) decreased by a unit in the fourth decimal if $0{\cdot}005 \leqslant P \leqslant 0{\cdot}40$ and $0{\cdot}1 \leqslant \theta \leqslant 0{\cdot}9$ and (b) increased by a unit in the fourth decimal if $0{\cdot}5 \leqslant P \leqslant 0{\cdot}97$ and $0{\cdot}1 \leqslant \theta \leqslant 0{\cdot}9$. When $n = 2$, instead of (a) and (b), add a unit in the fourth decimal if $0{\cdot}40 \leqslant P \leqslant 0{\cdot}92$ and $0{\cdot}1 \leqslant \theta \leqslant 0{\cdot}9$. This procedure will normally give results to within a unit in the fourth decimal.

12·4 *The percentage points of the range* (*Table* 22)

Inverse interpolation in Table 23 provides the percentage points of the range given in Table 22 for $n = 2$ (1) 20. This table is clearly useful in problems of industrial quality control where a check is being kept on the variability of the product.

Example 21. The same example may be taken as was used in § 10·2 to illustrate the table of the percentage points for the mean deviation. The critical standard deviation in time of fuse-running being fixed at 0·5 sec., inner and outer control limits for the range of observed times, in a sample of eight fuses, may be obtained as follows from the columns headed 2·5 % and 0·1 %.

Inner control limits at

$$0{\cdot}5 \times 1{\cdot}41 = 0{\cdot}70 \text{ sec.} \quad \text{and} \quad 0{\cdot}5 \times 4{\cdot}61 = 2{\cdot}30 \text{ sec.}$$

Outer control limits at

$$0{\cdot}5 \times 0{\cdot}83 = 0{\cdot}41 \text{ sec.} \quad \text{and} \quad 0{\cdot}5 \times 5{\cdot}82 = 2{\cdot}91 \text{ sec.}$$

12·5 *The moments of the distribution of the range (Tables* 20 *and* 27)

If $P(X) = 1 - Q(X)$ is the probability integral of a standardized random variable X, where $-\infty \leqslant X \leqslant \infty$, then it is known (see Karl Pearson, 1902; and Tippett, 1925) that the expectation and variance of the distribution of the range, W, in samples of size n may be written in the form

$$\left.\begin{aligned} \mathscr{E}(W) &= \int_{-\infty}^{\infty} \{1 - [Q(u)]^n - [P(u)]^n\}\, du, \\ \operatorname{var}(W) &= 2\int_{-\infty}^{\infty} du \int_{-\infty}^{u} \{1 - [P(u)]^n - [Q(v)]^n + [P(u) - P(v)]^n\}\, dv - \{\mathscr{E}(W)\}^2. \end{aligned}\right\} \quad (68)$$

Expressions of increasing complexity, which must be evaluated by quadrature, may be obtained for the higher moments. If the population sampled is normal, the probability integral of equation (2) will be substituted for $P(X)$.

Simpler formulae for the moments $\mu_r^*(n)$ of range about a working origin W^* can, of course, be based on its probability integral $P(W \mid n)$ in the form

$$\mu_r^*(n) = (-W^*)^r + r\int_0^{\infty} (1 - P(W \mid n))\,(W - W^*)^{r-1}\, dW, \quad (69)$$

but the accuracy obtainable from these is restricted by that available for $P(W \mid n)$.

Tippett (1925) used the first relation (68) to compute the expectation of $W_n = w_n/\sigma$ in samples from a normal population, for $n = 2\,(1)\,1000$. This table, given in his paper, was reproduced in *Tables for Statisticians and Biometricians*, vol. II, Table XXII. In the present reissue we have included as Table 27 only the first half of this table, for $n = 2\,(1)\,499$.† Tippett also calculated a few values of $\sigma(W)$ and of the moment ratios $\beta_1(W)$ and $\beta_2(W)$ using an approximate process, and on the basis of these provided charts for these statistics for sample sizes up to 1000. Much more accurate values of the moments for $n \leqslant 100$ have since been obtained by Harter *et al.* (1959) and also, for the single case of $n = 200$, by Harley & Pearson (1957). The following figures have been taken from these publications.

n	$\mathscr{E}(W)$	$\sigma(W)$	$\beta_1(W)$	$\beta_2(W)$
60	4·63856	0·639	0·201	3·35
100	5·01519	0·605	0·223	3·39
200	5·49209	0·566	0·252	3·44
500	6·07340	0·524	0·285	3·50
1000	6·48287	0·497	0·309	3·54

† Additional values for $n = 500\,(10)\,1000$ have been printed below the main table.

Table 20 of the present volume contains for $n=2\ (1)\ 20$ values of $\mathscr{E}(W)$ or in common notation d_n, var (W) and its square root $\sigma(W)$, $\beta_1(W)$ and $\beta_2(W)$, as well as certain special quantities derived from d_n and var (W), whose use is described on p. 54 below. The values have been checked against Harter's table and a considerable number of last figure errors in the original table were found and corrected.

12·6 *Applications*

(*a*) *Estimation of the standard deviation*

Using again the common notation, we may write

$$\mathscr{E}(W_n)=\mathscr{E}(w_n/\sigma)=d_n;$$

it follows that the range between the extreme observations in a sample, divided by the appropriate numerical factor d_n, has an expectation of σ, i.e. w_n/d_n is an unbiased estimate of σ. Further, the standard error of this estimate will be $\sigma(W_n)\times\sigma/d_n$, i.e. is obtained by multiplying σ by the coefficient of variation of the range. To assist in the computation of the range estimate in small samples the reciprocal of d_n is given in the second column of Table 22. It is now possible to compare the reliability, measured by the standard error, of three alternative estimators of the standard deviation, σ, in a normal population, namely, those obtained from the sample standard deviation, the mean deviation and the range. As in the case of the range, the first and second estimators can be adjusted so that their expectations are σ, by division by factors (depending on the sample size n) obtainable from equations (30) and (64). Thus using Tables 35 (for s) and 20 (for m and for w), we reach the comparison shown in the following table, where figures in columns 2, 3 and 4 are the coefficients of variation of s, m and w, respectively.

Standard errors of alternative unbiased estimators of σ (expressed in terms of σ as unit)

Sample size n	S.D. estimate	M.D. estimate	Range estimate	Range estimate ÷ S.D. estimate
2	0·756	0·756	0·756	1·00
3	·523	·525	·525	1·00
4	·422	·430	·427	1·01
5	·363	·373	·372	1·02
6	0·323	0·334	0·335	1·04
7	·294	·306	·308	1·05
8	·272	·283	·288	1·06
9	·254	·265	·272	1·07
10	0·239	0·250	0·259	1·08
12	·215	·227	·239	1·11
15	·191	·201	·218	1·14
20	·163	·173	·195	1·20

It is seen that, up to samples of 6, the range estimator is as reliable as that based on the mean deviation and up to $n=10$ there is very little to choose between them. Beyond this, the relative accuracy of the range estimator falls off progressively. It will be seen from Table 20 that the distribution of mean deviation tends to normality with increasing n, its β_1 and β_2 coefficients tending to 0 and 3; on the other hand, the distribution of range approaches the normal most closely in the neighbourhood of $n=8-10$, and afterwards diverges, β_1 and β_2 tending to limiting values of 0·6493 and 4·2000 (Gumbel, 1949).

(*b*) *The use of mean range*

As the sample size increases, the distribution of the extreme observations in a sample, and therefore that of the range, will depend more and more on the form of the tails of the population frequency curve. The distribution of range will therefore be rather sensitive to departure from normality of a kind which it is very difficult to detect in practice. For this reason and because the ratio of the coefficient of variation of w to that of s increases without limit as $n \to \infty$, it is common to estimate σ from the mean range of the observations in a number of small groups. If k samples of n observations are available, and we write the mean value of their ranges as $\overline{w}_n$, we may use $\overline{w}_n/d_n$ as an estimator of σ. This will have a standard error of $\sigma(W_n) \times \sigma/(\sqrt{k}\, d_n)$. Further, the sampling distribution of this estimator will have moment ratios

$$\beta_1(\overline{w}_n) = \beta_1(W_n)/k, \quad \beta_2(\overline{w}_n) = 3 + \{\beta_2(W_n) - 3\}/k,$$

and its approach to the normal form may be examined with the help of Table 20.

12·7 *Illustrative examples*

Example 22 (use of Tables 20 and 35). Example 15 (p. 37) contains measurements of tensile strength of six specimens of rubber drawn from each of five batches. These data may be used to obtain a range estimate of the within-batch standard deviation, σ, and also to make comparisons of alternative estimators of σ. It is necessary to assume that the within-batch variation is approximately normal and has a variance which is common from batch to batch. The following table shows the five sample variance estimates s^2 (and their square roots, s), the mean deviations (m) and the ranges (w). While s^2 is an unbiased estimator of σ^2, the expectation of s is less than σ. All three statistics therefore need adjustment before they give unbiased estimates of σ:

(i) Since $n = 6$, $\nu = 5$, s must be divided by $(\frac{2}{5})^{\frac{1}{2}}\,\Gamma(3)/\Gamma(2{\cdot}5)$ (putting $r = 1$ in equation (30)), whose numerical value col. 2 of Table 35 shows to be 0·9515.

(ii) The value of m must be divided by $\{2 \times 5/(6\pi)\}^{\frac{1}{2}}$ (see equation (64)); Table 20, col. 2, gives the value 0·7284.

(iii) The value of w must be divided by d_6 given as 2·5344 in Table 20, col. 7.

Estimator	Sample no.					Mean value
	1	2	3	4	5	
Variance, s^2	468·6	653·1	406·0	196·3	111·5	367·1
Standard deviation, s	21·6	25·6	20·1	14·0	10·6	
Unbiased estimate	**22·7**	**26·9**	**21·1**	**14·7**	**11·1**	**19·3**
Mean deviation, m	16·6	20·0	15·7	10·2	8·6	
Unbiased estimate	**22·8**	**27·5**	**21·6**	**14·0**	**11·8**	**19·5**
Range, w	59	66	56	38	27	
Unbiased estimate	**23·3**	**26·0**	**22·1**	**15·0**	**10·7**	**19·4**

N.B. The unit of measurement for the estimates of σ is kg./cm.^{2}.

The high correlation between the corresponding sample values of the three estimators is particularly noticeable. If common estimates are calculated by taking the mean of the five sample values, we obtain the figures in the last column of the table, namely, 19·34, 19·54 and 19·42, in close agreement.

(iv) If the five variance estimates are summed and averaged, we obtain 367·1, with square root 19·16. The latter will be a slightly biased estimate of σ; as the total degrees of freedom are $\nu = 25$, the appropriate corrective factor taken from Table 35, col. 2, is 0·9901, giving as an unbiased estimate $19{\cdot}16/0{\cdot}9901 = 19{\cdot}3\,(5)$.

The standard errors of these four alternative pooled estimates of σ may be obtained as follows:

(i) For estimate based on mean s,

$$\sigma \times \text{S.D.}(s/\sigma)/\{\mathscr{E}(s/\sigma) \times \sqrt{5}\} = 0{\cdot}145\sigma \quad \text{(Table 35).}$$

(ii) For estimate based on mean m,

$$\sigma \times \text{S.D.}(m/\sigma)/\{\mathscr{E}(m/\sigma) \times \sqrt{5}\} = 0{\cdot}150\sigma \quad \text{(Table 20).}$$

(iii) For estimate based on mean w,

$$\sigma \times \text{S.D.}(w/\sigma)/\{\mathscr{E}(w/\sigma) \times \sqrt{5}\} = 0{\cdot}150\sigma \quad \text{(Table 20).}$$

(iv) For estimate based on the total within-sample sum of squares ($\nu = 25$), Table 35 shows that the standard error when corrected for bias will be

$$\sigma \times 0{\cdot}1407/0{\cdot}9901 = 0{\cdot}142\sigma.$$

If we substitute for the unknown σ a value of, say, 19·4 based on the sample estimates, it is seen that these standard errors lie between 2·7 and 2·9 kg./cm.^{2}. While, therefore, the fourth method, based on the usual analysis of variance procedure of averaging the individual sample sums of squared deviations before taking the square root, provides the unbiased estimate of σ with the smallest standard error, it is clear that the differences between the four estimates are very much smaller than the standard error of any one of them. This is due to the very high correlation which exists between the alternative estimators.

Example 23 (use of Table 27). Tippett's table of the expectation of w/σ, i.e. d_n, in medium or large samples from a normal population may be of value in certain theoretical investigations, but is more likely to be useful in rough assessments of the homogeneity or accuracy of data. For example, it has been suggested that a useful check against major arithmetical errors in calculating a standard deviation is provided by comparing the figure obtained for s with the range of the observations, divided by the appropriate d_n factor. The following discussion illustrates this suggestion.

Hooke (1926) gives tables of the individual measurements of a large number of cranial characters, made on a series of skulls (probably of the seventeenth century) dug up during excavations in Farringdon Street, London, in 1924–5. The table on p. 49 gives information regarding four only of the cranial characters:

L = glabellar-occipital length, B = maximum parietal breadth,

LB = length from basion to nasion, S = arc from nasion to opisthion.

The values of d_n in col. 4 are taken from Table 27. The standard errors of s, in col. 6, are estimated from $s/\sqrt{(2n)}$, and those in col. 5 from $\text{S.D.}(w/\sigma) \times s/d_n$. The four values of $\text{S.D.}(w/\sigma)$ were read from Diagram VI of Tippett's (1925) paper.

A comparison of the standard errors of the two estimates of σ shows at once the inefficiency of the range estimator computed in this way from a single large sample. As the two estimates are correlated,* care is needed in testing whether the differences between them are significant. A more satisfactory check is obtained by calculating the ratios of range to standard deviation w_n/s_n and referring these values to Table 29*c*, discussed more fully on p. 59. It is seen that all the four values of the ratio given in the last column of the table below lie between the lower and upper 10% significance levels obtainable from Table 29*c*. No abnormalities of the distributions of the four cranial characters are therefore indicated by the w/s criterion.

Character	n	Range	d_n	Estimates of σ and their S.E.'s		$\frac{w}{s}$
				From range	From sample S.D.	
(1)	(2)	(3)	(4)	(5)	(6)	(7)
L	141	30·5	5·256	5·80 ± 0·72	6·46 ± 0·38	4·72
B	142	28·0	5·261	5·32 ± 0·65	5·90 ± 0·35	4·75
LB	119	23·0	5·138	4·48 ± 0·52	4·47 ± 0·29	5·15
S	129	75·5	5·195	14·53 ± 1·61	14·20 ± 0·88	5·32

N.B. The unit is 1 mm.

13. The extreme normal deviate measured from the sample mean and its studentized form† (Tables 25, 26, 26*a* and 26*b*)

The determination of the extreme standardized deviate $X_n = (x_n - \mu)/\sigma$ from the observed extreme variate x_n, depends on a knowledge of both μ and σ and the computation of the standardized mean deviation m/σ and range $W = w/\sigma$, on a knowledge of σ. Indeed, it is this relation to σ which make these latter statistics suitable estimators of σ. However, in the applications (such as Examples 19 (p. 43), 21 (p. 45) and 24 (p. 50)), where X_n and W are used as indicators of departures from a basic normal distribution, we require, for the computation of the standardized statistics X and W, a knowledge of μ and/or σ. Since these will often not be known, it is desirable to have statistics in which μ only, μ and σ, or σ only, are replaced by appropriate estimates, and Tables 25 and 26 are concerned with such statistics.

Table 25 gives percentage points of $u = (x_n - \bar{x})/\sigma$ or $u = -(x_1 - \bar{x})/\sigma$, i.e. of the deviation of the largest (smallest) observation in a sample of n from the sample mean $\bar{x}$, divided by the standard deviation of the sampled normal population. This statistic, although independent of μ, still depends on σ. The probability integral, originally derived by McKay (1935) and tabulated by K. R. Nair (1948*b*), is given by

$$P(u \mid n) = \sqrt{n}\,(2\pi)^{-\frac{1}{2}(n-1)}\,G_{n-1}(nu),$$

where the rth G-function is defined by the recurrence relation

$$G_r(x) = \int_0^x \exp\left\{-\frac{t^2}{2r(r+1)}\right\} G_{r-1}(t)\,dt, \quad G_0(x) = 1. \tag{70}$$

The criterion will be useful in deciding whether to reject an outlying observation as anomalous when no knowledge about the population mean μ is available, but σ is known from previous experience.

* It can be shown that this correlation is given by [S.D. $(s)/\mathscr{E}(s)$]/[S.D. $(w)/\mathscr{E}(w)$].

† For use of the term 'studentized', see p. 52, footnote.

Example 24. The following illustration has been given by McKay (1935). In the course of routine testing of a standard leather product from a tannery, five parallel tests yielded the following values for the hide substance content of the leather specimens: 32·44, 36·45, 39·64, 40·13, 41·09. The first observation appears rather low. Long experience with the product in question has established a value of 2·2 for the standard deviation σ. We find

$$u=(37{\cdot}95-32{\cdot}44)/2{\cdot}2=2{\cdot}5;$$

this exceeds the 5 % value of Table 25, and we conclude that the first observation may well be anomalous.

More often than not, however, σ will also be unknown. Suppose first that there is available a root-mean-square estimate s of σ based on ν degrees of freedom and independent of the sample under consideration.

Table 26 gives six upper percentage points of the studentized extreme deviate

$$t_n=(x_n-\bar{x})/s \quad \text{or} \quad t_1=(\bar{x}-x_1)/s.$$

It is based on the original tabulations by K. R. Nair (1948*b*, 1952), later revised and slightly extended by H. A. David (1956). For the upper 5 and 1 % points values for $\nu=5$ (1) 9 have been added from K. C. S. Pillai (1959).

Lower 5 % and 1 % points were also computed by Nair for $\nu=10$, 15 and 30 but are not reproduced here. For $\nu \geqslant 10$ these lower percentage points only change slightly with increasing ν. The values for $\nu=\infty$ are those given in the 4th and 6th columns, respectively, of Table 25 while for $\nu=10$ we have:

n	3	4	5	6	7	8	9
Lower 1 % point	0·09	0·19	0·29	0·37	0·43	0·49	0·54
Lower 5 % point	·20	·35	·46	·55	·62	·69	·74

In the application of this statistic to an analysis of variance of, say, variety trials, the sample would be represented by the n varietal means $\bar{x}_t$, of which $\bar{x}_n$, 'the best yielder', would be tested against the general mean $\bar{x}$, whilst the independent estimate of standard deviation would be provided by $s_\nu=s/\sqrt{k}$, where k is the number of replicates and s^2 is the error mean square.

Example 25. Snedecor (1946, p. 266) gives the results of a randomized block experiment for four strains of wheat, replicated in five blocks, with data and analysis of variance set out below:

Block	Strain			
	A	B	C	D
1	32·3	33·3	30·8	29·3
2	34·0	33·0	34·3	26·0
3	34·3	36·3	35·3	29·8
4	35·0	36·8	32·3	28·0
5	36·5	34·5	35·8	28·8
Mean	34·4	34·8	33·7	28·4

Grand mean 32·8

Analysis of variance of data

	Sum of squares	D.F.	Mean square
Blocks	21·46	4	5·36
Strains	134·45	3	44·82
Error	26·26	12	2·19
Total	182·17	19	

In testing for the significance of a strain effect, we find $F = 44{\cdot}82/2{\cdot}19 = 20{\cdot}5$, $\nu_1 = 3$ and $\nu_2 = 12$ degrees of freedom. To quote Snedecor: 'one suspects that the highly significant F is attributable largely to the small yield of strain D.' To test this we compute

$$t_n = (32{\cdot}8 - 28{\cdot}4)/\sqrt{(\tfrac{1}{5} \times 2{\cdot}19)} = 6{\cdot}6,$$

which is well beyond the 0·1 % significance level given in Table 26.

Had the error mean square been 13 instead of 2·19 (an error of 11 % per plot which is frequently encountered), the F-ratio would become insignificant, but on testing the difference between D and the general mean we obtain $t_n = 2{\cdot}7$. Entering Table 26 with $n = 4$ and $\nu = 12$, we find that this value is significant at the 2·5 % level, so that this test would have been able to detect the low-yielding strain, D, in spite of the F-test failing to detect any difference between strains in general.

Under the foregoing conditions one may wish to base the estimation of σ not only on the 'external' estimate s_ν but also on the 'internal' sum of squares $\Sigma(x_i - \bar{x})^2$, i.e. the sum of squares of the sample for which x_n is the largest and x_1 the smallest item. Table 26a (Quesenberry & David, 1961) provides upper 5 % and 1 % points of

$$b_n = (x_n - \bar{x})/S \quad \text{or} \quad b_1 = (\bar{x} - x_1)/S,$$

where

$$S^2 = \sum_{i=1}^{n} (x_i - \bar{x})^2 + \nu s_\nu^2.$$

Percentage points of the corresponding two-sided statistic

$$b^* = \max |x_i - \bar{x}|/S$$

are given in Table 26b, a condensed version of that in Quesenberry & David (1961). These b-statistics are known to have desirable optimal properties for the detection of a *single* outlier or, in the analysis of variance situation, of a single outlying population (the latter being the so-called slippage problem).

It is believed that Tables 26a and b will rarely be in error by more than a unit in the third decimal place.

In *Example* 25 we have

$$b_1 = (32{\cdot}8 - 28{\cdot}4)/\sqrt{[(134{\cdot}45 + 26{\cdot}26)/5]} = 0{\cdot}78.$$

Entering Table 26a with $n = 4$, $\nu = 15$ we see that b_1 is highly significant. With error mean square 13 instead of 2·19 we find $b_1 = 0{\cdot}58$ which is still significant at the 1 % level. In the absence of outside knowledge a two-sided test is perhaps more appropriate in the present example. Table 26b shows that our value of b^* (which here equals $b_1 = 0{\cdot}58$) is just significant at the 1 % level.

So far we have supposed that an independent estimate s_ν of σ is available. In the absence of such external information one has to base the estimation of σ purely on the internal component computed from the sample under consideration. This case was originally investigated by Thompson (1935) and later in more detail and in amended form by Pearson & Chandra Sekar (1936). It is covered by Tables 26*a* and *b* as the special case $\nu = 0$. Thus in *Example* 24, in the absence of knowledge on σ, we compute the sum of squares about their mean for the five observations, $S^2 = 50{\cdot}08$, and hence have

$$b_1 = (37{\cdot}95 - 32{\cdot}44)/7{\cdot}08 = 0{\cdot}78.$$

Here $n = 5$, $\nu = 0$, and this value of b_1 is not significant according to Table 26*a*.

The different conclusion reached lies in the fact that we are now estimating the standard deviation from the five observations in the sample, which include possible abnormal values, i.e. we take as our estimate of σ, $s = \sqrt{(\frac{1}{4} \times 50{\cdot}08)} = 3{\cdot}54$. Before, under the conditions of *Example* 24, we supposed a reliable value of $\sigma = 2{\cdot}2$ was available from past experience.

14. Further tables based on range (Tables 29, 29*a*, *b*, *c*, 30, 31*b* and *c*)

In this section we shall discuss a number of tables, several of which were not included in the original edition of this volume. All are dependent on the distribution of the range in samples from a normal population. In a variety of situations they will provide the statistician with quick answers which can later be confirmed if necessary by a more orthodox technique.

Table 29. Upper percentage points of the studentized range, $q = w_n/s_\nu$.

Table 29*a*. Upper percentage points of the ratio $u = |\bar{x}_1 - \bar{x}_2|/(w_1 + w_2)$.

Table 29*b*. Upper percentage points of the ratio of two independent ranges, $F' = w_1/w_2$.

Table 29*c*. Percentage points of the ratio of the range to the standard deviation, w/s, where w and s are derived from the same sample of n observations.

Tables 30 (A *and* B). Tables for the analysis of variance based on range.

Table 31*b*. Upper percentage points of the ratio $w_{\text{max.}} \Big/ \sum_{t=1}^{k} w_t$.

Table 31*c*. Upper percentage points of the ratio $w_{\text{max.}}/w_{\text{min.}}$

It has been established from a number of investigations (e.g. Lord, 1950; Moore, 1957; Harter, 1963) that provided the ranges have been determined from small samples—preferably $n \leqslant 10$—these 'quick' tests based on the range are not seriously less effective in establishing departure from hypothesis (i.e. less powerful) than the more orthodox tests using root-mean-square estimators.

14·1 *The studentized range: definition and tables* (*Tables* 29 *and* 30)

If in the standardized range, $W_n = w_n/\sigma$, we replace the unknown population standard deviation σ by the square root, s_ν, of an independent mean-square estimate of σ^2 based on ν degrees of freedom, we obtain the *studentized range**

$$q = w_n/s_\nu.$$

* The term *studentized* has been gradually accepted into current usuage and we have ventured to drop both the capital S and the inverted commas. The word may well have been first coined by W. S. Gosset himself; he certainly used it in correspondence as early as 1932 in connexion with this same problem of range (see letter quoted by E. S. Pearson, 1938, pp. 245–6).

The probability integral of this statistic is given by the integral

$$\Pr\{w_n/s_\nu \leqslant q\} = \int_0^\infty [\Gamma(\tfrac{1}{2}\nu)]^{-1}\, 2^{-\frac{1}{2}\nu+1}\, \nu^{\frac{1}{2}\nu}\, s^{\nu-1} \exp\{-\tfrac{1}{2}\nu s^2\}\, P(qs \mid n)\, ds, \tag{71}$$

where $P(w \mid n)$ is the probability integral of range (Table 23). For large ν, this integral tends to $P(w \mid n)$ taken at argument $w=q$ and can, in fact, be evaluated as a series in powers of ν^{-1} (Hartley, 1944); thus

$$\left.\begin{aligned} \Pr\{w_n/s_\nu \leqslant q\} &= P + a_1\nu^{-1} + a_2\nu^{-2} + \dots, \\ \text{where} \qquad a_1 &= \tfrac{1}{4}\{q^2P'' - qP'\}, \\ a_2 &= \tfrac{1}{96}\{3q^4P^{\text{iv}} - 2q^3P'''\} - \tfrac{1}{8}a_1 \end{aligned}\right\} \tag{72}$$

and P and its derivatives $P', \dots, P^{\text{iv}}$ are taken at argument $w=q$. Table 29, as published in the earlier editions of this volume, was obtained partly from an evaluation of (72) and, when this became inadequate for small ν and large q, by numerical quadrature of (71). More accurate later work by Pachares (1959) and Harter *et al.* (1959) has made it possible to remove a number of last figure errors in our original table and to add a table of upper 10% points for q. At the same time we have omitted the tables of lower 5% and 1% points. Harter's (1959) Tables give: (*a*) the probability integral of q to six decimal places for $n=2$ (1) 20 (2) 40 (10) 100; $\nu=1$ (1) 20, 24, 30, 40, 60, 120; (*b*) a table to four significant figures of the percentage points of q for $P=0{\cdot}001$, $0{\cdot}005$, $0{\cdot}010$, $0{\cdot}025$, $0{\cdot}05$, $0{\cdot}10$, $(0{\cdot}1)$, $0{\cdot}90$, $0{\cdot}95$, $0{\cdot}975$, $0{\cdot}99$, $0{\cdot}995$, $0{\cdot}999$.

14·2 *The studentized range: applications of Tables* 29 *and* 30. *Short-cut tests in the analysis of variance*

Although the control limits for range used in quality control are usually computed from *estimates* of σ, these are commonly based on so many observations that the use of the studentized range in place of range is not required and the procedure given in Example 21 (p. 45) is adequate.

The main use of this statistic is, therefore, in analysis of variance where q:

(*a*) is the basic statistic used in various 'multiple contrast' comparison procedures concerned with contrasts between treatment means, and in the construction of simultaneous confidence intervals for such contrasts;

(*b*) may be used as an alternative, short-cut, statistic in place of the F-test.

In the simplest form we can take q as the ratio of (i) the range of k 'treatment' means each based on n observations, to (ii) an estimate of the error standard deviation divided by $\sqrt{n}$, where this estimate is determined in the usual way from the within-treatment sum of squares. In this form q may provide a more powerful means than F of detecting, say, the difference between the best yielding treatment and the poorest treatment (which is often the experimental 'control' treatment), particularly in situations in which the gain of most of the treatments over the control is insignificant.

However, we may go further than this and use Table 29 and its companion Table 30 to provide short-cut tests in which both of the variance components are estimated from ranges or mean ranges instead of mean squares, so that the formation of sums of squares is avoided altogether. Patnaik (1950) has described the most straightforward of such tests. The procedure makes use of the fact that the mean value, $\bar{w}_n$, of the ranges in k samples, each containing n

observations, from normal populations having a common variance σ^2, is distributed approximately (see E. S. Pearson, 1952) as $c\chi\sigma/\sqrt{\nu}$, where the *scale factor* c and the *equivalent degrees of freedom* ν are functions of n and k. These functions are shown in Table 30 A for $n = 2$ (1) 10, $k = 1$ (1) 5, 10, (and ∞ for c). Extrapolation for ν beyond $k = 5$ (or 10 for Table B) may be carried out by adding to the value of ν at $k = 5$ the 'constant difference' (C.D.) multiplied by $(k-5)$.

Patnaik applied the method to the case of a simple classification, such as the analysis of tensile strength records of rubber specimens (Example 15, p. 37) into between- and within-batch components. In this case, we proceed as follows:

(*a*) Estimate the within-batch variance from the mean of the batch ranges, $\bar{w}_n$.

(*b*) Estimate the between-batch variance from the range, say, $\bar{x}_{\max.} - \bar{x}_{\min.}$ of the batch means $\bar{x}_t$.

(*c*) Calculate the ratio
$$q = (\bar{x}_{\max.} - \bar{x}_{\min.})\, c\sqrt{n}/\bar{w}_n. \tag{73}$$

Since the estimates (*a*) and (*b*) are independent, the expression q of equation (73) is approximately distributed as a studentized range for sample size k (in place of the tabular n of Table 29) and having the equivalent degrees of freedom ν, of Table 30 A, for the s-estimate.

Example 26. Applying this test to the data of Example 15 (p. 37), the five-batch ranges give a mean $\bar{w}_6 = 246/5 = 49{\cdot}2$. Entering Table 30 A with 'sample size', $n = 6$ and 'number of samples', $k = 5$, we find $\nu = 22{\cdot}6$ and $c = 2{\cdot}56$, so that

$$q = (194{\cdot}3 - 158{\cdot}3)\, 2{\cdot}56\sqrt{6}/49{\cdot}2 = 4{\cdot}59.$$

This value can be referred directly to the upper percentage points of q (Table 29). Entering the table with a 'sample size' of 5 (the number of batch means, $k = 5$, plays the role of the 'sample size' in the numerator of q) and degrees of freedom $\nu = 22{\cdot}6$, we find that the observed value of 4·59 falls between the 5 % point at 4·2 and the 1 % point at 5·2. A similar result was obtained in the application of the F-test. It will also be noted that our estimate of the within-batch standard deviation, $\bar{w}_6/c = 19{\cdot}2$, based on the equivalent of 22·6 degrees of freedom, is close to the square root of the 'within-batch mean square', $s_2 = \sqrt{367{\cdot}1} = 19{\cdot}16$, based on 25 degrees of freedom quoted on p. 38; the loss of the equivalent of about 2 degrees of freedom is due to the short-cut procedure. It should be remembered that the squares of both these estimates, $19{\cdot}2^2$ and $19{\cdot}16^2$, are unbiased estimates of σ^2, in contrast to the estimate

$$\bar{w}_6/d_6 = 49{\cdot}2/2{\cdot}534 = 19{\cdot}42,$$

which is an unbiased estimate of σ ($d_6 = 2{\cdot}534$ from Table 20).

In the case of k samples with ranges w_t based on unequal sample sizes n_t ($t = 1, 2, \ldots, k$), an estimate of σ may be obtained (H. A. David, 1951) from the weighted mean range

$$\frac{\sum_t w_t(d_{n_t}/V_{n_t})}{\sum_t (d^2_{n_t}/V_{n_t}) + \frac{1}{2}}.$$

The factors d_{n_t}/V_{n_t} and $d^2_{n_t}/V_{n_t}$ are given in Table 20.* This quantity is distributed approximately as the root-mean-square estimator, s, for $\nu = \frac{1}{2}\Sigma d^2_{n_t}/V_{n_t}$ degrees of freedom.

In the case of a double classification the procedure has to be modified (Hartley, 1950) as shown by the following example.

* V_{n_t} is the variance of the standardized range w/σ for sample size n_t.

Example 27. We again use Snedecor's data tabled in Example 25 (p. 50), giving yields of four strains of wheat planted in five randomized blocks. In order to estimate the error standard deviation, we first form differences of individual yields from their respective strain means (strain residuals) and then form ranges of these residuals for each of the five blocks. This work is shown below:

Strain residuals and their block ranges

Block	Strain				Range
	A	B	C	D	
1	−2·1	−1·5	−2·9	0·9	3·8
2	−0·4	−1·8	0·6	−2·4	3·0
3	−0·1	1·5	1·6	1·4	1·7
4	0·6	2·0	−1·4	−0·4	3·4
5	2·1	−0·3	2·1	0·4	2·4
Total	—	—	—	—	14·3

The procedure is now exactly the same as in the single classification except that Table 30B must be used,† with

number of ranges = 5, sample size (for range) = 4.

We find: mean range $\bar{w} = 14{\cdot}3/5 = 2{\cdot}86$, $\nu = 10{\cdot}9$, $c = 1{\cdot}88$,

and hence $q = (34{\cdot}8 - 28{\cdot}4)\, 1{\cdot}88 \sqrt{5}/2{\cdot}86 = 9{\cdot}4$.

Referring this value to the upper percentage points of q in Table 29, with sample size (for range) = 4 and degrees of freedom 10·9, we find that it exceeds the 1 % point which is 5·6, a highly significant result similar to that obtained previously from the F-test. The estimate of the error standard deviation is $2{\cdot}86/1{\cdot}88 = 1{\cdot}52$, which agrees well with that obtained from the analysis of variance table given on p. 51, namely, $\sqrt{2{\cdot}19} = 1{\cdot}48$.

The method of estimation in the present double classification is, of course, unsymmetrical, as we have taken residuals from strain means and the five ranges of four such residuals in each of five blocks. Alternatively, we could have taken four ranges of five residuals from block means. This latter method would here have given the same estimate of 1·52, but may in general differ slightly, the one based on the larger sample sizes usually having a larger ν (see Table 30B), and therefore being more precise. However, one would normally take residuals about 'treatment' means as these means are usually required for individual comparison of treatments (see § 14·3 below).

If a test for block differences is desired, we would have to form the means of the five blocks which are found to be 31·4, 31·8, 33·9, 33·0 and 33·9 respectively. The range of these is $33{\cdot}9 - 31{\cdot}4 = 2{\cdot}5$, so that $q = 2{\cdot}5 \times 1{\cdot}88 \sqrt{4}/2{\cdot}86 = 3{\cdot}3$. Entering Table 29 with a sample size of 5 and $\nu = 10{\cdot}9$, we find for q a 5 % value of 4·6 which is not exceeded, so that there are no significant block differences.

Generalizations to higher designs. A table making provision for applying the range method to the split plot experiment is given by H. A. David (1951), who also discusses other designs.

* Table 30B takes account of the correlation between the ranges of residuals. For details, see Hartley (1950).

14·3 *Use of range estimators in the comparison of two means* (*Table* 29*a*)

A natural extension of the use of range is to introduce this estimator of the population standard deviation in place of the root-mean-square estimator, s, in Student's t-tests, remembering that w, just as s, is independent of the mean in sampling from a normal population. Lord (1947) published tables of the percentage points of

$$(a)\ (\bar{x}-\mu)/w \quad \text{and} \quad (b)\ (\bar{x}_1-\bar{x}_2)/\{\tfrac{1}{2}(w_1+w_2)\},$$

where w is the range and $\bar{x}$ the mean in a sample of n observations from a normal population having mean μ. Similarly $\bar{x}_t$, w_t ($t=1$, 2) are the means and ranges in independent samples of equal size n from two normal populations with a common mean and variance. The first ratio, (a), provides a test of significance for the mean in a single sample, while the second ratio, (b), is appropriately used in comparing the means of two independent samples, each containing the same number of observations. Further, in the case of n paired observations the ratio (a) may be applied to the mean $\bar{x}$ and the range w of the n paired differences.

(a) *Single sample case.* On noting that the range in a sample of two observations is distributed as half a normal curve having standard deviation $\sqrt{2}\sigma$, it will be seen that the upper percentage points of $|\bar{x}-\mu|/w$ may be obtained from Table 29*b* described below, by entering the table with $n_1=2$, $n_2=n$ and dividing the tabular entry by $\sqrt{(2n)}$. For a single-tail test at level α', Table 29*b* should be entered in the column $\alpha=2\alpha'$.

As an illustration we may go back to Examples 5 and 6 (pp. 18, 20) and ask whether the 12 values of breaking strength of cotton fabric are consistent with the hypothesis that the average strength of this material is $\mu=171{\cdot}5$. We find that

$$|\bar{x}-\mu|/w=|167{\cdot}08-171{\cdot}5|/27=0{\cdot}164.$$

Table 29*b* entered with $n_1=2$, $n_2=12$ gives a 10% point for w_1/w_2 of 0·7726 and a 5% point of 0·9513; dividing these values by $\sqrt{24}=4{\cdot}899$ we obtain the corresponding values 0·158 and 0·194 for a double-tail test of the ratio $|\bar{x}-\mu|/w$. Thus the observed value falls between the 10% and 5% points as was the case when applying the standard Student t-test.

(b) *Two sample case.* (Table 29*a*). In the general case where $n_1 \neq n_2$, Moore (1957) suggested that the ratio $u=(\bar{x}_1-\bar{x}_2)/(w_1+w_2)$ might still be used as a test statistic although in theory the most effective test would now give different weights to the sample ranges, w_1 and w_2, depending on n_1 and n_2. He compared his test with one in which the denominator of the ratio had the value $\sum_{t=1}^{2}(w_t d_{n_t}/V_{n_t})$, the multiplying factors being the weights defined in §14·2 and given in Table 20. As a result of the comparison, he did not consider that the small gain in power when using the weighted range estimator justified the additional complication introduced into the test.

Table 29*a*, gives upper significance points for Moore's ratio, u, for all combinations of sample sizes $2 \leqslant n_1$, $n_2 \leqslant 20$. When divided by two, the table entries for $n_1=n_2$ correspond to those given by Lord (1947). Apart from the case $n_1=n_2$, the tabular entries are not exact as Moore used a χ-approximation for the distribution of $(w_1+w_2)/\sigma$ on the lines suggested in §14·2. The few comparisons made with exact results obtained by quadrature did not show errors in the approximation of more than two units in the third decimal place.

Example 28. The following illustration is taken from Moore's (1957) paper. Two operators,

A and B, make determinations of the percentage of ammonia in plant gas, with the following results:

Operator A	43	44	56	63	46			
Operator B	39	35	43	32	36	48	33	33

From these figures can it be said that the operators are consistently measuring the same thing? The data give

	n_t	$\bar{x}_t$	w_t	w_t/d_{n_t}	$\sum_i (x_{ti}-\bar{x}_t)^2$	s_t^2	s_t
Operator A ($t=1$)	5	50·4	20	8·60	305·2	76·30	8·73
Operator B ($t=2$)	8	37·375	16	5·62	221·875	31·70	5·63

It follows that $|\bar{x}_1-\bar{x}_2|/(w_1+w_2)=13{\cdot}025/36=0{\cdot}362$, a value which Table 29*a* shows to be just beyond the 1 % level of significance, remembering that we are here using a double-tail test. The usual form of t-test gives $t=3{\cdot}301$ which for $\nu=11$ degrees of freedom is found from Table 12 to be also significant at the 1 % level. In view of the unusually close correspondence between the range and root-mean-square estimates of σ shown in the 4th and 7th columns of the table above, agreement between the two tests was to be expected.

14·4 *Use of range estimators in comparing the variances in two populations* (*Table* 29*b*)

It is natural to proceed from a range-based 'substitute' t-test to a range-based F-test. Table 29*b* prepared by Harter (1963), considerably extending earlier tables by Link (1950) and Pillai & Buenaventura (1961), gives upper percentage points of the distribution of the ratio $F'=w_1/w_2$ of ranges determined from two independent samples of n_1 and n_2 observations drawn from normal populations having a common variance. To find the lower percentage points for w_1/w_2, we enter the table with n_1 and n_2 reversed and take the reciprocal of the figure extracted. For example, if $n_1=12$, $n_2=8$, the lower and upper 2·5 % points for w_1/w_2 are $1/1{\cdot}807=0{\cdot}553$ and 2·563, respectively.

When dealing with two small samples the use of the range ratio clearly provides an extremely rapid method of assessing the significance of differences in variance. The test is of course somewhat less powerful for the detection of differences between variances than that based on $F=s_1^2/s_2^2$,* but if observations are not difficult to come by, the inclusion of between one and three additional observations in each sample, followed by the use of w_1/w_2 rather than F, might be considered worthwhile.

Harter made some detailed computations comparing the power of the two tests for the case of equal sample sizes, $n_1=n_2$. The figures in the following table are taken from his results. It is supposed that in the two populations sampled, $\sigma_1=\theta\sigma_2$, with $\theta>1$; then the table shows the chances of establishing significance at the 5 % and 1 % levels ($\alpha=0{\cdot}05$ and 0·01) for six different values of θ.

* It should be noted that the loss in power when using range estimates of variability is a good deal less in tests for detecting 'non-centralities', i.e. differences between population means as in the linear comparisons of § 14·2.

$n_1=n_2$	$\alpha=0{\cdot}05$						$\alpha=0{\cdot}01$					
	$\theta=2$	$\theta=3$	$\theta=4$	$\theta=6$	$\theta=8$	$\theta=10$	$\theta=2$	$\theta=3$	$\theta=4$	$\theta=6$	$\theta=8$	$\theta=10$
3	0·174 ·174	0·321 ·321	0·456 ·457	0·653 ·655	0·769 ·771	0·839 ·840	0·039 ·039	0·083 ·083	0·139 ·139	0·266 ·267	0·391 ·393	0·501 ·503
6	0·387 ·402	0·707 ·729	0·868 ·884	0·969 ·975	0·991 ·993	0·997 ·997	0·141 ·146	0·396 ·417	0·629 ·656	0·873 ·891	0·954 ·962	0·981 ·985
9	0·537 ·582	0·869 ·902	0·966 ·978	0·997 ·998	1·000 1·000	1·000 1·000	0·257 ·288	0·649 ·708	0·866 ·905	0·981 ·990	0·997 ·998	0·999 1·000
12	0·642 ·714	0·938 ·967	0·990 ·996	1·000 1·000	1·000 1·000	1·000 1·000	0·360 ·430	0·796 ·870	0·950 ·978	0·997 ·999	1·000 1·000	1·000 1·000
15	0·717 ·808	0·969 ·989	0·997 ·999	1·000 1·000	1·000 1·000	1·000 1·000	0·447 ·558	0·879 ·946	0·981 ·995	0·999 1·000	1·000 1·000	1·000 1·000

(The first line of each pair gives the power of the F' test, the second line that of the F test; n_1 and n_2 are sample sizes.)

Taking Harter's results as a whole, we find the following 'equivalent' sample sizes:

n for F-test	2	3–9	10–11	12
n' for 'equivalent' range ratio F'-test	2	$n+1$	$n+2$	15

In other words, the power of the F'-test with $n_1=n_2=n'$ is at least as great as that of the F-test with sample sizes $n_1=n_2=n$, n' and n being as above.

As an illustration consider the data of *Example* 28. Here $w_1/w_2=20/16=1{\cdot}25$. Entering Table 29*b* with $n_1=5$, $n_2=8$ we see that the observed ratio lies between the upper 25% and 10% points, which fall at 1·112 and 1·475, respectively. There is therefore no evidence that Operator A's measurements are more variable than those of Operator B. If we use the standard test, $F=s_1^2/s_2^2=76{\cdot}30/31{\cdot}70=2{\cdot}41$; Table 18 entered with $\nu_1=4$, $\nu_2=7$ shows again that this value lies between the 25% and 10% points which fall at 1·72 and 2·96.

14·5 *Range tests for heterogeneity of variance* (*Table* 31*b and* 31*c*)

It was suggested by Bliss, Cochran & Tukey (1956) that a very simple test based on the observed ranges in several samples of equal size might be used as a quick substitute for Cochran's test based on variances, described in §16·5 below. If w_t ($t=1, 2, \ldots, k$) are the ranges in k independent samples, each containing n observations and drawn from normal populations having a common variance, then Table 31*b* gives the upper 5% points of the ratio of the largest range to the sum of all the ranges, i.e. of $w_{\text{max.}}/\sum w_t$. The test was originally developed to meet the need for a simple rejection criterion for use with several bioassays in the U.S. Pharmacopeia XV (1955). If the observed ratio exceeded the tabular value, the set of assays represented by this largest range was assumed to contain an aberrant observation or outlier, which was identified by inspection and rejected.

This test is exactly parallel with that suggested by Cochran (1941) in which sample variances are used instead of ranges and we calculate the ratio $s^2_{\text{max.}}/\sum s_t^2$ for which upper 5% and 1% points are given in Table 31*a*. A further alternative test is based on the ratio $w_{\text{max.}}/w_{\text{min.}}$, for which upper percentages are given in Table 31*c* (see pp. 202, 264–5).

As an illustration we may use the data of *Example* 15 (pp. 37–8) for measurements on tensile strength of rubber. Here

$$w_{max.} = 66, \quad \sum w_t = 246 \quad \text{and} \quad w_{max.}/\sum w_t = 0{\cdot}268; \quad w_{min.} = 27, \quad w_{max.}/w_{min.} = 2{\cdot}44;$$

for $k=5$, $n=6$ Table 31*b* and 31*c* give upper 5% points of 0·351 and 4·187, respectively. Similarly

$$s^2_{max.} = 653{\cdot}1, \quad \sum s_t^2 = 1835{\cdot}5 \quad \text{and} \quad s^2_{max.}/\sum s_t^2 = 0{\cdot}356;$$

this is a value well short of the 5% point (0·506) of Table 31*a*, taking $\nu = n-1 = 5$.

The three tests are therefore in agreement in providing no evidence suggesting a significant difference between batch variabilities. The same conclusion is reached in § 16·3 using the *M*-test and in § 16·5 using the ratio $s^2_{max.}/s^2_{min.}$. Of course these five tests will not always give the same verdict, their comparative powers depending on the alternative to variance homogeneity that holds good (see Pearson, 1966, *Biometrika*, **53**, 229–34).

For $k > 10$, the authors of the test suggest using the following supplementary table from which approximate 5% points can be obtained by dividing the tabled factors by $k+2$:

k \ n	2	3	4	5	6	7	8	9	10
10	4·06	3·04	2·65	$2{\cdot}44_5$	$2{\cdot}30_5$	2·21	$2{\cdot}14_5$	2·09	2·05
12	4·06	3·03	$2{\cdot}63_5$	$2{\cdot}42_5$	2·29	2·20	2·13	$2{\cdot}07_5$	2·04
15	4·06	3·02	$2{\cdot}62_5$	$2{\cdot}41_5$	2·28	$2{\cdot}18_5$	2·12	$2{\cdot}06_5$	$2{\cdot}02_5$
20	4·13	3·03	2·62	$2{\cdot}41_5$	2·28	$2{\cdot}18_5$	2·11	2·05	2·01
50	4·26	3·11	2·67	2·44	2·29	2·19	2·11	2·06	2·01

Bliss *et al.* (1956) give an example from the turbidimetric assay of vitamin B_{12}. For purpose of illustration the essential data are the ranges of percentage transmission, read in a photometer, among 32 sets of 3 tubes, as tabled below:

Observed range for $n = 3$	0	1	2	3	4	5	6	7	8	13
No. of sets of tubes, total 32	1	5	6	8	4	2	3	1	1	1

The total of ranges is 113, w (max.) is 13 and the ratio is 0·115. Since $k = 32$ the auxiliary table is entered with $(k+2) \times 0{\cdot}115 = 3{\cdot}91$ which clearly exceeds the value 3·08 obtained by interpolation in the table. Thus on the basis of all 32 ranges, the range of 13 was attributed to an outlier which was then identified and rejected.

14·6 *A test of internal consistency within a single univariate sample* (*Table* 29*c*)

The studentized range, w_n/s_ν, discussed in §§ 14·1 and 14·2, is the ratio of the range in a sample of n observations to a completely independent estimate of σ based on ν degrees of freedom. The ratio dealt with now, whose percentage points are given in Table 29*c*, is that of the range, w, to the standard deviation, s, where both are calculated from the same normal sample and are therefore highly correlated. The statistic can be useful in several ways in detecting what might be described as abnormality of sample pattern. Such abnormality might result from:

(*a*) various types of heterogeneity, e.g. if what has been termed slippage has occurred or if there has been a change in variability during the sampling;

(*b*) the presence of one or more rogue observations;

(*c*) a gross computing error in calculating the standard deviation;

(*d*) The population not following a normal distribution, being either platykurtic ($\beta_2 < 3$), when w/s will tend to be too small, or leptokurtic ($\beta_2 > 3$) when w/s will tend to be too great.

The methods employed in calculating the percentage points have been described by David, Hartley & Pearson (1954) and Pearson & Stephens (1964). The values given to three decimal places are exact; the remainder were obtained by fitting Pearson curves with the correct first four moments. These latter values should rarely be in error by more than two units in the second decimal place. The columns headed 0 % give the extreme lower and upper bounds for the ratio in samples of n observations from *any* distribution.

Example 28*a*. A report by Mahalanobis *et al.* (1949) describes an anthropometric survey of the United Provinces of India undertaken in 1941. From these data the frequency distributions of head length were taken for each of 23 racial groups and the values of w/s calculated. The sample sizes, n, varied between 57 and 197 and Fig. 2*a* shows the values of w/s plotted against n. Also drawn are the upper and lower 5 % and 1 % levels taken from Table 29*c*. It will be seen that for five of the 23 groups w/s falls beyond the upper 5 % level, for two of these, well beyond the 1 % level.

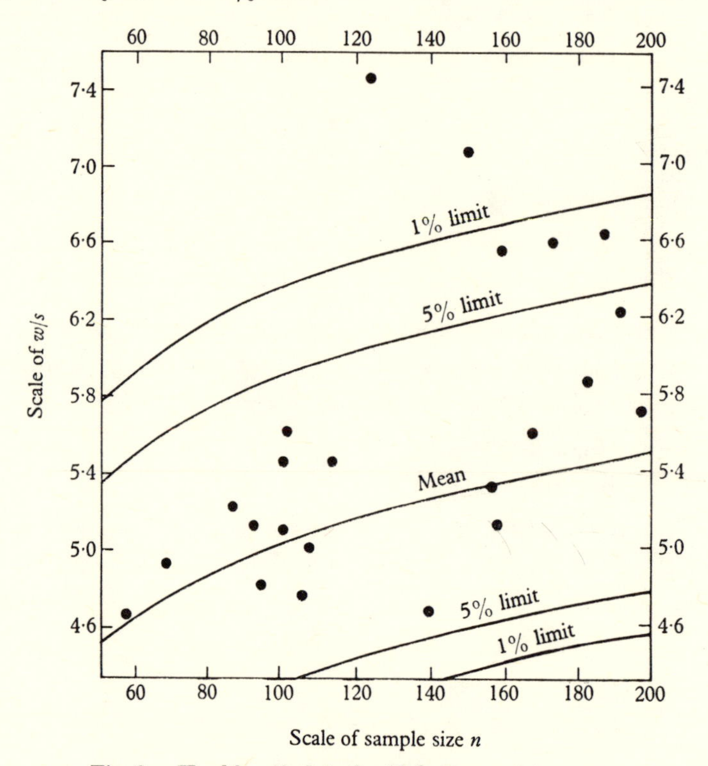

Fig. 2*a*. Head length data for 23 Indian racial groups.

On looking at the original frequency distributions it was found that in each of these five cases there was one or more than one divergent measurement, always in the sense of a small head length. Without further information it is not possible to say whether these outlying readings were due to faulty recording, but it is clear that the routine application of this simple check-test will from time to time provide a *prima facie* case for further scrutiny of data.

Example 28*b*. By considering the moment ratios $\sqrt{b_1}$ and b_2, it is shown below that the sample of 1000 measures of warp strength listed in the table of *Example* 35 (p. 69) are most unlikely to have been drawn randomly from a normal population. Could a quick pointer to this conclusion have been obtained from the ratio w/s?

As the data have been grouped, we cannot make a precise test. However, assuming that each observation falls at the centre of its group, we have $w = 95$ lb., $s = 17{\cdot}42$ lb. (not using Sheppard's correction for grouping). Hence $w/s = 5{\cdot}45$. Reference to Table 29c, with $n = 1000$, shows that the observed ratio falls below the lower 0·1 % point (at 5·50), certainly suggesting non-normality.

15. THE MEAN POSITION OF RANKED NORMAL DEVIATES (MEANS OF NORMAL ORDER STATISTICS) (TABLE 28)

15·1 *Definitions*

If we draw a sample of n observations from a normal population with zero mean and standard deviation σ, we may arrange these in ascending order of magnitude, so that x_1 is the smallest and x_n is the largest value. The ith value in the sequence, $X_i = x_i/\sigma$, may be termed the *ith ranked normal deviate*;* its expectation is

$$\xi(i \mid n) = \frac{n!}{(i-1)!\,(n-i)!} \int_{-\infty}^{\infty} Z P^{i-1} Q^{n-i} X \, dX, \tag{74}$$

where Z, P and Q are as defined in § 1·1. Table 28 gives $\xi(i \mid n)$ to 3 decimals for $n = 2\,(1)\,20$ and to 2 decimals for $n = 21\,(1)\,26\,(2)\,50$.

Since

$$\xi(i \mid n) = -\xi(n+1-i \mid n),$$

only positive values are tabled which may be regarded as the $[\frac{1}{2}n]$† largest ranks $\xi(n+1-i \mid n)$ or, with signs reversed, the $[\frac{1}{2}n]$ smallest ranks.

Reference to Table 27 will confirm that $|\xi(1 \mid n)| = \frac{1}{2}d_n$, i.e. the mean extreme deviate equals half the mean range.

For odd values of n between 26 and 50 linear interpolation between the neighbouring columns will give the value required with an error which should not be more than a unit in the second decimal. Note that for n odd $\xi(\frac{1}{2}(n+1) \mid n) = 0$.

15·2 *Applications*

The table, in providing what may be called the average pattern of a normal sample, has applications in problems where ranked observations are selected for special study or where ranks must be translated into quantitative scores.

Example 29. In rearing a flock of thirty lambs it is decided to market the five heaviest at $2\frac{1}{2}$ months. We may wish to estimate their total weight. Taking from previous experience the information that the weights at $2\frac{1}{2}$ months follow a normal distribution with a mean of about 60 lb. and a standard deviation of about 4 lb., what would be the expected weight of the five heaviest lambs?

* The term '*i*th normal order statistic' is current in American usage.

† $[x]$ stands for the largest integer smaller than or equal to x.

Entering Table 28 for $n=30$, we find for this weight the estimate

$$5 \times 60+4(2{\cdot}04+1{\cdot}62+1{\cdot}36+1{\cdot}18+1{\cdot}03)=329\text{ lb.}$$

In certain situations Table 28 may be used when estimating the parameters, μ and σ, of a normal population from the ordered observations of a random sample drawn from it. For example, the range estimate of σ, namely, $(x_n - x_1)/d_n$, may be calculated, but this would not normally be used for samples exceeding 20. For larger samples we may sometimes use the difference between the total of the r largest observations $x_{n-r+1}, \ldots, x_n$ and that of the r smallest observations $x_1, \ldots, x_r$ as an estimate of σ (see Jones, 1946). Thus

$$S=\frac{x_{n-r+1}+\ldots+x_n-(x_1+\ldots+x_r)}{2\{\xi(n \mid n)+\ldots+\xi(n-r+1 \mid n)\}} \tag{75}$$

is an unbiased estimator of σ.

Example 30. Out of the fifty numbers in the first column of Wold's *Random Normal Deviates* (1948), the four largest are found to be 2·82, 2·12, 1·98 and 1·39, while the four smallest are −2·30, −2·28, −1·77 and −1·00. An estimate of σ may therefore be computed from

$$S=\frac{2{\cdot}82+2{\cdot}12+1{\cdot}98+1{\cdot}39+1{\cdot}00+1{\cdot}77+2{\cdot}28+2{\cdot}30}{2(2{\cdot}25+1{\cdot}85+1{\cdot}63+1{\cdot}46)}$$

$$=15{\cdot}66/(2 \times 7{\cdot}19)=1{\cdot}089.$$

The usual root-mean-square estimate based on the whole fifty numbers is $s=1{\cdot}003$. The population standard deviation for Wold's numbers is, of course, unity.

Jones (1946) has evaluated the required divisors $-2\{\xi(1 \mid n)+\ldots+\xi(r \mid n)\}$ for selected values of n and r as well as the variances of the estimates of σ. All his n are $\geqslant 100$, but for $n \leqslant 10$ the variances may be obtained from the tables of variances and covariances of the x_i calculated by Godwin (1949*b*). Figures for the efficiency of Jones's type of estimator were given by K. R. Nair (1950), while Godwin (1949*a*) discussed the problem of optimum 'linear estimators' of σ based on the ordered sample values x_i. Similar questions were also considered by Mosteller (1946).

A further application of Table 28 is of the type described in Fisher & Yates's *Statistical Tables* (1963), p. 31 and Tables XX, XXI. If n objects have been ranked (say in order of preference), but no quantitative values measuring this preference have been assigned by the experimenter, the score $\xi(i \mid n)$ may be given to the ith object and these scores may then be subjected to analysis. For example, they may form the basis for calculating a correlation coefficient (say, between two judges' rankings) or analysis of variance technique may be applied. In such applications it should be noted that whilst linear aggregates of the scores are unbiased estimators of the corresponding aggregates from observations drawn from the standard normal distribution, quadratic forms of the scores do not necessarily have this unbiased property. For example

$$\mathscr{E}\Big(\sum_i x_i^2/\sigma^2\Big)=n, \quad \text{while} \quad \sum_i \{\xi(i \mid n)\}^2=n-\sum_i(\text{variance of } x_i/\sigma).$$

16. Test for heterogeneity of variance among normally distributed observations (Tables 31, 32, 32*a* and 33)

16·1 *Definitions*

The statistical analysis of data often leads to the calculation of a number, say k, of independent mean-square estimates of variance, s_t^2, of which it is desired to test the homogeneity. In particular, one of the basic assumptions of the usual analysis of variance tests is that the population variances of all experimental groups can be taken as equal, and it is often desirable to check the justification for this assumption. Mathematically, we may define the problem as follows: the s_t^2 are distributed independently as $\chi^2\sigma_t^2/\nu_t$ with ν_t degrees of freedom, respectively, and we wish to test the hypothesis that the σ_t^2 have a common, though unknown, value σ^2.

The most useful test employs Bartlett's (1937) modification of Neyman & Pearson's (1931) likelihood ratio statistic. We use as criterion*

$$M = \mathrm{N}\log_e\{\mathrm{N}^{-1}\sum_{t=1}^{k}\nu_t s_t^2\} - \sum_{t=1}^{k}\nu_t \log_e s_t^2, \tag{76}$$

where $\mathrm{N} = \sum_t \nu_t$. Provided that $\sigma_t^2 = \sigma^2$ for all t and none of the degrees of freedom ν_t is small, M is distributed approximately as χ^2 with $k-1$ degrees of freedom.

For small samples, Bartlett showed that M/C, with

$$C = 1 + \frac{1}{3(k-1)}\left\{\sum_t \frac{1}{\nu_t} - \frac{1}{\mathrm{N}}\right\}, \tag{77}$$

is more closely approximated by χ^2 for $k-1$ degrees of freedom. Investigations carried out by U. S. Nair (1938) and Bishop & Nair (1939) showed that the correction (77) is not always adequate if some of the ν_t are 1, 2 or 3.

In applying this test as well as the short-cut test described in § 16·5 below, it must be remembered that the theory is based on the assumption that the random variation within each of the k groups follows the normal law. If this is not true a significant value of M, or of the ratio $s^2_{\text{max.}}/s^2_{\text{min.}}$ of § 16·5, may indicate departure from normality rather than heterogeneity of variance. Tests of this kind are, indeed, more sensitive to departure from normality than the ordinary tests of the analysis of variance.

16·2 *Description of Tables* 32, 32*a and* 33

Table 32 is based on Hartley's (1940) χ^2-series expansion of M, which is slightly more accurate than the expansion leading to (77) and permits the ν_t to drop as low as 2.† The truncated series used in this approximation depends on the three parameters

$$k, \quad c_1 = \sum_t \frac{1}{\nu_t} - \frac{1}{\mathrm{N}} \quad \text{and} \quad c_3 = \sum_t \frac{1}{\nu_t^3} - \frac{1}{\mathrm{N}^3}. \tag{78}$$

The 5 and 1 % significance points of M are given in Table 32, whilst the auxiliary Table 33 is only required for accurate interpolation in Table 32. In Table 32, for each combination of

* M corresponds to Bartlett's $-2\log_e\mu$.

† Recently Box (1949) has re-examined this approximation. Whilst considerably improving current approximations to other likelihood criteria, Box did not materially improve on that for M.

k and c_1 there are two percentage points denoted by (a) and (b). These are approximately maximum and minimum values of the true percentage point which will normally have an intermediate value, dependent on c_3. Provided the degrees of freedom are not very unequal, the correct value of M will be close to the entry opposite (a).

The table has been arranged to make its use as simple as possible. If in the table of 5 % points, say, all entries in the lines for the appropriate k are greater than the value of M derived from the data, this value is not significant at the 5 % level. On the other hand, if the calculated M is larger than all entries for that k, then M is significant at the 5 % level. In neither case is it necessary to calculate c_1 or c_3. When, however, M falls within the range of values shown in the lines for the particular k, it is necessary to calculate c_1. Knowing this value, it will usually be possible to form an opinion on the significance of M without proceeding to the calculation of c_3 needed for interpolation between the entries (a) and (b).

It will be noted that the entries in the tables under $c_1 = 0$ are simply the 5 and 1 % probability levels of χ^2 with $k-1$ degrees of freedom, this being the limiting form approached when all the ν_t are large. On the other hand, for a given k, c_1 has its maximum value when all ν_t are unity and therefore $c_1 = k - 1/k$.*

Table 32a, due to Harsaae (*Biometrika*, 1969, **56**, 276-7) gives exact percentage points of M in the case where all $\nu_t = \nu'$ for $\nu = 1(1)10$, $k = 3(1)12$. Harsaae has confirmed that Table 32 should not be used when $\nu = 1$ and that there may be slight inaccuracy when $\nu = 2$.

16·3 *Applications of Tables* 31 *and* 32

Example 31. Column (3) of the table below gives a set of ten estimates of variance, calculated from ten samples of weight records of schoolboys of similar age, but from different forms. It is desired to test whether there are any real 'form differences' in the weight dispersion of the boys. The calculations for M is as follows, using Table 53 to obtain natural logarithms:

(1) Form no. t	(2) No. of boys n_t	(3) Weight variance s_t^2 (lb.²)	(4) ν_t	(5) $\log_e s_t^2$	(6) $\nu_t \log_e s_t^2$	(7) $1/\nu_t$
1	10	51	9	3·93	35·4	0·111
2	15	78	14	4·36	61·0	0·071
3	21	91	20	4·51	90·2	0·050
4	23	52	22	3·95	86·9	0·045
5	15	101	14	4·62	64·7	0·071
6	11	36	10	3·58	35·8	0·100
7	31	41	30	3·71	111·3	0·033
8	15	76	14	4·33	60·6	0·071
9	3	64	2	4·16	8·3	0·500
10	6	93	5	4·53	22·6	0·200
Totals	150	—	140 (=N)	—	576·8	1·252

We obtain further

$$\Sigma\nu_t s_t^2 = 9176, \quad \Sigma(\nu_t s_t^2)/\mathrm{N} = 65{\cdot}54, \quad \log_e\{\Sigma\nu_t s_t^2/\mathrm{N}\} = 4{\cdot}183.$$

Hence

$$M = 140 \times 4{\cdot}183 - 576{\cdot}8 = 8{\cdot}8.$$

* Actually, the last entry in each line has been computed from the approximating function, putting $c_1 = k$.

The observed value of M is therefore 8·8. This has to be compared with the appropriate tabulated 5% (or 1%) point. It is seen from Table 32 (5% points) that all entries opposite $k=10$ are greater than 8·8. Without further calculation it may therefore be concluded that M is not significant at the 5% level, and we may infer that no real differences are indicated in the weight dispersion among the ten forms of schoolboys.

Had the observed value of M been 18·8 (instead of 8·8), the decision as to its significance would not have been obvious, since some of the 5% points tabulated in the lines for $k=10$ are smaller than 18·8.* It is now necessary to calculate c_1, defined by (78). Using the reciprocals of ν_t given in col. 7 of the table above, it is found that

$$c_1 = 1{\cdot}25.$$

Since the percentage points (a) and (b) for $k=10$ and both $c_1 = 1{\cdot}0$ and 1·5 are less than 18·8, we can say that M would now be significant at the 5% level.

Had the data given a value of 17·6 for M, which lies between the four appropriate tabled entries, it would normally have satisfied our purpose merely to note that M was on the border line of 5% significance. If more precise information is needed, it will be necessary to proceed further by calculating c_3 and interpolating as follows.

It can be shown that, for a given k and c_1, the range of c_3 is (to a first order of approximation)

$$c_3(a) = c_1^3/k^2 < c_3 < c_1 = c_3(b). \tag{79}$$

The lower bound is approached when all values of ν_t are equal and the upper bound when j, say, of the ν_t are each equal to unity and the $k-j$ remaining values all tend to infinity. In Table 32 the entry for the percentage point opposite (a) is that for $c_3 = c_3(a)$; that opposite (b) is for $c_3 = c_3(b)$. It will be seen that, at any rate throughout the tabulated range of values, the entry (a) is greater than or equal to (b). In using the former, therefore, we shall in rare cases fail to detect the significance of M.

If interpolation for c_3 is decided on, use may be made of the auxiliary Table 33. This gives for all the marginal entries k and c_1 of Table 32 the two quantities

$$C = c_3(a) = c_1^3/k^2, \quad \Delta C = c_3(b) - c_3(a) = c_1 - c_1^3/k^2. \tag{80}$$

The procedure would then be first to interpolate linearly in the two nearest c_1 columns between the two percentage points (a) and (b), using the formula

Percentage point corresponding to c_3

$$= \frac{1}{\Delta C}\{(c_1 - c_3) \times \text{entry } (a) + (c_3 - C) \times \text{entry } (b)\}, \tag{81}$$

and then interpolating to the correct value of c_1.

Illustration. Suppose that $k=10$ and the degrees of freedom are the ten values of ν_t given in col. 4 of the illustrative data tabled above. Here

$$c_1 = 1{\cdot}25, \quad c_3 = 0{\cdot}14.$$

For the interpolation process, we need the following entries:

		$c_1 = 1{\cdot}0$	$c_1 = 1{\cdot}5$
From Table 33	C	0·010	0·034
	ΔC	0·990	1·466
From Table 32	Entry (a)	17·54	17·83
	Entry (b)	17·17	17·29

* Reference to Table 32 (1% points) shows, however, that M cannot be significant at the 1% level.

PBT

9

Hence the 5 % point corresponding to $c_1 = 1{\cdot}0$, $c_3 = 0{\cdot}14$ is approximately, from equation (81),

$$\frac{1}{0{\cdot}99}\{0{\cdot}86 \times 17{\cdot}54 + 0{\cdot}13 \times 17{\cdot}17\} = 17{\cdot}49.$$

The 5 % point corresponding to $c_1 = 1{\cdot}5$ and $c_3 = 0{\cdot}14$ will be approximately

$$\frac{1}{1{\cdot}47}\{1{\cdot}36 \times 17{\cdot}83 + 0{\cdot}11 + 17{\cdot}29\} = 17{\cdot}79.$$

Interpolating between these two values for $c_1 = 1{\cdot}25$, we find finally a 5 % point for M at 17·64. It will be seen that this value differs very little from that of 17·68 obtained by using the (*a*) entries only.

Example 32. As a second example, we may test the homogeneity of the $k = 5$ batch-variances quoted at the bottom of the table of tensile strength data presented in Example 15 (p. 38). Here, $\nu_t = \nu = 5$ for all t, so that the procedure used in the preceding example may be simplified by omitting the three columns headed ν_t, $\nu_t \log_e s_t^2$ and $1/\nu_t$.

A summary of the calculation is as follows:

$$M = k\nu \log_e\{\Sigma s_t^2/k\} - \nu\Sigma \log_e s_t^2 = 25 \log_e 367 - 5 \times 28{\cdot}63.$$

We find $M = 4{\cdot}5$ which, with $k = 5$, is clearly not significant.

16·4 *The relation between the M- and F-tests in the case of two groups*

It will be seen that Table 32 does not contain any entry for $k = 2$. This is because when two variance estimates s_1^2 and s_2^2 alone are available, we should normally test for the significance of their difference by calculating the variance ratio $F = s_1^2/s_2^2$, and using a double-tail F-test as described in § 9·3. It is, however, of interest to consider the relation between the two tests in this special case. It is easily seen that

$$e^{-M} = \left(\frac{\nu_1 + \nu_2}{\nu_2 + \nu_1 F}\right)^{\nu_1+\nu_2} F^{\nu_1}. \tag{82}$$

To $M = 0$ (when $s_1^2 = s_2^2$) corresponds the value $F = 1$, and for any value of $M > 0$ there will correspond two roots of (82), F_1 and F_2 (say), with $F_1 > 1$ and $F_2 < 1$. Thus the upper tail of the M-distribution for which $M \geqslant M(\alpha \mid \nu_t)$, where α is the significance level, will correspond to the two tails in the F-distribution, $F \geqslant F_1 = F(\alpha_1 \mid \nu_1, \nu_2)$ and $F \leqslant F_2 = F(1 - \alpha_2 \mid \nu_1, \nu_2)$. The total probability of falling in these tails will be $\alpha = \alpha_1 + \alpha_2$, e.g. 0·05 for the 5 % point of M, but this will not necessarily be divided equally between the tails. Only in the special case $\nu_1 = \nu_2$, where $F_2 = 1/F_1$ does the M-test correspond to the F-test with equal tails. It can be shown, however, that in the general case where $\nu_1 \neq \nu_2$, the M-test is approximated by the double-tail F-test for equal degrees of freedom ν, given by $\nu = 1{\cdot}5 \Big/ \left(\frac{1}{\nu_1} + \frac{1}{\nu_2} - \frac{1}{\nu_1 + \nu_2}\right)$. This approximation is based on the assumption that the distribution of M for $k = 2$ will depend only on c_1, and that we may therefore use an 'equivalent' common ν, such that

$$\frac{2}{\nu} - \frac{1}{2\nu} = c_1 = \frac{1}{\nu_1} + \frac{1}{\nu_2} - \frac{1}{\nu_1 + \nu_2}.$$

16·5 *Short-cut tests for heterogeneity of variance* (*Tables* 31 *and* 31*a*)

The computation of the criterion M as set out in *Example* 31 (p. 64) is comparatively laborious and often a deterrent in a rapid survey of data. For a quick assessment of heterogeneity in a set of mean squares we may compute the ratio of the largest to the smallest mean square $s^2_{\text{max.}}/s^2_{\text{min.}}$ or, alternatively, the ratio of the largest to the sum of the mean squares, $s^2_{\text{max.}}\Big/\sum_{t=1}^{k} s_t^2$ (Cochran, 1941). Tables 31 and 31a give 5% and 1% points of these ratios for the case where all k mean squares are based on ν degrees of freedom.

Example 33. Suppose we take the five batch variances of *Example* 15 (p. 37), i.e. the mean squares 469, 653, 406, 196, 111, for which $s^2_{\text{max.}} = 653$ and $s^2_{\text{min.}} = 111$. Their ratio is about 6 and clearly does not exceed the 5% value of 16·3 tabled for $k=5$, $\nu=5$. The application of Cochran's test to the same data has been given in § 14·5 above. Both results confirm that of the M-test (*Example* 32, p. 66).

As an approximate procedure the tables may be used when the ν_t differ by entering them with the average of the ν_t. In cases of doubt the M-criterion should be computed.

17. Tests for departure from normality (Tables 34A, B and C)

17·1 *Discussion of the problem*

If n independent observations x_i $(i=1,2,\ldots,n)$ of a random variable x are available, we may wish to test the hypothesis that the population frequency function follows the normal law, i.e. that

$$f(x) = \frac{1}{\sqrt{(2\pi)}\,\sigma} \exp\left[-\frac{1}{2}\left(\frac{x-\mu}{\sigma}\right)^2\right].$$

Usually μ and σ will be unknown and must be estimated from the data. It should be observed that by the very nature of the problem it is impossible to conclude definitely whether a given sample has been drawn from a particular kind of universe, normal or otherwise. The best that we can hope for is to be in a position to determine when the population is probably not normal, so that it would perhaps be more accurate to describe the tests considered below as *tests of non-normality*. Two methods of dealing with the problem are in common use:

(i) A normal curve is fitted to the sample data and the χ^2-test for goodness of fit applied to the grouped frequencies.

(ii) Certain functions of the moments of the sample are calculated and the significance of their departure from the expected values for a normal population is examined.

While the procedure (i) has certain advantages, it may prove somewhat insensitive to real departures from normality, partly because the χ^2-test only takes into account the magnitude of the differences between observed and expected group frequencies and not their sign and arrangement, and partly owing to the necessity of grouping together small tail frequencies when applying the test. If there are less than 50–100 observations, this loss of sensitivity through grouping will apply, indeed, to the whole range of the distribution. In the present discussion we are concerned only with procedure (ii).

If we define the moments of the observations as

$$\bar{x} = \sum_{i=1}^{n} \frac{x_i}{n}, \quad m_r = \sum_{i=1}^{n} \frac{(x_i-\bar{x})^r}{n} \qquad (r \geqslant 2),$$

the well-known moment ratios are

$$\sqrt{b_1} = m_3/m_2^{\frac{3}{2}}, \quad b_2 = m_4/m_2^2.$$

Departure of $\sqrt{b_1}$ from the 'normal' value of zero is an indication of skewness in the frequency function of the sampled population, while departure in b_2 from the 'normal' value of 3 is an indication of kurtosis.* For large samples, rough tests of normality may be obtained by comparing $\sqrt{b_1}$ and b_2-3 with the approximate values of their standard errors, viz. $\sqrt{(6/n)}$ and $\sqrt{(24/n)}$ respectively. Work by R. A. Fisher (1929, 1930) and others has made it possible to calculate the higher sampling moments of $\sqrt{b_1}$ and b_2 for a normal universe. The upper and lower 5 and 1 % points (single-tail test) given in Tables 34 B and C, were computed by assuming that the distributions of $\sqrt{b_1}$ and b_2 can be represented approximately by Pearson-type curves having the correct first four moment coefficients (E. S. Pearson, 1930, 1931; and P. Williams, 1935). Further work (E. S. Pearson, 1963, 1965) has made it possible to improve the accuracy of the approximation and to provide significance levels for b_2 when $n < 200$.

R. C. Geary (1935, 1936) has suggested an alternative statistic which may be used for detecting changes in kurtosis, particularly when samples contain less than 50 observations, where owing to the asymmetry of the b_2-distribution the method employed in deriving Table 34C could not safely be used. This statistic is

$$a = \frac{\text{mean deviation}}{\text{standard deviation}} = \frac{\sum_i |x_i - \bar{x}|}{\{n \sum_i (x_i - \bar{x})^2\}^{\frac{1}{2}}}. \tag{83}$$

For the normal population itself, this ratio has the value $\sqrt{(2/\pi)} = 0{\cdot}7979$; the ratio will be higher for platykurtic and lower for leptokurtic distributions.

Table 34A gives upper and lower 10, 5 and 1 % points for a (single-tail test) as well as the expectation and standard deviation in sampling from a normal population. As $n \to \infty$ the expectation of a tends to its normal population value of 0·7979.

[Charts based on the tables for a and $\sqrt{b_1}$ were published in *Biometrika* (1936), **28**, 304–7; these together with a similar chart of significance points for b_2, were reproduced by *Biometrika* as a 'separate', but they are not reproduced here.]

17·2 *Applications*

Example 34. The following table shows fifty observations of a variable x. Investigate whether it is likely that they have been drawn from a normal population. A rough plot in a dot diagram suggests a population distribution with marked positive skewness. However, it is desirable to establish a significant departure from normality on an objective basis. The sample size being small, we shall calculate a and $\sqrt{b_1}$, not b_2.

−6	−14	19	−5	−15	4	−9	−7	5	29
0	−11	2	−12	−1	−16	−2	9	−8	6
−17	11	15	0	−10	−3	14	1	0	29
−9	−2	−13	−13	−14	−11	−7	−4	20	−8
11	−8	−13	−11	0	−15	−4	−2	−11	−6

The mean deviation is perhaps calculated most readily, using the formula

$$\sum_{i=1}^{n} |x_i - \bar{x}| = 2\{(\textit{sum} \text{ of observations greater than } \bar{x}) - \bar{x} \times (\textit{number} \text{ of observations greater than } \bar{x})\}.$$

* The symmetrical platykurtic distribution, with $\beta_2 < 3$, is characterized by a flatter top and more abrupt terminals than the normal curve; the symmetrical leptokurtic distribution, with $\beta_2 > 3$ has a sharper peak at the mean and more extended tails.

The computation will then proceed as follows:

For all observations $\Sigma(x_i) = -112, \quad \bar{x} = -2\cdot24,$

$$\Sigma(x_i^2) = 6494, \quad \Sigma(x_i - \bar{x})^2 = 6243\cdot12,$$

$$\Sigma(x_i^3) = 33254, \quad \Sigma(x_i - \bar{x})^3 = 75769\cdot7376,$$

$$\sqrt{b_1} = 1\cdot086.$$

For observations greater than $\bar{x}$

$$\Sigma(x_i) = 168, \quad \text{number greater than } \bar{x} = 22.*$$

Hence $\sum_i |x_i - \bar{x}| = 2\{168 + 2\cdot24 \times 22\} = 434\cdot56,$

$$a = 0\cdot778.$$

Table 34B shows that, for $n=50$, $\sqrt{b_1}$ is well beyond the upper 1% point at 0·787. Table 34A indicates that a is below the expected value for a normal population (0·802) but not significantly so. We should therefore conclude that the population has almost certainly considerable skewness, in the positive sense, i.e. with the longer tail towards increasing x.

The observations actually formed part of a sampling experiment (E. S. Pearson, 1935), the population distribution following a Pearson Type I curve for which $\sqrt{\beta_1} = 1\cdot0$, $\beta_2 = 3\cdot8$ and (mean deviation)/(standard deviation) = 0·797.

Example 35. Bayes (1937) has given the following distribution of 1000 observations made on the strength of the warp of a 12/20's duck cloth:

Central value (lb.)	400	405	410	415	420	425	430	435	440	445
Observed frequency	2	4	4	17	27	27	50	49	68	76
Graduation	2·6	3·8	7·4	13·0	20·7	30·8	43·1	57·0	71·7	85·9

Central value (lb.)	450	455	460	465	470	475	480	485	490	495
Observed frequency	95	113	125	105	97	73	38	19	10	1
Graduation	98·1	106·6	109·5	105·5	93·6	74·2	49·2	23·1	4·2	0·0

For these data Pearson & Welch (1937) give the following moment coefficients (using Sheppard's corrections for grouping):

$$\text{Mean} = 454\cdot09\text{ lb.}, \quad \text{standard deviation} = 17\cdot358\text{ lb.}$$

$$\sqrt{b_1} = -0\cdot421, \quad b_2 = 2\cdot748.$$

Tables 34B and C, entered with $n=1000$, show that $\sqrt{b_1}$ is clearly significant while b_2 is close to its lower 5% point. In their further analysis of the problem, the authors represent the distribution by a Pearson Type I curve, the corresponding graduated frequencies obtained from which are shown in the bottom rows of the table. These data will be used again, below (see Examples 47 and 50, pp. 86, 89).

IV. TABLES RELATING TO CERTAIN DISCRETE DISTRIBUTIONS (TABLES 36–41)

Included under this general heading is a group of six tables relating to the binomial series, the hypergeometric series and Poisson's limit to the binomial. The common element lies in the fact that the random variable is a 'count' which can only assume integral values 0, 1, 2, ...

* Note that there are three values of -2 and one of -1, all greater than $\bar{x} = -2\cdot24$, although negative.

Table 7, which gives the cumulative sum of the Poisson series, as well as the probability integral of χ^2, and Table 16 giving the cumulative sum of binomial terms in the form of the incomplete B-function, have been printed more appropriately at an earlier stage.

18. Tables giving the individual terms of selected binomial series (Table 37)

The $(i+1)$th term in the expansion of the binomial $(q+p)^n$ may be written

$$f(i\,|\,n,p)=\binom{n}{i}p^i q^{n-i}$$

The table gives to 5 decimal places the values of $f(i\,|\,n,p)$, $i=0,1,\ldots,n$, for

$$n=5,\ 10,\ 15,\ 20,\ 25,\ 30$$

and
$$p=0{\cdot}01,0{\cdot}02\,(0{\cdot}02)\,0{\cdot}10\,(0{\cdot}10)\,0{\cdot}50.$$

For $p>0{\cdot}5$ the terms are obtained in reverse, using the arguments i and p printed in the last column but one and bottom row of each page.

The table is intended primarily for exploratory and illustrative purposes, e.g. in lecture or class work; the statistician who wishes to use a working table giving binomial terms over a wide range of values of n and p, should consult *Tables of the Binomial Probability Distribution*, National Bureau of Standards, Applied Mathematics Series, no. 6 (1950). Alternatively, he may obtain the cumulative sums, and hence obtain the individual terms by differencing either from the *Tables of the Incomplete B-function* (Karl Pearson, 1934) or more approximately, from Table 16 (see § 8·1 above), noting in both cases that

$$\sum_{i=0}^{a-1} f(i\,|\,n,p)=1-I_p(a,n-a+1)=1-\int_0^p u^{a-1}(1-u)^{n-a}\,du\Big/\!\!\int_0^1 u^{a-1}(1-u)^{n-a}\,du. \quad (84)$$

Example 36. Suppose that we wish to find out what kind of errors are involved in using the normal probability integral as an approximation to the sums of the tail terms of certain binomial series in the neighbourhood of the 5 % points. Take the case of the binomial with $n=30$, $p=0{\cdot}2$. Using the correction for continuity, we compare:

(i) *For the lower tail*, the sums of terms from Table 37 corresponding to $i=0,1,\ldots,a$ with the integral of the normal distribution (obtained from Table 1) up to

$$X=(a+0{\cdot}5-np)/\sqrt{(npq)}.$$

(ii) *For the upper tail*, the sums of terms corresponding to $i=a,a+1,\ldots,n$ with the integral beyond

$$X=(a-0{\cdot}5-np)/\sqrt{(npq)},$$

where $np=6{\cdot}0$, $\sqrt{(npq)}=2{\cdot}191$. Results are shown in the following table:

Terms		Binomial	Normal approx.
Up to and including	$a=2$	0·0442	0·0551
	$a=3$	0·1227	0·1269
Including and beyond	$a=10$	0·0611	0·0551
	$a=11$	0·0256	0·0200

The approximation is clearly rather rough, though it will be adequate for some practical purposes.

19. THE TEST OF SIGNIFICANCE IN A 2×2 CONTINGENCY TABLE CONTAINING SMALL FREQUENCIES (TABLE 38)

19·1 *Description of the problem**

Suppose that in a group of $N=A+B$ individuals, r possess a certain character, say Y, and $N-r$ do not. If now a sample of A individuals is drawn randomly from the N, *without replacement*, the probability that a of these possess the character Y is

$$f(a \mid r, A, B)=\frac{A!\,B!\,r!\,(N-r)!}{a!\,b!\,(A-a)!\,(B-b)!\,N!}, \tag{85}$$

where $b=r-a$. This expression is proportional to the coefficient of x^a in the hypergeometric series

$$F(\alpha,\beta,\gamma,x)=F(-r,\,-A,\,B-r+1,\,x).$$

If $A\geqslant B$, then a can assume values

$$\begin{array}{lll} \text{(i)} & 0,1,\ldots,r & \text{if} \quad r\leqslant B,\\ \text{(ii)} & r-B, r-B+1,\ldots,r & \text{if} \quad B<r\leqslant A,\\ \text{(iii)} & r-B, r-B+1,\ldots,A & \text{if} \quad r>A. \end{array}$$

For this probability distribution,

$$\text{Mean } a=rA/N, \tag{86}$$

$$\text{Variance of } a=\sigma_a^2=ABr(N-r)/\{N^2(N-1)\}. \tag{87}$$

The result of the sampling described, which is equivalent to a random partition of the N individuals into two groups, one of A and the other of B, may be set down in the following 2×2 table:

	With Y	Without Y	Total
Sample or 1st group	a	$A-a$	A
Remainder or 2nd group	b	$B-b$	B
Total	r	$N-r$	N

If the partition is made at random, without any relation to the presence or absence of the character Y, then the expectation of a is rA/N and that of b is rB/N; or, in other words, the expectation of the difference in proportions, $a/A-b/B$, is zero. In so far as some effect operates to make the presence of Y more probable in one group than the other, a will be more likely to differ from rA/N and $a/A-b/B$ from zero, than would be expected under the random distribution law of equation (85). Table 38† is designed to assist in the decision as to whether this departure from expectation on the random hypothesis is significant. The form of the table may be explained with the help of Fig. 3, which shows for the particular values $A=15$, $B=12$ the relation between possible values of a, b and $r=a+b$. For a given value of r, e.g. $r=9$, the probabilities $f(a \mid 9, 15, 12)$ of equation (85) are associated with the (a,b) values marked by circles along the line $r=9$. Taking the upper tail of this distribution, the probabilities that a is

* For rules applicable in routine tests of significance, see §19·2.

† This table is due to D. J. Finney (1948), and we are indebted to him for permission to reproduce it and to make very full use in this and the following section (19·2) of the explanatory comment which accompanied his original publication of the table.

7, 8 and 9 (and therefore that $b=2$, 1 and 0) are respectively 0·091, 0·017 and 0·001. Thus using a single-tail test, i.e. asking whether a/A is significantly greater than b/B, and working to a 5% level, we should conclude that there was a significant difference if $a=8$, $b=1$ ($\Pr\{a \geq 8\} = 0{\cdot}018$) but not if $a=7$, $b=2$ ($\Pr\{a \geq 7\} = 0{\cdot}109$).

If similar calculations are made for the cumulative sums of the terms in the upper tail of each of the hypergeometric series, for $r = 1, 2, \ldots, 26$, all the pairs of values (a, b) to the right and below the stepped line in Fig. 3 will be judged significantly different at the 5% level.

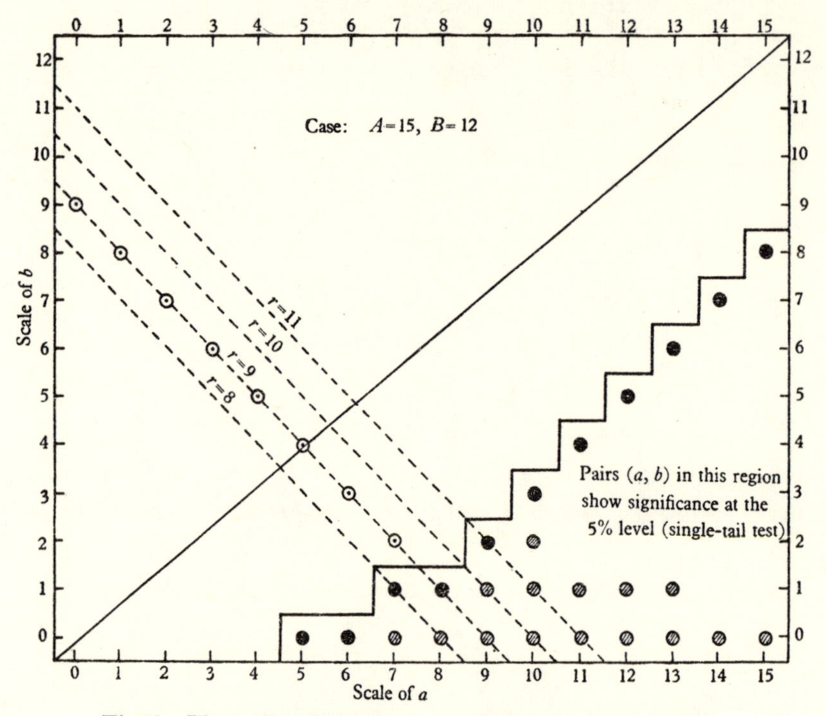

Fig. 3. Illustrating the test of significance in a 2×2 table.

Formally, this bounding line is placed above and to the left of points $(a_0, r - a_0)$, where

$$\sum_{a=a_0}^{k} f(a \mid r, A, B) \leq 0{\cdot}05 \quad \text{and} \quad \sum_{a=a_0-1}^{k} f(a \mid r, A, B) > 0{\cdot}05,$$

k being the smaller of the quantities r and A. Similar significance limits could be inserted in the diagram for other percentage levels, and lattice diagrams of this type could be drawn for other combinations of A and B.

As arranged, Table 38 does not give significance levels for a for fixed $r = a + b$, A and B, but shows in bold type, for given a, A and B, the highest value of b ($< a$) which is just significant at the 5, 2·5, 1 and 0·5% levels. Thus for the special case illustrated in Fig. 3, the 5% values tabled for (a, b) are those corresponding to the solid black circles. It will be seen that on this basis some entries, e.g. those for $a=7$, $b=0$ and for $a=9$, $b=1$ on the diagonals $r=7$ and $r=10$ respectively, are unnecessary and space is saved in tabulation.

The figures in small type following an entry, say $b = b_0$, are the sums along the diagonal of the hypergeometric terms for which $b \leq b_0$.

An extension of Table 38 for sample size combinations $B \leq A \leq 45$ has now been issued.*

* *Tables for Testing Significance in a 2 × 2 Contingency Table.* Compiled by D. J. Finney, R. Latscha, B. M. Bennett & P. Hsu with an Introduction by E. S. Pearson, 1963. Cambridge University Press.

Provided that none of the marginal totals A, B, r or $N-r$ are very small, a good approximation to the sum of the tail terms of the hypergeometric series of equation (85) may be obtained from the integral under a normal curve, having the mean, $\bar{a}$, and variance, σ_a^2, of equations (86) and (87). Thus if $r \leqslant B \leqslant A$ and we want the sum of the first a_0+1 terms of $f(a \mid r, A, B)$, we use the following approximation

$$\sum_{a=0}^{a_0} f(a \mid r, A, B) = \int_{-\infty}^{X} \frac{1}{\sqrt{(2\pi)}} e^{-\frac{1}{2}u^2} du = P(X),$$

where $$X = (a_0 + 0{\cdot}5 - \bar{a})/\sigma_a.$$

If we need an approximation to the sum of the terms corresponding to a_0 or *larger* integers, we use the complementary normal integral, $Q(X) = 1 - P(X)$ with $X = (a_0 - 0{\cdot}5 - \bar{a})/\sigma_a$. The half-unit added to or subtracted from a_0 is a correction introduced in approximating to the sum of a discrete distribution by a continuous integral.* Its use in connexion with the interpretation of 2×2 tables was first suggested by Yates (1934).

19·2 *Rules to follow in using Table* 38

Points to be noted are:

(i) The 'first group' in the sense of the 2×2 table on p. 71 is defined to be that for which $A \geqslant B$.

(ii) The type of observation which has been described conventionally in that table as possessing 'character Y', is that which makes $a/A \geqslant b/B$ (or $aB \geqslant bA$).

(iii) If b is equal to or less than the integer in the column headed 0·05, 0·025, 0·01 or 0·005, a/A is significantly greater than b/B (single-tail test) at these probability levels. On the other hand, for the double-tail test, if b is equal to or less than the integer in a given column, a/A is significantly *different from* b/B at a probability level equal to twice the figure heading that column, i.e. at the 0·10, 0·05, 0·02 and 0·01 levels, respectively.

(iv) A dash, or absence of an entry, for some combination of A, B and a indicates that no 2×2 table in that class can show a significant effect at that level.

(v) Owing to the discontinuous character of the hypergeometric distribution, the probability that, for given $r = a + b$, b will be equal to or less than the value specified in bold type will generally be less, and often very considerably less, than that shown at the head of the column; the true numerical values are shown in small type.

(vi) From the mathematical point of view, the two pairs of marginal totals play equivalent roles in the test. Consequently, if only one set satisfies the condition that neither total exceeds 15, that margin may be conventionally regarded as giving A and B. If both margins satisfy the condition, the two possible arrangements will lead to identical conclusions.

(vii) The use of the table presented here may be extended by noting that any 2×2 table more extreme than one known to show significant deviation from proportionality must itself be significant. Thus even though one of each pair of marginal totals exceeds 15, a test of significance without calculation may still be possible. In particular, if the standard arrangement of the table gives (1) below with $A > 15 \geqslant B$ and $aB \geqslant bA$, the deviation from proportionality will be significant if arrangement (2) is significant (but not necessarily non-significant

* This correction has also been made in Example 36 (p. 70), when using the normal probability integral to approximate to the sum of binomial terms.

otherwise), and the deviation will be non-significant if arrangement (3) is not significant (but not necessarily significant otherwise).

(1)

a	$A-a$	A
b	$B-b$	B
$a+b$	$A+B-a-b$	$A+B$

(2)

$15+a-A$	$A-a$	15
b	$B-b$	B
$15+a+b-A$	$A+B-a-b$	$15+B$

(3)

a	$15-a$	15
b	$B-b$	B
$a+b$	$15+B-a-b$	$15+B$

19·3 *Applications of Table* 38

These may be usefully classified under several heads. While the basic mathematical distribution underlying the test (i.e. that given by equation (85)) is the same in all cases, there are differences of some importance in the way in which the random element may be introduced in collecting the data. We think it useful to accompany the illustrative examples with some account of these differences.

Case 1

The most direct application of the theory is to the type of experiment where (i) two 'treatments' are being compared, the response being a quantal one so that each individual either 'reacts' or 'fails to react'; (ii) the $N=A+B$ individuals to be used in the experiments are divided by a random partition (e.g. by using random numbers) into a group of A receiving the first treatment and a group of B receiving the second. It is then observed that proportions a/A and b/B react; the question asked is whether the difference in these proportions is statistically significant.

Note that in this situation where the random selection has been applied to a specified set of N individuals, the experiment is self-contained and the reference set is strictly that which would be generated had different allocations of the treatments been made among the N individuals.

Example 37. Suppose that it is desired to test the value of a period of 'preliminary training' in improving the subsequent performance of newly recruited operatives who are to be employed on a particular industrial process. Out of twenty-nine new recruits, a group of fourteen is selected randomly for this training. Subsequently all twenty-nine operatives are tested and the performance of each is classed as 'faulty' or 'correct', with the result shown below:

	Performance		Total
	Faulty	Correct	
No preliminary training	9 $(=a)$	6	15 $(=A)$
Preliminary training	3 $(=b)$	11	14 $(=B)$
Total	12	17	29

The single-tail test is appropriate here; reference to Table 38 with $A=15$, $B=14$ and $a=9$, shows that $b=3$ is significant at the 5 % but not at the 2·5 % level. The figure in smaller type indicates that if the preliminary training had no effect on performance, and twelve out of the twenty-nine operators would have shown a faulty performance whether they had a preliminary training or not, then the probability of a random partition putting $b \leq 3$ into one group and $a \geq 9$ into the other is 0·041.

As none of the marginal frequencies in the table are very small, we may expect to get a reasonable approximation to this probability of 0·041, using the normal probability integral as described at the end of § 19·1. We have

$$\bar{a} = 15.12/29 = 6{\cdot}207, \quad \sigma_a^2 = 15.14.12.17/(29^2.28) = 1{\cdot}819, \quad \sigma_a = 1{\cdot}349.$$

Hence
$$X = (9 - 0{\cdot}5 - 6{\cdot}207)/1{\cdot}349 = 1{\cdot}70.$$

Since Table 1 shows that $Q(1{\cdot}70) = 1 - P(1{\cdot}70) = 0{\cdot}045$, the approximation can be regarded as satisfactory for our purpose.

Case 2

A second important type of application arises when we inquire whether the probability of an individual possessing a specified character Y is the same in two large populations from which random samples of size A and B, respectively, have been drawn. Here two separate random selections are involved, and the most natural reference set is one in which a and b (the number of individuals in the samples having the character Y) are no longer restricted to give a fixed total of r. Mathematically, the position may be defined briefly as follows.

If $p_1(Y)$ and $p_2(Y)$ are the probabilities in the two populations, the probability of obtaining a pair of samples with a and b individuals bearing the character Y, respectively, is

$$f(a, b \mid p_1, p_2) = \binom{A}{a} p_1^a (1-p_1)^{A-a} \times \binom{B}{b} p_2^b (1-p_2)^{B-b}.$$

If the hypothesis tested is true and $p_1 = p_2 = p$, then

$$f(a, b \mid p) = \binom{N}{r} p^r (1-p)^{N-r} \times \frac{A!\,B!\,r!\,(N-r)!}{a!\,b!\,(A-a)!\,(B-b)!\,N!}. \tag{88}$$

The first term in (88) is the binomial probability of obtaining $r = a + b$ Y's in a sample of $N = A + B$ when $p(Y) = p$; the second expression is the multinomial term already considered. Turning back to Fig. 3, the first term represents the probability that the point (a, b) falls on a particular diagonal, r = constant; this probability cannot be determined, since p is essentially unknown. The second term is the conditional probability of a (or b), given r and is independent of the unknown p.

If we use the same test procedure as before, based on the sum of the tail terms of the hypergeometric series, e.g. reject the hypothesis of a common probability in the two populations when (a, b) falls beyond the appropriate 5 % limit, we know that we shall be on the safe side, even in using the extended reference set, where r is regarded as varying from one sampling to another. Under these conditions, using the nominal 5 % level, the proportion of occasions when we should reject the hypothesis $p_1(Y) = p_2(Y)$ when it is true cannot be determined. It is the sum of true tail terms (always $\leq 0{\cdot}05$) such as those given in small print in the table, weighted with the unknown binomial terms of equation (88). All that we can say is that this proportion will be considerably below the nominal 0·05.

Barnard (1947) has considered how a test might be developed providing an upper limit to this unknown proportion.

Example 38. The relative sensitivity of two mining explosives, C and D, was tested under controlled conditions. Ten cartridges of explosive C and six of explosive D were fired in turn in a standard gas chamber and a record was made of whether each shot resulted in an ignition or not. The results (taken from data quoted by E. S. Pearson, 1950) were as follows:

	Ignitions	Non-ignitions	Total
Explosive C	0	10 $(=a)$	10 $(=A)$
Explosive D	4	2 $(=b)$	6 $(=B)$
Total	4	12	16

Is there clear evidence of a difference between the two explosives in liability to cause ignition?

Table 38 shows the following figures for $A=10$, $B=6$, $a=10$:

Probability	0·05	0·025	0·01	0·005
b	3	2	2	1

Regarding the double-tail test as appropriate and remembering the rule (iii) of § 19·2 above, the observed result with $b=2$ is significant at the 5 % level; it is also significant at the 2 % level.

Case 3

A third type of application is to the situation where the individuals in a population may possess two characters, Y and Z, and we wish to determine, on the basis of a single sample of N, small compared to the population and randomly selected from it, whether the characters are independent or not. Writing $\bar{Y}$ and $\bar{Z}$ for the absence of Y and Z, every individual may be classed in one or other of four categories and the probability scheme is as follows:

	Y	$\bar{Y}$	Total
Z	$p(Y, Z)$	$p(\bar{Y}, Z)$	$p(Z)$
$\bar{Z}$	$p(Y, \bar{Z})$	$p(\bar{Y}, \bar{Z})$	$p(\bar{Z})$
Total	$p(Y)$	$p(\bar{Y})$	1

The condition for independence is that $p(YZ)=p(Y)\times p(Z)$. It follows at once that if this is satisfied and if the sample data are arranged in a 2×2 contingency table with the same notation as before, then

$$f(a,b\mid p(Y),p(Z))=\binom{N}{r}p(Y)^r(1-p(Y))^{N-r}\times\binom{N}{A}p(Z)^A(1-p(Z))^{N-A}\times\frac{A!\,B!\,r!\,(N-r)!}{a!\,b!\,(A-a)!\,(B-b)!\,N!}. \quad (89)$$

Thus the conditional probability distribution for a, given r and A, is the same hypergeometric distribution as before. If we use the significance levels of this series, obtained from Table 38, rejecting the hypothesis of independence when b for given a, A and B falls, say, beyond the appropriate 5 % level, we shall again be on the safe side, i.e. the risk of wrong rejection will be less than 0·05 in relation to the more extended reference set in which both r and A are regarded as varying from one sample to another.

Example 39. In an extensive inquiry into the relationship between intelligence and physical and mental characters (Karl Pearson, 1906, p. 144), data were given showing the relationship between teachers' estimates of intelligence and athleticism among 1708 schoolboys. Suppose that in a preliminary survey of the data, the records randomly selected for 30 boys had provided the following figures:

Intelligence	Athletic	Not athletic	Total
Above average Below average	11 8	3 $(=b)$ 8 $(=a)$	14 $(=B)$ 16 $(=A)$
Total	19	11	30

Do these figures show any significant relationship between the two qualities? Since $A > 15$, we follow rule (vii) of § 19·2 and consider the tables

$$\begin{array}{cc|c} 11 & 3 & 14 \\ 8 & 7 & 15 \\ \hline 19 & 10 & 29 \end{array} \quad \text{and} \quad \begin{array}{cc|c} 11 & 3 & 14 \\ 7 & 8 & 15 \\ \hline 18 & 11 & 29 \end{array}$$

When Table 38 is entered with $A = 15$, $B = 14$, it is seen that in neither case is the arrangement significant, even at the 10 % level (double-tail test). This means that in the original table a/A is not significantly different from b/B at this level.

19·4 *The case of larger samples*

The scope of Table 38 is of course very limited.* When dealing with medium or large samples, using the mean and standard deviation of the hypergeometric series, we note that the ratio

$$u = \frac{a - \mathscr{E}(a \mid r, A)}{\sigma(a \mid r, A)} = \frac{a - rA/N}{\left\{\dfrac{ABr(N-r)}{N^2(N-1)}\right\}^{\frac{1}{2}}} \tag{90}$$

has expectation zero and unit standard deviation. This is true for the reference set of each of the three forms of application discussed, i.e. whether both margins, one margin or neither margin of the 2×2 table is regarded as fixed. The usual approximation is to regard the overall distribution of this ratio as roughly normal, i.e. to judge the significance of u by reference to the normal probability scale of Table 1. Under these conditions it is clearly justifiable to substitute N for $N-1$, and the ratio is quoted under various alternative, but mathematically identical, forms, e.g. as

$$\frac{a/A - b/B}{\left\{\dfrac{r}{N}\left(1 - \dfrac{r}{N}\right)\left(\dfrac{1}{A} + \dfrac{1}{B}\right)\right\}^{\frac{1}{2}}} \quad \text{or} \quad \frac{\{a(B-b) - b(A-a)\}\sqrt{N}}{\{ABr(N-r)\}^{\frac{1}{2}}}. \tag{90a}$$

* For a reference to a more extensive table covering the range $B \leqslant A \leqslant 45$, see p. 72.

Assuming, on the hypothesis tested, that the ratio is a unit normal deviate, its square may be regarded as a value of χ^2 based on $\nu=1$ degree of freedom, and the test is often applied, making use of the χ^2 tables. It should, however, be noted that in this form the test is automatically a double-tail one; by working with the square root as in equations (90) or (90a) and referring to the Normal tables, we do not lose sight of the sign of $a/A-b/B$ which may be particularly informative.

It is common practice to insert in the numerator of u (equation (90)) the '$\frac{1}{2}$' correction for continuity referred to on p. 73 above, or to make the equivalent correction to the numerator of the ratios in equations (90a). In this way we shall in general get a more accurate approximation to the sum of the tail terms of the hypergeometric series from the integrals of the normal curve or of the χ^2-distribution having one degree of freedom. However, it should be remembered that the expectation of u^2, when so corrected, is now less than unity, and as Lancaster (1949) has pointed out, this may result in considerable bias in any form of analysis which depends upon the combination of a number of independent values of u^2.

Example 40. When the full data for the 1708 schoolboys referred to in Example 39 (p. 77) were tabled, the following results were obtained:

Intelligence	Athletic	Not athletic	Total
Above average	581	209	790
Below average	567	351	918
Total	1148	560	1708

Using equation (90), or either of the alternative forms (90a), we find that $u=5\cdot17$. Regarded as a standardized normal deviate, this value is clearly highly significant, so that there is little doubt that in the category of boys considered a positive relationship existed between 'intelligence' and skill in athletics.

20. Test of the difference between two Poisson expectations (Table 36)

Definition

If a and b are two variables distributed independently in accordance with the Poisson law, with expectations m_1 and m_2 respectively, their joint probability function is

$$f(a,b\mid m_1,m_2)=m_1^a e^{-m_1}/a!\times m_2^b e^{-m_2}/b!$$

$$=M^r e^{-M}/r!\times\binom{r}{a}\theta^a(1-\theta)^b, \tag{91}$$

where $$r=a+b,\quad M=m_1+m_2,\quad \theta=m_1/(m_1+m_2).$$

The first term in (91) represents the probability that a Poisson variable with expectation M has a value r, while the second expression is the term of a binomial. Thus the conditional probability distribution of a (and therefore of $b=r-a$) for given r and θ is a binomial. The position is very similar to that described in § 19·1 above, and illustrated in Fig. 3, except that the lattice is no longer closed, A and B having become infinite. Further, the conditional distribution, along each diagonal, $r =$ constant, is now a binomial in place of a hypergeometric series.

In practical application, when $\theta = m_1/(m_1 + m_2)$ is specified by hypothesis, significance levels may be obtained for the conditional distribution from the tail sums of binomials, e.g. by use of the incomplete B-function relation (§ 8·1 and 18). In the special case where we wish to test whether the two Poisson expectations are the same, i.e. whether $m_1 = m_2$, so that $\theta = \frac{1}{2}$, we are concerned with the significance levels of the binomial $(\frac{1}{2} + \frac{1}{2})^r$.

Formally, Table 36 A shows for values of $r = a + b$ in the range 1 (1) 80, the values of b for which

$$\sum_{i=0}^{b} \binom{r}{i} \left(\frac{1}{2}\right)^r \leqslant \alpha \quad \text{and} \quad \sum_{i=0}^{b+1} \binom{r}{i} \left(\frac{1}{2}\right)^r > \alpha,$$

for $\alpha = 0{\cdot}10$, 0·05, 0·025, 0·01 and 0·005.

If we denote by b the smaller of the two observed frequencies a and b, then if b is equal to or less than the integer in the column headed 0·10, 0·05, 0·025, 0·01 or 0·005, b is significantly *less than a* (single-tail test) at these levels. Alternatively, if a double-tail test is appropriate, we should say that b was significantly *different from a* at the probability levels 0·20, 0·10, 0·05, 0·02 or 0·01 respectively. Absence of an entry indicates that at this level significance is not attainable for that value of r.

As in the case of Table 38, owing to the discontinuous character of the binomial distribution, the tail probability corresponding to a given critical value of b will usually be considerably less than that shown at the head of the column. As a result, in the long run application of the test the proportion of times in which significance will be claimed, when in fact $m_1 = m_2$, will be considerably less than the nominal value defined by the probability level used. The extent of this over-estimation of the risk of error involved will depend on the value of the common expectation $m = m_1 = m_2$. To give some idea of the position, Table 36 B shows (for the single-tail test) for certain values of m the probability of rejecting the hypothesis $m_1 = m_2$, when it is true, using the five nominal significance levels of Table 36 A. These probabilities were obtained from the double summation

$$\sum_{r=r_0}^{\infty} \left\{ (2m)^r e^{-2m}/r! \times \sum_{i=0}^{b} \binom{r}{i} \left(\frac{1}{2}\right)^r \right\},$$

where for given r, b has the critical value given in Table 36 A, and where r_0 is the value of r opposite the first entry of 0 in the appropriate α-column.

It is seen that these true significance levels change systematically with m. In the nature of the problem, m cannot be known, but within broad limits the table indicates how much on the safe side we shall be in using the nominal or upper bound significance values. The exceptionally low values of the probabilities in Table 36 B for $m = 2{\cdot}5$ occur, of course, because with an expectation of this magnitude the two samples will often show frequencies a and b for which $a + b = r < r_0$. Significance is then not attainable, however different are a and b. This effect, however, scarcely remains when $m = 5{\cdot}0$.

The tables have been based on those given in a paper by Przyborowski & Wilenski (1940). These authors also gave tables and charts, not reproduced here, from which it is possible to assess the power of the test, i.e. the probability of establishing significance for given unequal values of the expectations m_1 and m_2.

Example 41. Maguire, Pearson & Wynn (1952) have quoted figures for the time intervals between explosions in coal mines in the United Kingdom involving more than ten men killed. If, starting from the beginning of the record, we take successive periods of 10 years, it is found that eight accidents occurred in the third of these periods and twenty accidents in the fourth period. If accidents occur randomly in time with a constant expectation per unit time, the frequency of accidents in a period of given length should follow the Poisson law. If it is questioned whether the difference between accident frequencies of eight and twenty is significant, we may use Table 36A to test whether these two figures are consistent with sampling subject to a common Poisson law with expectation m.

Entering the table with $r = a + b = 28$ and using a double-tail test, i.e. doubling the probability at the top of the columns, we find that $b = 8$ is significant at the 5 % level. Thus there is some evidence for an increase in accident risk during the 20 years in question, although, of course, this is most unlikely to have taken the form of an increase from a constant expectation m_1 in the first period to another constant value m_2 in the second period.

It is seen from Table 36B that when m is in the neighbourhood of 12·5–15, the nominal 5 % level (double-tail test) corresponds to a true level of 0·030.

21. Individual terms of Poisson's limit to the binomial (Table 39)

If a variable x follows the Poisson distribution law, with expectation m, the probability that $x = i$ will be

$$f(i \mid m) = m^i e^{-m}/i! \quad (i = 0, 1, \ldots, \infty). \tag{92}$$

As described in § 3·3 above, use has been made of the relation between the χ^2-integral and the cumulative sum of the terms of the Poisson series, to arrange the table of the former (Table 7) so that it can be used for the latter. Table 7, in fact, gives $\sum_{i=0}^{c-1} f(i \mid m)$ to 5-decimal accuracy. The present Table 39, calculated at an earlier date, gives to 6-decimal accuracy the individual terms in the series (92) for $m = 0{\cdot}1\,(0{\cdot}1)\,15{\cdot}0$. The two tables are complementary, and it will be found that sometimes one, sometimes the other is the more appropriate to use. The choice will depend partly on whether it is the cumulative sum or the individual probabilities that are required. But apart from this, from the point of view of interpolation between tabled values of m, the argument interval may be the controlling factor.

For $0 < m \leqslant 1{\cdot}0$: Table 7 has the smaller interval, giving figures for 38 values of m in this range, whereas Table 39 gives figures for only 10 values.

For $1{\cdot}0 < m < 5{\cdot}0$: both tables use the same argument interval, 0·1.

For $5{\cdot}0 < m < 15{\cdot}0$: Table 39 has the advantage, retaining the smaller interval of 0·1.

For $m > 15{\cdot}0$: we are beyond the range of Table 39. While Table 7 gives cumulative sums up to $m = 67$ these are not carried further than the term $c = 35$, a fact which limits the usefulness of the extension.

An illustration of the use of Table 7 in fitting a Poisson distribution to a series of observed frequencies has already been given (Example 3, p. 11), and no further illustration is needed here.

22. The determination of confidence limits for the Poisson and binomial parameters

22·1 *General comment on the theory*

If observations of random variables, $x_1, x_2, \ldots, x_n$, are used to provide an estimate T of a population parameter θ, it is common practice to record with T an estimate s_T of its standard error. When dealing with samples containing many observations, the sampling distribution of T

will generally be approximately normal and the difference between s_T and its true value σ_T will usually be of smaller order than the difference between T and θ. Hence little ambiguity results from the various forms of putting on record the accuracy of the estimate T, e.g. from the statement that the odds are roughly 19 to 1 that the numerical value of the difference $T-\theta$ is less than $2s_T$.

The position, however, is different when the estimate is based on relatively few observations, since the sampling error in s_T cannot then be overlooked. The fiducial theory of R. A. Fisher and the confidence interval theory of J. Neyman were developed to meet this situation. Both aim at providing means of calculating from the data intervals within which the unknown parameter θ may be expected to lie within a given measure of probability. While from the practical standpoint there is in most (though not in all) problems no difference between the result of applying the two methods, Fisher's approach introduces the concept of fiducial probability, while Neyman's employs only the classical concept of direct probability. The application of theory described here is in terms of Neyman's approach.

Several illustrations have been given earlier in this Introduction of the calculation of confidence limits: viz. (i) limits for the standard deviation, σ, of a normal population, based on the χ^2 distribution (§ 4·3); (ii) limits for the population mean μ, based on the t-distribution (§ 5·5); (iii) limits for the difference between the means of two normal populations having unequal variances, based on Welch's method (§ 6·2); (iv) limits for the correlation coefficient ρ between two normal variables, based on the distribution of r (§ 7·3). In all these cases the distributions of the sample statistics employed in the estimation were continuous. In the two following sections we are concerned with discontinuous distributions, and the meaning to be attached to the probability statement associated with the confidence interval needs rather careful definition; this is our reason for developing the argument in connexion with the Poisson parameter, m, in some detail.

22·2 *Confidence limits for the expectation, m, of a Poisson variable* (*Table* 40)

The general term of a Poisson series may be written as $e^{-m}m^i/i!$ $(i=0,1,2,\ldots)$. For any given value c and $\alpha<0{\cdot}5$, we may determine two values of m, say $m_A(c\mid\alpha)<m_B(c\mid\alpha)$, such that

$$\sum_{i=c}^{\infty}\frac{e^{-m_A}m_A^i}{i!}=\alpha,\quad \sum_{i=0}^{c}\frac{e^{-m_B}m_B^i}{i!}=\alpha. \tag{93}$$

Within the range of tabulation $m_A(c)$ and $m_B(c)$ may be readily determined from the table of percentage points of the χ^2-distribution (Table 8) using the relation (18) of p. 10. Thus

$$m_B(c\mid\alpha)=\tfrac{1}{2}\chi_1^2,\quad \text{where}\quad Q(\chi_1^2\mid\nu)=\alpha,\quad \nu=2(c+1), \tag{94}$$

$$m_A(c\mid\alpha)=\tfrac{1}{2}\chi_2^2,\quad \text{where}\quad 1-Q(\chi_2^2\mid\nu)=\alpha,\quad \nu=2c. \tag{95}$$

If, for example, we take $\alpha=0{\cdot}005$ and $c=6$, we see from Table 8 that

(*a*) the upper 0·5 % point ($Q=0{\cdot}005$) for χ^2 with $\nu=14$ is 31·3193 so that

$$m_B(6\mid 0{\cdot}005)=15{\cdot}66;$$

(*b*) the lower 0·5 % point ($Q=0{\cdot}995$) for χ^2 with $\nu=12$ is 3·07382 so that

$$m_A(6\mid 0{\cdot}005)=1{\cdot}54.$$

If now we determine $m_B(c\mid\alpha)$ and $m_A(c\mid\alpha)$ for $c=0,1,2,\ldots$ and plot their values as ordinates against c as abscissa, we obtain the diagram shown in Fig. 4, where the two series of points have been connected by arbitrarily drawn smooth curves. The area between the curves is the so-called *confidence belt.* Suppose that for a particular observed value of c we determine the limits $m_B(c)$ and $m_A(c)$, either from a chart or appropriate table, then the probability that the interval between these limits includes the true expectation m is at least

$$1-2\alpha = 1-2\times 0{\cdot}005 = 0{\cdot}99,$$

or

$$\Pr\{m_A(c \mid 0{\cdot}005) \leqslant m \leqslant m_B(c \mid 0{\cdot}005)\} \geqslant 0{\cdot}99. \tag{96}$$

This result may be demonstrated as follows. If repeated observations are made of a Poisson variable having fixed but unknown expectation, say $m = m_0$, the points (c, m_0) will lie along a line such as *CDEF* in Fig. 4 (where, for illustration, $m_0 = 12$). This line is divided into three parts by the boundaries of the confidence belt:

(i) For observations with $c = 0, 1, \ldots$ on *CD* the statement (96) will be incorrect because $m_B < m_0$; but from the nature of the construction the probability of c falling in *CD* is $\leqslant \alpha = 0{\cdot}005$.

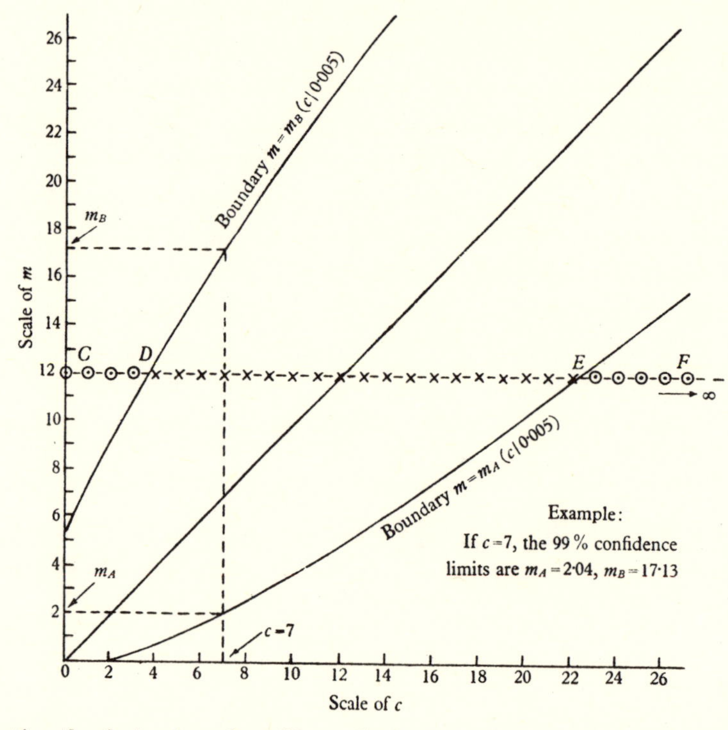

Fig. 4. Illustrating the derivation of confidence limits for a Poisson expectation with $1-2\alpha = 0{\cdot}99$.

(ii) For observations with c on *EF* the statement (96) will be incorrect because $m_A > m_0$; the probability of c falling in *EF* is also $\leqslant \alpha = 0{\cdot}005$.

(iii) For observations with c on *DE*, the statement (96) will be correct; the probability of this occurring is therefore clearly $\geqslant 1-2\alpha = 0{\cdot}99$.

This position will hold whatever be the expectation m_0, and thus in general (96) provides a relation in probability between two functions of any Poisson variable and its expectation.

Owing to the discontinuity in distribution, it is clear that in repeated sampling, particularly when m is small, the proportion of occasions on which the interval includes the true expectation may be considerably in excess of the lower bound $1-2\alpha$. Thus it should be possible to find alternative methods of solution which would fix the confidence coefficient at exactly $1-2\alpha$ and at the same time narrow the interval. This matter has been discussed by W. L. Stevens (1950).

Table 40 gives values of $m_A(c \mid \alpha)$ and $m_B(c \mid \alpha)$ for $c = 0\,(1)\,30\,(5)\,50$ and $\alpha = 0{\cdot}001, 0{\cdot}005, 0{\cdot}01, 0{\cdot}025$ and $0{\cdot}05$. Beyond $c = 50$, the values of m_A and m_B may be computed using the relations (94) and (95) of p. 81, the required percentage points of χ^2 being obtained from one of the approximate relations quoted under Table 8.

Example 42. A spot sample shows eighteen incoming calls received during an interval of 1 min. on a particular section of the equipment at a telephone exchange. What limits may be reasonably placed on the expected number of calls per $\frac{1}{4}$ hour?

Assuming that the rate of traffic remains reasonably constant during the period considered and that the distribution of calls received in an interval of constant length follows the Poisson law, we may obtain confidence limits from Table 40. For example, for $c = 18$ the lower and upper 95 % confidence limits are 10·67 and 28·45. For the average number of calls per $\frac{1}{4}$ hour we have, therefore, as limits

$$15 \times 10{\cdot}67 = 160 \quad \text{and} \quad 15 \times 28{\cdot}45 = 427.$$

In determining these values we are using a procedure giving limits which will include the true expectation, in the long run, at least 19 times out of 20. Clearly these are very wide limits, but it is a function of statistical method to emphasize that precise conclusions cannot be drawn from inadequate data.

22·3 *Confidence limits for p in binomial sampling* (*Table* 41, *two charts*)

The general term of the binomial expansion $(q+p)^n$ may be written as before

$$f(i \mid n, p) = \binom{n}{i} p^i (1-p)^{n-i}.$$

The value of n is supposed known, and it is required to find confidence limits $p_A(c \mid n, \alpha)$ and $p_B(c \mid n, \alpha)$ for p (where $p_A < p_B$) on the basis of an observed value $i = c$. The derivation proceeds on lines similar to those followed for the Poisson expectation (see § 22·2 above), except that the confidence belt of Fig. 4 is now 'closed', since $0 \leqslant c \leqslant n$ and $0 \leqslant p \leqslant 1$. For a given value of $i = c$, the values of p_A and p_B are defined by

$$\sum_{i=c}^{n} f(i \mid n, p_A) = \alpha, \quad \sum_{i=0}^{c} f(i \mid n, p_B) = \alpha, \tag{97}$$

where the cumulative sums may be conveniently expressed as in equation (84) in terms of the incomplete B-function integral.

Since there are now two parameters, p (to be estimated) and n (supposed known), no detailed tables have been given, but sufficient points c, $p_A(c \mid n, \alpha)$ and c, $p_B(c \mid n, \alpha)$ were calculated to draw smooth curves through them. By taking c/n as abscissa and p as ordinate these curves can be reproduced in a single diagram. The charts of Table 41 show confidence belts prepared in this way for

$$\alpha = 0{\cdot}025 \quad \text{and} \quad 0{\cdot}005 \qquad \text{or} \qquad 1 - 2\alpha = 0{\cdot}95 \quad \text{and} \quad 0{\cdot}99,$$

and for $\qquad n = 8, 10, 12, 16, 20, 24, 30, 40, 60, 100, 200, 400, 1000.$

The same limitation applies as in the case of the Poisson confidence limits; namely, the probability that the statement

$$p_A(c \mid n, \alpha) \leqslant p \leqslant p_B(c \mid n, \alpha) \tag{98}$$

is correct is likely to be considerably in excess of the lower bound, $1 - 2\alpha$.

The charts cannot and are not intended to provide very precise readings. The broad picture which they give of the relation between p and c/n will, however, be found of value in answering a variety of questions regarding probabilities, proportions and percentages. Some of these are illustrated in the following examples.

Example 43. The toxicity of a drug may be measured by the proportion, p, of mice in a standard laboratory population that will die after injection with a dose of given strength. Out of a sample of thirty mice randomly selected from the population, eight die after injection; within what limits may we expect that p lies? Turning to the first chart of Table 41, and taking $n=30$, $c/n=8/30=0{\cdot}267$, it will be seen that we may say that $0{\cdot}12<p<0{\cdot}46$, if we are prepared to accept a risk of error of not more than 1 in 20. To obtain greater confidence in prediction (risk of error 1 in 100) we must turn to the second chart and obtain $0{\cdot}095<p<0{\cdot}52$.*

Example 44. In a manufacturing process a crude index of quality, P, has been the percentage of articles which pass a certain test. This index has fluctuated in the past round $P=60$, but it is proposed to make an intensive effort to improve quality (which will mean the raising of this percentage) by tightening the control of manufacture. Improvement is to be judged by studying the changes in the proportion of articles (c/n) passing the test in a random sample of n articles. How large would n need to be to obtain from the sample an estimate of P, with a range of uncertainty of not more than 5?

At the start, the value of $p=P/100$ in the material sampled is not more than 0·60, and we wish to determine n so that the confidence belt will be of breadth about 0·05. On the assumption that a confidence coefficient of 0·95 is adequate, we may use the first chart. It will be seen that for c/n having values between 0·6 and 0·8, n must be more than 1000 for the interval p_2-p_1 to be as small as 0·05.† In many cases the testing of so large a sample would be quite out of the question, and this result points to the fact that an index of this type is not an efficient measure of quality. Much more information of changes could probably be drawn from a smaller sample, if the index could be based on the mean value of some measured character determined for each article of the sample.

Example 45. There are two alternative hypotheses regarding the chance of an individual in a certain population bearing a given character; the alternatives are that $p=\frac{1}{4}$ or $p=\frac{1}{2}$. Such might be the case in some genetic investigation. How large a sample must be planned to make it practically certain that we can discriminate between the two hypotheses?

In this case we are concerned with the sampling variation of c for $p=\frac{1}{4}$ and $p=\frac{1}{2}$, and n should be chosen so large that there is no 'overlap' of any consequence between the two distributions. Suppose we choose n so that the upper 0·005 point of the c distribution for $p=\frac{1}{4}$, as judged from the curves of the second chart, corresponds to the lower 0·005 point of the distribution for $p=\frac{1}{2}$. This will occur when n is slightly over 100, say 110.

If we were prepared to accept a greater risk of an inconclusive result, which we might well be prepared to do if the sample could be readily increased in size in a doubtful case, then we might choose n so that the upper and lower 0·025 points of the c distributions correspond. Turning to the first chart it is found that this occurs when n is about 60.

* If interpolation is carried out in Table 16, it is found more precisely that the 95 % confidence limits fall at 0·123 and 0·459 while the 99 % limits fall at 0·093 and 0·516.

† Since for large values of n the upper and lower bounds of the confidence belt are very nearly parallel lines making an angle of 45° with the axes, and the binomial may be represented by a normal curve, the breadth of the belt is approximately $4\{p(1-p)/n\}^{\frac{1}{2}}$, which if equated to 0·05 gives roughly $n=1600$ for $p=0{\cdot}60$ and $n=1000$ for $p=0{\cdot}80$.

V. MISCELLANEOUS TABLES (PEARSON-TYPE CURVES, RANK CORRELATION, ORTHOGONAL POLYNOMIALS)

23. Chart relating the type of Pearson frequency curve to the values of β_1 and β_2 (Table 43)

23·1 *Specification of the curves*

Karl Pearson's system of frequency curves (1895, 1901, 1916), $y=f(x)$, are obtained as solutions of the differential equation

$$\frac{1}{y}\frac{dy}{dx}=\frac{-(x+c_1)}{c_0+c_1x+c_2x^2}, \tag{99}$$

where the origin for x is at the mean. The form of the solution depends on the values of the constants c_0, c_1 and c_2, which may be shown to be related to the moments of the curve. If we write

$$\mu_r=\int_{l_1}^{l_2} x^r f(x)\,dx$$

for the rth moment about the mean ($\mu_0=1$, $\mu_1=0$), where l_1 and l_2 are the permissible lower and upper limits for x, and if

$$\sigma^2=\mu_2,\quad \beta_1=\mu_3^2/\mu_2^3,\quad \beta_2=\mu_4/\mu_2^2,$$

then

$$c_0=\frac{\sigma^2(4\beta_2-3\beta_1)}{2(5\beta_2-6\beta_1-9)},\quad c_1=\frac{\sigma\sqrt{\beta_1}(\beta_2+3)}{2(5\beta_2-6\beta_1-9)},\quad c_2=\frac{2\beta_2-3\beta_1-6}{2(5\beta_2-6\beta_1-9)}. \tag{100}$$

It follows that if a diagram is drawn plotting β_1, β_2 to rectangular axes, different areas, curves or points in the plane can be associated with particular types of curve. A chart of this kind is presented as Table 43, covering the main types of Pearson curves, i.e. Types I–VII. The line $\beta_2-\beta_1-1=0$ forms an upper boundary above which, in the sense of the chart, the β_1, β_2 point of no real frequency distribution can fall, while for the Pearson system, μ_8 is infinite on and below the line $8\beta_2-15\beta_1-36=0$ shown at the bottom of the chart. Using a standard notation, the equations to these seven curves may be expressed as follows:

Type	Equation	Origin for x	Limits for x
I*	$y=y_0\left(1+\frac{x}{a_1}\right)^{m_1}\left(1-\frac{x}{a_2}\right)^{m_2}$	Mode	$-a_1\leqslant x\leqslant a_2$
II	$y=y_0\left(1-\frac{x^2}{a^2}\right)^{m}$	Mean (=mode)	$-a\leqslant x\leqslant a$
III	$y=y_0e^{-\gamma x}\left(1+\frac{x}{a}\right)^{\gamma a}$	Mode	$-a\leqslant x<\infty$
IV	$y=y_0e^{-\nu\tan^{-1}x/a}\left(1+\frac{x^2}{a^2}\right)^{-m}$	Mean $+\frac{\nu a}{r}$†	$-\infty<x<\infty$
V	$y=y_0e^{-\gamma/x}x^{-p}$	At start of curve	$0\leqslant x<\infty$
VI	$y=y_0(x-a)^{q_2}x^{-q_1}$	At a before start of curve	$a\leqslant x<\infty$
VII	$y=y_0\left(1+\frac{x^2}{a^2}\right)^{-m}$	Mean (=mode)	$-\infty<x<\infty$

The notation used is that of W. P. Elderton (1953), whose book, *Frequency Curves and Correlation*, or the original papers of Pearson should be consulted for the relations between the moments and the parameters of the curves.

* For the most common form of Type I curve of 'bell' shape (area I in chart), both m_1 and m_2 are positive; for J-shaped curves (area I(J) in chart), either m_1 or m_2 is negative; for U-shaped curves (area I(U) in chart), both are negative.

† $r=6(\beta_2-\beta_1-1)/(2\beta_2-3\beta_1-6)=2m-2$; this origin is chosen to simplify the form of the equation.

Historically this system of curves was developed to graduate frequency distributions of observational data which could not be represented by the Normal curve. Latterly, however, the system has been used increasingly to represent probability distributions whose moments are known but for which the mathematical equations are either undetermined or not expressible in simple form. Illustrations of this application are given in § 24·3 below.

Example 46. Moment ratios for the distribution of range in samples of n observations from a normal population were evaluated approximately by L. H. C. Tippett (1925) and E. S. Pearson (1926) as follows:

Sample size, n	6	10	20	60	100
$\beta_1(w)$	0·189	0·156	0·161	0·201	0·223
$\beta_2(w)$	3·17	3·22	3·26	3·35	3·39
Type areas in which points fall	I	I	VI	VI	VI

Determine which types of Pearson curve are likely to be most appropriate to use in approximating to the distribution of range. The first two β_1, β_2 points fall in the Type I area and the last three in the Type VI area, the Type III line, on which $2\beta_2 - 3\beta_1 - 6 = 0$, being crossed between $n = 10$ and 20.

The Type I and VI approximations were in fact used by E. S. Pearson (1932) to obtain rough percentage points for the distribution of range. The practical utility of this method is illustrated by the fact that the values so obtained were found to be scarcely in error when correct values for $n \leqslant 20$ were later derived from the true probability integral (see § 12·4 above). The discrepancy arose as much from the approximate nature of the β_1, β_2 values available in 1932 (compare the figures given above with those in Table 20) as from the use of Type I and VI curves.

Example 47. On p. 69 (Example 35) a frequency distribution has been given for 1000 observations on the warp strength of a duck cloth. The values of the moments and moment ratios for this distribution are

$$\text{Mean} = 454{\cdot}09\,\text{lb.}, \quad \text{Standard deviation} = 17{\cdot}358\,\text{lb.},$$

$$\sqrt{b_1} = -0{\cdot}421, \quad b_1 = 0{\cdot}1772, \quad b_2 = 2{\cdot}748.$$

It has already been shown that the value of $\sqrt{b_1}$ establishes a significant departure from normality. Turning to the chart of Table 43 it will be seen that the point (0·177, 2·748) lies clearly in the Type I area. Pearson & Welch (1937) fitted a Type I curve to the observed distribution, using the method of moments, and obtained the following equation, the origin being at zero lb.:

$$y = 21{\cdot}9662(x - 370{\cdot}02)^{5{\cdot}1218}(495{\cdot}59 - x)^{1{\cdot}9805}.$$

The graduated frequencies obtained from this equation are shown in the table on p. 69. The fit appears good except towards the upper tail, where the curve comes down steeply to the terminal at $x = 495{\cdot}59$ lb. An improved graduation could be obtained, no doubt, by applying corrections to the moment solution, so as to secure either a maximum-likelihood or minimum χ^2 fit. Pearson & Welch were, however, concerned with the lower tail of the distribution corresponding to low breaking strengths and regarded the graduation as adequate. If the three frequency groups centred at 485, 490 and 495 lb. are combined into one group and those centred at 400 and 405 lb. into another, the value of χ^2 based on seventeen groups is 14·45; for $\nu = 17 - 5 = 12$ degrees of freedom,* the probability of obtaining a worse fit through chance fluctuations (when the true law is of Type I form) equals 0·27.

24. Tables of certain percentage points of Pearson curves, for given β_1 and β_2, expressed in standardized measure (Table 42)

24·1 *Definition*

If $y = f(x)$, $l_1 \leqslant x \leqslant l_2$, is a frequency curve of the Pearson system (see § 23·1 above), having mean μ and standard deviation σ, so that $X = (x-\mu)/\sigma$ is a standardized deviate, then Table 42 gives values of

$$X_P = (x_P - \mu)/\sigma,$$

where

$$P = \int_{l_1}^{x_P} f(x)\,dx,$$

for $P = 0{\cdot}005$, 0·01, 0·025, 0·05, 0·95, 0·975, 0·99 and 0·995 and for values of β_1 and β_2 in the area of Fig. 5, lying below the line labelled Type IX. This diagram, taken from the paper by

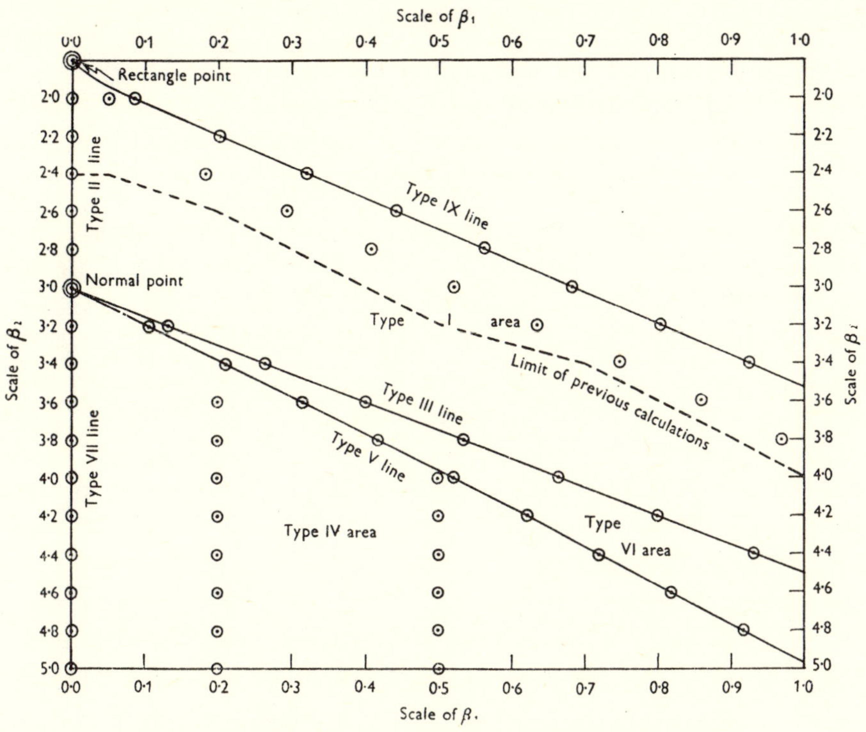

Fig. 5. Showing area in β_1, β_2 plane covered by Table 42.

Pearson & Merrington (1951), shows the points for which framework values of X_P were computed in preparing the table.† The Type IX line, on which lie the (β_1, β_2) points of the curves

$$y = y_0\left(1 + \frac{x}{a}\right)^m,$$

forms an upper boundary for the table entries; above this line (in the sense of the diagram) the (β_1, β_2) points correspond to J-shaped and U-shaped curves.

* This involves an application of the χ^2 test for goodness of fit referred to in § 3·2 above in connexion with Table 7. Since four moments have been used in the fitting and the total observed and theoretical frequencies are equal, five relations exist among the seventeen differences $(n_t - m_t)$. Hence $\nu = 12$. In theory, the fit should have been adjusted to obtain a 'minimum' χ^2 before applying this test; in the present application this was unnecessary.

† A much fuller table has since been computed by Johnson *et al.* (1963). This gives the limits to three and sometimes four decimal places, adds 0·25, 10 and 25 % points and much extends the range of β_1 and β_2 covered.

24·2 *Interpolation*

To the accuracy justified, linear interpolation for β_2 and for β_1, except in the panel 0·00–0·01, is adequate. If at a given (β_1, β_2) point the values of X_P for several P values are required, it is worth proceeding systematically using the formula

$$X(\theta,\phi) = X(0,0)\times(1-\theta-\phi+\theta\phi) + X(1,0)\times(\theta-\theta\phi) + X(1,1)\times\theta\phi + X(0,1)\times(\phi-\theta\phi). \quad (101)$$

Here, as shown in Fig. 6, the four X-values are the four tabular entries at the corners of the panel in which the interpolate falls, and θ and ϕ are the appropriate fractions of the tabular intervals for β_1 and β_2 at the part of the table considered. For an illustration, see Example 48 (below).

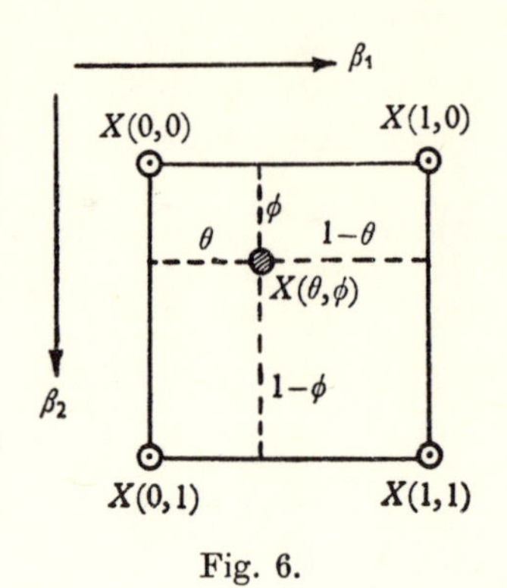

Fig. 6.

In the interval for β_1, 0·00–0·01, for a given β_2 a three-point Lagrangian formula may be used. Thus

$$X(\theta) = X(0)\times\tfrac{1}{3}(1-\theta)(3-\theta) + X(1)\times\tfrac{1}{2}\theta(3-\theta) - X(3)\times\tfrac{1}{6}\theta(1-\theta),$$

where $\theta = 100\beta_1$ and $X(0)$, $X(1)$ and $X(3)$ are the tabular entries for $\beta_1 = 0·00$, 0·01 and 0·03 respectively.

24·3 *Applications*

A considerable proportion of current statistical research is concerned with the determination of the sampling distributions of statistics required either as estimators or for use in tests of significance. It is often the case that while the distribution itself cannot be expressed in any simple form, the sampling moments can be derived and numerical values calculated either precisely or as approximations. In a great number of cases such investigation shows that the distribution tends to the normal form as the sample size is increased. However, this information regarding the limit is often not sufficient, and we require an answer to the question: what error is involved in assuming normality when the sample size has a specified value, n? In other words, to what extent in practice will the knowledge of the expectation and standard error of a particular statistic suffice?

In many instances experience has shown the value of the Pearson curves in approximating, on the basis of known moments, to the distributions of frequency functions which are either undetermined or are not readily expressed in simple form. In addition, these curves, of course, represent exactly the distributions of a number of the statistics in common use, e.g. χ^2, t and F.

Example 48. Bailey (1950), in considering the mathematical theory of what he terms a simple stochastic epidemic, gives a table of moments for the time of completion of an epidemic, when all available susceptibles have been exhausted. Thus if τ is the completion time in a community containing n susceptibles, he shows that appreciable skewness and kurtosis remain in the distribution of τ, even when n is very large. With $n = 80$, he gives:

γ_1	γ_2	$\beta_1 = \gamma_1^2$	$\beta_2 = \gamma_2 + 3$
0·771	1·114	0·594	4·114

The following figures for percentage levels of $(\tau - \bar{\tau})/\sigma_\tau$ may be derived by interpolation in Table 42:

	Lower		Upper	
	0·5 %	5·0 %	5 %	0·5 %
Pearson curve (Type IV) approx.	−1·95	−1·41	1·82	3·33
From 'normal' probability scale	−2·58	−1·64	1·64	2·58

The figures give added meaning to Bailey's measures of skewness and kurtosis, and a comparison with the normal scale percentage levels is instructive.

The interpolation was carried out systematically, using the formula (101). Thus $\theta = 0{\cdot}94$, $\phi = 0{\cdot}57$ and for $P = 0{\cdot}005$

$$-X_{0\cdot005}(0{\cdot}94, 0{\cdot}57) = 2{\cdot}02 \times 0{\cdot}026 + 1{\cdot}90 \times 0{\cdot}404 + 1{\cdot}97 \times 0{\cdot}536 + 2{\cdot}09 \times 0{\cdot}034 = 1{\cdot}947.$$

Example 49. If we write $m = \sum_{i=1}^{n} \frac{|x_i - \bar{x}|}{n}$ for the mean deviation in a random sample of observations $x_1, x_2, \ldots, x_n$ drawn from a normal population, the sampling moments of m are known (see § 10·1 and Table 20). The following are the appropriate values for the cases $n = 10$ and 20:

n	$\bar{m}$	σ_m	$\beta_1(m)$	$\beta_2(m)$
10	0·75694	0·1894	0·106	3·093
20	0·77768	0·1344	0·0513	3·045

We may now determine approximations to the lower and upper 0·5 % points in these two sampling distributions, expressed in terms of the population standard deviation as unit. These will be given by $\bar{m} + X_P \sigma_m$, where the factors X_P for $P = 0{\cdot}005$ and 0·995 are obtained by interpolation from Table 42. The results are as follows:

n	$X_{0\cdot005}$	$X_{0\cdot995}$	Lower 0·5 % point	Upper 0·5 % point
10	−2·24	2·86	0·333	1·299
20	−2·35	2·78	0·462	1·151

The values for $n = 10$ correspond, to 3-decimal accuracy, with the exact percentage points derived from the probability integral of m and given in Table 21. This integral has not been calculated beyond $n = 15$, but the accuracy of the approximation at $n = 10$ suggests that the values obtained at $n = 20$ should also be satisfactory, and that this method could be used to extend the table of percentage points of mean deviation (Table 21) beyond $n = 10$.

Example 50. The distribution of the warp strength, given in the table of Example 35 (p. 69), and discussed again in Example 47 (p. 86), is clearly not normal. Suppose it was desired to construct a control chart giving lower limits for the mean $\bar{x}_5$, of determinations of strength in $n=5$ randomly selected test pieces of this type of cloth. We may ask how much error would be involved in subtracting from μ multiples of $\sigma(\bar{x}_5)=\sigma_x/\sqrt{5}$ obtained in the ordinary way from the *normal* probability scale.

Since $\sqrt{b_1}$ is negative, we shall obtain from Table 42 the upper 0·5 and 5 % values of X_P for

$$\beta_1(\bar{x}_5)=\beta_1(x)/5=0{\cdot}1772/5=0{\cdot}035,$$

$$\beta_2(\bar{x}_5)=3+(\beta_2(x)-3)/5=3+(2{\cdot}748-3)/5=2{\cdot}95.$$

The appropriate panels of values of X_P, which will have negative signs, are:

for $P=0{\cdot}05$:

	0·03	0·05
2·8	−1·70	−1·72
3·0	−1·69	−1·71

and for $P=0{\cdot}005$:

	0·03	0·05
2·8	−2·64	−2·68
3·0	−2·72	−2·76

It is seen by inspection that $X_{0{\cdot}05}=-1{\cdot}70$, while linear interpolation for the two variables gives $X_{0{\cdot}005}=-2{\cdot}71$. The corresponding standardized deviates for a normal distribution are −1·64 and −2·58 respectively. Using these factors to obtain outer (0·005) and inner (0·05) limits in a control chart for $\bar{x}_5$, taking the mean level as 454·09 lb. and the standard error of $\bar{x}_5$ as $17{\cdot}358/\sqrt{5}=7{\cdot}76$ lb., we find:

	Outer limit	Inner limit
Using observed β_1, β_2 for the 1000 values of x	433·1	440·9
Assuming that x is normally distributed	434·1	441·4

As the limits would probably be required to the nearest 1 lb. only, it is seen that there would be no error involved in basing the inner limit on normal theory and only an error of 1 lb. in the case of the outer limit.

25. Tables for use in connexion with rank correlation (Tables 44–46)

25·1 *Definition of the coefficients*

The whole subject of rank correlation has been fully discussed in M. G. Kendall's *Rank Correlation Methods* (1962), and we are indebted to Professor Kendall and his publishers for permission to use some of his examples as well as extracts from his tables of probability values. Here we shall confine ourselves to a brief definition of the measures and some illustration of the use of the tables.

When a number of individuals are arranged in order according to some quality which they all possess to a varying degree, they are said to be *ranked*. In cases where each individual is ranked for two or more different qualities, it may be important to determine whether there is any association or correlation between the qualities and, if so, to calculate and compare measures of this relationship.

Spearman's coefficient of rank correlation, which has often been denoted by the letter ρ, and Kendall's coefficient τ, both measure the correspondence between two rankings. Kendall and Babington Smith's coefficient of concordance W is a measure of the correspondence between any number of rankings. To avoid confusion in notation, we shall here denote Spearman's and Kendall's coefficients by r_s and t_k, respectively, retaining W for the coefficient of concordance. The coefficients are most readily defined with the help of simple examples.

(a) The coefficient r_s (Spearman's ρ)

Suppose that a number of boys have been ranked according to their ability in mathematics and in music. Such a pair of rankings for $n = 10$ boys, denoted by the letters A to J, might be as follows:

Boy ...	*A*	*B*	*C*	*D*	*E*	*F*	*G*	*H*	*I*	*J*
Rank for mathematics	7	4	3	10	6	2	9	8	1	5
Rank for music	5	7	3	10	1	9	6	2	8	4
Difference in ranks, d	2	−3	0	0	5	−7	3	6	−7	1
Squared difference, d^2	4	9	0	0	25	49	9	36	49	1

Spearman's coefficient, r_s, is obtained by treating the two series of ordinal numbers in the rankings as though they were cardinal numbers representing quality on two numerical scales, and then calculating the usual product-moment coefficient of correlation between the corresponding numbers in the two series. If d_i is the difference in the ranking of the ith individual for the two qualities, then this coefficient may be shown to be given in simplest form by

$$\begin{aligned} r_s &= 1 - 6\sum_{i=1}^{n} (d_i^2)/(n^3 - n) \\ &= 1 - 6S_r/(n^3 - n), \end{aligned} \tag{102}$$

where $S_r = \sum_i d_i^2$. For the figures given above, $n = 10$ and $\Sigma d_i^2 = 182$, so that $r_s = -0{\cdot}103$.

(b) The coefficient t_k (Kendall's τ)

Suppose we consider any one of the $\frac{1}{2}n(n-1)$ possible pairs of the n individuals and associate with this pair, (1) a score of $+1$ if the ranking for both qualities is the same order or (2) a score of -1 if the ranking is in different order. Thus in the illustration given above, the pair AB scores -1, while AC scores $+1$. Denote the sum of these scores for the $\frac{1}{2}n(n-1)$ pairs by S_t. It is seen that S_t has a maximum value of $\frac{1}{2}n(n-1)$ if the two rankings are identical and a minimum value of $-\frac{1}{2}n(n-1)$ if the individuals are ranked in exactly opposite order. Kendall defines his coefficient as

$$t_k = \frac{S_t}{\frac{1}{2}n(n-1)}. \tag{103}$$

It is clear that $-1 \leqslant t_k \leqslant 1$.

Short-cut methods are available for calculating S_t; one of these commonly used is as follows. Since the scores will not be affected by any rearrangement of the individuals, we may order these so that one ranking is in the natural order. Thus the table on the preceding page may be written:

Boy ...	I	F	C	B	J	E	A	H	G	D
Rank for mathematics	1	2	3	4	5	6	7	8	9	10
Rank for music	8	9	3	7	4	1	5	2	6	10

Each pair of numbers in the second ranking which is in the right order will now contribute $+1$ to S_t. It is unnecessary to count the pairs in the wrong order, so that

$$S_t = N_+ \times (+1) + \{\tfrac{1}{2}n(n-1) - N_+\} \times (-1) = 2N_+ - \tfrac{1}{2}n(n-1),$$

$$t_k = \frac{4N_+}{n(n-1)} - 1,$$

where N_+ is the number of pairs in the second ranking which are in the right order.

For the illustration given, the computation of N_+ proceeds as follows. Consider first the pairs associated with the first member of the second ranking, namely 8; two members greater than 8 are on the right of it, so the contribution to N_+ is 2. Taking now the pairs associated with the next member 9 (other than 8, 9, which has already been taken into account), we find a contribution of 1. Similarly, the contribution from pairs associated with 3, arising from members to the right of it, is 5. Proceeding in this way, we find that

$$N_+ = 2+1+5+1+3+4+2+2+1 = 21.$$

Hence $$S_t = 2 \times 21 - 45 = -3, \quad t_k = -0{\cdot}067.$$

(c) *The coefficient of concordance W*

We now consider the case where there are several rankings, say m in number, of n individuals. Suppose, for example, that four observers rank six objects as follows (i.e. $m=4$, $n=6$):

	Object					
	A	B	C	D	E	F
Observer I	5	4	1	6	3	2
Observer II	2	3	1	5	6	4
Observer III	4	1	6	3	2	5
Observer IV	4	3	2	5	1	6
Sum of ranks	15	11	10	19	12	17
Deviation of sum from average	1	−3	−4	5	−2	3

We may wish to measure the degree of consistency or concordance between the four observers' judgements. It would be possible to calculate a rank correlation coefficient between each pair of observers, but a single measure of the concordance of the observers, taken as a group, will usually be needed. The coefficient of concordance as defined by Kendall and Babington Smith is obtained as follows.

(1) The n sums of the ranks allotted by each observer to an individual are found. These sums are shown in the bottom row of the table given above; they must add up to $\frac{1}{2}mn(n+1)$, or 84 in the case illustrated.

(2) The deviations of these sums from their mean value of $\frac{1}{2}m(n+1)$ are then found: the sum of the squares of these deviations will be denoted by S_W. In the example, the mean is 14 and $S_W = 64$.

(3) S_W cannot be less than zero and has a maximum value of $\frac{1}{12}m^2(n^3-n)$ when all the rankings are identical; the coefficient of concordance is obtained by dividing S_W by its maximum value, i.e.

$$W = \frac{12S_W}{m^2(n^3-n)}. \tag{104}$$

In the example, $W = 12 \times 64/(16 \times 210) = 0{\cdot}229$. It may be shown that if the Spearman coefficient, r_s, is calculated for each of the $\frac{1}{2}m(m-1)$ possible pairs of observers, then

$$\text{Average } r_s = \frac{mW-1}{m-1}.$$

In the special case when there are only two observers,

$$r_s = 2W-1.$$

(d) The case of tied ranks

In the practical application of ranking methods cases will arise in which two or more individuals are so similar that no preference can be expressed between them. The members are then said to be *tied*. For the appropriate modification of the coefficients and the effect on the tests of significance described below, the reader may consult Kendall (1962), where the subject is fully discussed.

25·2 *Tests of significance* (*Tables* 44–46)

If the sample of individuals whose qualities are ranked has been drawn randomly from a much larger population, the question will arise as to whether the two or more qualities are correlated in this population. To test for significance, the observed value of the coefficient is referred to the distribution of values it would have in random samples from a population in which the qualities are completely uncorrelated. The scores S_r, S_t and S_W, which can assume integer values only, are the basic quantities in the three coefficients, and for small values of n (and m) it has been found possible to calculate the exact probabilities associated with each of their possible values, on the hypothesis of no correlation. These calculations form the basis of the tables. A brief statement follows regarding the distribution and the probability table for each score, S; some indication will also be given of how a test may be carried out when the numbers are beyond the range of the table.

(a) The coefficient r_s (Spearman's ρ)

When the qualities ranked are independent, S_r, which can assume only even integer values, is distributed symmetrically about $\frac{1}{6}(n^3-n)$. The figures in Table 44, covering the range $n = 4\,(1)\,10$, have been taken from the complete tables given by Kendall (1948) and David, Kendall & Stuart (1951). $Q = 1 - P(S_r \mid n)$ is the probability that S_r is equalled or exceeded. Since the lower and upper limits for S_r are 0 and $\frac{1}{3}(n^3-n)$ respectively, the probability is also Q that the sum of squares of the rank differences is less than or equal to $\frac{1}{3}(n^3-n) - S_r$. To facilitate computation of the lower tail significance levels of the distribution of S_r, values of $\frac{1}{3}(n^3-n)$ are given in the bottom row of the table.

Owing to the rapid increase with n of the range of the distribution, the argument interval for S_r has been changed progressively across the table. Complete results are given for $n=4, 5$; afterwards selected values only are quoted, ranging from near the 20 % level to near the 0·1 % level. It should be remembered that the upper percentage points for S_r correspond to the lower points for r_s and vice versa.

Example 51. We may ask whether there is any significant association between the rankings for mathematics and music for the ten boys considered in the table on p. 91. Here

$$r_s = -0{\cdot}103, \quad S_r = 182, \quad n = 10.$$

Turning to the columns headed $n=10$ in Table 44, we see that if the two qualities were independent, the expectation of S_r is 165, half the figure entered in the bottom line of the table, while there is a probability of 0·235 that $S_r \geqslant 208$. Clearly r_s does not differ significantly from zero.

Had the score for two other subjects, say mathematics and physics, been found to be $S_r = 52$, giving a rank correlation $r_s = 0{\cdot}685$, we might again apply the test. S_r is now below expectation, so that

$$\Pr\{r_s \geqslant 0{\cdot}685\} = \Pr\{S_r \leqslant 52\} = \Pr\{S_r \geqslant 330 - 52 = 278\}.$$

The table shows this probability to be 0·017. Whether we regard a single-tail or double-tail test as appropriate, the correlation is certainly significant at the 5 % level.

Tests for the significance of r_s when $n > 10$. If the two qualities are independent, the expectation of r_s is zero and its standard error is $(n-1)^{-\frac{1}{2}}$. In samples with $n > 20$ an adequate test of significance may generally be obtained by referring the ratio of r_s to its standard error, i.e. $r_s\sqrt{(n-1)}$, to the normal probability scale. For $10 < n < 20$, Kendall has shown that a good approximation will be obtained by treating r_s as a product moment coefficient of correlation between normally distributed variables (§ 7·1) and making use of the significance points of Table 13, with degrees of freedom $\nu = n-2$. Alternatively the use of the transformation $t = r_s\{(n-2)/(1-r_s^2)\}^{\frac{1}{2}}$, obtained by inverting equation (46), makes it possible to use Table 9.

To illustrate the accuracy of the t-approximation, we may use it in the case just considered, where it was supposed that a coefficient of $r_s = 0{\cdot}685$ was found for ten rankings in mathematics and physics. We compute

$$t = 0{\cdot}685\{8/(1-0{\cdot}685^2)\}^{\frac{1}{2}} = 2{\cdot}66.$$

Entering Table 9 with $\nu = n-2 = 8$, we find that $\Pr\{t \leqslant 2{\cdot}66\} = 0{\cdot}9855$. Thus $\Pr\{r_s \geqslant 0{\cdot}685\}$ approximately equals 0·0145. The correct value, we have seen, is 0·017. If we used the cruder normal approximation, we should refer $r_s\sqrt{(n-1)} = 0{\cdot}685\sqrt{9} = 2{\cdot}055$ to the table of the normal probability integral (Table 1) and find a probability of 0·020. In this case there is little to choose between the two approximations, but Kendall has shown that the t-approximation will in general be the more accurate.

(*b*) *The coefficient t_k (Kendall's τ)*

When the qualities ranked are independent, S_t, which can assume only integer values,* is distributed symmetrically about zero. Table 45, which is reproduced in full from Kendall (1962), shows for $S_t \geqslant 0$ the probability $Q = 1 - P(S_t \mid n)$, i.e. the probability that S_t is equalled or exceeded. Q is also the probability that the score is less than or equal to $-S_t$. The upper percentage points for S_t correspond to the upper percentage points for t_k. The table deals with sample sizes for which $4 \leqslant n \leqslant 10$.

* If $\frac{1}{2}n(n-1)$ is even, S_t can take only even values; otherwise it can take only odd values.

For the illustration of mathematics and music ranking already considered, we found

$$t_s = -0{\cdot}067, \quad S_t = -3, \quad \text{with} \quad n = 10.$$

It is at once clear from Table 45 that this result is quite insignificant, since $\Pr\{S_t \leqslant -3\} = 0{\cdot}431$.

Tests for significance of t_k when $n > 10$. In the case of independence, t_k and S_t vary about zero with standard errors

$$\sigma(t_k) = \left\{\frac{2(2n+5)}{9n(n-1)}\right\}^{\frac{1}{2}}, \quad \sigma(S_t) = \{\tfrac{1}{18}n(n-1)(2n+5)\}^{\frac{1}{2}}.$$

As n increases, the sampling distribution approaches normality more rapidly than does the distribution of r_s. A test of significance may therefore be obtained by comparing the ratio of the observed t_k (or S_t) to its standard error with the normal probability scale.

For values of n between 10 and 30, some improvement in accuracy will be obtained by introducing a correction for continuity of the type used when the cumulative sum of binomial terms is approximated by the integral of the normal curve. Since S_t increases by discrete steps of 2 units, this correction is applied by subtracting 1 from the observed S_t if $S_t > 0$ (and adding 1 if $S_t < 0$), and then dividing the result by the standard error, $\sigma(S_t)$.

Illustration of the use of correction for continuity. Suppose $n = 10$ and the observed value of S_t is 19. From the expression given above we see that

$$\sigma(S_t) = \{10 \times 9 \times 25/18\}^{\frac{1}{2}} = 11{\cdot}18.$$

Our approximation, therefore, consists in determining the integral under the normal curve beyond a standardized deviate of

$$X = (S_t - 1)/\sigma(S_t) = (19-1)/11{\cdot}18 = 1{\cdot}61.$$

Table 1 gives this probability as 0·054, a figure which to 3 decimals corresponds exactly with the true probability given in Table 45. Had no correction for continuity been used, the ratio would have been 1·70, giving a probability of 0·045.

(c) *The coefficient of concordance, W* (*Table* 46)

For the case of independence, exact values of the probability $Q = 1 - P(S_W \mid n, m)$ that a specified value of S_W will be attained or exceeded are given in Table 46 for $n = 3$, $m = 3\,(1)\,10$; $n = 4$, $m = 3\,(1)\,6$; $n = 5$, $m = 3$. The figures are taken from Kendall (1962). S_W can assume only certain integral values, the range of possibilities increasing rapidly as m increases; the distributions are not symmetrical. For $n = 3$, $m = 3\,(1)\,6$ and for $n = 4$, $m = 3$, we have given the complete results for S_W greater than the lowest value shown in the table. In other cases a selection has been made, covering a range from near the upper 20 % level to near the upper 0·1 % level.

Approximation beyond the range of the tables. It may be shown that for independence

$$\mathscr{E}(W) = 1/m, \quad \sigma_W = \{2(m-1)/(m^3(n-1))\}^{\frac{1}{2}}. \tag{105}$$

Since $0 \leqslant W \leqslant 1$ an approximation to the distribution of W may be obtained as suggested in § 8·4 above, that is, by using a Pearson Type I curve having the correct terminals and the mean and standard deviation given in (105). Thus

$$\Pr\{W \geqslant W_0\} \sim I_{1-W_0}(a, b), \tag{106}$$

where

$$b = \tfrac{1}{2}(n-1) - 1/m, \quad a = (m-1)\,b, \tag{107}$$

and I is the incomplete B-function integral for which the percentage points are given in Table 16 for

$$\nu_1 = 2b, \quad \nu_2 = 2a.^*$$

* Alternatively, the chart of Table 17 could be used.

It has been suggested by Kendall & Babington Smith (1939) that an improved approximation will be obtained by using a correction for continuity, which involves setting

$$\Pr\{W \geq W_0\} \sim I_{1-W_0'}(a, b), \tag{108}$$

where a and b are as given in (107) and

$$W_0' = (S_{W_0} - 1)/\{\tfrac{1}{12}m^2(n^3 - n) + 2\}, \tag{109}$$

S_{W_0} being the observed score.*

An alternative approximation has been suggested by Friedman (1940) which makes use of the fact that the distribution of a Type I or 'B' variable tends to that of χ^2 as $a = 2\nu_2$ increases, for a given $b = 2\nu_1$. Equating W to $c\chi^2$, and determining the *scale factor* c and the *equivalent degrees of freedom*, ν, so that the first two moments of W and $c\chi^2$ agree, we find that

$$m(n-1)\,W = S_W/\{\tfrac{1}{12}mn(n+1)\} \tag{110}$$

will be distributed approximately as χ^2, with degrees of freedom $\nu = n - 1$. In reaching this simplest form of the result a factor $m/(m-1)$ has been written as 1.

Example 52. Consider first the results given on p. 93, where

$$m = 4, \quad n = 6, \quad S_{W_0} = 64, \quad W_0 = 0{\cdot}229.$$

Is the coefficient of concordance significant? Since for independence, $\mathscr{E}(W) = 1/m = 0{\cdot}25$, it is clear that the result is not significant and no further analysis is necessary.

Suppose, however, that we required an approximation to the upper 5 % point for W in this case. We find from (107) that $b = 2{\cdot}25$, $a = 6{\cdot}75$. Interpolating in the 5 % points of Table 16 with $\nu_1 = 2b = 4{\cdot}5$, $\nu_2 = 2a = 13{\cdot}5$, we find $x = 0{\cdot}496$. This puts the 5 % point for W at 0·504. If we use the correction for continuity, the observed coefficient to be compared with this limit would be calculated from equation (109).

Example 53. Suppose that three assessors, I–III, had ranked five objects, *A–E*, in the following orders:

	A	*B*	*C*	*D*	*E*
I	1	3	2	4	5
II	2	4	1	5	3
III	1	3	2	4	5
Sum of ranks	4	10	5	13	13
Deviation of sum from 9	−5	1	−4	4	4

We have

$$m = 3, \quad n = 5, \quad S_W = 74, \quad W = 0{\cdot}822.$$

Is the concordance significant? We can here use Table 46, and see that $\Pr\{S_W \geq 74\} = 0{\cdot}015$. The result is therefore significant at the 5 % level, but not at the 1 % level.

Now consider the approximation. We find

$$\nu_1 = 2b = 3\tfrac{1}{3}, \quad \nu_2 = 2a = 6\tfrac{2}{3}.$$

Interpolating in the 1% points of Table 16, we find that $x = 0{\cdot}189$, suggesting an approximate 1 % significance point for W of 0·811. The observed $W_0 = 0{\cdot}822$ falls beyond this point, but if we apply the correction for continuity, we obtain from equation (109)

$$W_0' = (74 - 1)/\{\tfrac{1}{12} \times 9 \times 120 + 2\} = 73/92 = 0{\cdot}793.$$

This value falls short of 0·811, and we should conclude that the concordance was just not significant at the 1 % level, an answer agreeing with that obtained exactly from Table 46.

* Strictly speaking this correction for continuity is only correct when the interval in S_W is 2, which is the interval most frequently met with when $n \geq 4$.

26. ORTHOGONAL POLYNOMIALS (TABLE 47)

26·1 *Description of the method of fitting and definition of the functions tabled*

When determining the regression law of a 'dependent' variable y on an 'independent' variable x, a linear relationship between the two variables is often found to be inadequate. Unless some special regression law is suggested by theoretical considerations, a polynomial of the form

$$Y(x) = a_0 + a_1 x + a_2 x^2 + \ldots + a_k x^k \tag{111}$$

is often found to be a convenient curvilinear representation of y as a function of x and is fitted to observed pairs of associated values y_t, x_t. With such data the x_t are often equidistant, as, for example, when n observations y_t $(t = 1, 2, \ldots, n)$ are made at times x_t progressing by constant intervals. In such cases it is convenient to standardize the x-scale so that

$$x_t = t - \tfrac{1}{2}(n+1) \quad (t = 1, 2, \ldots, n),$$

and to fit $Y(x)$ as a weighted sum of orthogonal polynomials,

$$Y(x) = A_0\phi_0(x) + A_1\phi_1(x) + \ldots + A_k\phi_k(x). \tag{112}$$

Here $\phi_i(x)$ is, apart from a constant factor, the ith degree Tchebycheff polynomial.*

Any pair of these polynomials $\phi_i(x)$, $\phi_j(x)$ $(i \neq j)$ satisfies the orthogonality condition

$$\sum_{t=1}^{n} \phi_i(x_t)\,\phi_j(x_t) = 0, \tag{113}$$

which uniquely determines all $\phi_i(x)$, apart from a constant factor. In particular, we have

$$\left.\begin{aligned}
&\phi_0(x) = 1, \quad \phi_1(x) = \lambda_1 x, \quad \phi_2(x) = \lambda_2\{x^2 - \tfrac{1}{12}(n^2-1)\},\\
&\phi_3(x) = \lambda_3\{x^3 - \tfrac{1}{20}(3n^2-7)\,x\}, \quad \phi_4(x) = \lambda_4\{x^4 - \tfrac{1}{14}(3n^2-13)\,x^2 + \tfrac{3}{560}(n^2-1)(n^2-9)\},\\
&\phi_5(x) = \lambda_5\{x^5 - \tfrac{5}{18}(n^2-7)\,x^3 + \tfrac{1}{1008}(15n^4 - 230n^2 + 407)\,x\},\\
&\phi_6(x) = \lambda_6\{x^6 - \tfrac{5}{44}(3n^2-31)\,x^4 + \tfrac{1}{176}(5n^4 - 110n^2 + 329)\,x^2 - \tfrac{5}{14784}(n^2-1)(n^2-9)(n^2-25)\},
\end{aligned}\right\} \tag{114}$$

where the λ_i are chosen such that the $\phi_i(x_t)$ are positive or negative integers throughout.†

If we assume that the true regression law is a polynomial

$$\eta(x) = \sum_{i=0}^{k} \alpha_i \phi_i(x),$$

from which the observed y_t differ by independent normal deviates z_t having a common variance σ^2, i.e. if we assume that $y_t = \eta(x_t) + z_t$, then the orthogonality condition (113) implies that the least-square estimators A_i of the α_i are given by the familiar formulae for ordinary regression coefficients, viz.

$$A_i = \sum_t y_t \phi_i(x_t) \Big/ \sum_t \{\phi_i(x_t)\}^2. \tag{115}$$

Further, the coefficients A_i, which can be computed from (115) for any $i \leq n-1$, are independent normal variates with means (i) α_i for $i \leq k$, and (ii) 0 for $i > k$ and with variances

* P. L. Tchebycheff's classical papers on the problem of interpolation by means of orthogonal functions appeared between 1854 and 1875. Statisticians will be familiar with R. A. Fisher's (1921 *b*) application of these polynomials to the statistical field and with A. C. Aitken's (1933) concise proof of their mathematical properties.

† The choice of these computationally most convenient factors was first introduced by R. A. Fisher, who denoted the Tchebycheff polynomials, so modified, by ξ_i' in contrast to his earlier polynomials ξ_i in which the factors λ_i were omitted.

$\sigma^2/\sum_t \{\phi_i(x_t)\}^2$. A mean-square estimate of σ^2 (independent of the A_i for $i \leqslant k$) is provided by the residual mean square

$$s^2 = \left(\sum_{t=1}^{n} y_t^2 - \sum_{i=0}^{k} \left\{A_i^2 \sum_{t=1}^{n} [\phi_i(x_t)]^2\right\}\right) \Big/ (n-k-1). \tag{116}$$

Thus the ratios $(A_i - \alpha_i)\sqrt{\sum_t \{\phi_i(x_t)\}^2}/s$, with $\alpha_i = 0$ for $i > k$, follow Student's t-distribution (§ 5·1) for $\nu = n-k-1$ degrees of freedom. To facilitate these calculations, Table 47 provides the exact values of $\phi_i(x_t)$ for $n = 3\,(1)\,52$ and $i = 1\,(1)\,6$, except for small n ($n < 7$) when some of the higher order $\phi_i(x_t)$ are omitted. Since for i even, $\phi_i(x_t) = \phi_i(x_{n-t+1})$ and for i odd, $\phi_i(x_t) = -\phi_i(x_{n-t+1})$, values of ϕ for $n > 12$ are tabled only for $x_t \leqslant 0$. The two values given at the foot of each column are, respectively, the sum of squares $\sum_{t=1}^{n} \{\phi_i(x_t)\}^2$ which is required for equations (115) and (116) and the factor λ_i which is the coefficient of the highest power in each polynomial.

Values of the functions $\phi_i(x_t)$ have been tabled by: Fisher & Yates (1938, 1953) for $i = 1\,(1)\,5$, $n = 3\,(1)\,75$; Anderson & Houseman (1942) for $i = 1\,(1)\,5$, $n = 3\,(1)\,104$; van der Reyden (1943) for $i = 1\,(1)\,9$, $n = 5\,(1)\,52$.* The present Table 47 has been reproduced from van der Reyden's table with permission of the author and the Editor of the *Onderstepoort Journal of Veterinary Science and Animal Husbandry*. We have therefore used van der Reyden's notation of ϕ_i in place of Fisher's ξ_i'.

26·2 *Illustration of the use of Table* 47

Example 54. To illustrate the use of the table we choose the example given by van der Reyden (1943, p. 377) dealing with the average diurnal variation of atmospheric temperature, y_t, during December 1940 at Armoedsvlakte, Bechuanaland. A plot of the data suggests that at least a cubic will be required to represent the temperature trend, and in order to investigate the matter fully, all terms up to the sixth degree were computed. The work is set out in the table following.

Fitting of sixth-degree polynomial to twenty-four observations of temperature (in degrees Fahrenheit)

(1) t	(2) y_t	(3) y_t	(4) t	(5) Sum y_t	(6) ϕ_2	(7) ϕ_4	(8) ϕ_6	(9) Diff. y_t	(10) ϕ_1	(11) ϕ_3	(12) ϕ_5
(1)	65·25	65·88	(24)	131·13	253	253	4807	−0·63	−23	−1771	−4807
(2)	69·37	66·71	(23)	136·08	187	33	−3971	2·66	−21	−847	1463
(3)	74·44	67·09	(22)	141·53	127	−97	−4769	7·35	−19	−133	3743
(4)	79·00	68·12	(21)	147·12	73	−157	−2147	10·88	−17	391	3553
(5)	82·72	69·08	(20)	151·80	25	−165	1045	13·64	−15	745	2071
(6)	85·97	70·57	(19)	156·54	−17	−137	3271	15·40	−13	949	169
(7)	88·11	71·90	(18)	160·01	−53	−87	3957	16·21	−11	1023	−1551
(8)	89·67	73·13	(17)	162·80	−83	−27	3183	16·54	−9	987	−2721
(9)	90·36	74·07	(16)	164·43	−107	33	1419	16·29	−7	861	−3171
(10)	90·35	76·15	(15)	166·50	−125	85	−695	14·20	−5	665	−2893
(11)	88·92	80·41	(14)	169·33	−137	123	−2525	8·51	−3	419	−2005
(12)	87·81	84·95	(13)	172·76	−143	143	−3575	2·86	−1	143	−715
Total	991·97	868·06		1860·03				123·91			

* In all cases, for small values of n the range of i is restricted.

The series of 24 observed temperature values is given in cols. 2 and 3; the first 12 values are found by descending col. 2 and the last 12 values by ascending col. 3. This arrangement facilitates the formation of the sums $y_t + y_{25-t}$ (entered in col. 5) and differences $y_t - y_{25-}$ (entered in col. 9). When computing A_i from equation (115), the former are required as multipliers for ϕ_2, ϕ_4 and ϕ_6 and the latter as multipliers for ϕ_1, ϕ_3 and ϕ_5. The required values of the ϕ_i for $t = 1, 2, \ldots, 12$ are shown in cols. 6, 7, 8 and 10, 11, 12, but would normally not be copied out from Table 47.

Calculating the product sums and dividing by the appropriate values of $\Sigma\phi_i^2$, we obtain the seven values of A_i as follows:

$$A_0 = 77{\cdot}50125, \qquad A_4 = 0{\cdot}0028930,$$
$$A_1 = -0{\cdot}2850804, \qquad A_5 = -0{\cdot}00043867,$$
$$A_2 = -0{\cdot}05024726, \qquad A_6 = -0{\cdot}000025491,$$
$$A_3 = 0{\cdot}004913506,$$

To obtain the actual values, $Y(x)$, from the fitted polynomial, we would have to multiply these values of A_i by the corresponding $\phi_i(x_t)$ tabled above in accordance with formula (112).

26·3 *Tests of significance*

With the assumptions made regarding the random errors

$$z_t = y_t - \eta(x_t),$$

a variety of tests regarding polynomial regressions will follow on similar lines to those used elsewhere in the analysis of variance.

(*a*) *A general test*

If we are *prepared to assume* that

$$\eta(x) = \sum_{i=0}^{k+l} \alpha_i \phi_i(x), \quad \text{with} \quad \alpha_i = 0 \text{ for } i > k+l,$$

and wish to *test the hypothesis* that

$$\alpha_i = 0, \quad i = k+1,\ k+2,\ \ldots,\ k+l,$$

we may obtain our residual estimate of σ^2 from

$$s^2 = \left(\sum_t y_t^2 - \sum_{i=0}^{k+l} \{A_i^2 \sum_t [\phi_i(x_t)]^2\}\right) \Big/ (n-k-l-1). \tag{117}$$

Since on the hypothesis tested $\displaystyle\sum_{i=k+1}^{k+l} \{A_i^2 \sum_t [\phi_i(x_t)]^2\}$

is distributed as the sum of l independent normal deviates each with variance σ^2, we may complete the test by calculating the variance ratio

$$F = \sum_{i=k+1}^{k+l} \frac{A_i^2 \sum_t [\phi_i(x_t)]^2}{ls^2}, \tag{118}$$

and entering Table 18 with $\nu_1 = l$, $\nu_2 = n-k-l-1$.

If, for example, in the case of the temperature data of Example **54** (p. 98), we were prepared to assume that the trend could certainly be represented by a polynomial of fifth degree

($\alpha_i = 0$ for $i > 5$), and wished to test whether a third-degree polynomial sufficed (i.e. to test whether $\alpha_4 = 0 = \alpha_5$), we could use the F of equation (118) with $k = 3$, $l = 2$ as a comprehensive test statistic. Since from van der Reyder (1943, p. 378) we have $s^2 = 12{\cdot}738/18$, it is found that

$$F = 18 \times 37{\cdot}542/(2 \times 12{\cdot}738) = 26{\cdot}53,$$

a value which clearly exceeds the 0·1 % point of F given in Table 18 for $\nu_1 = 2$, $\nu_2 = 18$ degrees of freedom. Thus a third-degree polynomial is not adequate in this case.

(*b*) *Case* $k = 0$

A special case of the test described under (*a*) occurs when $k = 0$. This case arises when, from theoretical considerations, it is known that if the data have a trend at all, this must be some polynomial of the lth degree and the test is for the significance of this polynomial as a whole.

For example, if for the above data we suspected from theoretical considerations that the diurnal temperature variation might follow a law representable by a fifth-degree polynomial, we could test this hypothesis by computing the F ratio of (118) for $k = 0$, $l = 5$. We then obtain

$$18 \times 1836{\cdot}66/(5 \times 12{\cdot}738) = 519{\cdot}1,$$

which clearly exceeds all tabulated percentage points of F for $\nu_1 = 5$ and $\nu_2 = 18$ degrees of freedom; thus the theory would be confirmed.

(*c*) *Case* $l = 1$

In this case we assume that the significance of a polynomial regression of degree k has already been established and we desire to test whether the next higher term, i.e. that of $(k+1)$th degree, is required for the representation of the real law.

This test is often used in a hierarchical sequence, starting from $k = 0$ and testing in turn the significance of each term until an insignificant coefficient is reached. With such a procedure it should be remembered that: (i) A coefficient α_i of $\phi_i(x)$ in the real law may well be 0, or small although some higher coefficient is large; if this is the case, the residual variance s^2 calculated after fitting the ith degree polynomial will no longer be a correct estimate of σ^2. (ii) The chance of returning an observed A_i as significant by the ith test in the sequence is no longer equal to the nominal level of significance in the ith F-test employed and may, in fact, be considerably smaller. For details of this test procedure in the present example see van der Reyden (1943).

It is clear that, in the case of equidistant x_t, the use of orthogonal polynomials considerably simplifies both the estimation of the polynomial regression of a given degree and also the tests of significance which may be required to decide between alternative polynomial regressions of different degree. In such tests, no recalculation of the estimates A_i ($i \leqslant k$) is required if the degree of the fitted polynomial is increased from k to $k + l$.

26·4 *Fitting polynomials to series with missing values*

It is worth noting that the present tables can still be used if the x_t are equidistant except for a number of gaps, as, for instance, in experiments in which it was planned to make observations at a regular time interval but a number of omissions have occurred. The procedure is similar to the missing-plot technique of the analysis of variance, but we shall describe it in detail by means of an example.

Illustration of fitting a fifth-degree polynomial for a series of $n=24$ values, with gaps at $t=6$ and $t=18$

	Polynomial values for the gaps						y-estimates and corrections			
(1)	ϕ_0 (2)	ϕ_1 (3)	ϕ_2 (4)	ϕ_3 (5)	ϕ_4 (6)	ϕ_5 (7)	${}_0y_t$ (8)	${}_1\Delta y_t$ (9)	${}_2\Delta y_t$ (10)	Final y_t (11)
(1) Gap at $t=6$	1	−13	−17	949	−137	169	85	1·08	0·20	86·28
(2) Gap at $t=18$	1	11	−53	−1023	− 87	1551	72	−0·90	−0·16	70·94
(3) $\Sigma\phi_i^2$ (all t)	24	4 600	394 680	$10^3\times$ 17 761	394 680	$10^3\times$ 177 929				
(4) ${}_0A_i$	77·465	−0·2821	−0·050 219	0·004 856	0·003 208	−0·000 4387				
(5) ${}_1\Delta A_i$	0·0075	−0·0052	0·000 074	0·000 110	−0·000 176	−0·000 0068				
(6) ${}_2\Delta A_i$	0·002	−0·0009	0·000 013	0·000 020	−0·000 034	−0·000 0012				
(7) Final A_i	77·47	−0·288	−0·050 13	0·004 99	0·003 00	−0·000 447				

Example 55. Let us suppose that the sixth and eighteenth temperature values were missing from the data used in Example 54, and that we wish to fit a fifth-degree polynomial to the remaining values. We first 'guess' values (${}_0y_t$, say) for the two missing temperatures; let these be ${}_0y_6=85$ and ${}_0y_{18}=72$. Using these ${}_0y_t$ in (115), we compute trial values ${}_0A_i$ for the A_i. These are shown in line 4 of the table above. Using these trial ${}_0A_i$ we compute corrected values, ${}_1y_t$, from (112), but only at the gaps, $t=6$ and $t=18$. The corrections

$$ {}_1\Delta y_6={}_1y_6-{}_0y_6, \quad {}_1\Delta y_{18}={}_1y_{18}-{}_0y_{18} $$

are shown in col. 9 of the table. These corrections are now used in turn to obtain corrections, ${}_1\Delta A_i$, for the A_i from

$$ {}_1\Delta A_i=\sum_{t=6,18}\{{}_1\Delta y_t\phi_i(x_t)\}\Big/\left\{\sum_{t=1}^{24}[\phi_i(x_t)]^2\right\} $$

and are shown in line 5 of the table; the remainder of the iteration process will be apparent. The final values obtained after the second iteration, viz. 86·28 and 70·94, should be compared with the graduated values, 86·25 and 71·07, obtained by van der Reyden (1943, p. 379) after fitting a fifth-degree polynomial to the complete data.

Usually one or two cycles of the iteration will suffice. The appropriate tests of significance are described by Hartley (1951) who also shows that the solution of the iteration described above is identical with the least squares fit to the observed y_t.

VI. AUXILIARY TABLES

27. Tables of miscellaneous mathematical quantities and conversion factors (Tables 48–54)

In this final section we have assembled a number of auxiliary tables giving some of the more general mathematical quantities which are used incidentally in statistical work. The applications are extremely varied and we shall select only a few examples to illustrate the use of the tables.

27·1 *Powers and sums of powers; squares and square roots; reciprocals*

Table 48 gives the powers of natural numbers $n = 1\,(1)\,100$ up to n^7, and Table 49 gives the cumulative sums of these powers, $\sum_{n=1}^{s} n^r$.* There are obvious uses of the former in the evaluation of moments, polynomials and power series. The computation of special tables to assist in fitting orthogonal polynomials to ordinates spaced at equal intervals has removed one of the more common uses for Table 49, but this will still be useful when fitting laws involving isolated powers, e.g. the law of heat loss through radiation $y = ct^4$.

Table 48 is also helpful in the computation of the general roots (fractional powers) of numbers. Thus if we need to find

$$x = y^{1/r}$$

from a given y, we may proceed as follows:

(*a*) Determine a first approximation, x_0, to x by selecting the argument $x_0 = n$ such that n^r is the nearest entry to y in the appropriate column.

(*b*) Compute an improved approximation from

$$x_1 = \{(r-1)\,x_0 + y/x_0^{r-1}\}/r, \tag{119}$$

where $x_0^{r-1} = n^{r-1}$ is the number immediately preceding n^r in the same line of the table.

This method gives as a rule 4-figure accuracy and, certainly, 3-figure accuracy.

Illustration. Find $(30068)^{\frac{1}{3}}$.

Entering the column headed n^3, we find that the entry $29791 = 31^3$ is nearest to $y = x^3 = 30068$, so that $x_0 = 31$. An improved value is then computed from

$$x_1 = \tfrac{1}{3}\{2 \times 31 + 30068/961\} = 31{\cdot}096.$$

This happens to be accurate to 5 figures.

Table 50 giving the squares of natural numbers, n^2, for $n = 1\,(1)\,999$, has been included mainly to assist in the evaluation of the sums of squares required in computing standard deviations or components in the analysis of variance when a calculating machine is not to hand. It may also be used for obtaining square roots (beyond the range of those given in Table 51) by the method just described for finding $x = y^{1/r}$, when $r = 2$. Equation (119) now becomes

$$x_1 = \tfrac{1}{2}\{x_0 + y/x_0\}, \tag{120}$$

and the use of Table 50 rather than Table 48 is now preferable, since it gives squares for 3-figure arguments. Equation (120) will yield square roots of 5–6-figure accuracy.

Table 51 gives, for $n = 1\,(1)\,100$, the three quantities $\sqrt{n}$, $1/\sqrt{n}$ and $1/n$, expressions frequently needed in the analysis of small samples.

27·2 *Factorials, Γ-functions and their logarithms to base* 10

Factorials are required in the evaluation of the terms of numerous statistical series of which it is sufficient to mention the binomial and hypergeometric. The Γ-function appears in the moments of many well-known probability distributions; in particular, it arises in the calculation of parameters for all the Pearson frequency curves (§ 23·1).

Table 51 gives $n!$ as an exact integer for $n = 1\,(1)\,9$; beyond this, for $n = 10\,(1)\,250$, the 5-decimal quantity tabled must be multiplied by 10^c, where c is the characteristic of $\log n!$

* For $r > 7$ Dr John Wishart has suggested that the relation $\sum_{n=1}^{s} n^r = \sum_{i=1}^{r} \binom{s+1}{i+1} \Delta^i 0^r$ may be found useful, particularly if tables of the binomial coefficients and of the differences of the powers of zero are available.

shown alongside in the next column. Similarly the entry for $1/n!$ must be multiplied by 10^{-c} for all n. For $n = 251\,(1)\,1000$, only $\log n!$ is tabled. Throughout, this function is tabled to 7 decimals.

Table 52 gives $\Gamma(1+p)$ and $\log\Gamma(1+p)$ to 7 decimals for $p = 0{\cdot}00\,(0{\cdot}01)\,1{\cdot}00$. More generally, $\log\Gamma(n+p)$ with n an integer and $0 < p < 1$ may be found as follows:*

(*a*) For $n = 1$; linear interpolation in Table 52 yields a maximum error of a unit in the fifth decimal.

(*b*) For $n = 2\,(1)\,6$; values may be obtained from the formula

$$\log\Gamma(n+p) = \log(n-1+p) + \log(n-2+p) + \ldots + \log(1+p) + \log\Gamma(1+p),$$

using ordinary tables of logarithms and Table 52.

(*c*) For $n > 6$; remembering that for integer n, $\log\Gamma(n) = \log(n-1)!$, we may interpolate in Table 51. Four-point Lagrangian interpolation yields 4-decimal accuracy for small n with a progressive improvement until, near $n = 1000$, 7 accurate decimals are obtained.

27·3 *Functions of p and $q = 1-p$ $(0 < p < 1)$*

The later columns of Table 52 give to 4 or 5 decimal places values of the auxiliary functions $1-p^2$, $(1-p^2)^{\frac{1}{2}}$, $(1-p^2)^{-\frac{1}{2}}$, pq, $(pq)^{\frac{1}{2}}$ and p^2+q^2. Much fuller tables of some of these functions have been given by Kelley (1948). The following are some illustrations of the use of these functions:

$1-p^2$. Problems of correlation, with $p = \rho$; for reducing a variance to a residual variance after removing a linear regression term; for use in a large sample approximation to the sampling variance of a product moment correlation coefficient, r (normally distributed variables).

$(1-p^2)^{\frac{1}{2}}$. As above, for standard deviation instead of variance.

$(1-p^2)^{-\frac{1}{2}}$. For the evaluation of partial correlation coefficients by continued multiplication.

pq and $(pq)^{\frac{1}{2}}$. For computing the binomial variance and standard deviation.

p^2+q^2. For evaluating the variance of a weighted mean from

$$\operatorname{var}(px_1 + qx_2) = (p^2+q^2)\,\sigma^2,$$

where σ^2 is the variance of independent variables x_1 and x_2.

27·4 *Natural logarithms* (*Table* 53)

Natural logarithms are provided for the decade $1 \leqslant x < 10$; if x lies outside this range, the logarithm may be found by adding the appropriate value of $\log_e 10^n$ shown at the bottom of the table. However, in many calculations it is unnecessary to add the constant $\log_e 10^n$, as this cancels out in the course of the computation; such is the case when calculating the statistic M (§ 16·1) used in testing for variance heterogeneity.

If proportional parts are used for interpolation, they should be added to, or subtracted from, the nearest tabular value. Examples are given at the foot of the table. This method will yield 5-decimal accuracy for $x > 5{\cdot}0$, but for $1{\cdot}0 < x < 5{\cdot}0$ errors of up to 4 units in the fifth decimal may occur and to obtain 5-decimal accuracy, more exact interpolation must be used.

27·5 *Miscellaneous constants* (*Table* 54)

Only mathematical constants and conversion factors for weights and measures are given here; for other physical constants, etc., the reader must consult standard works of reference.

* For fuller tables see Brownlee (1923), Davis (1933), Pearson, E. S. (1922).

APPENDIX I

LIST OF REFERENCES

Aitken, A. C. (1933). *Proc. Roy. Soc. Edinb.* **53**, 54–78.
Anderson, R. L. & Houseman, E. E. (1942). *Res. Bull., Iowa Agric. Exp. Sta.* no. 297.
Aspin, A. A. (1949). *Biometrika*, **36**, 290–3.
Bailey, N. T. J. (1950). *Biometrika*, **37**, 193–202.
Barnard, G. A. (1947). *Biometrika*, **34**, 123–38.
Bartlett, M. S. (1937). *Proc. Roy. Soc.* A, **160**, 268–82.
Bayes, A. W. (1937). *Suppl. J.R. Statist. Soc.* **4**, 61–80.
Bishop, D. J. (1939). *Biometrika*, **31**, 31–55.
Bishop, D. J. & Nair, U. S. (1939). *Suppl. J.R. Statist. Soc.* **6**, 89–99.
Bliss, C. I. (1952). *The Statistics of Bio-assay, with special reference to Vitamins.* New York: Academic Press.
Bliss, C. I., Cochran, W. G. & Tukey, J. W. (1956). *Biometrika*, **43**, 418–22.
Box, G. E. P. (1949). *Biometrika*, **36**, 317–46.
Brownlee, J. (1923). Log $\Gamma(x)$ *from* $x=1$ *to* 50·9 *by intervals of* 0·01. Tracts for Computers, no. IX. Cambridge University Press.
Cadwell, J. H. (1953). *Biometrika*, **40**, 336–46.
Carter, A. H. (1947). *Biometrika*, **34**, 352–8.
Cochran, W. G. (1940). *Ann. Math. Statist.* **11**, 93–5.
Cochran, W. G. (1941). *Ann. Eugen., Lond.*, **11**, 47–52.
Colcord, C. G. & Deming, L. S. (1935). *Sankhyā*, **2**, 423–4.
Comrie, L. J. (1936). *Interpolation and Allied Tables.* London: H.M. Stationery Office.
Comrie, L. J. (1950). *Chambers' Shorter Six-figure Mathematical Tables.*
Comrie, L. J. & Hartley, H. O. (1941). *Biometrika*, **32**, 183–6.
David, F. N. (1938). *Tables of the Ordinates and Probability Integral of the Distribution of the Correlation Coefficient in Small Samples.* Cambridge University Press for the Biometrika Trustees.
David, H. A. (1951). *Biometrika*, **38**, 393–409.
David, H. A. (1952). *Biometrika*, **39**, 422–4.
David, H. A. (1956). *Biometrika*, **43**, 449–51.
David, H. A., Hartley, H. O. & Pearson, E. S. (1954). *Biometrika*, **41**, 482–93.
David, S. T., Kendall, M. G. & Stuart, A. (1951). *Biometrika*, **38**, 131–40.
Davis, H. T. (1933). *Tables of the Higher Mathematical Functions*, **1**. Bloomington, Indiana: Principia Press.
Eisenhart, C., Hastay, M. W. & Wallis, W. A. (1947). *Techniques of Statistical Analysis.* New York: McGraw Hill Book Company.
Elderton, W. P. (1953). *Frequency Curves and Correlation*, 4th ed. Cambridge University Press.
Finney, D. J. (1941). *Ann. Eugen., Lond.*, **11**, 136–40.
Finney, D. J. (1948). *Biometrika*, **35**, 145–56.
Finney, D. J. (1952). *Probit Analysis. A Statistical Treatment of the Sigmoid Response Curve*, 2nd ed. Cambridge University Press.
Finney, D. J. & Stevens, W. L. (1948). *Biometrika*, **35**, 191–201.
Fisher, R. A. (1915). *Biometrika*, **10**, 507–21.
Fisher, R. A. (1920). *Mon. Not. R. Astr. Soc.* **80**, 758–70.
Fisher, R. A. (1921*a*). *Metron.* **1**, Part 4, 3–32.
Fisher, R. A. (1921*b*). *J. Agric. Sci.* **11**, 107–35.
Fisher, R. A. (1929). *Proc. Lond. Math. Soc.*, Series 2, **30**, 199–238.
Fisher, R. A. (1930). *Proc. Roy. Soc.* A, **130**, 16–28.
Fisher, R. A. & Yates, F. (1963). *Statistical Tables for Biological, Agricultural and Medical Research*, 6th ed. (1st ed. 1938). Edinburgh: Oliver and Boyd.
Friedman, M. (1940). *Ann. Math. Statist.* **11**, 86–92.
Gayen, A. K. (1951). *Biometrika*, **38**, 219–47.
Geary, R. C. (1935). *Biometrika*, **27**, 310–32.
Geary, R. C. (1936). *Biometrika*, **28**, 295–307.
Godwin, H. J. (1945). *Biometrika*, **33**, 254–6.
Godwin, H. J. (1949*a*). *Biometrika*, **36**, 92–100.

Godwin, H. J. (1949b). *Ann. Math. Statist.* **20**, 279–85.
Gosset, W. S. ('Student') (1908). *Biometrika*, **6**, 1–25.
Greenwood, J. A. & Hartley, H. O. (1962). *Guide to Tables in Mathematical Statistics.* Princeton University Press.
Gumbel, E. J. (1949). *Biometrika*, **36**, 142–8.
Halton Thomson, D. (1947). *Biometrika*, **34**, 368–72.
Harley, B. I. & Pearson, E. S. (1957). *Biometrika*, **44**, 257–60.
Harter, H. L. (1963). *Biometrika*, **50**, 187–94.
Harter, H. L. (1964). *Biometrika*, **51**, 231–39.
Harter, H. L., Clemm, D. S. & Guthrie, E. H. (1959). *The Probability Integrals of the Range and of the Studentized Range.* **1** and **2**. Wright Air Development Center, Ohio, Tech. Report 58–484.
Hartley, H. O. (1938). *Suppl. J.R. Statist. Soc.* **5**, 80–8.
Hartley, H. O. (1940). *Biometrika*, **31**, 249–55.
Hartley, H. O. (1941). *Biometrika*, **32**, 161–7.
Hartley, H. O. (1942). *Biometrika*, **32**, 309–10.
Hartley, H. O. (1944). *Biometrika*, **33**, 173–80.
Hartley, H. O. (1945). *Biometrika*, **33**, 257–8.
Hartley, H. O. (1950). *Biometrika*, **37**, 271–80.
Hartley, H. O. (1951). *Biometrika*, **38**, 410–13.
Helmert, F. R. (1876). *Astr. Nachr.* **88**, no. 2096.
Hooke, B. G. E. (1926). *Biometrika*, **18**, 1–55.
Johnson, N. L., Nixon, Eric, Amos, D. E. & Pearson, E. S. (1963). *Biometrika*, **50**, 459–98.
Jones, A. E. (1946). *Biometrika*, **33**, 274–82.
Kelley, T. L. (1948). *The Kelley Statistical Tables*, 2nd ed. Harvard University Press.
Kendall, M. G. (1962). *Rank Correlation Methods*, 3rd ed. London: Chas. Griffin and Co. Ltd.
Kendall, M. G. & Babington Smith, B. (1939). *Ann. Math. Statist.* **10**, 275–87.
Lancaster, H. O. (1949). *Biometrika*, **36**, 370–82.
Latter, O. H. (1906). *Biometrika*, **4**, 363–73.
Leggatt, C. W. (1935). *Proc. Int. Seed Test. Ass.* **7**, 27–37.
Link, R. F. (1950). *Ann. Math. Statist.* **21**, 112–16.
Lord, E. (1947). *Biometrika*, **34**, 41–67.
Lord, E. (1950). *Biometrika*, **37**, 64–77.
Maguire, B. A., Pearson, E. S. & Wynn, A. H. A. (1952). *Biometrika*, **39**, 168–80.
Mahalanobis, P. C., Majumdar, D. N. & Rao, C. R. (1949). *Sankhyā*, **9**, 89.
McKay, A. T. (1935). *Biometrika*, **27**, 466–71.
Merrington, M. & Thompson, C. M. (1943). *Biometrika*, **33**, 73–88.
Moore, P. G. (1957). *Biometrika*, **44**, 482–89.
Mosteller, F. (1946). *Ann. Math. Statist.* **17**, 377–408.
Nair, K. R. (1948a). *Biometrika*, **35**, 16–31.
Nair, K. R. (1948b). *Biometrika*, **35**, 118–44.
Nair, K. R. (1950). *Biometrika*, **37**, 182–3.
Nair, K. R. (1952). *Biometrika*, **39**, 189–91.
Nair, U. S. (1938). *Biometrika*, **30**, 274–94.
National Bureau of Standards (1950). *Tables of the Binomial Probability Distribution.* Applied Maths. series no. 6.
Neyman, J. & Pearson, E. S. (1931). *Bull. Int. Acad. Cracovie*, A, 460–81.
Norton, H. W. (1952). *Math. Tab., Wash.*, **6**, 35–8.
Pachares, J. (1959). *Biometrika*, **46**, 461–4.
Patnaik, P. B. (1949). *Biometrika*, **36**, 202–32.
Patnaik, P. B. (1950). *Biometrika*, **37**, 78–87.
Pearson, E. S. (1922). *Table of Logarithms of the Complete Gamma-function* (*to* 10 *decimal places*) *for Argument* 2–1200. Tracts for Computers, no. VIII. Cambridge University Press.
Pearson, E. S. (1930). *Biometrika*, **22**, 239–49.
Pearson, E. S. (1931). *Biometrika*, **22**, 423–4.
Pearson, E. S. (1932). *Biometrika*, **24**, 404–17.
Pearson, E. S. (1935a). *Biometrika*, **27**, 333–47.
Pearson, E. S. (1935b). *The Application of Statistical Methods to Industrial Standardisation and Quality Control.* British Standards Institution No. 600. 1935.
Pearson, E. S. (1938). *Biometrika*, **30**, 210–50.
Pearson, E. S. (1945). *Biometrika*, **33**, 252–3.
Pearson, E. S. (1948). *Biometrika*, **35**, 424.
Pearson, E. S. (1950). *Biometrika*, **37**, 383–98.
Pearson, E. S. (1952). *Biometrika*, **39**, 130–6.

Pearson, E. S. (1963). *Biometrika*, **50**, 95–112.
Pearson, E. S. (1965), *Biometrika*, **52**, 282–5.
Pearson, E. S. & Chandra Sekar, C. (1936). *Biometrika*, **28**, 308–20.
Pearson, E. S. & Hartley, H. O. (1951). *Biometrika*, **38**, 112–30.
Pearson, E. S. & Merrington, M. (1951). *Biometrika*, **38**, 4–10.
Pearson, E. S. & Stephens, M. A. (1964). *Biometrika*, **51**, 484–6.
Pearson, E. S. & Welch, B. L. (1937). *Suppl. J.R. Statist. Soc.* **4**, 94–101.
Pearson, K. (1895). *Phil. Trans.* A, **186**, 343–414.
Pearson, K. (1901). *Phil. Trans.* A, **197**, 443–59.
Pearson, K. (1902). *Biometrika*, **1**, 390–9.
Pearson, K. (1906). *Biometrika*, **5**, 105–46.
Pearson, K. (1914). *Tables for Statisticians and Biometricians*, Part 1. Cambridge University Press for the Biometrika Trustees.
Pearson, K. (1916). *Phil. Trans.* A, **216**, 429–57.
Pearson, K. (1922). *Tables of the Incomplete* Γ-*function*. Cambridge University Press for the Biometrika Trustees.
Pearson, K. (1931). *Tables for Statisticians and Biometricians*, Part 2. Cambridge University Press for the Biometrika Trustees.
Pearson, K. (1934). *Tables of the Incomplete* B-*function*. Cambridge University Press for the Biometrika Trustees.
Pillai, K. C. S. (1959). *Biometrika*, **46**, 473–4.
Pillai, K. C. S. & Buenaventura, A. R. (1961). *Biometrika*, **48**, 195–6.
Przyborowski, J. & Wilenski, H. (1940). *Biometrika*, **31**, 313–23.
Quesenberry, C. P. & David, H. A. (1961). *Biometrika*, **48**, 379–90.
Reyden, van der, D. (1943). *Onderstepoort J. Vet. Sci.* **18**, 355–404.
Sheppard, W. F. (1902). *Biometrika*, **2**, 174–90.
Simaika, J. B. (1942). *Biometrika*, **32**, 263–76.
Snedecor, G. W. (1946). *Statistical Methods*, 4th ed. Ames, Iowa: Collegiate Press, Inc.
Soper, H. E. (1921). *The Numerical Evaluation of the Incomplete* B-*function*. Tracts for Computers, no. VII. Cambridge University Press.
Stevens, W. L. (1950). *Biometrika*, **37**, 117–29.
Thompson, W. R. (1935). *Ann. Math. Statist.* **6**, 214–19.
Tippett, L. H. C. (1925). *Biometrika*, **17**, 364–87.
Tippett, L. H. C. (1952). *The Methods of Statistics*, 4th ed. Williams and Norgate Ltd.
Tschepourkowsky, E. (1905). *Biometrika*, **4**, 286–312.
Welch, B. L. (1947). *Biometrika*, **34**, 28–35.
Williams, P. (1935). *Biometrika*, **27**, 269–71.
Wise, M. E. (1950). *Biometrika*, **37**, 208–18.
Wishart, J. (1927). *Biometrika*, **19**, 1–38.
Wishart, J. (1938). *J. Agric. Sci.* **28**, 299–306.
Wold, H. (1948). *Random Normal Deviates*. Tracts for Computers, no. XXV. Cambridge University Press.
Yates, F. (1934). *Suppl. J.R. Statist. Soc.* **1**, 217–35.

APPENDIX II

ORIGIN OF TABLES

The following list shows the source or sources from which each of the tables in the present volume has been taken as well as the names of their computers where these are known to us. As statisticians, as well as editors, we should like to put on record our gratitude for these labours, spread over fifty years. The statistician's debt of thanks is, indeed, due not only to the contributors whose tables we are reissuing or expect to reissue, but to all those others whose work we are omitting because changes in statistical techniques have called for new forms of tabular aid. With few exceptions, their work can be found if required in the past volumes of *Biometrika*.

The great majority of the present tables have also appeared in *Biometrika* or have been computed recently under *Biometrika* auspices, but in a few cases referred to below we have drawn from outside sources, as stated in the Preface. The page numbers of a journal in the 'source' column is that of the first page of a table.

ORIGIN OF INDIVIDUAL TABLES

	Source	Computer
1	*Biometrika*, **2**, 182. *T.S.B.* **1**, II*	W. F. Sheppard
2	*Drapers Co. Research Memoirs*, Biometric Series, **8**, 27. *T.S.B.* **1**, IV	Julia Bell
3	*The Kelley Statistical Tables* (1948), Table I	
4	*Biometrika*, **5**, 405. *T.S.B.* **1**, I	W. F. Sheppard
5	*Biometrika*, **13**, 428. *T.S.B.* **2**, I	Derived from work of W. F. Sheppard by H. E. Soper and others
6	*Biometrika*, **35**, 193	D. J. Finney and W. L. Stevens
7	*Biometrika*, **37**, 318	Editorial and Mathematics Division, National Physical Laboratory
8	*Biometrika*, **32**, 188 (with the addition of a 0·1% column). Corrections: Harter (1963); D. B. Owen	Catherine M. Thompson (Jean H. Thompson)
9	*Biometrika*, **37**, 170	Editorial and Mathematics Division, National Physical Laboratory
10	*Biometrika*, **38**, 115	Jean H. Thompson and Mathematics Division, National Physical Laboratory
11	*Biometrika*, **36**, 291; *Biometrika*, **43**, 204	Alice A. Aspin; W. H. Trickett, B. L. Welch and G. S. James
12, 13	Freshly compiled, partly using *Biometrika*, **32**, 168, 300 and partly from Fisher & Yates, *Statistical Tables* (1953), Tables III and VI	
14	Freshly compiled from standard sources	
15	*Tables of the Ordinates and Probability Integral of the Distribution of the Correlation Coefficient* (1938), Charts II and IV	Florence N. David
16	*Biometrika*, **32**, 168; *Biometrika*, **50**, 452	Catherine M. Thompson; D. E. Amos
17	*Biometrika*, **38**, 424	H. O. Hartley and Elaine R. Fitch
18	Reduced from *Biometrika*, **33**, 74 with 0·1% points compiled from various sources	Maxime Merrington and Catherine M. Thompson
19	*Biometrika*, **35**, 26	K. R. Nair
20	Mean deviation: freshly compiled	Joyce M. May
	Range: *Biometrika*, **38**, 463 (derived from 5 decimal MS. of Table 23)†	Mathematics Division, National Physical Laboratory

* *T.S.B.* **1** or 2 indicates *Tables for Statisticians and Biometricians*, parts 1 or 2, with table number following in Roman figures.

† The later and fuller tables computed by Harter *et al.* (1959) have made it possible to remove a number of last-figure errors from these three tables.

	Source	Computer
21	*Biometrika*, **33**, 265	Scientific Computing Service Ltd., M. Sumner
22	*Biometrika*, **32**, 308 (freshly compiled for $n > 12$)*	Scientific Computing Service Ltd.
23	*Biometrika*, **32**, 302*	Scientific Computing Service Ltd.
24	Freshly compiled	Jean H. Thompson and Joyce M. May
25	*Biometrika*, **35**, 140	K. R. Nair
26	*Biometrika*, **35**, 143; **43**, 450; **46**, 473	K. R. Nair; H. A. David; K. C. S. Pillai
26*a*	*Biometrika*, **48**, 388	C. P. Quesenberry and H. A. David
26*b*	*Biometrika*, **48**, 389 (after condensation)	C. P. Quesenberry and H. A. David
27	*Biometrika*, **17**, 386. *T.S.B.* **2**, XXII	L. H. C. Tippett
28	Compiled from *Biometrika*, **36**, 98 ($n \leqslant 10$); Fisher & Yates, *Statistical Tables* (1953) ($n > 20$); with some fresh calculations ($10 < n < 20$)	In part, H. J. Godwin, Jean H. Thompson
29	*Biometrika*, **46**, 464	H. L. Harter; J. Pachares
29*a*	*Biometrika*, **44**, 487	P. G. Moore
29*b*	*Biometrika*, **50**, 191	H. L. Harter
29*c*	*Biometrika*, **41**, 491; *Biometrika*, **51**, 486	H. A. David, H. O. Hartley and E. S. Pearson; E. S. Pearson and M. A. Stephens
30	*Biometrika*, **38**, 408	H. A. David
31	*Biometrika*, **37**, 310; **39**, 424	H. O. Hartley; H. A. David and Joyce M. May
31*a*	Abridged from C. Eisenhart, M. W. Hastay and W. A. Wallis (1947). *Techniques of Statistical Analysis*, Table 15·1. See also W. G. Cochran (1941)	
31*b*	*Biometrika*, **43**, 420	C. I. Bliss, W. G. Cochran and J. W. Tukey
32, 33	*Biometrika*, **33**, 302, 304	Catherine M. Thompson and Maxine Merrington
34 (A)	a *Biometrika*, **28**, 303	R. C. Geary
34 (B)	$\sqrt{b_1}$ *Biometrika*, **28**, 306. *T.S.B.* **2**, XXXVII *bis*	E. S. Pearson and P. Williams
34 (C)	b_2 *Biometrika*, **22**, 248. *T.S.B.* **2**, XXXVII *bis*	E. S. Pearson
35	Freshly compiled, using *Biometrika*, **10**, 529. *T.S.B.* **2**, XVII	Joyce M. May and H. O. Hartley
36	Enlarged, from *Biometrika*, **31**, 320	J. Przyborowski and H. Wilenski; E. S. Pearson
37	Derived from *Tables of the Incomplete* B-*function*; see also National Bureau of Standards, Applied Mathematics Series, no. 6	
38	*Biometrika*, **35**, 149	D. J. Finney
39	*Biometrika*, **10**, 27. *T.S.B.* **1**, LI	H. E. Soper
40	Enlarged from *Biometrika*, **28**, 439	F. Garwood
41	Enlarged from *Biometrika*, **26**, 410	J. C. Clopper and E. S. Pearson
42	*Biometrika*, **38**, 6, with fresh calculations for 2·5 and 1 % points	Maxine Merrington and E. S. Pearson
43	Redrawn from *Biometrika*, **7**, 131. *T.S.B.* **1**, XXXV	A. Rhind
44, 45, 46	Reduced from M. G. Kendall, *Rank Correlation Methods* (1962), Tables 1, 2 and 5, and *Biometrika*, **38**, 132	M. G. Kendall
47	Based on tables published by D. van der Reyden, *Onderstepoort J. Vet. Sci.* (1943), **18**, 383	D. F. I. van Heerden
48	*Biometrika*, **2**, 477. *T.S.B.* **1**, XXVII	W. P. Elderton
49	*Biometrika*, **2**, 479. *T.S.B.* **1**, XXVIII	W. P. Elderton
50	From standard sources	
51	Enlarged from an earlier table. *T.S.B.* **1**, XLIX	Julia Bell
52	Freshly compiled; partly based on *The Kelley Statistical Tables* (1948), Table I	
53	Freshly compiled from standard sources	
54	Freshly compiled; largely based on L. J. Comrie, *Chambers' Shorter Six-figure Mathematical Tables* (1950)	

* See footnote † on page 107.

TABLES

Table 1. *The Normal probability function. The integral P(X) and ordinate Z(X) in terms of the standardized deviate X*

X	$P(X)$	δ +	δ^2 −	$Z(X)$	δ −	δ^2 −
·00	·5000000	39894	0	·3989423	199	399
·01	·5039894	39890	4	·3989223	598	399
·02	·5079783	39882	8	·3988625	997	399
·03	·5119665	39870	12	·3987628	1395	398
·04	·5159534	39854	16	·3986233	1793	398
·05	·5199388	39834	20	·3984439	2191	397
·06	·5239222	39810	24	·3982248	2588	397
·07	·5279032	39782	28	·3979661	2984	396
·08	·5318814	39750	32	·3976677	3379	395
·09	·5358564	39714	36	·3973298	3773	394
·10	·5398278	39675	40	·3969525	4166	393
·11	·5437953	39631	44	·3965360	4558	392
·12	·5477584	39584	48	·3960802	4948	390
·13	·5517168	39532	51	·3955854	5337	389
·14	·5556700	39477	55	·3950517	5724	387
·15	·5596177	39418	59	·3944793	6110	386
·16	·5635595	39355	63	·3938684	6493	384
·17	·5674949	39288	67	·3932190	6875	382
·18	·5714237	39217	71	·3925315	7255	380
·19	·5753454	39143	74	·3918060	7633	378
·20	·5792597	39065	78	·3910427	8008	375
·21	·5831662	38983	82	·3902419	8381	373
·22	·5870644	38897	86	·3894038	8752	371
·23	·5909541	38808	89	·3885286	9120	368
·24	·5948349	38715	93	·3876166	9485	365
·25	·5987063	38618	97	·3866681	9847	362
·26	·6025681	38518	100	·3856834	10207	360
·27	·6064199	38414	104	·3846627	10564	357
·28	·6102612	38306	107	·3836063	10917	354
·29	·6140919	38195	111	·3825146	11268	350
·30	·6179114	38081	114	·3813878	11615	347
·31	·6217195	37963	118	·3802264	11958	344
·32	·6255158	37842	121	·3790305	12298	340
·33	·6293000	37717	125	·3778007	12635	337
·34	·6330717	37589	128	·3765372	12968	333
·35	·6368307	37458	131	·3752403	13297	329
·36	·6405764	37323	135	·3739106	13623	325
·37	·6443088	37185	138	·3725483	13944	322
·38	·6480273	37044	141	·3711539	14262	318
·39	·6517317	36900	144	·3697277	14575	313
·40	·6554217	36753	147	·3682701	14885	309
·41	·6590970	36602	150	·3667817	15190	305
·42	·6627573	36449	153	·3652627	15491	301
·43	·6664022	36293	156	·3637136	15787	296
·44	·6700314	36133	159	·3621349	16079	292
·45	·6736448	35971	162	·3605270	16367	288
·46	·6772419	35806	165	·3588903	16650	283
·47	·6808225	35638	168	·3572253	16928	278
·48	·6843863	35467	171	·3555325	17202	274
·49	·6879331	35294	173	·3538124	17470	269
·50	·6914625		176	·3520653		264

X	$P(X)$	δ +	δ^2 −
·50	·6914625	35118	176
·51	·6949743	34939	179
·52	·6984682	34758	181
·53	·7019440	34574	184
·54	·7054015	34388	186
·55	·7088403	34200	189
·56	·7122603	34009	191
·57	·7156612	33815	193
·58	·7190427	33620	196
·59	·7224047	33422	198
·60	·7257469	33222	200
·61	·7290691	33020	202
·62	·7323711	32816	204
·63	·7356527	32610	206
·64	·7389137	32402	208
·65	·7421539	32192	210
·66	·7453731	31980	212
·67	·7485711	31767	214
·68	·7517478	31551	215
·69	·7549029	31334	217
·70	·7580363	31116	219
·71	·7611479	30896	220
·72	·7642375	30674	222
·73	·7673049	30451	223
·74	·7703500	30226	225
·75	·7733726	30001	226
·76	·7763727	29773	227
·77	·7793501	29545	228
·78	·7823046	29316	230
·79	·7852361	29085	231
·80	·7881446	28853	232
·81	·7910299	28620	233
·82	·7938919	28387	234
·83	·7967306	28152	235
·84	·7995458	27917	235
·85	·8023375	27680	236
·86	·8051055	27443	237
·87	·8078498	27205	238
·88	·8105703	26967	238
·89	·8132671	26728	239
·90	·8159399	26489	239
·91	·8185887	26249	240
·92	·8212136	26008	240
·93	·8238145	25768	241
·94	·8263912	25527	241
·95	·8289439	25285	241
·96	·8314724	25044	242
·97	·8339768	24802	242
·98	·8364569	24560	242
·99	·8389129	24318	242
1·00	·8413447		242

$$Z(X)=e^{-\frac{1}{2}X^2}/\sqrt{(2\pi)},\quad P(X)=1-Q(X)=\int_{-\infty}^{X} Z(u)\,du.$$

Table 1 (*continued*)

$Z(X)$	δ −	δ^2 −
·3520653	17734	264
·3502919	17994	259
·3484925	18248	254
·3466677	18497	249
·3448180	18741	244
·3429439	18981	239
·3410458	19215	234
·3391243	19444	229
·3371799	19667	224
·3352132	19886	219
·3332246	20099	213
·3312147	20307	208
·3291840	20510	203
·3271330	20707	197
·3250623	20899	192
·3229724	21086	187
·3208638	21267	181
·3187371	21442	176
·3165929	21613	170
·3144317	21777	165
·3122539	21936	159
·3100603	22090	154
·3078513	22239	148
·3056274	22381	143
·3033893	22519	137
·3011374	22650	132
·2988724	22777	126
·2965948	22897	121
·2943050	23013	115
·2920038	23122	110
·2896916	23227	104
·2873689	23325	99
·2850364	23419	93
·2826945	23507	88
·2803438	23589	83
·2779849	23666	77
·2756182	23738	72
·2732444	23805	66
·2708640	23866	61
·2684774	23922	56
·2660852	23972	51
·2636880	24017	45
·2612863	24058	40
·2588805	24093	35
·2564713	24122	30
·2540591	24147	25
·2516443	24167	20
·2492277	24182	15
·2468095	24191	10
·2443904	24196	5
·2419707		0

X	$P(X)$	δ +	δ^2 −	$Z(X)$	δ −	δ^2 +
1·00	·8413447	24076	242	·2419707	24196	0
1·01	·8437524	23834	242	·2395511	24191	5
1·02	·8461358	23592	242	·2371320	24182	10
1·03	·8484950	23351	242	·2347138	24168	14
1·04	·8508300	23109	242	·2322970	24149	19
1·05	·8531409	22868	241	·2298821	24125	24
1·06	·8554277	22626	241	·2274696	24097	28
1·07	·8576903	22386	241	·2250599	24064	33
1·08	·8599289	22145	240	·2226535	24027	37
1·09	·8621434	21905	240	·2202508	23986	41
1·10	·8643339	21665	240	·2178522	23940	46
1·11	·8665005	21426	239	·2154582	23890	50
1·12	·8686431	21188	239	·2130691	23836	54
1·13	·8707619	20950	238	·2106856	23778	58
1·14	·8728568	20712	237	·2083078	23715	62
1·15	·8749281	20475	237	·2059363	23649	66
1·16	·8769756	20239	236	·2035714	23578	70
1·17	·8789995	20004	235	·2012135	23504	74
1·18	·8809999	19769	235	·1988631	23426	78
1·19	·8829768	19535	234	·1965205	23344	82
1·20	·8849303	19302	233	·1941861	23259	85
1·21	·8868606	19070	232	·1918602	23170	89
1·22	·8887676	18839	231	·1895432	23077	93
1·23	·8906514	18609	230	·1872354	22981	96
1·24	·8925123	18379	229	·1849373	22882	99
1·25	·8943502	18151	228	·1826491	22779	103
1·26	·8961653	17924	227	·1803712	22673	106
1·27	·8979577	17697	226	·1781038	22564	109
1·28	·8997274	17472	225	·1758474	22452	112
1·29	·9014747	17248	224	·1736022	22337	115
1·30	·9031995	17026	223	·1713686	22218	118
1·31	·9049021	16804	222	·1691468	22097	121
1·32	·9065825	16584	220	·1669370	21973	124
1·33	·9082409	16365	219	·1647397	21847	127
1·34	·9098773	16147	218	·1625551	21717	129
1·35	·9114920	15930	217	·1603833	21585	132
1·36	·9130850	15715	215	·1582248	21451	134
1·37	·9146565	15501	214	·1560797	21314	137
1·38	·9162067	15289	212	·1539483	21175	139
1·39	·9177356	15078	211	·1518308	21033	142
1·40	·9192433	14868	210	·1497275	20890	144
1·41	·9207302	14660	208	·1476385	20744	146
1·42	·9221962	14453	207	·1455641	20596	148
1·43	·9236415	14248	205	·1435046	20446	150
1·44	·9250663	14044	204	·1414600	20294	152
1·45	·9264707	13842	202	·1394306	20140	154
1·46	·9278550	13642	201	·1374165	19985	155
1·47	·9292191	13443	199	·1354181	19828	157
1·48	·9305634	13245	197	·1334353	19669	159
1·49	·9318879	13049	196	·1314684	19508	160
1·50	·9331928		194	·1295176		162

Note sign of second difference, δ^2.

Table 1. *The Normal probability function (continued)*

X	$P(X)$	δ +	δ^2 −	$Z(X)$	δ −	δ^2 +
1·50	·9331928	12855	194	·1295176	19346	162
1·51	·9344783	12662	193	·1275830	19183	163
1·52	·9357445	12471	191	·1256646	19018	165
1·53	·9369916	12282	189	·1237628	18853	166
1·54	·9382198	12094	188	·1218775	18685	167
1·55	·9394292	11908	186	·1200090	18517	168
1·56	·9406201	11724	184	·1181573	18348	169
1·57	·9417924	11541	183	·1163225	18177	170
1·58	·9429466	11360	181	·1145048	18006	171
1·59	·9440826	11181	179	·1127042	17834	172
1·60	·9452007	11004	177	·1109208	17661	173
1·61	·9463011	10828	176	·1091548	17487	174
1·62	·9473839	10654	174	·1074061	17312	174
1·63	·9484493	10482	172	·1056748	17137	175
1·64	·9494974	10311	170	·1039611	16962	176
1·65	·9505285	10142	169	·1022649	16786	176
1·66	·9515428	9975	167	·1005864	16609	177
1·67	·9525403	9810	165	·0989255	16432	177
1·68	·9535213	9647	163	·0972823	16255	177
1·69	·9544860	9485	162	·0956568	16077	178
1·70	·9554345	9326	160	·0940491	15899	178
1·71	·9563671	9167	158	·0924591	15722	178
1·72	·9572838	9011	156	·0908870	15544	178
1·73	·9581849	8856	155	·0893326	15366	178
1·74	·9590705	8704	153	·0877961	15188	178
1·75	·9599408	8553	151	·0862773	15010	178
1·76	·9607961	8403	149	·0847764	14832	178
1·77	·9616364	8256	147	·0832932	14654	178
1·78	·9624620	8110	146	·0818278	14477	177
1·79	·9632730	7966	144	·0803801	14300	177
1·80	·9640697	7824	142	·0789502	14123	177
1·81	·9648521	7684	140	·0775379	13946	176
1·82	·9656205	7545	139	·0761433	13770	176
1·83	·9663750	7409	137	·0747663	13594	176
1·84	·9671159	7273	135	·0734068	13419	175
1·85	·9678432	7140	133	·0720649	13245	175
1·86	·9685572	7009	132	·0707404	13071	174
1·87	·9692581	6879	130	·0694333	12897	173
1·88	·9699460	6751	128	·0681436	12725	173
1·89	·9706210	6624	126	·0668711	12553	172
1·90	·9712834	6500	125	·0656158	12382	171
1·91	·9719334	6377	123	·0643777	12211	170
1·92	·9725711	6255	121	·0631566	12041	170
1·93	·9731966	6136	120	·0619524	11873	169
1·94	·9738102	6018	118	·0607652	11705	168
1·95	·9744119	5902	116	·0595947	11538	167
1·96	·9750021	5787	115	·0584409	11372	166
1·97	·9755808	5674	113	·0573038	11206	165
1·98	·9761482	5563	111	·0561831	11042	164
1·99	·9767045	5453	110	·0550789	10879	163
2·00	·9772499		108	·0539910		162

X	$P(X)$	δ +	δ^2 −
2·00	·9772499	5345	108
2·01	·9777844	5239	106
2·02	·9783083	5134	105
2·03	·9788217	5031	103
2·04	·9793248	4929	102
2·05	·9798178	4829	100
2·06	·9803007	4731	98
2·07	·9807738	4634	97
2·08	·9812372	4539	95
2·09	·9816911	4445	94
2·10	·9821356	4352	92
2·11	·9825708	4262	91
2·12	·9829970	4172	89
2·13	·9834142	4084	88
2·14	·9838226	3998	86
2·15	·9842224	3913	85
2·16	·9846137	3829	84
2·17	·9849966	3747	82
2·18	·9853713	3666	81
2·19	·9857379	3587	79
2·20	·9860966	3509	78
2·21	·9864474	3432	77
2·22	·9867906	3357	75
2·23	·9871263	3283	74
2·24	·9874545	3210	73
2·25	·9877755	3138	71
2·26	·9880894	3068	70
2·27	·9883962	2999	69
2·28	·9886962	2932	68
2·29	·9889893	2865	66
2·30	·9892759	2800	65
2·31	·9895559	2736	64
2·32	·9898296	2674	63
2·33	·9900969	2612	62
2·34	·9903581	2552	60
2·35	·9906133	2492	59
2·36	·9908625	2434	58
2·37	·9911060	2377	57
2·38	·9913437	2321	56
2·39	·9915758	2267	55
2·40	·9918025	2213	54
2·41	·9920237	2160	53
2·42	·9922397	2108	52
2·43	·9924506	2058	51
2·44	·9926564	2008	50
2·45	·9928572	1960	49
2·46	·9930531	1912	48
2·47	·9932443	1865	47
2·48	·9934309	1820	46
2·49	·9936128	1775	45
2·50	·9937903		44

$$Z(X) = e^{-\frac{1}{2}X^2}/\sqrt{(2\pi)}, \quad P(X) = 1 - Q(X) = \int_{-\infty}^{X} Z(u)\,du.$$

Table 1 (*continued*)

$Z(X)$	δ −	δ^2 +
·0539910	10717	162
·0529192	10557	161
·0518636	10397	160
·0508239	10238	159
·0498001	10081	157
·0487920	9924	156
·0477996	9769	155
·0468226	9616	154
·0458611	9463	153
·0449148	9312	151
·0439836	9162	150
·0430674	9013	149
·0421661	8866	147
·0412795	8720	146
·0404076	8575	145
·0395500	8432	143
·0387069	8290	142
·0378779	8149	140
·0370629	8010	139
·0362619	7873	138
·0354746	7737	136
·0347009	7602	135
·0339408	7468	133
·0331939	7337	132
·0324603	7206	130
·0317397	7077	129
·0310319	6950	127
·0303370	6824	126
·0296546	6699	125
·0289847	6576	123
·0283270	6455	122
·0276816	6335	120
·0270481	6216	119
·0264265	6099	117
·0258166	5984	116
·0252182	5870	114
·0246313	5757	113
·0240556	5646	111
·0234910	5536	110
·0229374	5428	108
·0223945	5322	107
·0218624	5217	105
·0213407	5113	104
·0208294	5011	102
·0203284	4910	101
·0198374	4811	99
·0193563	4713	98
·0188850	4617	96
·0184233	4522	95
·0179711	4428	93
·0175283		92

X	$P(X)$	δ +	δ^2 −	$Z(X)$	δ −	δ^2 +
2·50	·9937903	1731	44	·0175283	4336	92
2·51	·9939634	1688	43	·0170947	4246	91
2·52	·9941323	1646	42	·0166701	4157	89
2·53	·9942969	1605	41	·0162545	4069	88
2·54	·9944574	1565	40	·0158476	3982	86
2·55	·9946139	1525	39	·0154493	3897	85
2·56	·9947664	1487	39	·0150596	3814	84
2·57	·9949151	1449	38	·0146782	3731	82
2·58	·9950600	1412	37	·0143051	3650	81
2·59	·9952012	1376	36	·0139401	3571	80
2·60	·9953388	1341	35	·0135830	3493	78
2·61	·9954729	1306	35	·0132337	3416	77
2·62	·9956035	1272	34	·0128921	3340	76
2·63	·9957308	1239	33	·0125581	3266	74
2·64	·9958547	1207	32	·0122315	3193	73
2·65	·9959754	1176	32	·0119122	3121	72
2·66	·9960930	1145	31	·0116001	3051	70
2·67	·9962074	1115	30	·0112951	2981	69
2·68	·9963189	1085	29	·0109969	2913	68
2·69	·9964274	1056	29	·0107056	2847	67
2·70	·9965330	1028	28	·0104209	2781	66
2·71	·9966358	1001	27	·0101428	2717	64
2·72	·9967359	974	27	·0098712	2654	63
2·73	·9968333	948	26	·0096058	2592	62
2·74	·9969280	922	26	·0093466	2531	61
2·75	·9970202	897	25	·0090936	2471	60
2·76	·9971099	873	24	·0088465	2413	59
2·77	·9971972	849	24	·0086052	2355	57
2·78	·9972821	825	23	·0083697	2299	56
2·79	·9973646	803	23	·0081398	2244	55
2·80	·9974449	781	22	·0079155	2189	54
2·81	·9975229	759	22	·0076965	2136	53
2·82	·9975988	738	21	·0074829	2084	52
2·83	·9976726	717	21	·0072744	2033	51
2·84	·9977443	697	20	·0070711	1983	50
2·85	·9978140	678	20	·0068728	1934	49
2·86	·9978818	658	19	·0066793	1886	48
2·87	·9979476	640	19	·0064907	1839	47
2·88	·9980116	622	18	·0063067	1793	46
2·89	·9980738	604	18	·0061274	1748	45
2·90	·9981342	587	17	·0059525	1704	44
2·91	·9981929	570	17	·0057821	1661	43
2·92	·9982498	553	16	·0056160	1619	42
2·93	·9983052	537	16	·0054541	1578	41
2·94	·9983589	522	16	·0052963	1537	40
2·95	·9984111	507	15	·0051426	1497	40
2·96	·9984618	492	15	·0049929	1459	39
2·97	·9985110	478	14	·0048470	1421	38
2·98	·9985588	464	14	·0047050	1384	37
2·99	·9986051	450	14	·0045666	1347	36
3·00	·9986501		13	·0044318		35

Note sign of second difference, δ^2.

Table 1. *The Normal probability function (continued)*

X	$P(X)$	δ +	δ^2 −	$Z(X)$	δ −	δ^2 +
3·00	·9986501	437	13	·0044318	1312	35
3·01	·9986938	424	13	·0043007	1277	35
3·02	·9987361	411	13	·0041729	1243	34
3·03	·9987772	399	12	·0040486	1210	33
3·04	·9988171	387	12	·0039276	1178	32
3·05	·9988558	375	12	·0038098	1146	32
3·06	·9988933	364	11	·0036951	1115	31
3·07	·9989297	353	11	·0035836	1085	30
3·08	·9989650	342	11	·0034751	1056	29
3·09	·9989992	332	10	·0033695	1027	29
3·10	·9990324	322	10	·0032668	999	28
3·11	·9990646	312	10	·0031669	971	27
3·12	·9990957	302	10	·0030698	944	27
3·13	·9991260	293	9	·0029754	918	26
3·14	·9991553	284	9	·0028835	893	26
3·15	·9991836	275	9	·0027943	868	25
3·16	·9992112	267	9	·0027075	843	24
3·17	·9992378	258	8	·0026231	820	24
3·18	·9992636	250	8	·0025412	797	23
3·19	·9992886	242	8	·0024615	774	23
3·20	·9993129	235	8	·0023841	752	22
3·21	·9993363	227	7	·0023089	731	21
3·22	·9993590	220	7	·0022358	710	21
3·23	·9993810	213	7	·0021649	689	20
3·24	·9994024	206	7	·0020960	669	20
3·25	·9994230	200	7	·0020290	650	19
3·26	·9994429	193	6	·0019641	631	19
3·27	·9994623	187	6	·0019010	612	18
3·28	·9994810	181	6	·0018397	595	18
3·29	·9994991	175	6	·0017803	577	17
3·30	·9995166	169	6	·0017226	560	17
3·31	·9995335	164	6	·0016666	543	17
3·32	·9995499	159	5	·0016122	527	16
3·33	·9995658	153	5	·0015595	512	16
3·34	·9995811	148	5	·0015084	496	15
3·35	·9995959	143	5	·0014587	481	15
3·36	·9996103	139	5	·0014106	467	15
3·37	·9996242	134	5	·0013639	453	14
3·38	·9996376	130	4	·0013187	439	14
3·39	·9996505	125	4	·0012748	426	13
3·40	·9996631	121	4	·0012322	413	13
3·41	·9996752	117	4	·0011910	400	13
3·42	·9996869	113	4	·0011510	388	12
3·43	·9996982	109	4	·0011122	376	12
3·44	·9997091	106	4	·0010747	364	12
3·45	·9997197	102	4	·0010383	353	11
3·46	·9997299	99	3	·0010030	342	11
3·47	·9997398	95	3	·0009689	331	11
3·48	·9997493	92	3	·0009358	320	10
3·49	·9997585	89	3	·0009037	310	10
3·50	·9997674		3	·0008727		10

X	$P(X)$	δ +	δ^2 −
3·50	·9997674	86	3
3·51	·9997759	83	3
3·52	·9997842	80	3
3·53	·9997922	77	3
3·54	·9997999	74	3
3·55	·9998074	72	3
3·56	·9998146	69	3
3·57	·9998215	67	2
3·58	·9998282	65	2
3·59	·9998347	62	2
3·60	9998409	60	2
3·61	9998469	58	2
3·62	·9998527	56	2
3·63	·9998583	54	2
3·64	·9998637	52	2
3·65	·9998689	50	2
3·66	·9998739	48	2
3·67	·9998787	47	2
3·68	·9998834	45	2
3·69	·9998879	43	2
3·70	·9998922	42	2
3·71	·9998964	40	2
3·72	·9999004	39	1
3·73	·9999043	37	1
3·74	·9999080	36	1
3·75	·9999116	35	1
3·76	·9999150	33	1
3·77	·9999184	32	1
3·78	·9999216	31	1
3·79	·9999247	30	1
3·80	·9999277	29	1
3·81	·9999305	28	1
3·82	·9999333	27	1
3·83	·9999359	26	1
3·84	·9999385	25	1
3·85	·9999409	24	1
3·86	·9999433	23	1
3·87	·9999456	22	1
3·88	·9999478	21	1
3·89	·9999499	20	1
3·90	·9999519	19	1
3·91	·9999539	19	1
3·92	·9999557	18	1
3·93	·9999575	17	1
3·94	·9999593	17	1
3·95	·9999609	16	1
3·96	·9999625	15	1
3·97	·9999641	15	1
3·98	·9999655	14	1
3·99	·9999670	14	1
4·00	·9999683		1

$$Z(X) = e^{-\frac{1}{2}X^2}/\sqrt{(2\pi)}, \quad P(X) = 1 - Q(X) = \int_{-\infty}^{X} Z(u)\, du.$$

Table 1 (*continued*)

Z(X)	δ −	δ² +
·0008727	301	10
·0008426	291	10
·0008135	282	9
·0007853	273	9
·0007581	264	9
·0007317	256	8
·0007061	247	8
·0006814	239	8
·0006575	232	8
·0006343	224	8
·0006119	217	7
·0005902	210	7
·0005693	203	7
·0005490	196	7
·0005294	189	6
·0005105	183	6
·0004921	177	6
·0004744	171	6
·0004573	165	6
·0004408	160	6
·0004248	155	5
·0004093	149	5
·0003944	144	5
·0003800	139	5
·0003661	135	5
·0003526	130	5
·0003396	125	4
·0003271	121	4
·0003149	117	4
·0003032	113	4
·0002919	109	4
·0002810	105	4
·0002705	102	4
·0002604	98	4
·0002506	95	3
·0002411	91	3
·0002320	88	3
·0002232	85	3
·0002147	82	3
·0002065	79	3
·0001987	76	3
·0001910	73	3
·0001837	71	3
·0001766	68	3
·0001698	66	2
·0001633	63	2
·0001569	61	2
·0001508	59	2
·0001449	57	2
·0001393	55	2
·0001338		2

X	P(X)	δ +	δ² −	Z(X)	δ −	δ² +
4·00	·9999683	13	1	·0001338	53	2
4·01	·9999696	13	1	·0001286	51	2
4·02	·9999709	12	0	·0001235	49	2
4·03	·9999721	12		·0001186	47	2
4·04	·9999733	11		·0001140	45	2
4·05	·9999744	11		·0001094	43	2
4·06	·9999755	10		·0001051	42	2
4·07	·9999765	10		·0001009	40	2
4·08	·9999775	9		·0000969	39	2
4·09	·9999784	9		·0000930	37	1
4·10	·9999793	9		·0000893	36	1
4·11	·9999802	8		·0000857	35	1
4·12	·9999811	8		·0000822	33	1
4·13	·9999819	8		·0000789	32	1
4·14	·9999826	7		·0000757	31	1
4·15	·9999834	7		·0000726	30	1
4·16	·9999841	7		·0000697	28	1
4·17	·9999848	7		·0000668	27	1
4·18	·9999854	6		·0000641	26	1
4·19	·9999861	6		·0000615	25	1
4·20	·9999867	6		·0000589	24	1
4·21	·9999872	6		·0000565	23	1
4·22	·9999878	5		·0000542	22	1
4·23	·9999883	5		·0000519	22	1
4·24	·9999888	5		·0000498	21	1
4·25	·9999893	5		·0000477	20	1
4·26	·9999898	4		·0000457	19	1
4·27	·9999902	4		·0000438	18	1
4·28	·9999907	4		·0000420	18	1
4·29	·9999911	4		·0000402	17	1
4·30	·9999915	4		·0000385	16	1
4·31	·9999918	4		·0000369	16	1
4·32	·9999922	3		·0000354	15	1
4·33	·9999925	3		·0000339	14	1
4·34	·9999929	3		·0000324	14	1
4·35	·9999932	3		·0000310	13	1
4·36	·9999935	3		·0000297	13	1
4·37	·9999938	3		·0000284	12	1
4·38	·9999941	3		·0000272	12	0
4·39	·9999943	3		·0000261	11	
4·40	·9999946	2		·0000249	11	
4·41	·9999948	2		·0000239	10	
4·42	·9999951	2		·0000228	10	
4·43	·9999953	2		·0000218	9	
4·44	·9999955	2		·0000209	9	
4·45	·9999957	2		·0000200	9	
4·46	·9999959	2		·0000191	8	
4·47	·9999961	2		·0000183	8	
4·48	·9999963	2		·0000175	8	
4·49	·9999964	2		·0000167	7	
4·50	·9999966			·0000160		

Note sign of second difference, δ^2.

Table 1. *The Normal probability function (continued)*

X	$P(X)$*	$Z(X)$*
4·50	66023	159837
4·51	67586	152797
4·52	69080	146051
4·53	70508	139590
4·54	71873	133401
4·55	73177	127473
4·56	74423	121797
4·57	75614	116362
4·58	76751	111159
4·59	77838	106177
4·60	78875	101409
4·61	79867	96845
4·62	80813	92477
4·63	81717	88297
4·64	82580	84298
4·65	83403	80472
4·66	84190	76812
4·67	84940	73311
4·68	85656	69962
4·69	86340	66760
4·70	86992	63698
4·71	87614	60771
4·72	88208	57972
4·73	88774	55296
4·74	89314	52739
4·75	89829	50295
4·76	90320	47960
4·77	90789	45728
4·78	91235	43596
4·79	91661	41559
4·80	92067	39613
4·81	92453	37755
4·82	92822	35980
4·83	93173	34285
4·84	93508	32667
4·85	93827	31122
4·86	94131	29647
4·87	94420	28239
4·88	94696	26895
4·89	94958	25613
4·90	95208	24390
4·91	95446	23222
4·92	95673	22108
4·93	95889	21046
4·94	96094	20033
4·95	96289	19066
4·96	96475	18144
4·97	96652	17265
4·98	96821	16428
4·99	96981	15629

X	$P(X)$*	$Z(X)$*
5·00	97133	14867
5·01	97278	14141
5·02	97416	13450
5·03	97548	12791
5·04	97672	12162
5·05	97791	11564
5·06	97904	10994
5·07	98011	10451
5·08	98113	9934
5·09	98210	9441
5·10	98302	8972
5·11	98389	8526
5·12	98472	8101
5·13	98551	7696
5·14	98626	7311
5·15	98698	6944
5·16	98765	6595
5·17	98830	6263
5·18	98891	5947
5·19	98949	5647
5·20	99004	5361
5·21	99056	5089
5·22	99105	4831
5·23	99152	4585
5·24	99197	4351
5·25	99240	4128
5·26	99280	3917
5·27	99318	3716
5·28	99354	3525
5·29	99388	3344
5·30	99421	3171
5·31	99452	3007
5·32	99481	2852
5·33	99509	2704
5·34	99535	2563
5·35	99560	2430
5·36	99584	2303
5·37	99606	2183
5·38	99628	2069
5·39	99648	1960
5·40	99667	1857
5·41	99685	1760
5·42	99702	1667
5·43	99718	1579
5·44	99734	1495
5·45	99748	1416
5·46	99762	1341
5·47	99775	1270
5·48	99787	1202
5·49	99799	1138

X	$P(X)$*	$Z(X)$*
5·50	99810	1077
5·51	99821	1019
5·52	99831	965
5·53	99840	913
5·54	99849	864
5·55	99857	817
5·56	99865	773
5·57	99873	731
5·58	99880	691
5·59	99886	654
5·60	99893	618
5·61	99899	585
5·62	99905	553
5·63	99910	522
5·64	99915	494
5·65	99920	467
5·66	99924	441
5·67	99929	417
5·68	99933	394
5·69	99936	372
5·70	99940	351
5·71	99944	332
5·72	99947	313
5·73	99950	296
5·74	99953	280
5·75	99955	264
5·76	99958	249
5·77	99960	235
5·78	99963	222
5·79	99965	210
5·80	99967	198
5·81	99969	187
5·82	99971	176
5·83	99972	166
5·84	99974	157
5·85	99975	148
5·86	99977	139
5·87	99978	131
5·88	99979	124
5·89	99981	117
5·90	99982	110
5·91	99983	104
5·92	99984	98
5·93	99985	92
5·94	99986	87
5·95	99987	82
5·96	99987	77
5·97	99988	73
5·98	99989	68
5·99	99990	65
6·00	99990	61

$$Z(X) = e^{-\frac{1}{2}X^2}/\sqrt{(2\pi)}, \quad P(X) = 1 - Q(X) = \int_{-\infty}^{X} Z(u)\,du.$$

* The entries for $P(X)$ and $Z(X)$ on this page are given to 10 decimal places; thus 0·99999 should be prefixed to each entry for $P(X)$ and a decimal point, followed by four, five, ..., eight zeros, as appropriate, to $Z(X)$.

Table 2. *The Normal probability function. Values of* $-\log Q(X) = -\log\{1-P(X)\}$ *for large values of* X (*extension of Table* 1)

X	$-\log Q(X)$	X	$-\log Q(X)$	X	$-\log Q(X)$
5	6·54265	30	197·30921	50	544·96634
6	9·00586	31	210·56940	60	783·90743
7	11·89285	32	224·26344	70	1066·26576
8	15·20614	33	238·39135	80	1392·04459
9	18·94746	34	252·95315	90	1761·24604
10	23·11805	35	267·94888	100	2173·87154
11	27·71882	36	283·37855	150	4888·38812
12	32·75044	37	299·24218	200	8688·58977
13	38·21345	38	315·53979	250	13574·49960
14	44·10827	39	332·27139	300	19546·12790
15	50·43522	40	349·43701	350	26603·48018
16	57·19458	41	367·03664	400	34746·55970
17	64·38658	42	385·07032	450	43975·36860
18	72·01140	43	403·53804	500	54289·90830
19	80·06919	44	422·43983		
20	88·56010	45	441·77568		
21	97·48422	46	461·54561		
22	106·84167	47	481·74964		
23	116·63253	48	502·38776		
24	126·85686	49	523·45999		
25	137·51475	50	544·96634		
26	148·60624				
27	160·13139				
28	172·09024				
29	184·48283				
30	197·30921				

N.B. To obtain anything but a rough appreciation after $X=50$, the table would require much extension, but for many practical problems it suffices to take after $X=50$:

$$Q(X) = \frac{1}{\sqrt{(2\pi)}}\frac{1}{X}e^{-\frac{1}{2}X^2}$$

Table 3. *The Normal probability function. Values of* X *for extreme values of* Q *and* P (*extension of Table* 4)

Q	0·0000	0·0001	0·0002	0·0003	0·0004	0·0005	0·0006	0·0007	0·0008	0·0009	0·0010	
0·000	∞	3·7190	3·5401	3·4316	3·3528	3·2905	3·2389	3·1947	3·1559	3·1214	3·0902	0·999
·001	3·0902	3·0618	3·0357	3·0115	2·9889	2·9677	2·9478	2·9290	2·9112	2·8943	2·8782	·998
·002	2·8782	2·8627	2·8480	2·8338	2·8202	2·8070	2·7944	2·7822	2·7703	2·7589	2·7478	·997
·003	2·7478	2·7370	2·7266	2·7164	2·7065	2·6968	2·6874	2·6783	2·6693	2·6606	2·6521	·996
·004	2·6521	2·6437	2·6356	2·6276	2·6197	2·6121	2·6045	2·5972	2·5899	2·5828	2·5758	·995
0·005	2·5758	2·5690	2·5622	2·5556	2·5491	2·5427	2·5364	2·5302	2·5241	2·5181	2·5121	0·994
·006	2·5121	2·5063	2·5006	2·4949	2·4893	2·4838	2·4783	2·4730	2·4677	2·4624	2·4573	·993
·007	2·4573	2·4522	2·4471	2·4422	2·4372	2·4324	2·4276	2·4228	2·4181	2·4135	2·4089	·992
·008	2·4089	2·4044	2·3999	2·3954	2·3911	2·3867	2·3824	2·3781	2·3739	2·3698	2·3656	·991
·009	2·3656	2·3615	2·3575	2·3535	2·3495	2·3455	2·3416	2·3378	2·3339	2·3301	2·3263	·990
0·010	2·3263	2·3226	2·3189	2·3152	2·3116	2·3080	2·3044	2·3009	2·2973	2·2938	2·2904	0·989
·011	2·2904	2·2869	2·2835	2·2801	2·2768	2·2734	2·2701	2·2668	2·2636	2·2603	2·2571	·988
·012	2·2571	2·2539	2·2508	2·2476	2·2445	2·2414	2·2383	2·2353	2·2322	2·2292	2·2262	·987
·013	2·2262	2·2232	2·2203	2·2173	2·2144	2·2115	2·2086	2·2058	2·2029	2·2001	2·1973	·986
·014	2·1973	2·1945	2·1917	2·1890	2·1862	2·1835	2·1808	2·1781	2·1754	2·1727	2·1701	·985
0·015	2·1701	2·1675	2·1648	2·1622	2·1596	2·1571	2·1545	2·1520	2·1494	2·1469	2·1444	0·984
·016	2·1444	2·1419	2·1394	2·1370	2·1345	2·1321	2·1297	2·1272	2·1248	2·1224	2·1201	·983
·017	2·1201	2·1177	2·1154	2·1130	2·1107	2·1084	2·1061	2·1038	2·1015	2·0992	2·0969	·982
·018	2·0969	2·0947	2·0924	2·0902	2·0880	2·0858	2·0836	2·0814	2·0792	2·0770	2·0749	·981
·019	2·0749	2·0727	2·0706	2·0684	2·0663	2·0642	2·0621	2·0600	2·0579	2·0558	2·0537	·980
	0·0010	0·0009	0·0008	0·0007	0·0006	0·0005	0·0004	0·0003	0·0002	0·0001	0·0000	P

Table 4. *The Normal probability function. Values of X in terms of Q and P*

Q	·000	·001	·002	·003	·004	·005	·006	·007	·008	·009	·010	
·00	∞	3·0902	2·8782	2·7478	2·6521	2·5758	2·5121	2·4573	2·4089	2·3656	2·3263	·99
·01	2·3263	2·2904	2·2571	2·2262	2·1973	2·1701	2·1444	2·1201	2·0969	2·0749	2·0537	·98
·02	2·0537	2·0335	2·0141	1·9954	1·9774	1·9600	1·9431	1·9268	1·9110	1·8957	1·8808	·97
·03	1·8808	1·8663	1·8522	1·8384	1·8250	1·8119	1·7991	1·7866	1·7744	1·7624	1·7507	·96
·04	1·7507	1·7392	1·7279	1·7169	1·7060	1·6954	1·6849	1·6747	1·6646	1·6546	1·6449	·95
·05	1·6449	1·6352	1·6258	1·6164	1·6072	1·5982	1·5893	1·5805	1·5718	1·5632	1·5548	·94
·06	1·5548	1·5464	1·5382	1·5301	1·5220	1·5141	1·5063	1·4985	1·4909	1·4833	1·4758	·93
·07	1·4758	1·4684	1·4611	1·4538	1·4466	1·4395	1·4325	1·4255	1·4187	1·4118	1·4051	·92
·08	1·4051	1·3984	1·3917	1·3852	1·3787	1·3722	1·3658	1·3595	1·3532	1·3469	1·3408	·91
·09	1·3408	1·3346	1·3285	1·3225	1·3165	1·3106	1·3047	1·2988	1·2930	1·2873	1·2816	·90
·10	1·2816	1·2759	1·2702	1·2646	1·2591	1·2536	1·2481	1·2426	1·2372	1·2319	1·2265	·89
·11	1·2265	1·2212	1·2160	1·2107	1·2055	1·2004	1·1952	1·1901	1·1850	1·1800	1·1750	·88
·12	1·1750	1·1700	1·1650	1·1601	1·1552	1·1503	1·1455	1·1407	1·1359	1·1311	1·1264	·87
·13	1·1264	1·1217	1·1170	1·1123	1·1077	1·1031	1·0985	1·0939	1·0893	1·0848	1·0803	·86
·14	1·0803	1·0758	1·0714	1·0669	1·0625	1·0581	1·0537	1·0494	1·0450	1·0407	1·0364	·85
·15	1·0364	1·0322	1·0279	1·0237	1·0194	1·0152	1·0110	1·0069	1·0027	0·9986	0·9945	·84
·16	0·9945	0·9904	0·9863	0·9822	0·9782	0·9741	0·9701	0·9661	0·9621	0·9581	0·9542	·83
·17	0·9542	0·9502	0·9463	0·9424	0·9385	0·9346	0·9307	0·9269	0·9230	0·9192	0·9154	·82
·18	0·9154	0·9116	0·9078	0·9040	0·9002	0·8965	0·8927	0·8890	0·8853	0·8816	0·8779	·81
·19	0·8779	0·8742	0·8705	0·8669	0·8633	0·8596	0·8560	0·8524	0·8488	0·8452	0·8416	·80
·20	0·8416	0·8381	0·8345	0·8310	0·8274	0·8239	0·8204	0·8169	0·8134	0·8099	0·8064	·79
·21	0·8064	0·8030	0·7995	0·7961	0·7926	0·7892	0·7858	0·7824	0·7790	0·7756	0·7722	·78
·22	0·7722	0·7688	0·7655	0·7621	0·7588	0·7554	0·7521	0·7488	0·7454	0·7421	0·7388	·77
·23	0·7388	0·7356	0·7323	0·7290	0·7257	0·7225	0·7192	0·7160	0·7128	0·7095	0·7063	·76
·24	0·7063	0·7031	0·6999	0·6967	0·6935	0·6903	0·6871	0·6840	0·6808	0·6776	0·6745	·75
·25	0·6745	0·6713	0·6682	0·6651	0·6620	0·6588	0·6557	0·6526	0·6495	0·6464	0·6433	·74
·26	0·6433	0·6403	0·6372	0·6341	0·6311	0·6280	0·6250	0·6219	0·6189	0·6158	0·6128	·73
·27	0·6128	0·6098	0·6068	0·6038	0·6008	0·5978	0·5948	0·5918	0·5888	0·5858	0·5828	·72
·28	0·5828	0·5799	0·5769	0·5740	0·5710	0·5681	0·5651	0·5622	0·5592	0·5563	0·5534	·71
·29	0·5534	0·5505	0·5476	0·5446	0·5417	0·5388	0·5359	0·5330	0·5302	0·5273	0·5244	·70
·30	0·5244	0·5215	0·5187	0·5158	0·5129	0·5101	0·5072	0·5044	0·5015	0·4987	0·4959	·69
·31	0·4959	0·4930	0·4902	0·4874	0·4845	0·4817	0·4789	0·4761	0·4733	0·4705	0·4677	·68
·32	0·4677	0·4649	0·4621	0·4593	0·4565	0·4538	0·4510	0·4482	0·4454	0·4427	0·4399	·67
·33	0·4399	0·4372	0·4344	0·4316	0·4289	0·4261	0·4234	0·4207	0·4179	0·4152	0·4125	·66
·34	0·4125	0·4097	0·4070	0·4043	0·4016	0·3989	0·3961	0·3934	0·3907	0·3880	0·3853	·65
·35	0·3853	0·3826	0·3799	0·3772	0·3745	0·3719	0·3692	0·3665	0·3638	0·3611	0·3585	·64
·36	0·3585	0·3558	0·3531	0·3505	0·3478	0·3451	0·3425	0·3398	0·3372	0·3345	0·3319	·63
·37	0·3319	0·3292	0·3266	0·3239	0·3213	0·3186	0·3160	0·3134	0·3107	0·3081	0·3055	·62
·38	0·3055	0·3029	0·3002	0·2976	0·2950	0·2924	0·2898	0·2871	0·2845	0·2819	0·2793	·61
·39	0·2793	0·2767	0·2741	0·2715	0·2689	0·2663	0·2637	0·2611	0·2585	0·2559	0·2533	·60
·40	0·2533	0·2508	0·2482	0·2456	0·2430	0·2404	0·2378	0·2353	0·2327	0·2301	0·2275	·59
·41	0·2275	0·2250	0·2224	0·2198	0·2173	0·2147	0·2121	0·2096	0·2070	0·2045	0·2019	·58
·42	0·2019	0·1993	0·1968	0·1942	0·1917	0·1891	0·1866	0·1840	0·1815	0·1789	0·1764	·57
·43	0·1764	0·1738	0·1713	0·1687	0·1662	0·1637	0·1611	0·1586	0·1560	0·1535	0·1510	·56
·44	0·1510	0·1484	0·1459	0·1434	0·1408	0·1383	0·1358	0·1332	0·1307	0·1282	0·1257	·55
·45	0·1257	0·1231	0·1206	0·1181	0·1156	0·1130	0·1105	0·1080	0·1055	0·1030	0·1004	·54
·46	0·1004	0·0979	0·0954	0·0929	0·0904	0·0878	0·0853	0·0828	0·0803	0·0778	0·0753	·53
·47	0·0753	0·0728	0·0702	0·0677	0·0652	0·0627	0·0602	0·0577	0·0552	0·0527	0·0502	·52
·48	0·0502	0·0476	0·0451	0·0426	0·0401	0·0376	0·0351	0·0326	0·0301	0·0276	0·0251	·51
·49	0·0251	0·0226	0·0201	0·0175	0·0150	0·0125	0·0100	0·0075	0·0050	0·0025	0·0000	·50
	·010	·009	·008	·007	·006	·005	·004	·003	·002	·001	·000	P

Extreme values:

Q	10^{-4}	10^{-5}	10^{-6}	10^{-7}	10^{-8}	10^{-9}
X	3·7190	4·2649	4·7534	5·1993	5·6120	5·9978

For an extension in the range $Q = 0{\cdot}0000$ (0·0001) 0·0200, see Table 3; also see Table 2.

Table 5. *The Normal probability function. Values of Z in terms of Q and P*

Q	·000	·001	·002	·003	·004	·005	·006	·007	·008	·009	·010	
·00	·00000	·00337	·00634	·00915⁻	·01185⁻	·01446	·01700	·01949	·02192	·02431	·02665⁺	·99
·01	·02665⁺	·02896	·03123	·03348	·03569	·03787	·04003	·04216	·04427	·04635⁺	·04842	·98
·02	·04842	·05046	·05249	·05449	·05648	·05845⁺	·06040	·06233	·06425⁺	·06615⁺	·06804	·97
·03	·06804	·06992	·07177	·07362	·07545⁺	·07727	·07908	·08087	·08265⁻	·08442	·08617	·96
·04	·08617	·08792	·08965⁺	·09137	·09309	·09479	·09648	·09816	·09983	·10149	·10314	·95
·05	·10314	·10478	·10641	·10803	·10964	·11124	·11284	·11442	·11600	·11756	·11912	·94
·06	·11912	·12067	·12222	·12375⁻	·12528	·12679	·12830	·12981	·13130	·13279	·13427	·93
·07	·13427	·13574	·13720	·13866	·14011	·14156	·14299	·14442	·14584	·14726	·14867	·92
·08	·14867	·15007	·15146	·15285⁺	·15423	·15561	·15698	·15834	·15970	·16105⁻	·16239	·91
·09	·16239	·16373	·16506	·16639	·16770	·16902	·17033	·17163	·17292	·17421	·17550	·90
·10	·17550	·17678	·17805⁻	·17932	·18058	·18184	·18309	·18433	·18557	·18681	·18804	·89
·11	·18804	·18926	·19048	·19169	·19290	·19410	·19530	·19649	·19768	·19886	·20004	·88
·12	·20004	·20121	·20238	·20354	·20470	·20585⁺	·20700	·20814	·20928	·21042	·21155⁻	·87
·13	·21155⁻	·21267	·21379	·21490	·21601	·21712	·21822	·21932	·22041	·22149	·22258	·86
·14	·22258	·22365⁺	·22473	·22580	·22686	·22792	·22898	·23003	·23108	·23212	·23316	·85
·15	·23316	·23419	·23522	·23625⁻	·23727	·23829	·23930	·24031	·24131	·24232	·24331	·84
·16	·24331	·24430	·24529	·24627	·24726	·24823	·24921	·25017	·25114	·25210	·25305⁺	·83
·17	·25305⁺	·25401	·25495⁺	·25590	·25684	·25778	·25871	·25964	·26056	·26148	·26240	·82
·18	·26240	·26331	·26422	·26513	·26603	·26693	·26782	·26871	·26960	·27049	·27137	·81
·19	·27137	·27224	·27311	·27398	·27485⁻	·27571	·27657	·27742	·27827	·27912	·27996	·80
·20	·27996	·28080	·28164	·28247	·28330	·28413	·28495⁻	·28577	·28658	·28739	·28820	·79
·21	·28820	·28901	·28981	·29060	·29140	·29219	·29298	·29376	·29454	·29532	·29609	·78
·22	·29609	·29686	·29763	·29840	·29916	·29991	·30067	·30142	·30216	·30291	·30365⁻	·77
·23	·30365⁻	·30439	·30512	·30585⁻	·30658	·30730	·30802	·30874	·30945⁺	·31017	·31087	·76
·24	·31087	·31158	·31228	·31298	·31367	·31436	·31505⁺	·31574	·31642	·31710	·31778	·75
·25	·31778	·31845⁻	·31912	·31979	·32045⁻	·32111	·32177	·32242	·32307	·32372	·32437	·74
·26	·32437	·32501	·32565⁻	·32628	·32691	·32754	·32817	·32879	·32941	·33003	·33065⁻	·73
·27	·33065⁻	·33126	·33187	·33247	·33307	·33367	·33427	·33486	·33545⁻	·33604	·33662	·72
·28	·33662	·33720	·33778	·33836	·33893	·33950	·34007	·34063	·34119	·34175⁻	·34230	·71
·29	·34230	·34286	·34341	·34395⁺	·34449	·34503	·34557	·34611	·34664	·34717	·34769	·70
·30	·34769	·34822	·34874	·34925⁺	·34977	·35028	·35079	·35129	·35180	·35230	·35279	·69
·31	·35279	·35329	·35378	·35427	·35475⁺	·35524	·35572	·35620	·35667	·35714	·35761	·68
·32	·35761	·35808	·35854	·35900	·35946	·35991	·36037	·36082	·36126	·36171	·36215⁻	·67
·33	·36215⁻	·36259	·36302	·36346	·36389	·36431	·36474	·36516	·36558	·36600	·36641	·66
·34	·36641	·36682	·36723	·36764	·36804	·36844	·36884	·36923	·36962	·37001	·37040	·65
·35	·37040	·37078	·37116	·37154	·37192	·37229	·37266	·37303	·37340	·37376	·37412	·64
·36	·37412	·37447	·37483	·37518	·37553	·37588	·37622	·37656	·37690	·37724	·37757	·63
·37	·37757	·37790	·37823	·37855⁺	·37888	·37920	·37951	·37983	·38014	·38045⁻	·38076	·62
·38	·38076	·38106	·38136	·38166	·38196	·38225⁺	·38254	·38283	·38312	·38340	·38368	·61
·39	·38368	·38396	·38423	·38451	·38478	·38504	·38531	·38557	·38583	·38609	·38634	·60
·40	·38634	·38659	·38684	·38709	·38734	·38758	·38782	·38805⁺	·38829	·38852	·38875⁻	·59
·41	·38875⁻	·38897	·38920	·38942	·38964	·38986	·39007	·39028	·39048	·39069	·39089	·58
·42	·39089	·39109	·39129	·39149	·39168	·39187	·39206	·39224	·39243	·39261	·39279	·57
·43	·39279	·39296	·39313	·39330	·39347	·39364	·39380	·39396	·39411	·39427	·39442	·56
·44	·39442	·39457	·39472	·39486	·39501	·39514	·39528	·39542	·39555⁻	·39568	·39580	·55
·45	·39580	·39593	·39605⁺	·39617	·39629	·39640	·39651	·39662	·39673	·39683	·39694	·54
·46	·39694	·39703	·39713	·39723	·39732	·39741	·39749	·39758	·39766	·39774	·39781	·53
·47	·39781	·39789	·39796	·39803	·39809	·39816	·39822	·39828	·39834	·39839	·39844	·52
·48	·39844	·39849	·39854	·39858	·39862	·39866	·39870	·39873	·39876	·39879	·39882	·51
·49	·39882	·39884	·39886	·39888	·39890	·39891	·39892	·39893	·39894	·39894	·39894	·50
	·010	·009	·008	·007	·006	·005	·004	·003	·002	·001	·000	P

Table 6. *The Normal probability function. Table for probit analysis*

Proportion P	Expected probit Y	Maximum working probit $Y+Q/Z$	Minimum working probit $Y-P/Z$	Range $1/Z$	Weighting coefficient Z^2/PQ	Maximum working probit $Y+Q/Z$	Minimum working probit $Y-P/Z$	Expected probit Y
0·500	5·00	6·2533	3·7467	2·5066	0·6366	6·2533	3·7467	5·00
·504	·01	·2534	·7466	·5068	·6366	·2534	·7466	4·99
·508	·02	·2536	·7465	·5071	·6365	·2535	·7464	·98
·512	·03	·2539	·7461	·5078	·6364	·2539	·7461	·97
·516	·04	·2543	·7457	·5086	·6362	·2543	·7457	·96
0·520	5·05	6·2548	3·7450	2·5098	0·6360	6·2550	3·7452	4·95
·524	·06	·2555	·7444	·5111	·6358	·2556	·7445	·94
·528	·07	·2563	·7435	·5128	·6355	·2565	·7437	·93
·532	·08	·2572	·7425	·5147	·6351	·2575	·7428	·92
·536	·09	·2582	·7414	·5168	·6347	·2586	·7418	·91
0·540	5·10	6·2593	3·7401	2·5192	0·6343	6·2599	3·7407	4·90
·544	·11	·2605	·7387	·5218	·6338	·2613	·7395	·89
·548	·12	·2618	·7371	·5247	·6333	·2629	·7382	·88
·552	·13	·2632	·7353	·5279	·6327	·2647	·7368	·87
·556	·14	·2647	·7334	·5313	·6321	·2666	·7353	·86
0·560	5·15	6·2664	3·7314	2·5350	0·6314	6·2686	3·7336	4·85
·564	·16	·2681	·7292	·5389	·6307	·2708	·7319	·84
·567	·17	·2699	·7268	·5431	·6300	·2732	·7301	·83
·571	·18	·2718	·7242	·5476	·6292	·2758	·7282	·82
·575	·19	·2738	·7215	·5523	·6283	·2785	·7262	·81
0·579	5·20	6·2759	3·7186	2·5573	0·6274	6·2814	3·7241	4·80
·583	·21	·2781	·7156	·5625	·6265	·2844	·7219	·79
·587	·22	·2804	·7124	·5680	·6255	·2876	·7196	·78
·591	·23	·2828	·7090	·5738	·6245	·2910	·7172	·77
·595	·24	·2853	·7054	·5799	·6234	·2946	·7147	·76
0·599	5·25	6·2878	3·7016	2·5862	0·6223	6·2984	3·7122	4·75
·603	·26	·2905	·6977	·5928	·6211	·3023	·7095	·74
·606	·27	·2932	·6935	·5997	·6199	·3065	·7068	·73
·610	·28	·2960	·6892	·6068	·6187	·3108	·7040	·72
·614	·29	·2989	·6846	·6143	·6174	·3154	·7011	·71
0·618	5·30	6·3018	3·6798	2·6220	0·6161	6·3202	3·6982	4·70
·622	·31	·3049	·6749	·6300	·6147	·3251	·6951	·69
·626	·32	·3080	·6697	·6383	·6133	·3303	·6920	·68
·629	·33	·3112	·6643	·6469	·6119	·3357	·6888	·67
·633	·34	·3145	·6587	·6558	·6104	·3413	·6855	·66
0·637	5·35	6·3178	3·6528	2·6650	0·6088	6·3472	3·6822	4·65
·641	·36	·3213	·6469	·6744	·6072	·3531	·6787	·64
·644	·37	·3248	·6406	·6842	·6056	·3594	·6752	·63
·648	·38	·3283	·6340	·6943	·6040	·3660	·6717	·62
·652	·39	·3320	·6273	·7047	·6023	·3727	·6680	·61
0·655	5·40	6·3357	3·6203	2·7154	0·6005	6·3797	3·6643	4·60
·659	·41	·3394	·6130	·7264	·5987	·3870	·6606	·59
·663	·42	·3433	·6055	·7378	·5969	·3945	·6567	·58
·666	·43	·3472	·5978	·7494	·5951	·4022	·6528	·57
·670	·44	·3512	·5898	·7614	·5932	·4102	·6488	·56
0·674	5·45	6·3552	3·5815	2·7737	0·5912	6·4185	3·6448	4·55
·677	·46	·3593	·5729	·7864	·5893	·4271	·6407	·54
·681	·47	·3635	·5641	·7994	·5872	·4359	·6365	·53
·684	·48	·3677	·5550	·8127	·5852	·4450	·6323	·52
·688	·49	·3720	·5456	·8264	·5831	·4544	·6280	·51
0·691	5·50	6·3764	3·5360	2·8404	0·5810	6·4640	3·6236	4·50
				$1/Z$ Range	Z^2/PQ Weighting coefficient	$Y+Q/Z$ Maximum working probit	$Y-P/Z$ Minimum working probit	Y Expected probit

$$Z = e^{-\frac{1}{2}X^2}/\sqrt{(2\pi)}$$

$$Y = X + 5$$

$$P = 1 - Q = \int_{-\infty}^{X} Z(u)\,du$$

Table 6 (*continued*)

Proportion P	Expected probit Y	Maximum working probit $Y+Q/Z$	Minimum working probit $Y-P/Z$	Range $1/Z$	Weighting coefficient Z^2/PQ			
0·691	5·50	6·3764	3·5360	2·8404	0·5810	6·4640	3·6236	4·50
·695	·51	·3808	·5260	·8548	·5788	·4740	·6192	·49
·698	·52	·3852	·5157	·8695	·5766	·4843	·6148	·48
·702	·53	·3898	·5052	·8846	·5744	·4948	·6102	·47
·705	·54	·3944	·4943	·9001	·5722	·5057	·6056	·46
0·709	5·55	6·3990	3·4831	2·9159	0·5699	6·5169	3·6010	4·45
·712	·56	·4037	·4715	·9322	·5675	·5285	·5963	·44
·716	·57	·4085	·4597	·9488	·5652	·5403	·5915	·43
·719	·58	·4133	·4475	·9658	·5628	·5525	·5867	·42
·722	·59	·4181	·4349	·9832	·5603	·5651	·5819	·41
0·726	5·60	6·4230	3·4220	3·0010	0·5579	6·5780	3·5770	4·40
·729	·61	·4280	·4088	·0192	·5554	·5912	·5720	·39
·732	·62	·4330	·3952	·0378	·5529	·6048	·5670	·38
·736	·63	·4381	·3812	·0569	·5503	·6188	·5619	·37
·739	·64	·4432	·3669	·0763	·5477	·6331	·5568	·36
0·742	5·65	6·4484	3·3522	3·0962	0·5451	6·6478	3·5516	4·35
·745	·66	·4536	·3370	·1166	·5425	·6630	·5464	·34
·749	·67	·4588	·3214	·1374	·5398	·6786	·5412	·33
·752	·68	·4641	·3055	·1586	·5371	·6945	·5359	·32
·755	·69	·4695	·2892	·1803	·5343	·7108	·5305	·31
0·758	5·70	6·4749	3·2724	3·2025	0·5316	6·7276	3·5251	4·30
·761	·71	·4803	·2551	·2252	·5288	·7449	·5197	·29
·764	·72	·4858	·2375	·2483	·5260	·7625	·5142	·28
·767	·73	·4914	·2194	·2720	·5232	·7806	·5086	·27
·770	·74	·4969	·2008	·2961	·5203	·7992	·5031	·26
0·773	5·75	6·5026	3·1819	3·3207	0·5174	6·8181	3·4974	4·25
·776	·76	·5082	·1623	·3459	·5145	·8377	·4918	·24
·779	·77	·5139	·1423	·3716	·5116	·8577	·4861	·23
·782	·78	·5197	·1219	·3978	·5086	·8781	·4803	·22
·785	·79	·5255	·1009	·4246	·5056	·8991	·4745	·21
0·788	5·80	6·5313	3·0794	3·4519	0·5026	6·9206	3·4687	4·20
·791	·81	·5372	·0574	·4798	·4996	·9426	·4628	·19
·794	·82	·5431	·0348	·5083	·4965	·9652	·4569	·18
·797	·83	·5490	·0116	·5374	·4935	·9884	·4510	·17
·800	·84	·5550	2·9880	·5670	·4904	7·0120	·4450	·16
0·802	5·85	6·5611	2·9638	3·5973	0·4873	7·0362	3·4389	4·15
·805	·86	·5671	·9389	·6282	·4841	·0611	·4329	·14
·808	·87	·5732	·9135	·6597	·4810	·0865	·4268	·13
·811	·88	·5794	·8875	·6919	·4778	·1125	·4206	·12
·813	·89	·5855	·8608	·7247	·4746	·1392	·4145	·11
0·816	5·90	6·5917	2·8335	3·7582	0·4714	7·1665	3·4083	4·10
·819	·91	·5980	·8056	·7924	·4682	·1944	·4020	·09
·821	·92	·6043	·7771	·8272	·4650	·2229	·3957	·08
·824	·93	·6106	·7478	·8628	·4617	·2522	·3894	·07
·826	·94	·6169	·7178	·8991	·4585	·2822	·3831	·06
0·829	5·95	6·6233	2·6872	3·9361	0·4552	7·3128	3·3767	4·05
·831	·96	·6297	·6558	·9739	·4519	·3442	·3703	·04
·834	·97	·6362	·6238	4·0124	·4486	·3762	·3638	·03
·836	·98	·6426	·5909	·0517	·4453	·4091	·3574	·02
·839	·99	·6491	·5573	·0918	·4420	·4427	·3509	·01
0·841	6·00	6·6557	2·5230	4·1327	0·4386	7·4770	3·3443	4·00
				$1/Z$ Range	Z^2/PQ Weighting coefficient	$Y+Q/Z$ Maximum working probit	$Y-P/Z$ Minimum working probit	Y Expected probit

$$Z = e^{-\frac{1}{2}X^2}/\sqrt{(2\pi)}$$
$$Y = X + 5$$
$$P = 1 - Q = \int_{-\infty}^{X} Z(u)\, du$$

Table 6. *Table for probit analysis (continued)*

Proportion P	Expected probit Y	Maximum working probit $Y+Q/Z$	Minimum working probit $Y-P/Z$	Range $1/Z$	Weighting coefficient Z^2/PQ			
0·841	6·00	6·6557	2·5230	4·1327	0·4386	7·4770	3·3443	4·00
·844	·01	·6623	·4878	·1745	·4353	·5122	·3377	3·99
·846	·02	·6689	·4518	·2171	·4319	·5482	·3311	·98
·848	·03	·6755	·4150	·2605	·4285	·5850	·3245	·97
·851	·04	·6822	·3774	·3048	·4252	·6226	·3178	·96
0·853	6·05	6·6888	2·3387	4·3501	0·4218	7·6613	3·3112	3·95
·855	·06	·6956	·2994	·3962	·4184	·7006	·3044	·94
·858	·07	·7023	·2590	·4433	·4150	·7410	·2977	·93
·860	·08	·7091	·2178	·4913	·4116	·7822	·2909	·92
·862	·09	·7159	·1756	·5403	·4082	·8244	·2841	·91
0·864	6·10	6·7227	2·1324	4·5903	0·4047	7·8676	3·2773	3·90
·867	·11	·7296	·0883	·6413	·4013	·9117	·2704	·89
·869	·12	·7365	·0432	·6933	·3979	·9568	·2635	·88
·871	·13	·7434	1·9970	·7464	·3944	8·0030	·2566	·87
·873	·14	·7504	·9498	·8006	·3910	·0502	·2496	·86
0·875	6·15	6·7573	1·9014	4·8559	0·3876	8·0986	3·2427	3·85
·877	·16	·7643	·8520	·9123	·3841	·1480	·2357	·84
·879	·17	·7714	·8016	·9698	·3807	·1984	·2286	·83
·881	·18	·7784	·7498	5·0286	·3772	·2502	·2216	·82
·883	·19	·7855	·6970	·0885	·3738	·3030	·2145	·81
0·885	6·20	6·7926	1·6429	5·1497	0·3703	8·3571	3·2074	3·80
·887	·21	·7997	·5876	·2121	·3669	·4124	·2003	·79
·889	·22	·8068	·5310	·2758	·3634	·4690	·1932	·78
·891	·23	·8140	·4731	·3409	·3600	·5269	·1860	·77
·893	·24	·8212	·4140	·4072	·3565	·5860	·1788	·76
0·894	6·25	6·8284	1·3534	5·4750	0·3531	8·6466	3·1716	3·75
·896	·26	·8357	·2916	·5441	·3496	·7084	·1643	·74
·898	·27	·8429	·2282	·6147	·3462	·7718	·1571	·73
·900	·28	·8502	·1635	·6867	·3428	·8365	·1498	·72
·901	·29	·8575	·0972	·7603	·3393	·9028	·1425	·71
0·903	6·30	6·8649	1·0295	5·8354	0·3359	8·9705	3·1351	3·70
·905	·31	·8722	0·9602	·9120	·3325	9·0398	·1278	·69
·907	·32	·8796	·8893	·9903	·3291	·1107	·1204	·68
·908	·33	·8870	·8168	6·0702	·3256	·1832	·1130	·67
·910	·34	·8944	·7426	·1518	·3222	·2574	·1056	·66
0·911	6·35	6·9019	0·6668	6·2351	0·3188	9·3332	3·0981	3·65
·913	·36	·9093	·5892	·3201	·3155	·4108	·0907	·64
·915	·37	·9168	·5098	·4070	·3121	·4902	·0832	·63
·916	·38	·9243	·4286	·4957	·3087	·5714	·0757	·62
·918	·39	·9318	·3455	·5863	·3053	·6545	·0682	·61
0·919	6·40	6·9394	0·2606	6·6788	0·3020	9·7394	3·0606	3·60
·921	·41	·9469	·1736	·7733	·2986	·8264	·0531	·59
·922	·42	·9545	·0847	·8698	·2953	·9153	·0455	·58
·924	·43	·9621		·9684	·2920		·0379	·57
·925	·44	·9697		7·0691	·2887		·0303	·56
0·926	6·45	6·9774		7·1720	0·2854		3·0226	3·55
·928	·46	·9850		·2771	·2821		·0150	·54
·929	·47	6·9927		·3845	·2788		·0073	·53
·931	·48	7·0004		·4943	·2756		2·9996	·52
·932	·49	·0081		·6064	·2723		·9919	·51
0·933	6·50	7·0158		7·7210	0·2691		2·9842	3·50
				$1/Z$ Range	Z^2/PQ Weighting coefficient	$Y+Q/Z$ Maximum working probit	$Y-P/Z$ Minimum working probit	Y Expected probit

$Z = e^{-\frac{1}{2}X^2}/\sqrt{(2\pi)}$

$Y = X + 5$

$P = 1 - Q = \int_{-\infty}^{X} Z(u)\,du$

Table 6 (*continued*)

Proportion P	Expected probit Y	Maximum working probit $Y+Q/Z$	Range $1/Z$	Weighting coefficient Z^2/PQ		
0·9332	6·50	7·0158	7·7210	0·2691	2·9842	3·50
345	·51	·0236	·8380	·2658	·9764	·49
357	·52	·0313	·9577	·2626	·9687	·48
370	·53	·0391	8·0800	·2594	·9609	·47
382	·54	·0469	·2050	·2563	·9531	·46
0·9394	6·55	7·0547	8·3327	0·2531	2·9453	3·45
406	·56	·0625	·4633	·2500	·9375	·44
418	·57	·0704	·5968	·2468	·9296	·43
429	·58	·0783	·7333	·2437	·9217	·42
441	·59	·0861	·8728	·2406	·9139	·41
0·9452	6·60	7·0940	9·0154	0·2375	2·9060	3·40
463	·61	·1020	·1613	·2345	·8980	·39
474	·62	·1099	·3105	·2314	·8901	·38
484	·63	·1178	·4630	·2284	·8822	·37
495	·64	·1258	·6190	·2254	·8742	·36
0·9505	6·65	7·1338	9·7785	0·2224	2·8662	3·35
515	·66	·1417	·9417	·2194	·8583	·34
525	·67	·1498	10·1086	·2165	·8502	·33
535	·68	·1578	10·2794	·2135	·8422	·32
545	·69	·1658	10·4540	·2106	·8342	·31
0·9554	6·70	7·1739	10·6327	0·2077	2·8261	3·30
564	·71	·1819	10·8156	·2049	·8181	·29
573	·72	·1900	11·0027	·2020	·8100	·28
582	·73	·1981	11·1941	·1992	·8019	·27
591	·74	·2062	11·3900	·1964	·7938	·26
0·9599	6·75	7·2143	11·5905	0·1936	2·7857	3·25
608	·76	·2224	11·7957	·1908	·7776	·24
616	·77	·2306	12·0058	·1881	·7694	·23
625	·78	·2387	12·2208	·1853	·7613	·22
633	·79	·2469	12·4409	·1826	·7531	·21
0·9641	6·80	7·2551	12·6662	0·1799	2·7449	3·20
649	·81	·2633	12·8969	·1773	·7367	·19
656	·82	·2715	13·1331	·1746	·7285	·18
664	·83	·2797	13·3750	·1720	·7203	·17
671	·84	·2880	13·6227	·1694	·7120	·16
0·9678	6·85	7·2962	13·8764	0·1669	2·7038	3·15
686	·86	·3045	14·1362	·1643	·6955	·14
693	·87	·3128	14·4023	·1618	·6872	·13
699	·88	·3210	14·6749	·1593	·6790	·12
706	·89	·3293	14·9541	·1568	·6707	·11
0·9713	6·90	7·3376	15·2402	0·1544	2·6624	3·10
719	·91	·3460	15·5333	·1519	·6540	·09
726	·92	·3543	15·8337	·1495	·6457	·08
732	·93	·3626	16·1414	·1471	·6374	·07
738	·94	·3710	16·4568	·1448	·6290	·06
0·9744	6·95	7·3794	16·7800	0·1424	2·6206	3·05
750	·96	·3877	17·1113	·1401	·6123	·04
756	·97	·3961	17·4509	·1378	·6039	·03
761	·98	·4045	17·7989	·1356	·5955	·02
767	·99	·4129	18·1558	·1333	·5871	·01
0·9772	7·00	7·4214	18·5216	0·1311	2·5786	3·00
			$1/Z$ Range	Z^2/PQ Weighting coefficient	$Y-P/Z$ Minimum working probit	Y Expected probit

$Z = e^{-\frac{1}{2}X^2}/\sqrt{(2\pi)}$

$Y = X + 5$

$P = 1 - Q = \int_{-\infty}^{X} Z(u)\, du$

Table 6. *Table for probit analysis (continued)*

Proportion P	Expected probit Y	Maximum working probit $Y+Q/Z$	Range $1/Z$	Weighting coefficient Z^2/PQ		
0·9772	7·00	7·4214	18·5216	0·1311	2·5786	3·00
778	·01	·4298	18·8967	·1289	·5702	2·99
783	·02	·4382	19·2814	·1268	·5618	·98
788	·03	·4467	19·6758	·1246	·5533	·97
793	·04	·4552	20·0803	·1225	·5448	·96
0·9798	7·05	7·4636	20·4952	0·1204	2·5364	2·95
803	·06	·4721	20·9207	·1183	·5279	·94
808	·07	·4806	21·3572	·1163	·5194	·93
812	·08	·4891	21·8050	·1142	·5109	·92
817	·09	·4976	22·2644	·1122	·5024	·91
0·9821	7·10	7·5062	22·7357	0·1103	2·4938	2·90
826	·11	·5147	23·2194	·1083	·4853	·89
830	·12	·5232	23·7157	·1064	·4768	·88
834	·13	·5318	24·2251	·1045	·4682	·87
838	·14	·5404	24·7478	·1026	·4596	·86
0·9842	7·15	7·5489	25·2844	0·1007	2·4511	2·85
846	·16	·5575	25·8352	·0989	·4425	·84
850	·17	·5661	26·4006	·0971	·4339	·83
854	·18	·5747	26·9812	·0953	·4253	·82
857	·19	·5833	27·5772	·0935	·4167	·81
0·9861	7·20	7·5919	28·1892	0·0918	2·4081	2·80
864	·21	·6006	28·8177	·0901	·3994	·79
868	·22	·6092	29·4631	·0884	·3908	·78
871	·23	·6178	30·1260	·0867	·3822	·77
875	·24	·6265	30·8069	·0851	·3735	·76
0·9878	7·25	7·6351	31·5063	0·0834	2·3649	2·75
881	·26	·6438	32·2249	·0818	·3562	·74
884	·27	·6525	32·9631	·0802	·3475	·73
887	·28	·6612	33·7216	·0787	·3388	·72
890	·29	·6699	34·5010	·0771	·3301	·71
0·9893	7·30	7·6786	35·3020	0·0756	2·3214	2·70
896	·31	·6873	36·1251	·0741	·3127	·69
898	·32	·6960	36·9712	·0727	·3040	·68
901	·33	·7047	37·8408	·0712	·2953	·67
904	·34	·7135	38·7348	·0698	·2865	·66
0·9906	7·35	7·7222	39·6539	0·0684	2·2778	2·65
909	·36	·7310	40·5988	·0671	·2690	·64
911	·37	·7397	41·5704	·0656	·2603	·63
913	·38	·7485	42·5695	·0643	·2515	·62
916	·39	·7573	43·5970	·0630	·2427	·61
0·9918	7·40	7·7661	44·6538	0·0617	2·2339	2·60
920	·41	·7748	45·7407	·0604	·2252	·59
922	·42	·7836	46·8588	·0591	·2164	·58
925	·43	·7924	48·0090	·0579	·2076	·57
927	·44	·8013	49·1924	·0567	·1987	·56
0·9929	7·45	7·8101	50·4099	0·0555	2·1899	2·55
931	·46	·8189	51·6628	·0543	·1811	·54
932	·47	·8277	52·9521	·0532	·1723	·53
934	·48	·8366	54·2791	·0520	·1634	·52
936	·49	·8454	55·6448	·0509	·1546	·51
0·9938	7·50	7·8543	57·0506	0·0498	2·1457	2·50
$Z=e^{-\frac{1}{2}X^2}/\sqrt{(2\pi)}$ $Y=X+5$ $P=1-Q=\int_{-\infty}^{X} Z(u)\,du$			$1/Z$ Range	Z^2/PQ Weighting coefficient	$Y-P/Z$ Minimum working probit	Y Expected probit

Table 6 (*continued*)

Proportion P	Expected probit Y	Maximum working probit $Y+Q/Z$	Range $1/Z$	Weighting coefficient Z^2/PQ		
0·99379	7·50	7·8543	57·0506	0·0498	2·1457	2·50
396	·51	·8631	58·4978	·0487	·1369	·49
413	·52	·8720	59·9876	·0476	·1280	·48
430	·53	·8809	61·5216	·0466	·1191	·47
446	·54	·8897	63·1011	·0456	·1103	·46
0·99461	7·55	7·8986	64·7277	0·0446	2·1014	2·45
477	·56	·9075	66·4028	·0436	·0925	·44
492	·57	·9164	68·1280	·0426	·0836	·43
506	·58	·9253	69·9051	·0416	·0747	·42
520	·59	·9342	71·7357	·0407	·0658	·41
0·99534	7·60	7·9432	73·6216	0·0398	2·0568	2·40
547	·61	·9521	75·5646	·0389	·0479	·39
560	·62	·9610	77·5667	·0380	·0390	·38
573	·63	·9700	79·6298	·0371	·0300	·37
585	·64	·9789	81·7559	·0362	·0211	·36
0·99598	7·65	7·9879	83·9472	0·0354	2·0121	2·35
609	·66	·9968	86·2059	·0346	·0032	·34
621	·67	8·0058	88·5342	·0338	1·9942	·33
632	·68	·0147	90·9344	·0330	·9853	·32
643	·69	·0237	93·4091	·0322	·9763	·31
0·99653	7·70	8·0327	95·9607	0·0314	1·9673	2·30
664	·71	·0417	98·5918	·0307	·9583	·29
674	·72	·0507	101·3053	·0300	·9493	·28
683	·73	·0597	104·1038	·0292	·9403	·27
693	·74	·0687	106·9903	·0285	·9313	·26
0·99702	7·75	8·0777	109·9679	0·0278	1·9223	2·25
711	·76	·0867	113·0396	·0272	·9133	·24
720	·77	·0957	116·2088	·0265	·9043	·23
728	·78	·1047	119·4788	·0258	·8953	·22
736	·79	·1138	122·8530	·0252	·8862	·21
0·99744	7·80	8·1228	126·3352	0·0246	1·8772	2·20
752	·81	·1318	129·9290	·0240	·8682	·19
760	·82	·1409	133·6385	·0234	·8591	·18
767	·83	·1499	137·4676	·0228	·8501	·17
774	·84	·1590	141·4206	·0222	·8410	·16
0·99781	7·85	8·1681	145·5018	0·0217	1·8319	2·15
788	·86	·1771	149·7158	·0211	·8229	·14
795	·87	·1862	154·0671	·0206	·8138	·13
801	·88	·1953	158·5609	·0200	·8047	·12
807	·89	·2044	163·2020	·0195	·7956	·11
0·99813	7·90	8·2134	167·9957	0·0190	1·7866	2·10
819	·91	·2225	172·9476	·0185	·7775	·09
825	·92	·2316	178·0632	·0181	·7684	·08
831	·93	·2407	183·3485	·0176	·7593	·07
836	·94	·2498	188·8095	·0171	·7502	·06
0·99841	7·95	8·2590	194·4526	0·0167	1·7410	2·05
846	·96	·2681	200·2844	·0162	·7319	·04
851	·97	·2772	206·3118	·0158	·7228	·03
856	·98	·2863	212·5418	·0154	·7137	·02
861	·99	·2955	218·9818	·0150	·7045	·01
0·99865	8·00	8·3046	225·6395	0·0146	1·6954	2·00
			$1/Z$ Range	Z^2/PQ Weighting coefficient	$Y-P/Z$ Minimum working probit	Y Expected probit

$Z=e^{-\frac{1}{2}X^2}/\sqrt{(2\pi)}$

$Y=X+5$

$P=1-Q=\int_{-\infty}^{X} Z(u)\,du$

Table 6. *Table for probit analysis (continued)*

Proportion P	Expected probit Y	Maximum working probit $Y+Q/Z$	Range $1/Z$	Weighting coefficient Z^2/PQ		
0·99865	8·00	8·3046	225·6395	0·0146	1·6954	2·00
869	·01	·3137	232·5229	·0142	·6863	1·99
874	·02	·3229	239·6402	·0138	·6771	·98
878	·03	·3320	247·0000	·0134	·6680	·97
882	·04	·3412	254·6114	·0131	·6588	·96
0·99886	8·05	8·3503	262·4836	0·0127	1·6497	1·95
889	·06	·3595	270·6262	·0124	·6405	·94
893	·07	·3687	279·0493	·0120	·6313	·93
896	·08	·3778	287·7634	·0117	·6222	·92
900	·09	·3870	296·7792	·0114	·6130	·91
0·99903	8·10	8·3962	306·1082	0·0110	1·6038	1·90
906	·11	·4054	315·7619	·0107	·5946	·89
910	·12	·4146	325·7527	·0104	·5854	·88
913	·13	·4238	336·0932	·0101	·5762	·87
916	·14	·4330	346·7966	·0099	·5670	·86
0·99918	8·15	8·4422	357·8732	0·0096	1·5578	1·85
921	·16	·4514	369·3477	·0093	·5486	·84
924	·17	·4606	381·2245	·0090	·5394	·83
926	·18	·4698	393·5226	·0088	·5302	·82
929	·19	·4790	406·2580	·0085	·5210	·81
0·99931	8·20	8·4882	419·4476	0·0083	1·5118	1·80
934	·21	·4974	433·1086	·0080	·5026	·79
936	·22	·5067	447·2593	·0078	·4933	·78
938	·23	·5159	461·9185	·0076	·4841	·77
940	·24	·5251	477·1059	·0074	·4749	·76
0·99942	8·25	8·5344	492·8419	0·0071	1·4656	1·75
944	·26	·5436	509·1479	·0069	·4564	·74
946	·27	·5529	526·0459	·0067	·4471	·73
948	·28	·5621	543·5592	·0065	·4379	·72
950	·29	·5714	561·7116	·0063	·4286	·71
0·99952	8·30	8·5806	580·5283	0·0061	1·4194	1·70
953	·31	·5899	600·0353	·0060	·4101	·69
955	·32	·5992	620·2599	·0058	·4008	·68
957	·33	·6084	641·2302	·0056	·3916	·67
958	·34	·6177	662·9758	·0054	·3823	·66
0·99960	8·35	8·6270	685·5274	0·0053	1·3730	1·65
961	·36	·6363	708·9171	·0051	·3637	·64
962	·37	·6456	733·1780	·0050	·3544	·63
964	·38	·6548	758·3451	·0048	·3452	·62
965	·39	·6641	784·4545	·0047	·3359	·61
0·99966	8·40	8·6734	811·5439	0·0045	1·3266	1·60
968	·41	·6827	839·6528	·0044	·3173	·59
969	·42	·6920	868·8222	·0042	·3080	·58
970	·43	·7013	899·0948	·0041	·2987	·57
971	·44	·7106	930·5153	·0040	·2894	·56
0·99972	8·45	8·7200	963·1301	0·0038	1·2800	1·55
973	·46	·7293	996·9878	·0037	·2707	·54
974	·47	·7386	1032·1389	·0036	·2614	·53
975	·48	·7479	1068·6362	·0035	·2521	·52
976	·49	·7572	1106·5347	·0034	·2428	·51
0·99977	8·50	8·7666	1145·8919	0·0033	1·2334	1·50
			$1/Z$ Range	Z^2/PQ Weighting coefficient	$Y-P/Z$ Minimum working probit	Y Expected probit

$$Z = e^{-\frac{1}{2}X^2}/\sqrt{(2\pi)}$$
$$Y = X + 5$$
$$P = 1 - Q = \int_{-\infty}^{X} Z(u)\,du$$

Table 6 (*continued*)

Proportion P	Expected probit Y	Maximum working probit $Y+Q/Z$	Range $1/Z$	Weighting coefficient Z^2/PQ		
0·999767	8·50	8·7666	1145·8919	0·0033	1·2334	**1·50**
776	·51	·7759	1186·7675	·0032	·2241	**·49**
784	·52	·7852	1229·2242	·0031	·2148	**·48**
792	·53	·7946	1273·3271	·0030	·2054	**·47**
800	·54	·8039	1319·1443	·0029	·1961	**·46**
0·999807	8·55	8·8133	1366·7467	0·0028	1·1867	**1·45**
815	·56	·8226	1416·2085	·0027	·1774	**·44**
822	·57	·8320	1467·6071	·0026	·1680	**·43**
828	·58	·8413	1521·0232	·0025	·1587	**·42**
835	·59	·8507	1576·5411	·0024	·1493	**·41**
0·999841	8·60	8·8600	1634·2488	0·0024	1·1400	**1·40**
847	·61	·8694	1694·2383	·0023	·1306	**·39**
853	·62	·8788	1756·6055	·0022	·1212	**·38**
858	·63	·8881	1821·4507	·0021	·1119	**·37**
864	·64	·8975	1888·8785	·0021	·1025	**·36**
0·999869	8·65	8·9069	1958·9983	0·0020	1·0931	**1·35**
874	·66	·9162	2031·9243	·0019	·0838	**·34**
879	·67	·9256	2107·7758	·0019	·0744	**·33**
883	·68	·9350	2186·6775	·0018	·0650	**·32**
888	·69	·9444	2268·7596	·0017	·0556	**·31**
0·999892	8·70	8·9538	2354·1583	0·0017	1·0462	**1·30**
896	·71	·9632	2443·0158	·0016	·0368	**·29**
900	·72	·9726	2535·4807	·0016	·0274	**·28**
904	·73	·9820	2631·7085	·0015	·0180	**·27**
908	·74	·9914	2731·8615	·0015	·0086	**·26**
0·999912	8·75	9·0008	2836·1096	0·0014	0·9992	**1·25**
915	·76	·0102	2944·6302	·0014	·9898	**·24**
918	·77	·0196	3057·6091	·0013	·9804	**·23**
922	·78	·0290	3175·2401	·0013	·9710	**·22**
925	·79	·0384	3297·7264	·0012	·9616	**·21**
0·999928	8·80	9·0478	3425·2801	0·0012	0·9522	**1·20**
931	·81	·0572	3558·1233	·0011	·9428	**·19**
933	·82	·0667	3696·4883	·0011	·9333	**·18**
936	·83	·0761	3840·6179	·0011	·9239	**·17**
938	·84	·0855	3990·7662	·0010	·9145	**·16**
0·999941	8·85	9·0949	4147·1994	0·0010	0·9051	**1·15**
943	·86	·1044	4310·1955	·0010	·8956	**·14**
946	·87	·1138	4480·0457	·0009	·8862	**·13**
948	·88	·1232	4657·0549	·0009	·8768	**·12**
950	·89	·1327	4841·5419	·0009	·8673	**·11**
0·999952	8·90	9·1421	5033·8407	0·0008	0·8579	**1·10**
954	·91	·1516	5234·3007	·0008	·8484	**·09**
956	·92	·1610	5443·2878	·0008	·8390	**·08**
958	·93	·1704	5661·1851	·0007	·8296	**·07**
959	·94	·1799	5888·3938	·0007	·8201	**·06**
0·999961	8·95	9·1894	6125·3338	0·0007	0·8106	**1·05**
963	·96	·1988	6372·4452	·0007	·8012	**·04**
964	·97	·2083	6630·1886	·0006	·7917	**·03**
966	·98	·2177	6899·0468	·0006	·7823	**·02**
967	·99	·2272	7179·5252	·0006	·7728	**·01**
0·999968	9·00	9·2367	7472·1536	0·0006	0·7633	**1·00**
			$1/Z$ Range	**Z^2/PQ Weighting coefficient**	**$Y-P/Z$ Minimum working probit**	**Y Expected probit**

$Z=e^{-\frac{1}{2}X^2}/\sqrt{(2\pi)}$
$Y=X+5$
$P=1-Q=\int_{-\infty}^{X} Z(u)\,du$

Table 7. *Probability integral of the χ^2-distribution and the cumulative sum of the Poisson distribution*

ν	χ^2=0·001 m=0·0005	0·002 0·0010	0·003 0·0015	0·004 0·0020	0·005 0·0025	0·006 0·0030	0·007 0·0035	0·008 0·0040	0·009 0·0045	0·010 0·0050	c
1	0·97477	0·96433	0·95632	0·94957	0·94363	0·93826	0·93332	0·92873	0·92442	0·92034	
2	·99950	·99900	·99850	·99800	·99750	·99700	·99651	·99601	·99551	·99501	1
3	·99999	·99998	·99996	·99993	·99991	·99988	·99984	·99981	·99977	·99973	
4							·99999	·99999	·99999	·99999	2

ν	χ^2=10·5 m= 5·25	11·0 5·5	11·5 5·75	12·0 6·0	12·5 6·25	13·0 6·5	13·5 6·75	14·0 7·0	14·5 7·25	15·0 7·5	c
1	0·00119	0·00091	0·00070	0·00053	0·00041	0·00031	0·00024	0·00018	0·00014	0·00011	
2	·00525	·00409	·00318	·00248	·00193	·00150	·00117	·00091	·00071	·00055	1
3	·01476	·01173	·00931	·00738	·00585	·00464	·00367	·00291	·00230	·00182	
4	·03280	·02656	·02148	·01735	·01400	·01128	·00907	·00730	·00586	·00470	2
5	·06225	·05138	·04232	·03479	·02854	·02338	·01912	·01561	·01273	·01036	
6	·10511	·08838	·07410	·06197	·05170	·04304	·03575	·02964	·02452	·02026	3
7	0·16196	0·13862	0·11825	0·10056	0·08527	0·07211	0·06082	0·05118	0·04297	0·03600	
8	·23167	·20170	·17495	·15120	·13025	·11185	·09577	·08177	·06963	·05915	4
9	·31154	·27571	·24299	·21331	·18657	·16261	·14126	·12233	·10562	·09094	
10	·39777	·35752	·31991	·28506	·25299	·22367	·19704	·17299	·15138	·13206	5
11	·48605	·44326	·40237	·36364	·32726	·29333	·26190	·23299	·20655	·18250	
12	·57218	·52892	·48662	·44568	·40640	·36904	·33377	·30071	·26992	·24144	6
13	0·65263	0·61082	0·56901	0·52764	0·48713	0·44781	0·40997	0·37384	0·33960	0·30735	
14	·72479	·68604	·64639	·60630	·56622	·52652	·48759	·44971	·41316	·37815	7
15	·78717	·75259	·71641	·67903	·64086	·60230	·56374	·52553	·48800	·45142	
16	·83925	·80949	·77762	·74398	·70890	·67276	·63591	·59871	·56152	·52464	8
17	·88135	·85656	·82942	·80014	·76896	·73619	·70212	·66710	·63145	·59548	
18	·91436	·89436	·87195	·84724	·82038	·79157	·76106	·72909	·69596	·66197	9
19	0·93952	0·92384	0·90587	0·88562	0·86316	0·83857	0·81202	0·78369	0·75380	0·72260	
20	·95817	·94622	·93221	·91608	·89779	·87738	·85492	·83050	·80427	·77641	10
21	·97166	·96279	·95214	·93962	·92513	·90862	·89010	·86960	·84718	·82295	
22	·98118	·97475	·96686	·95738	·94618	·93316	·91827	·90148	·88279	·86224	11
23	·98773	·98319	·97748	·97047	·96201	·95199	·94030	·92687	·91165	·89463	
24	·99216	·98901	·98498	·97991	·97367	·96612	·95715	·94665	·93454	·92076	12
25	0·99507	0·99295	0·99015	0·98657	0·98206	0·97650	0·96976	0·96173	0·95230	0·94138	
26	·99696	·99555	·99366	·99117	·98798	·98397	·97902	·97300	·96581	·95733	13
27	·99815	·99724	·99598	·99429	·99208	·98925	·98567	·98125	·97588	·96943	
28	·99890	·99831	·99749	·99637	·99487	·99290	·99037	·98719	·98324	·97844	14
29	·99935	·99899	·99846	·99773	·99672	·99538	·99363	·99138	·98854	·98502	
30	0·99963	0·99940	0·99907	0·99860	0·99794	0·99704	0·99585	0·99428	0·99227	0·98974	15
32	·99988	·99980	·99968	·99949	·99922	·99884	·99831	·99759	·99664	·99539	16
34	·99996	·99994	·99989	·99983	·99972	·99957	·99935	·99904	·99862	·99804	17
36	·99999	·99998	·99997	·99994	·99991	·99985	·99976	·99964	·99946	·99921	18
38			·99999	·99998	·99997	·99995	·99992	·99987	·99980	·99970	19
40					0·99999	0·99998	0·99997	0·99996	0·99993	0·99989	20
42							·99999	·99999	·99998	·99996	21
44									·99999	·99999	22

The quantity tabled is

$$Q(\chi^2 \mid \nu) = 1 - P(\chi^2 \mid \nu) = 2^{-\frac{1}{2}\nu} \{\Gamma(\tfrac{1}{2}\nu)\}^{-1} \int_{\chi^2}^{\infty} e^{-\frac{1}{2}x} x^{\frac{1}{2}\nu - 1}\, dx,$$

which is equal to $\sum_{j=0}^{c-1} e^{-m} m^j / j!$, if ν is even and $c = \frac{1}{2}\nu$, $m = \frac{1}{2}\chi^2$.

Table 7 (*continued*)

ν	$\chi^2=0{\cdot}01$ m = 0·005	0·02 0·010	0·03 0·015	0·04 0·020	0·05 0·025	0·06 0·030	0·07 0·035	0·08 0·040	0·09 0·045	0·10 0·050	c
1	0·92034	0·88754	0·86249	0·84148	0·82306	0·80650	0·79134	0·77730	0·76418	0·75183	
2	·99501	·99005	·98511	·98020	·97531	·97045	·96561	·96079	·95600	·95123	1
3	·99973	·99925	·99863	·99790	·99707	·99616	·99518	·99412	·99301	·99184	
4	·99999	·99995	·99989	·99980	·99969	·99956	·99940	·99922	·99902	·99879	2
5			·99999	·99998	·99997	·99995	·99993	·99991	·99987	·99984	
6							·99999	·99999	·99999	·99998	3

ν	$\chi^2=15{\cdot}5$ m = 7·75	16·0 8·0	16·5 8·25	17·0 8·5	17·5 8·75	18·0 9·0	18·5 9·25	19·0 9·5	19·5 9·75	20·0 10·0	c
1	0·00008	0·00006	0·00005	0·00004	0·00003	0·00002	0·00002	0·00001	0·00001	0·00001	
2	·00043	·00034	·00026	·00020	·00016	·00012	·00010	·00008	·00006	·00005	1
3	·00144	·00113	·00090	·00071	·00056	·00044	·00035	·00027	·00022	·00017	
4	·00377	·00302	·00242	·00193	·00154	·00123	·00099	·00079	·00063	·00050	2
5	·00843	·00684	·00555	·00450	·00364	·00295	·00238	·00192	·00155	·00125	
6	·01670	·01375	·01131	·00928	·00761	·00623	·00510	·00416	·00340	·00277	3
7	0·03010	0·02512	0·02092	0·01740	0·01444	0·01197	0·00991	0·00819	0·00676	0·00557	
8	·05012	·04238	·03576	·03011	·02530	·02123	·01777	·01486	·01240	·01034	4
9	·07809	·06688	·05715	·04872	·04144	·03517	·02980	·02519	·02126	·01791	
10	·11487	·09963	·08619	·07436	·06401	·05496	·04709	·04026	·03435	·02925	5
11	·16073	·14113	·12356	·10788	·09393	·08158	·07068	·06109	·05269	·04534	
12	·21522	·19124	·16939	·14960	·13174	·11569	·10133	·08853	·07716	·06709	6
13	0·27719	0·24913	0·22318	0·19930	0·17744	0·15752	0·13944	0·12310	0·10840	0·09521	
14	·34485	·31337	·28380	·25618	·23051	·20678	·18495	·16495	·14671	·13014	7
15	·41604	·38205	·34962	·31886	·28986	·26267	·23729	·21373	·19196	·17193	
16	·48837	·45296	·41864	·38560	·35398	·32390	·29544	·26866	·24359	·22022	8
17	·55951	·52383	·48871	·45437	·42102	·38884	·35797	·32853	·30060	·27423	
18	·62740	·59255	·55770	·52311	·48902	·45565	·42320	·39182	·36166	·33282	9
19	0·69033	0·65728	0·62370	0·58987	0·55603	0·52244	0·48931	0·45684	0·42521	0·39458	
20	·74712	·71662	·68516	·65297	·62031	·58741	·55451	·52183	·48957	·45793	10
21	·79705	·76965	·74093	·71111	·68039	·64900	·61718	·58514	·55310	·52126	
22	·83990	·81589	·79032	·76336	·73519	·70599	·67597	·64533	·61428	·58304	11
23	·87582	·85527	·83304	·80925	·78402	·75749	·72983	·70122	·67185	·64191	
24	·90527	·88808	·86919	·84866	·82657	·80301	·77810	·75199	·72483	·69678	12
25	0·92891	0·91483	0·89912	0·88179	0·86287	0·84239	0·82044	0·79712	0·77254	0·74683	
26	·94749	·93620	·92341	·90908	·89320	·87577	·85683	·83643	·81464	·79156	13
27	·96182	·95295	·94274	·93112	·91806	·90352	·88750	·87000	·85107	·83076	
28	·97266	·96582	·95782	·94859	·93805	·92615	·91285	·89814	·88200	·86446	14
29	·98071	·97554	·96939	·96218	·95383	·94427	·93344	·92129	·90779	·89293	
30	0·98659	0·98274	0·97810	0·97258	0·96608	0·95853	0·94986	0·94001	0·92891	0·91654	15
32	·99379	·99177	·98925	·98617	·98243	·97796	·97269	·96653	·95941	·95126	16
34	·99728	·99628	·99500	·99339	·99137	·98889	·98588	·98227	·97799	·97296	17
36	·99887	·99841	·99779	·99700	·99597	·99468	·99306	·99107	·98864	·98572	18
38	·99955	·99935	·99907	·99870	·99821	·99757	·99675	·99572	·99442	·99281	19
40	0·99983	0·99975	0·99963	0·99947	0·99924	0·99894	0·99855	0·99804	0·99738	0·99655	20
42	·99994	·99991	·99986	·99979	·99969	·99956	·99938	·99914	·99882	·99841	21
44	·99998	·99997	·99995	·99992	·99988	·99983	·99975	·99964	·99949	·99930	22
46	·99999	·99999	·99998	·99997	·99996	·99993	·99990	·99986	·99979	·99970	23
48			·99999	·99999	·99998	·99998	·99996	·99994	·99992	·99988	24
50					0·99999	0·99999	0·99999	0·99998	0·99997	0·99995	25
52								·99999	·99999	·99998	26
54										·99999	27

Table 7. *Probability integral of the χ^2-distribution* (*continued*)

ν	χ^2=0·1 m=0·05	0·2 0·10	0·3 0·15	0·4 0·20	0·5 0·25	0·6 0·30	0·7 0·35	0·8 0·40	0·9 0·45	1·0 0·50	c
1	0·75183	0·65472	0·58388	0·52709	0·47950	0·43858	0·40278	0·37109	0·34278	0·31731	
2	·95123	·90484	·86071	·81873	·77880	·74082	·70469	·67032	·63763	·60653	1
3	·99184	·97759	·96003	·94024	·91889	·89643	·87320	·84947	·82543	·80125	
4	·99879	·99532	·98981	·98248	·97350	·96306	·95133	·93845	·92456	·90980	2
5	·99984	·99911	·99764	·99533	·99212	·98800	·98297	·97703	·97022	·96257	
6	·99998	·99985	·99950	·99885	·99784	·99640	·99449	·99207	·98912	·98561	3
7		0·99997	0·99990	0·99974	0·99945	0·99899	0·99834	0·99744	0·99628	0·99483	
8			·99998	·99994	·99987	·99973	·99953	·99922	·99880	·99825	4
9				·99999	·99997	·99993	·99987	·99978	·99964	·99944	
10					·99999	·99998	·99997	·99994	·99989	·99983	5
11							·99999	·99998	·99997	·99995	
12									·99999	·99999	6

ν	χ^2=21 m=10·5	22 11·0	23 11·5	24 12·0	25 12·5	26 13·0	27 13·5	28 14·0	29 14·5	30 15·0	c
1	0·00001										
2	·00003	0·00002	0·00001	0·00001							1
3	·00011	·00007	·00004	·00003	0·00002	0·00001	0·00001				
4	·00032	·00020	·00013	·00008	·00005	·00003	·00002	0·00001	0·00001	0·00001	2
5	·00081	·00052	·00034	·00022	·00014	·00009	·00006	·00004	·00002	·00002	
6	·00184	·00121	·00080	·00052	·00034	·00022	·00015	·00009	·00006	·00004	3
7	0·00377	0·00254	0·00171	0·00114	0·00076	0·00050	0·00033	0·00022	0·00015	0·00010	
8	·00715	·00492	·00336	·00229	·00155	·00105	·00071	·00047	·00032	·00021	4
9	·01265	·00888	·00620	·00430	·00297	·00204	·00140	·00095	·00065	·00044	
10	·02109	·01511	·01075	·00760	·00535	·00374	·00260	·00181	·00125	·00086	5
11	·03337	·02437	·01768	·01273	·00912	·00649	·00460	·00324	·00227	·00159	
12	·05038	·03752	·02773	·02034	·01482	·01073	·00773	·00553	·00394	·00279	6
13	0·07293	0·05536	0·04168	0·03113	0·02308	0·01700	0·01244	0·00905	0·00655	0·00471	
14	·10163	·07861	·06027	·04582	·03457	·02589	·01925	·01423	·01045	·00763	7
15	·13683	·10780	·08414	·06509	·04994	·03802	·02874	·02157	·01609	·01192	
16	·17851	·14319	·11374	·08950	·06982	·05403	·04148	·03162	·02394	·01800	8
17	·22629	·18472	·14925	·11944	·09471	·07446	·05807	·04494	·03453	·02635	
18	·27941	·23199	·19059	·15503	·12492	·09976	·07900	·06206	·04838	·03745	9
19	0·33680	0·28426	0·23734	0·19615	0·16054	0·13019	0·10465	0·08343	0·06599	0·05180	
20	·39713	·34051	·28880	·24239	·20143	·16581	·13526	·10940	·08776	·06985	10
21	·45894	·39951	·34398	·29306	·24716	·20645	·17085	·14015	·11400	·09199	
22	·52074	·45989	·40173	·34723	·29707	·25168	·21123	·17568	·14486	·11846	11
23	·58109	·52025	·46077	·40381	·35029	·30087	·25597	·21578	·18031	·14940	
24	·63873	·57927	·51980	·46160	·40576	·35317	·30445	·26004	·22013	·18475	12
25	0·69261	0·63574	0·57756	0·51937	0·46237	0·40760	0·35588	0·30785	0·26392	0·22429	
26	·74196	·68870	·63295	·57597	·51898	·46311	·40933	·35846	·31108	·26761	13
27	·78629	·73738	·68501	·63032	·57446	·51860	·46379	·41097	·36090	·31415	
28	·82535	·78129	·73304	·68154	·62784	·57305	·51825	·46445	·41253	·36322	14
29	·85915	·82019	·77654	·72893	·67825	·62549	·57171	·51791	·46507	·41400	
30	0·88789	0·85404	0·81526	0·77203	0·72503	0·67513	0·62327	0·57044	0·51760	0·46565	15
32	·93167	·90740	·87830	·84442	·80603	·76361	·71779	·66936	·61916	·56809	16
34	·96039	·94408	·92360	·89871	·86931	·83549	·79755	·75592	·71121	·66412	17
36	·97814	·96781	·95425	·93703	·91584	·89047	·86088	·82720	·78972	·74886	18
38	·98849	·98231	·97383	·96258	·94815	·93017	·90838	·88264	·85296	·81947	19
40	0·99421	0·99071	0·98568	0·97872	0·96941	0·95733	0·94213	0·92350	0·90122	0·87522	20
42	·99721	·99533	·99250	·98840	·98269	·97499	·96491	·95209	·93622	·91703	21
44	·99871	·99775	·99623	·99394	·99060	·98592	·97955	·97116	·96038	·94689	22
46	·99943	·99896	·99818	·99695	·99509	·99238	·98854	·98329	·97630	·96726	23
48	·99976	·99954	·99916	·99853	·99754	·99603	·99382	·99067	·98634	·98054	24
50	0·99990	0·99980	0·99962	0·99931	0·99881	0·99801	0·99678	0·99498	0·99241	0·98884	25
52	·99996	·99992	·99984	·99969	·99944	·99903	·99839	·99739	·99592	·99382	26
54	·99999	·99997	·99993	·99987	·99975	·99955	·99922	·99869	·99789	·99669	27
56		·99999	·99997	·99994	·99989	·99980	·99963	·99937	·99894	·99828	28
58			·99999	·99998	·99995	·99991	·99983	·99970	·99949	·99914	29
60				0·99999	0·99998	0·99996	0·99993	0·99986	0·99976	0·99958	30
62					·99999	·99998	·99997	·99994	·99989	·99980	31
64						·99999	·99999	·99997	·99995	·99991	32
66								·99999	·99998	·99996	33
68									·99999	·99998	34
70										0·99999	35

ν	$\chi^2=1{\cdot}1$ m=0·55	1·2 0·60	1·3 0·65	1·4 0·70	1·5 0·75	1·6 0·80	1·7 0·85	1·8 0·90	1·9 0·95	2·0 1·00	c
1	0·29427	0·27332	0·25421	0·23672	0·22067	0·20590	0·19229	0·17971	0·16808	0·15730	
2	·57695	·54881	·52205	·49659	·47237	·44933	·42741	·40657	·38674	·36788	1
3	·77707	·75300	·72913	·70553	·68227	·65939	·63693	·61493	·59342	·57241	
4	·89427	·87810	·86138	·84420	·82664	·80879	·79072	·77248	·75414	·73576	2
5	·95410	·94488	·93493	·92431	·91307	·90125	·88890	·87607	·86280	·84915	
6	·98154	·97689	·97166	·96586	·95949	·95258	·94512	·93714	·92866	·91970	3
7	0·99305	0·99093	0·98844	0·98557	0·98231	0·97864	0·97457	0·97008	0·96517	0·95984	
8	·99753	·99664	·99555	·99425	·99271	·99092	·98887	·98654	·98393	·98101	4
9	·99917	·99882	·99838	·99782	·99715	·99633	·99537	·99425	·99295	·99147	
10	·99973	·99961	·99944	·99921	·99894	·99859	·99817	·99766	·99705	·99634	5
11	·99992	·99987	·99981	·99973	·99962	·99948	·99930	·99908	·99882	·99850	
12	·99998	·99996	·99994	·99991	·99987	·99982	·99975	·99966	·99954	·99941	6
13	0·99999	0·99999	0·99998	0·99997	0·99996	0·99994	0·99991	0·99988	0·99983	0·99977	
14			·99999	·99999	·99999	·99998	·99997	·99996	·99994	·99992	7
15						·99999	·99999	·99999	·99998	·99997	
16									·99999	·99999	8

ν	$\chi^2=31$ m=15·5	32 16·0	33 16·5	34 17·0	35 17·5	36 18·0	37 18·5	38 19·0	39 19·5	40 20·0	c
5	0·00001	0·00001									
6	·00003	·00002	0·00001	0·00001							3
7	0·00006	0·00004	0·00003	0·00002	0·00001	0·00001					
8	·00014	·00009	·00006	·00004	·00003	·00002	0·00001	0·00001			4
9	·00030	·00020	·00013	·00009	·00006	·00004	·00003	·00002	0·00001	0·00001	
10	·00059	·00040	·00027	·00019	·00012	·00008	·00006	·00004	·00003	·00002	5
11	·00110	·00076	·00053	·00036	·00025	·00017	·00012	·00008	·00005	·00004	
12	·00197	·00138	·00097	·00068	·00047	·00032	·00022	·00015	·00011	·00007	6
13	0·00337	0·00240	0·00170	0·00120	0·00085	0·00059	0·00041	0·00029	0·00020	0·00014	
14	·00554	·00401	·00288	·00206	·00147	·00104	·00074	·00052	·00036	·00026	7
15	·00878	·00644	·00469	·00341	·00246	·00177	·00127	·00090	·00064	·00045	
16	·01346	·01000	·00739	·00543	·00397	·00289	·00210	·00151	·00109	·00078	8
17	·01997	·01505	·01127	·00840	·00622	·00459	·00337	·00246	·00179	·00129	
18	·02879	·02199	·01669	·01260	·00945	·00706	·00524	·00387	·00285	·00209	9
19	0·04037	0·03125	0·02404	0·01838	0·01397	0·01056	0·00793	0·00593	0·00442	0·00327	
20	·05519	·04330	·03374	·02613	·02010	·01538	·01170	·00886	·00667	·00500	10
21	·07366	·05855	·04622	·03624	·02824	·02187	·01683	·01289	·00981	·00744	
22	·09612	·07740	·06187	·04912	·03875	·03037	·02366	·01832	·01411	·01081	11
23	·12279	·10014	·08107	·06516	·05202	·04125	·03251	·02547	·01984	·01537	
24	·15378	·12699	·10407	·08467	·06840	·05489	·04376	·03467	·02731	·02139	12
25	0·18902	0·15801	0·13107	0·10791	0·08820	0·07160	0·05774	0·04626	0·03684	0·02916	
26	·22827	·19312	·16210	·13502	·11165	·09167	·07475	·06056	·04875	·03901	13
27	·27114	·23208	·19707	·16605	·13887	·11530	·09507	·07786	·06336	·05124	
28	·31708	·27451	·23574	·20087	·16987	·14260	·11886	·09840	·08092	·06613	14
29	·36542	·31987	·27774	·23926	·20454	·17356	·14622	·12234	·10166	·08394	
30	0·41541	0·36753	0·32254	0·28083	0·24264	0·20808	0·17714	0·14975	0·12573	0·10486	15
32	·51701	·46675	·41802	·37145	·32754	·28665	·24903	·21479	·18398	·15651	16
34	·61544	·56596	·51648	·46774	·42040	·37505	·33214	·29203	·25497	·22107	17
36	·70518	·65934	·61205	·56402	·51600	·46865	·42259	·37836	·33639	·29703	18
38	·78246	·74235	·69965	·65496	·60893	·56225	·51555	·46948	·42461	·38142	19
40	0·84551	0·81225	0·77572	0·73632	0·69453	0·65092	0·60607	0·56061	0·51514	0·47026	20
42	·89437	·86817	·83848	·80548	·76943	·73072	·68979	·64717	·60342	·55909	21
44	·93043	·91077	·88780	·86147	·83185	·79912	·76355	·72550	·68538	·64370	22
46	·95584	·94176	·92478	·90473	·88150	·85509	·82558	·79314	·75804	·72061	23
48	·97296	·96331	·95131	·93670	·91928	·89889	·87547	·84902	·81963	·78749	24
50	0·98402	0·97769	0·96955	0·95935	0·94682	0·93174	0·91392	0·89325	0·86968	0·84323	25
52	·99087	·98688	·98159	·97476	·96611	·95539	·94238	·92687	·90872	·88782	26
54	·99496	·99254	·98923	·98483	·97908	·97177	·96263	·95144	·93800	·92211	27
56	·99731	·99590	·99390	·99117	·98750	·98268	·97650	·96873	·95914	·94752	28
58	·99861	·99781	·99665	·99502	·99275	·98970	·98567	·98046	·97387	·96567	29
60	0·99930	0·99887	0·99822	0·99727	0·99593	0·99406	0·99152	0·98815	0·98377	0·97818	30
62	·99966	·99943	·99908	·99855	·99778	·99667	·99512	·99302	·99021	·98653	31
64	·99984	·99972	·99954	·99925	·99882	·99819	·99728	·99600	·99425	·99191	32
66	·99993	·99987	·99978	·99963	·99939	·99904	·99852	·99777	·99672	·99527	33
68	·99997	·99994	·99989	·99982	·99970	·99951	·99922	·99879	·99818	·99731	34
70	0·99999	0·99997	0·99995	0·99991	0·99985	0·99975	0·99960	0·99936	0·99902	0·99851	35

ν	$\chi^2=2{\cdot}2$ m=1·1	2·4 1·2	2·6 1·3	2·8 1·4	3·0 1·5	3·2 1·6	3·4 1·7	3·6 1·8	3·8 1·9	4·0 2·0	c
1	0·13801	0·12134	0·10686	0·09426	0·08327	0·07364	0·06520	0·05778	0·05125	0·04550	
2	·33287	·30119	·27253	·24660	·22313	·20190	·18268	·16530	·14957	·13534	1
3	·53195	·49363	·45749	·42350	·39163	·36181	·33397	·30802	·28389	·26146	
4	·69903	·66263	·62682	·59183	·55783	·52493	·49325	·46284	·43375	·40601	2
5	·82084	·79147	·76137	·73079	·69999	·66918	·63857	·60831	·57856	·54942	
6	·90042	·87949	·85711	·83350	·80885	·78336	·75722	·73062	·70372	·67668	3
7	0·94795	0·93444	0·91938	0·90287	0·88500	0·86590	0·84570	0·82452	0·80250	0·77978	
8	·97426	·96623	·95691	·94628	·93436	·92119	·90681	·89129	·87470	·85712	4
9	·98790	·98345	·97807	·97170	·96430	·95583	·94631	·93572	·92408	·91141	
10	·99457	·99225	·98934	·98575	·98142	·97632	·97039	·96359	·95592	·94735	5
11	·99766	·99652	·99503	·99311	·99073	·98781	·98431	·98019	·97541	·96992	
12	·99903	·99850	·99777	·99680	·99554	·99396	·99200	·98962	·98678	·98344	6
13	0·99961	0·99938	0·99903	0·99856	0·99793	0·99711	0·99606	0·99475	0·99314	0·99119	
14	·99985	·99975	·99960	·99938	·99907	·99866	·99813	·99743	·99655	·99547	7
15	·99994	·99990	·99984	·99974	·99960	·99940	·99913	·99878	·99832	·99774	
16	·99998	·99996	·99994	·99989	·99983	·99974	·99961	·99944	·99921	·99890	8
17	·99999	·99999	·99998	·99996	·99993	·99989	·99983	·99975	·99964	·99948	
18			·99999	·99998	·99997	·99995	·99993	·99989	·99984	·99976	9
19				0·99999	0·99999	0·99998	0·99997	0·99995	0·99993	0·99989	
20						·99999	·99999	·99998	·99997	·99995	10
21								·99999	·99999	·99998	
22										·99999	11

ν	$\chi^2=42$ m=21	44 22	46 23	48 24	50 25	52 26	54 27	56 28	58 29	60 30	c
10	0·00001										5
11	·00002	0·00001									
12	·00003	·00002	0·00001								6
13	0·00006	0·00003	0·00001	0·00001							
14	·00012	·00006	·00003	·00001	0·00001						7
15	·00023	·00011	·00005	·00003	·00001	0·00001					
16	·00040	·00020	·00010	·00005	·00002	·00001	0·00001				8
17	·00067	·00034	·00017	·00009	·00004	·00002	·00001	0·00001			
18	·00111	·00058	·00030	·00015	·00008	·00004	·00002	·00001			9
19	0·00177	0·00094	0·00050	0·00026	0·00013	0·00007	0·00003	0·00002	0·00001		
20	·00277	·00151	·00081	·00043	·00022	·00011	·00006	·00003	·00001	0·00001	10
21	·00421	·00234	·00128	·00069	·00036	·00019	·00010	·00005	·00003	·00001	
22	·00625	·00355	·00198	·00109	·00059	·00031	·00016	·00009	·00004	·00002	11
23	·00908	·00526	·00299	·00167	·00092	·00050	·00027	·00014	·00007	·00004	
24	·01291	·00763	·00443	·00252	·00142	·00078	·00043	·00023	·00012	·00006	12
25	0·01797	0·01085	0·00642	0·00373	0·00213	0·00120	0·00066	0·00036	0·00020	0·00011	
26	·02455	·01512	·00912	·00540	·00314	·00180	·00102	·00056	·00031	·00017	13
27	·03292	·02068	·01272	·00768	·00455	·00265	·00152	·00086	·00048	·00026	
28	·04336	·02779	·01743	·01072	·00647	·00384	·00224	·00129	·00073	·00041	14
29	·05616	·03670	·02346	·01470	·00903	·00545	·00324	·00189	·00109	·00062	
30	0·07157	0·04769	0·03107	0·01983	0·01240	0·00762	0·00460	0·00273	0·00160	0·00092	15
32	·11108	·07689	·05200	·03440	·02229	·01417	·00884	·00543	·00328	·00195	16
34	·16292	·11704	·08208	·05627	·03775	·02482	·01601	·01014	·00632	·00387	17
36	·22696	·16900	·12277	·08713	·06048	·04111	·02739	·01791	·01151	·00727	18
38	·30168	·23250	·17477	·12828	·09204	·06463	·04446	·03000	·01987	·01293	19
40	0·38426	0·30603	0·23771	0·18026	0·13358	0·09682	0·06872	0·04781	0·03263	0·02187	20
42	·47097	·38691	·31010	·24264	·18549	·13867	·10147	·07274	·05114	·03529	21
44	·55769	·47164	·38938	·31393	·24730	·19048	·14357	·10599	·07669	·05444	22
46	·64046	·55637	·47227	·39170	·31753	·25172	·19525	·14830	·11038	·08057	23
48	·71603	·63742	·55515	·47285	·39388	·32094	·25591	·19981	·15285	·11465	24
50	0·78216	0·71172	0·63458	0·55400	0·47340	0·39593	0·32416	0·25990	0·20417	0·15724	25
52	·83770	·77710	·70766	·63191	·55292	·47392	·39786	·32721	·26371	·20836	26
54	·88257	·83242	·77230	·70382	·62939	·55190	·47440	·39970	·33011	·26734	27
56	·91746	·87750	·82737	·76774	·70019	·62700	·55094	·47486	·40143	·33287	28
58	·94363	·91291	·87260	·82253	·76340	·69674	·62475	·55003	·47530	·40308	29
60	0·96258	0·93978	0·90848	0·86788	0·81790	0·75926	0·69347	0·62261	0·54917	0·47572	30
62	·97585	·95949	·93598	·90415	·86331	·81345	·75531	·69035	·62058	·54835	31
64	·98483	·97347	·95639	·93224	·89993	·85889	·80917	·75153	·68738	·61864	32
66	·99073	·98308	·97106	·95330	·92854	·89582	·85462	·80507	·74792	·68454	33
68	·99448	·98949	·98128	·96862	·95022	·92491	·89181	·85049	·80112	·74445	34
70	0·99680	0·99364	0·98819	0·97943	0·96616	0·94716	0·92134	0·88790	0·84649	0·79731	35

Table 7 (*continued*)

ν	χ^2=4·2 m=2·1	4·4 2·2	4·6 2·3	4·8 2·4	5·0 2·5	5·2 2·6	5·4 2·7	5·6 2·8	5·8 2·9	6·0 3·0	c
1	0·04042	0·03594	0·03197	0·02846	0·02535	0·02259	0·02014	0·01796	0·01603	0·01431	
2	·12246	·11080	·10026	·09072	·08208	·07427	·06721	·06081	·05502	·04979	1
3	·24066	·22139	·20354	·18704	·17180	·15772	·14474	·13278	·12176	·11161	
4	·37962	·35457	·33085	·30844	·28730	·26739	·24866	·23108	·21459	·19915	2
5	·52099	·49337	·46662	·44077	·41588	·39196	·36904	·34711	·32617	·30622	
6	·64963	·62271	·59604	·56971	·54381	·51843	·49363	·46945	·44596	·42319	3
7	0·75647	0·73272	0·70864	0·68435	0·65996	0·63557	0·61127	0·58715	0·56329	0·53975	
8	·83864	·81935	·79935	·77872	·75758	·73600	·71409	·69194	·66962	·64723	4
9	·89776	·88317	·86769	·85138	·83431	·81654	·79814	·77919	·75976	·73992	
10	·93787	·92750	·91625	·90413	·89118	·87742	·86291	·84768	·83178	·81526	5
11	·96370	·95672	·94898	·94046	·93117	·92109	·91026	·89868	·88637	·87337	
12	·97955	·97509	·97002	·96433	·95798	·95096	·94327	·93489	·92583	·91608	6
13	0·98887	0·98614	0·98298	0·97934	0·97519	0·97052	0·96530	0·95951	0·95313	0·94615	
14	·99414	·99254	·99064	·98841	·98531	·98283	·97943	·97559	·97128	·96649	7
15	·99701	·99610	·99501	·99369	·99213	·99029	·98816	·98571	·98291	·97975	
16	·99851	·99802	·99741	·99666	·99575	·99467	·99338	·99187	·99012	·98810	8
17	·99928	·99902	·99869	·99828	·99777	·99715	·99639	·99550	·99443	·99319	
18	·99966	·99953	·99936	·99914	·99886	·99851	·99809	·99757	·99694	·99620	9
19	0·99985	0·99978	0·99969	0·99958	0·99943	0·99924	0·99901	0·99872	0·99836	0·99793	
20	·99993	·99990	·99986	·99980	·99972	·99962	·99950	·99934	·99914	·99890	10
21	·99997	·99995	·99993	·99991	·99987	·99982	·99975	·99967	·99956	·99943	
22	·99999	·99998	·99997	·99996	·99994	·99991	·99988	·99984	·99978	·99971	11
23	·99999	·99999	·99999	·99998	·99997	·99996	·99994	·99992	·99989	·99986	
24			·99999	·99999	·99999	·99998	·99997	·99996	·99995	·99993	12
25					0·99999	0·99999	0·99999	0·99998	0·99998	0·99997	
26								·99999	·99999	·99998	13
27									·99999	·99999	

ν	χ^2=62 m=31	64 32	66 33	68 34	70 35	72 36	74 37	76 38	78 39	80 40	c
21	0·00001										
22	·00001	0·00001									11
23	·00002	·00001	0·00001								
24	·00003	·00002	·00001								12
25	0·00006	0·00003	0·00002	0·00001							
26	·00009	·00005	·00003	·00001	0·00001						13
27	·00014	·00008	·00004	·00002	·00001	0·00001					
28	·00023	·00012	·00007	·00004	·00002	·00001	0·00001				14
29	·00035	·00019	·00011	·00006	·00003	·00002	·00001				
30	0·00052	0·00029	0·00016	0·00009	0·00005	0·00003	0·00001	0·00001			15
32	·00114	·00066	·00038	·00021	·00012	·00007	·00004	·00002	0·00001	0·00001	16
34	·00234	·00139	·00082	·00047	·00027	·00015	·00009	·00005	·00003	·00001	17
36	·00452	·00277	·00167	·00100	·00059	·00034	·00020	·00011	·00006	·00004	18
38	·00828	·00522	·00324	·00198	·00120	·00071	·00042	·00025	·00014	·00008	19
40	0·01441	0·00934	0·00596	0·00375	0·00233	0·00142	0·00086	0·00051	0·00030	0·00018	20
42	·02392	·01594	·01045	·00675	·00430	·00270	·00167	·00102	·00062	·00037	21
44	·03795	·02600	·01751	·01161	·00758	·00488	·00310	·00194	·00120	·00073	22
46	·05772	·04062	·02810	·01912	·01281	·00846	·00550	·00353	·00224	·00140	23
48	·08437	·06097	·04329	·03023	·02077	·01405	·00937	·00616	·00399	·00256	24
50	0·11880	0·08810	0·06418	0·04596	0·03237	0·02245	0·01533	0·01032	0·00685	0·00448	25
52	·16148	·12283	·09175	·06736	·04862	·03453	·02415	·01664	·01130	·00757	26
54	·21237	·16557	·12675	·09533	·07049	·05127	·03670	·02587	·01797	·01231	27
56	·27080	·21623	·16953	·13057	·09884	·07358	·05390	·03888	·02762	·01934	28
58	·33550	·27412	·21994	·17335	·13428	·10227	·07663	·05652	·04105	·02938	29
60	0·40465	0·33801	0·27730	0·22351	0·17705	0·13789	0·10563	0·07964	0·05912	0·04323	30
62	·47611	·40615	·34040	·28035	·22694	·18063	·14140	·10893	·08260	·06169	31
64	·54757	·47649	·40758	·34270	·28328	·23026	·18409	·14482	·11215	·08552	32
66	·61680	·54683	·47685	·40894	·34490	·28609	·23346	·18745	·14816	·11530	33
68	·68183	·61504	·54612	·47719	·41025	·34700	·28880	·23654	·19071	·15140	34
70	0·74112	0·67923	0·61335	0·54544	0·47752	0·41150	0·34903	0·29141	0·23953	0·19388	35

Table 7. *Probability integral of the χ^2-distribution (continued)*

ν	$\chi^2=6{\cdot}2$ m = 3·1	6·4 3·2	6·6 3·3	6·8 3·4	7·0 3·5	7·2 3·6	7·4 3·7	7·6 3·8	7·8 3·9	8·0 4·0	c
1	0·01278	0·01141	0·01020	0·00912	0·00815	0·00729	0·00652	0·00584	0·00522	0·00468	
2	·04505	·04076	·03688	·03337	·03020	·02732	·02472	·02237	·02024	·01832	1
3	·10228	·09369	·08580	·07855	·07190	·06579	·06018	·05504	·05033	·04601	
4	·18470	·17120	·15860	·14684	·13589	·12569	·11620	·10738	·09919	·09158	2
5	·28724	·26922	·25213	·23595	·22064	·20619	·19255	·17970	·16761	·15624	
6	·40116	·37990	·35943	·33974	·32085	·30275	·28543	·26890	·25313	·23810	3
7	0·51660	0·49390	0·47168	0·45000	0·42888	0·40836	0·38845	0·36918	0·35056	0·33259	
8	·62484	·60252	·58034	·55836	·53663	·51522	·49415	·47349	·45325	·43347	4
9	·71975	·69931	·67869	·65793	·63712	·61631	·59555	·57490	·55442	·53415	
10	·79819	·78061	·76259	·74418	·72544	·70644	·68722	·66784	·64837	·62884	5
11	·85969	·84539	·83049	·81504	·79908	·78266	·76583	·74862	·73110	·71330	
12	·90567	·89459	·88288	·87054	·85761	·84412	·83009	·81556	·80056	·78513	6
13	0·93857	0·93038	0·92157	0·91216	0·90215	0·89155	0·88038	0·86865	0·85638	0·84360	
14	·96120	·95538	·94903	·94215	·93471	·92673	·91819	·90911	·89948	·88933	7
15	·97619	·97222	·96782	·96296	·95765	·95186	·94559	·93882	·93155	·92378	
16	·98579	·98317	·98022	·97693	·97326	·96921	·96476	·95989	·95460	·94887	8
17	·99174	·99007	·98816	·98599	·98355	·98081	·97775	·97437	·97064	·96655	
18	·99532	·99429	·99309	·99171	·99013	·98833	·98630	·98402	·98147	·97864	9
19	0·99741	0·99679	0·99606	0·99521	0·99421	0·99307	0·99176	0·99026	0·98857	0·98667	
20	·99860	·99824	·99781	·99729	·99669	·99598	·99515	·99420	·99311	·99187	10
21	·99926	·99905	·99880	·99850	·99814	·99771	·99721	·99662	·99594	·99514	
22	·99962	·99950	·99936	·99919	·99898	·99873	·99843	·99807	·99765	·99716	11
23	·99981	·99974	·99967	·99957	·99945	·99931	·99913	·99892	·99867	·99837	
24	·99990	·99987	·99983	·99978	·99971	·99963	·99953	·99941	·99926	·99908	12
25	0·99995	0·99994	0·99991	0·99989	0·99985	0·99981	0·99975	0·99968	0·99960	0·99949	
26	·99998	·99997	·99996	·99994	·99992	·99990	·99987	·99983	·99978	·99973	13
27	·99999	·99999	·99998	·99997	·99996	·99995	·99993	·99991	·99989	·99985	
28		·99999	·99999	·99999	·99998	·99998	·99997	·99996	·99994	·99992	14
29				·99999	·99999	·99999	·99998	·99998	·99997	·99996	
30						0·99999	0·99999	0·99999	0·99999	0·99998	15

ν	$\chi^2=82$ m = 41	84 42	86 43	88 44	90 45	92 46	94 47	96 48	98 49	100 50	c
34	0·00001										17
36	·00002	0·00001									18
38	·00004	·00002	0·00001	0·00001							19
40	0·00010	0·00006	0·00003	0·00002	0·00001	0·00001					20
42	·00022	·00013	·00007	·00004	·00002	·00001	0·00001				21
44	·00044	·00026	·00016	·00009	·00005	·00002	·00002	0·00001			22
46	·00086	·00053	·00032	·00019	·00011	·00007	·00004	·00002	0·00001	0·00001	23
48	·00162	·00101	·00062	·00038	·00023	·00014	·00008	·00005	·00003	·00002	24
50	0·00290	0·00185	0·00117	0·00073	0·00045	0·00027	0·00017	0·00010	0·00006	0·00003	25
52	·00500	·00326	·00210	·00134	·00084	·00053	·00032	·00020	·00012	·00007	26
54	·00832	·00555	·00365	·00238	·00153	·00097	·00061	·00038	·00023	·00014	27
56	·01335	·00910	·00612	·00406	·00267	·00173	·00111	·00070	·00044	·00027	28
58	·02073	·01442	·00990	·00671	·00450	·00298	·00195	·00126	·00081	·00051	29
60	0·03115	0·02214	0·01552	0·01074	0·00734	0·00495	0·00330	0·00218	0·00142	0·00092	30
62	·04540	·03294	·02357	·01664	·01160	·00798	·00543	·00365	·00242	·00159	31
64	·06425	·04757	·03473	·02502	·01778	·01248	·00865	·00592	·00401	·00269	32
66	·08839	·06678	·04974	·03653	·02648	·01894	·01338	·00934	·00644	·00439	33
68	·11839	·09122	·06928	·05189	·03834	·02795	·02012	·01431	·01005	·00698	34
70	0·15457	0·12142	0·09401	0·07176	0·05404	0·04015	0·02944	0·02132	0·01525	0·01078	35

Table 7 (*continued*)

ν	χ^2=8·2 m=4·1	8·4 4·2	8·6 4·3	8·8 4·4	9·0 4·5	9·2 4·6	9·4 4·7	9·6 4·8	9·8 4·9	10·0 5·0	c
1	0·00419	0·00375	0·00336	0·00301	0·00270	0·00242	0·00217	0·00195	0·00175	0·00157	
2	·01657	·01500	·01357	·01228	·01111	·01005	·00910	·00823	·00745	·00674	1
3	·04205	·03843	·03511	·03207	·02929	·02675	·02442	·02229	·02034	·01857	
4	·08452	·07798	·07191	·06630	·06110	·05629	·05184	·04773	·04394	·04043	2
5	·14555	·13553	·12612	·11731	·10906	·10135	·09413	·08740	·08110	·07524	
6	·22381	·21024	·19736	·18514	·17358	·16264	·15230	·14254	·13333	·12465	3
7	0·31529	0·29865	0·28266	0·26734	0·25266	0·23861	0·22520	0·21240	0·20019	0·18857	
8	·41418	·39540	·37715	·35945	·34230	·32571	·30968	·29423	·27935	·26503	4
9	·51412	·49439	·47499	·45594	·43727	·41902	·40120	·38383	·36692	·35049	
10	·60931	·58983	·57044	·55118	·53210	·51323	·49461	·47626	·45821	·44049	5
11	·69528	·67709	·65876	·64035	·62189	·60344	·58502	·56669	·54846	·53039	
12	·76931	·75314	·73666	·71991	·70293	·68576	·66844	·65101	·63350	·61596	6
13	0·83033	0·81660	0·80244	0·78788	0·77294	0·75768	0·74211	0·72627	0·71020	0·69393	
14	·87865	·86746	·85579	·84365	·83105	·81803	·80461	·79081	·77666	·76218	7
15	·91551	·90675	·89749	·88774	·87752	·86683	·85569	·84412	·83213	·81974	
16	·94269	·93606	·92897	·92142	·91341	·90495	·89603	·88667	·87686	·86663	8
17	·96208	·95723	·95198	·94633	·94026	·93378	·92687	·91954	·91179	·90361	
18	·97551	·97207	·96830	·96420	·95974	·95493	·94974	·94418	·93824	·93191	9
19	0·98454	0·98217	0·97955	0·97666	0·97348	0·97001	0·96623	0·96213	0·95771	0·95295	
20	·99046	·98887	·98709	·98511	·98291	·98047	·97779	·97486	·97166	·96817	10
21	·99424	·99320	·99203	·99070	·98921	·98755	·98570	·98365	·98139	·97891	
22	·99659	·99593	·99518	·99431	·99333	·99222	·99098	·98958	·98803	·98630	11
23	·99802	·99761	·99714	·99659	·99596	·99524	·99442	·99349	·99245	·99128	
24	·99888	·99863	·99833	·99799	·99760	·99714	·99661	·99601	·99532	·99455	12
25	0·99937	0·99922	0·99905	0·99884	0·99860	0·99831	0·99798	0·99760	0·99716	0·99665	
26	·99966	·99957	·99947	·99934	·99919	·99902	·99882	·99858	·99830	·99798	13
27	·99981	·99977	·99971	·99963	·99955	·99944	·99932	·99917	·99900	·99880	
28	·99990	·99987	·99984	·99980	·99975	·99969	·99962	·99953	·99942	·99930	14
29	·99995	·99993	·99991	·99989	·99986	·99983	·99979	·99973	·99967	·99960	
30	0·99997	0·99997	0·99996	0·99994	0·99993	0·99991	0·99988	0·99985	0·99982	·99977	15
32	·99999	·99999	·99999	·99998	·99998	·99997	·99997	·99996	·99995	·99993	16
34						·99999	·99999	·99999	·99999	·99998	17

ν	χ^2=102 m= 51	104 52	106 53	108 54	110 55	112 56	114 57	116 58	118 59	120 60	c
48	0·00001										24
50	0·00002	0·00001	0·00001								25
52	·00004	·00002	·00001	0·00001							26
54	·00009	·00005	·00003	·00002	0·00001	0·00001					27
56	·00017	·00010	·00006	·00004	·00002	·00001	0·00001				28
58	·00032	·00020	·00012	·00007	·00004	·00003	·00002	0·00001	0·00001		29
60	0·00059	0·00037	0·00023	0·00014	0·00009	0·00005	0·00003	0·00002	0·00001	0·00001	30
62	·00104	·00067	·00043	·00027	·00017	·00010	·00006	·00004	·00002	·00001	31
64	·00178	·00117	·00076	·00049	·00031	·00019	·00012	·00008	·00005	·00003	32
66	·00296	·00198	·00130	·00085	·00055	·00035	·00022	·00014	·00009	·00005	33
68	·00479	·00325	·00219	·00145	·00095	·00062	·00040	·00026	·00016	·00010	34
70	0·00753	0·00521	0·00356	0·00241	0·00161	0·00107	0·00070	0·00046	0·00029	0·00019	35

ν	χ^2=122 m= 61	124 62	126 63	128 64	130 65	132 66	134 67				
62	0·00001										31
64	·00002	0·00001	0·00001								32
66	·00003	·00002	·00001	0·00001							33
68	·00006	·00004	·00002	·00002	0·00001						34
70	0·00012	0·00008	0·00005	0·00003	0·00002	0·00001	0·00001				35

Table 8. *Percentage points of the χ^2-distribution*

ν \ Q	0·995	0·990	0·975	0·950	0·900	0·750	0·500
1	392704.10^{-10}	157088.10^{-9}	982069.10^{-9}	393214.10^{-8}	0·0157908	0·1015308	0·454936
2	0·0100251	0·0201007	0·0506356	0·102587	0·210721	0·575364	1·38629
3	0·0717218	0·114832	0·215795	0·351846	0·584374	1·212534	2·36597
4	0·206989	0·297109	0·484419	0·710723	1·063623	1·92256	3·35669
5	0·411742	0·554298	0·831212	1·145476	1·61031	2·67460	4·35146
6	0·675727	0·872090	1·23734	1·63538	2·20413	3·45460	5·34812
7	0·989256	1·239043	1·68987	2·16735	2·83311	4·25485	6·34581
8	1·34441	1·64650	2·17973	2·73264	3·48954	5·07064	7·34412
9	1·73493	2·08790	2·70039	3·32511	4·16816	5·89883	8·34283
10	2·15586	2·55821	3·24697	3·94030	4·86518	6·73720	9·34182
11	2·60322	3·05348	3·81575	4·57481	5·57778	7·58414	10·3410
12	3·07382	3·57057	4·40379	5·22603	6·30380	8·43842	11·3403
13	3·56503	4·10692	5·00875	5·89186	7·04150	9·29907	12·3398
14	4·07467	4·66043	5·62873	6·57063	7·78953	10·1653	13·3393
15	4·60092	5·22935	6·26214	7·26094	8·54676	11·0365	14·3389
16	5·14221	5·81221	6·90766	7·96165	9·31224	11·9122	15·3385
17	5·69722	6·40776	7·56419	8·67176	10·0852	12·7919	16·3382
18	6·26480	7·01491	8·23075	9·39046	10·8649	13·6753	17·3379
19	6·84397	7·63273	8·90652	10·1170	11·6509	14·5620	18·3377
20	7·43384	8·26040	9·59078	10·8508	12·4426	15·4518	19·3374
21	8·03365	8·89720	10·28293	11·5913	13·2396	16·3444	20·3372
22	8·64272	9·54249	10·9823	12·3380	14·0415	17·2396	21·3370
23	9·26043	10·19567	11·6886	13·0905	14·8480	18·1373	22·3369
24	9·88623	10·8564	12·4012	13·8484	15·6587	19·0373	23·3367
25	10·5197	11·5240	13·1197	14·6114	16·4734	19·9393	24·3366
26	11·1602	12·1981	13·8439	15·3792	17·2919	20·8434	25·3365
27	11·8076	12·8785	14·5734	16·1514	18·1139	21·7494	26·3363
28	12·4613	13·5647	15·3079	16·9279	18·9392	22·6572	27·3362
29	13·1211	14·2565	16·0471	17·7084	19·7677	23·5666	28·3361
30	13·7867	14·9535	16·7908	18·4927	20·5992	24·4776	29·3360
40	20·7065	22·1643	24·4330	26·5093	29·0505	33·6603	39·3353
50	27·9907	29·7067	32·3574	34·7643	37·6886	42·9421	49·3349
60	35·5345	37·4849	40·4817	43·1880	46·4589	52·2938	59·3347
70	43·2752	45·4417	48·7576	51·7393	55·3289	61·6983	69·3345
80	51·1719	53·5401	57·1532	60·3915	64·2778	71·1445	79·3343
90	59·1963	61·7541	65·6466	69·1260	73·2911	80·6247	89·3342
100	67·3276	70·0649	74·2219	77·9295	82·3581	90·1332	99·3341
X	−2·5758	−2·3263	−1·9600	−1·6449	−1·2816	−0·6745	0·0000

$$Q = Q(\chi^2 \mid \nu) = 1 - P(\chi^2 \mid \nu) = 2^{-\frac{1}{2}\nu}\{\Gamma(\tfrac{1}{2}\nu)\}^{-1}\int_{\chi^2}^{\infty} e^{-\frac{1}{2}x}\, x^{\frac{1}{2}\nu-1}\, dx.$$

Table 8 (*continued*)

ν \ Q	0·250	0·100	0·050	0·025	0·010	0·005	0·001
1	1·32330	2·70554	3·84146	5·02389	6·63490	7·87944	10·828
2	2·77259	4·60517	5·99146	7·37776	9·21034	10·5966	13·816
3	4·10834	6·25139	7·81473	9·34840	11·3449	12·8382	16·266
4	5·38527	7·77944	9·48773	11·1433	13·2767	14·8603	18·467
5	6·62568	9·23636	11·0705	12·8325	15·0863	16·7496	20·515
6	7·84080	10·6446	12·5916	14·4494	16·8119	18·5476	22·458
7	9·03715	12·0170	14·0671	16·0128	18·4753	20·2777	24·322
8	10·2189	13·3616	15·5073	17·5345	20·0902	21·9550	26·125
9	11·3888	14·6837	16·9190	19·0228	21·6660	23·5894	27·877
10	12·5489	15·9872	18·3070	20·4832	23·2093	25·1882	29·588
11	13·7007	17·2750	19·6751	21·9200	24·7250	26·7568	31·264
12	14·8454	18·5493	21·0261	23·3367	26·2170	28·2995	32·909
13	15·9839	19·8119	22·3620	24·7356	27·6882	29·8195	34·528
14	17·1169	21·0641	23·6848	26·1189	29·1412	31·3194	36·123
15	18·2451	22·3071	24·9958	27·4884	30·5779	32·8013	37·697
16	19·3689	23·5418	26·2962	28·8454	31·9999	34·2672	39·252
17	20·4887	24·7690	27·5871	30·1910	33·4087	35·7185	40·790
18	21·6049	25·9894	28·8693	31·5264	34·8053	37·1565	42·312
19	22·7178	27·2036	30·1435	32·8523	36·1909	38·5823	43·820
20	23·8277	28·4120	31·4104	34·1696	37·5662	39·9968	45·315
21	24·9348	29·6151	32·6706	35·4789	38·9322	41·4011	46·797
22	26·0393	30·8133	33·9244	36·7807	40·2894	42·7957	48·268
23	27·1413	32·0069	35·1725	38·0756	41·6384	44·1813	49·728
24	28·2412	33·1962	36·4150	39·3641	42·9798	45·5585	51·179
25	29·3389	34·3816	37·6525	40·6465	44·3141	46·9279	52·618
26	30·4346	35·5632	38·8851	41·9232	45·6417	48·2899	54·052
27	31·5284	36·7412	40·1133	43·1945	46·9629	49·6449	55·476
28	32·6205	37·9159	41·3371	44·4608	48·2782	50·9934	56·892
29	33·7109	39·0875	42·5570	45·7223	49·5879	52·3356	58·301
30	34·7997	40·2560	43·7730	46·9792	50·8922	53·6720	59·703
40	45·6160	51·8051	55·7585	59·3417	63·6907	66·7660	73·402
50	56·3336	63·1671	67·5048	71·4202	76·1539	79·4900	86·661
60	66·9815	74·3970	79·0819	83·2977	88·3794	91·9517	99·607
70	77·5767	85·5270	90·5312	95·0232	100·425	104·215	112·317
80	88·1303	96·5782	101·879	106·629	112·329	116·321	124·839
90	98·6499	107·565	113·145	118·136	124·116	128·299	137·208
100	109·141	118·498	124·342	129·561	135·807	140·169	149·449
X	+0·6745	+1·2816	+1·6449	+1·9600	+2·3263	+2·5758	+3·0902

For $\nu > 100$ take

$$\chi^2 = \nu\left\{1 - \frac{2}{9\nu} + X\sqrt{\frac{2}{9\nu}}\right\}^3 \quad \text{or} \quad \chi^2 = \tfrac{1}{2}\{X + \sqrt{(2\nu - 1)}\}^2,$$

according to the degree of accuracy required. X is the standardized normal deviate corresponding to $P = 1 - Q$, and is shown in the bottom line of the table.

Table 9. *Probability integral, $P(t \mid \nu)$, of the t-distribution*

t \ ν	1	2	3	4	5	6	7	8	9	10
0·0	0·50000	0·50000	0·50000	0·50000	0·50000	0·50000	0·50000	0·50000	0·50000	0·50000
0·1	·53173	·53527	·53667	·53742	·53788	·53820	·53843	·53860	·53873	·53884
0·2	·56283	·57002	·57286	·57438	·57532	·57596	·57642	·57676	·57704	·57726
0·3	·59277	·60376	·60812	·61044	·61188	·61285	·61356	·61409	·61450	·61484
0·4	·62112	·63608	·64203	·64520	·64716	·64850	·64946	·65019	·65076	·65122
0·5	0·64758	0·66667	0·67428	0·67834	0·68085	0·68256	0·68380	0·68473	0·68546	0·68605
0·6	·67202	·69529	·70460	·70958	·71267	·71477	·71629	·71745	·71835	·71907
0·7	·69440	·72181	·73284	·73875	·74243	·74493	·74674	·74811	·74919	·75006
0·8	·71478	·74618	·75890	·76574	·76999	·77289	·77500	·77659	·77784	·77885
0·9	·73326	·76845	·78277	·79050	·79531	·79860	·80099	·80280	·80422	·80536
1·0	0·75000	0·78868	0·80450	0·81305	0·81839	0·82204	0·82469	0·82670	0·82828	0·82955
1·1	·76515	·80698	·82416	·83346	·83927	·84325	·84614	·84834	·85006	·85145
1·2	·77886	·82349	·84187	·85182	·85805	·86232	·86541	·86777	·86961	·87110
1·3	·79129	·83838	·85777	·86827	·87485	·87935	·88262	·88510	·88705	·88862
1·4	·80257	·85177	·87200	·88295	·88980	·89448	·89788	·90046	·90249	·90412
1·5	0·81283	0·86380	0·88471	0·89600	0·90305	0·90786	0·91135	0·91400	0·91608	0·91775
1·6	·82219	·87464	·89605	·90758	·91475	·91964	·92318	·92587	·92797	·92966
1·7	·83075	·88439	·90615	·91782	·92506	·92998	·93354	·93622	·93833	·94002
1·8	·83859	·89317	·91516	·92688	·93412	·93902	·94256	·94522	·94731	·94897
1·9	·84579	·90109	·92318	·93488	·94207	·94691	·95040	·95302	·95506	·95669
2·0	0·85242	0·90825	0·93034	0·94194	0·94903	0·95379	0·95719	0·95974	0·96172	0·96331
2·1	·85854	·91473	·93672	·94817	·95512	·95976	·96306	·96553	·96744	·96896
2·2	·86420	·92060	·94241	·95367	·96045	·96495	·96813	·97050	·97233	·97378
2·3	·86945	·92593	·94751	·95853	·96511	·96945	·97250	·97476	·97650	·97787
2·4	·87433	·93077	·95206	·96282	·96919	·97335	·97627	·97841	·98005	·98134
2·5	0·87888	0·93519	0·95615	0·96662	0·97275	0·97674	0·97950	0·98153	0·98307	0·98428
2·6	·88313	·93923	·95981	·96998	·97587	·97967	·98229	·98419	·98563	·98675
2·7	·88709	·94292	·96311	·97295	·97861	·98221	·98468	·98646	·98780	·98884
2·8	·89081	·94630	·96607	·97559	·98100	·98442	·98674	·98840	·98964	·99060
2·9	·89430	·94941	·96875	·97794	·98310	·98633	·98851	·99005	·99120	·99208
3·0	0·89758	0·95227	0·97116	0·98003	0·98495	0·98800	0·99003	0·99146	0·99252	0·99333
3·1	·90067	·95490	·97335	·98189	·98657	·98944	·99134	·99267	·99364	·99437
3·2	·90359	·95733	·97533	·98355	·98800	·99070	·99247	·99369	·99459	·99525
3·3	·90634	·95958	·97713	·98503	·98926	·99180	·99344	·99457	·99539	·99599
3·4	·90895	·96166	·97877	·98636	·99037	·99275	·99428	·99532	·99606	·99661
3·5	0·91141	0·96358	0·98026	0·98755	0·99136	0·99359	0·99500	0·99596	0·99664	0·99714
3·6	·91376	·96538	·98162	·98862	·99223	·99432	·99563	·99651	·99713	·99758
3·7	·91598	·96705	·98286	·98958	·99300	·99496	·99617	·99698	·99754	·99795
3·8	·91809	·96860	·98400	·99045	·99369	·99552	·99664	·99738	·99789	·99826
3·9	·92010	·97005	·98504	·99123	·99430	·99601	·99705	·99773	·99819	·99852
4·0	0·92202	0·97141	0·98600	0·99193	0·99484	0·99644	0·99741	0·99803	0·99845	0·99874
4·2	·92560	·97386	·98768	·99315	·99575	·99716	·99798	·99850	·99885	·99909
4·4	·92887	·97602	·98912	·99415	·99649	·99772	·99842	·99886	·99914	·99933
4·6	·93186	·97792	·99034	·99498	·99708	·99815	·99876	·99912	·99936	·99951
4·8	·93462	·97962	·99140	·99568	·99756	·99850	·99902	·99932	·99951	·99964
5·0	0·93717	0·98113	0·99230	0·99625	0·99795	0·99877	0·99922	0·99947	0·99963	0·99973
5·2	·93952	·98248	·99309	·99674	·99827	·99899	·99937	·99959	·99972	·99980
5·4	·94171	·98369	·99378	·99715	·99853	·99917	·99950	·99968	·99978	·99985
5·6	·94375	·98478	·99437	·99750	·99875	·99931	·99959	·99975	·99983	·99989
5·8	·94565	·98577	·99490	·99780	·99893	·99942	·99967	·99980	·99987	·99991
6·0	0·94743	0·98666	0·99536	0·99806	0·99908	0·99952	0·99973	0·99984	0·99990	0·99993
6·2	·94910	·98748	·99577	·99828	·99920	·99959	·99978	·99987	·99992	·99995
6·4	·95066	·98822	·99614	·99847	·99931	·99966	·99982	·99990	·99994	·99996
6·6	·95214	·98890	·99646	·99863	·99940	·99971	·99985	·99992	·99995	·99997
6·8	·95352	·98953	·99675	·99878	·99948	·99975	·99987	·99993	·99996	·99998
7·0	0·95483	0·99010	0·99701	0·99890	0·99954	0·99979	0·99990	0·99994	0·99997	0·99998
7·2	·95607	·99063	·99724	·99901	·99960	·99982	·99991	·99995	·99997	·99999
7·4	·95724	·99111	·99745	·99911	·99964	·99984	·99993	·99996	·99998	·99999
7·6	·95836	·99156	·99764	·99920	·99969	·99986	·99994	·99997	·99998	·99999
7·8	·95941	·99198	·99781	·99927	·99972	·99988	·99995	·99997	·99999	·99999
8·0	0·96042	0·99237	0·99796	0·99934	0·99975	0·99990	0·99996	0·99998	0·99999	0·99999

Table 9 (*continued*)

t \ ν	11	12	13	14	15	16	17	18	19	20
0·0	0·50000	0·50000	0·50000	0·50000	0·50000	0·50000	0·50000	0·50000	0·50000	0·50000
0·1	·53893	·53900	·53907	·53912	·53917	·53921	·53924	·53928	·53930	·53933
0·2	·57744	·57759	·57771	·57782	·57792	·57800	·57807	·57814	·57820	·57825
0·3	·61511	·61534	·61554	·61571	·61585	·61598	·61609	·61619	·61628	·61636
0·4	·65159	·65191	·65217	·65240	·65260	·65278	·65293	·65307	·65319	·65330
0·5	0·68654	0·68694	0·68728	0·68758	0·68783	0·68806	0·68826	0·68843	0·68859	0·68873
0·6	·71967	·72017	·72059	·72095	·72127	·72155	·72179	·72201	·72220	·72238
0·7	·75077	·75136	·75187	·75230	·75268	·75301	·75330	·75356	·75380	·75400
0·8	·77968	·78037	·78096	·78146	·78190	·78229	·78263	·78293	·78320	·78344
0·9	·80630	·80709	·80776	·80833	·80883	·80927	·80965	·81000	·81031	·81058
1·0	0·83060	0·83148	0·83222	0·83286	0·83341	0·83390	0·83433	0·83472	0·83506	0·83537
1·1	·85259	·85355	·85436	·85506	·85566	·85620	·85667	·85709	·85746	·85780
1·2	·87233	·87335	·87422	·87497	·87562	·87620	·87670	·87715	·87756	·87792
1·3	·88991	·89099	·89191	·89270	·89339	·89399	·89452	·89500	·89542	·89581
1·4	·90546	·90658	·90754	·90836	·90907	·90970	·91025	·91074	·91118	·91158
1·5	0·91912	0·92027	0·92125	0·92209	0·92282	0·92346	0·92402	0·92452	0·92498	0·92538
1·6	·93105	·93221	·93320	·93404	·93478	·93542	·93599	·93650	·93695	·93736
1·7	·94140	·94256	·94354	·94439	·94512	·94576	·94632	·94683	·94728	·94768
1·8	·95034	·95148	·95245	·95328	·95400	·95463	·95518	·95568	·95612	·95652
1·9	·95802	·95914	·96008	·96089	·96158	·96220	·96273	·96321	·96364	·96403
2·0	0·96460	0·96567	0·96658	0·96736	0·96803	0·96861	0·96913	0·96959	0·97000	0·97037
2·1	·97020	·97123	·97209	·97283	·97347	·97403	·97452	·97495	·97534	·97569
2·2	·97496	·97593	·97675	·97745	·97805	·97858	·97904	·97945	·97981	·98014
2·3	·97898	·97990	·98067	·98132	·98189	·98238	·98281	·98319	·98352	·98383
2·4	·98238	·98324	·98396	·98457	·98509	·98554	·98594	·98629	·98660	·98688
2·5	0·98525	0·98604	0·98671	0·98727	0·98775	0·98816	0·98853	0·98885	0·98913	0·98938
2·6	·98765	·98839	·98900	·98951	·98995	·99033	·99066	·99095	·99121	·99144
2·7	·98967	·99035	·99090	·99137	·99177	·99211	·99241	·99267	·99290	·99311
2·8	·99136	·99198	·99249	·99291	·99327	·99358	·99385	·99408	·99429	·99447
2·9	·99278	·99334	·99380	·99418	·99450	·99478	·99502	·99523	·99541	·99557
3·0	0·99396	0·99447	0·99488	0·99522	0·99551	0·99576	0·99597	0·99616	0·99632	0·99646
3·1	·99495	·99541	·99578	·99608	·99634	·99656	·99675	·99691	·99705	·99718
3·2	·99577	·99618	·99652	·99679	·99702	·99721	·99738	·99752	·99764	·99775
3·3	·99646	·99683	·99713	·99737	·99757	·99774	·99789	·99801	·99812	·99821
3·4	·99703	·99737	·99763	·99784	·99802	·99817	·99830	·99840	·99850	·99858
3·5	0·99751	0·99781	0·99804	0·99823	0·99839	0·99852	0·99863	0·99872	0·99880	0·99887
3·6	·99791	·99818	·99838	·99855	·99869	·99880	·99890	·99898	·99905	·99911
3·7	·99825	·99848	·99867	·99881	·99893	·99903	·99911	·99918	·99924	·99929
3·8	·99853	·99874	·99890	·99902	·99913	·99921	·99928	·99934	·99939	·99944
3·9	·99876	·99895	·99909	·99920	·99929	·99936	·99942	·99948	·99952	·99956
4·0	0·99896	0·99912	0·99924	0·99934	0·99942	0·99948	0·99954	0·99958	0·99962	0·99965
4·2	·99926	·99938	·99948	·99955	·99961	·99966	·99970	·99973	·99976	·99978
4·4	·99947	·99957	·99964	·99970	·99974	·99978	·99980	·99983	·99985	·99986
4·6	·99962	·99969	·99975	·99979	·99983	·99985	·99987	·99989	·99990	·99991
4·8	·99972	·99978	·99983	·99986	·99988	·99990	·99992	·99993	·99994	·99995
5·0	0·99980	0·99985	0·99988	0·99990	0·99992	0·99993	0·99995	0·99995	0·99996	0·99997
5·2	·99985	·99989	·99992	·99993	·99995	·99996	·99996	·99997	·99997	·99998
5·4	·99989	·99992	·99994	·99995	·99996	·99997	·99998	·99998	·99998	·99999
5·6	·99992	·99994	·99996	·99997	·99997	·99998	·99998	·99999	·99999	·99999
5·8	·99994	·99996	·99997	·99998	·99998	·99999	·99999	·99999	·99999	·99999
6·0	0·99995	0·99997	0·99998	0·99998	0·99999	0·99999	0·99999	0·99999		
6·2	·99997	·99998	·99998	·99999	·99999	·99999				
6·4	·99997	·99998	·99999	·99999	·99999					
6·6	·99998	·99999	·99999	·99999						
6·8	·99998	·99999	·99999							
7·0	0·99999	0·99999								

Upper percentage points of t

$1-P(t\mid\nu)$	$\nu=1$	2	3	4	5	6	7	8	9	10
10^{-3}	318·3	22·33	10·21	7·17	5·89	5·21	4·79	4·50	4·30	4·14
10^{-4}	3183	70·7	22·20	13·03	9·68	8·02	7·06	6·44	6·01	5·69
10^{-5}	31831	224	47·91	23·33	15·54	12·03	10·11	8·90	8·10	7·53
5×10^{-6}	63652	316	60·40	27·82	17·89	13·55	11·22	9·79	8·83	8·15

Table 9. *Probability integral of the t-distribution (continued)*

t \ ν	20	21	22	23	24	30	40	60	120	∞
0·00	0·50000	0·50000	0·50000	0·50000	0·50000	0·50000	0·50000	0·50000	0·50000	0·50000
0·05	·51969	·51970	·51971	·51972	·51973	·51977	·51981	·51986	·51990	·51994
0·10	·53933	·53935	·53938	·53939	·53941	·53950	·53958	·53966	·53974	·53983
0·15	·55887	·55890	·55893	·55896	·55899	·55912	·55924	·55937	·55949	·55962
0·20	·57825	·57830	·57834	·57838	·57842	·57858	·57875	·57892	·57909	·57926
0·25	0·59743	0·59749	0·59755	0·59760	0·59764	0·59785	0·59807	0·59828	0·59849	0·59871
0·30	·61636	·61644	·61650	·61656	·61662	·61688	·61713	·61739	·61765	·61791
0·35	·63500	·63509	·63517	·63524	·63530	·63561	·63591	·63622	·63652	·63683
0·40	·65330	·65340	·65349	·65358	·65365	·65400	·65436	·65471	·65507	·65542
0·45	·67122	·67134	·67144	·67154	·67163	·67203	·67243	·67283	·67324	·67364
0·50	0·68873	0·68886	0·68898	0·68909	0·68919	0·68964	0·69009	0·69055	0·69100	0·69146
0·55	·70579	·70594	·70607	·70619	·70630	·70680	·70731	·70782	·70833	·70884
0·60	·72238	·72254	·72268	·72281	·72294	·72349	·72405	·72462	·72518	·72575
0·65	·73846	·73863	·73879	·73893	·73907	·73968	·74030	·74091	·74153	·74215
0·70	·75400	·75419	·75437	·75453	·75467	·75534	·75601	·75668	·75736	·75804
0·75	0·76901	0·76921	0·76940	0·76957	0·76973	0·77045	0·77118	0·77191	0·77264	0·77337
0·80	·78344	·78367	·78387	·78405	·78422	·78500	·78578	·78657	·78735	·78814
0·85	·79731	·79754	·79776	·79796	·79814	·79897	·79981	·80065	·80149	·80234
0·90	·81058	·81084	·81107	·81128	·81147	·81236	·81325	·81414	·81504	·81594
0·95	·82327	·82354	·82378	·82401	·82421	·82515	·82609	·82704	·82799	·82894
1·00	0·83537	0·83565	0·83591	0·83614	0·83636	0·83735	0·83834	0·83934	0·84034	0·84134
1·05	·84688	·84717	·84744	·84769	·84791	·84895	·84999	·85104	·85209	·85314
1·10	·85780	·85811	·85839	·85864	·85888	·85996	·86105	·86214	·86323	·86433
1·15	·86814	·86846	·86875	·86902	·86926	·87039	·87151	·87265	·87378	·87493
1·20	·87792	·87825	·87855	·87882	·87907	·88023	·88140	·88257	·88375	·88493
1·25	0·88714	0·88747	0·88778	0·88807	0·88832	0·88952	0·89072	0·89192	0·89313	0·89435
1·30	·89581	·89616	·89647	·89676	·89703	·89825	·89948	·90071	·90195	·90320
1·35	·90395	·90431	·90463	·90492	·90519	·90644	·90770	·90896	·91022	·91149
1·40	·91158	·91194	·91227	·91257	·91285	·91411	·91539	·91667	·91795	·91924
1·45	·91872	·91908	·91942	·91972	·92000	·92128	·92257	·92387	·92517	·92647
1·50	0·92538	0·92575	0·92608	0·92639	0·92667	0·92797	0·92927	0·93057	0·93188	0·93319
1·55	·93159	·93196	·93230	·93260	·93289	·93419	·93549	·93680	·93811	·93943
1·60	·93736	·93773	·93807	·93838	·93866	·93996	·94127	·94257	·94389	·94520
1·65	·94272	·94309	·94342	·94373	·94401	·94531	·94661	·94792	·94922	·95053
1·70	·94768	·94805	·94839	·94869	·94897	·95026	·95155	·95284	·95414	·95543
1·75	0·95228	0·95264	0·95297	0·95327	0·95355	0·95483	0·95611	0·95738	0·95866	0·95994
1·80	·95652	·95688	·95720	·95750	·95778	·95904	·96030	·96156	·96281	·96407
1·85	·96043	·96078	·96110	·96140	·96167	·96291	·96414	·96538	·96661	·96784
1·90	·96403	·96437	·96469	·96498	·96524	·96646	·96767	·96888	·97008	·97128
1·95	·96733	·96767	·96798	·96827	·96852	·96971	·97089	·97207	·97325	·97441
2·0	0·97037	0·97070	0·97100	0·97128	0·97153	0·97269	0·97384	0·97498	0·97612	0·97725
2·1	·97569	·97601	·97629	·97655	·97679	·97788	·97896	·98003	·98109	·98214
2·2	·98014	·98043	·98070	·98094	·98116	·98218	·98318	·98416	·98514	·98610
2·3	·98383	·98410	·98435	·98457	·98478	·98571	·98663	·98753	·98841	·98928
2·4	·98688	·98712	·98735	·98756	·98774	·98860	·98943	·99024	·99103	·99180
2·5	0·98938	0·98961	0·98982	0·99000	0·99017	0·99094	0·99169	0·99241	0·99312	0·99379
2·6	·99144	·99164	·99183	·99200	·99215	·99284	·99350	·99414	·99475	·99534
2·7	·99311	·99329	·99346	·99361	·99375	·99436	·99494	·99550	·99603	·99653
2·8	·99447	·99463	·99478	·99492	·99504	·99557	·99608	·99657	·99702	·99744
2·9	·99557	·99572	·99585	·99596	·99607	·99654	·99698	·99740	·99778	·99813
3·0	0·99646	0·99659	0·99670	0·99681	0·99690	0·99730	0·99768	0·99804	0·99836	0·99865
3·1	·99718	·99729	·99739	·99748	·99756	·99791	·99823	·99853	·99879	·99903
3·2	·99775	·99785	·99793	·99801	·99808	·99838	·99865	·99890	·99912	·99931
3·3	·99821	·99829	·99837	·99844	·99849	·99875	·99898	·99918	·99936	·99952
3·4	·99858	·99865	·99871	·99877	·99882	·99904	·99923	·99940	·99954	·99966
3·5	0·99887	0·99893	0·99899	0·99904	0·99908	0·99926	0·99942	0·99956	0·99967	0·99977
3·6	·99911	·99916	·99920	·99925	·99928	·99943	·99957	·99968	·99977	·99984
3·7	·99929	·99933	·99937	·99941	·99944	·99957	·99967	·99976	·99984	·99989
3·8	·99944	·99948	·99951	·99954	·99956	·99967	·99976	·99983	·99989	·99993
3·9	·99956	·99959	·99961	·99964	·99966	·99975	·99982	·99988	·99992	·99995
4·0	0·99965	0·99967	0·99970	0·99972	0·99974	0·99981	0·99987	0·99991	0·99995	0·99997
5·0	0·99997	0·99997	0·99998	0·99998	0·99998	0·99999	0·99999			

Table 10. *Chart for determining the power function of the t-test*

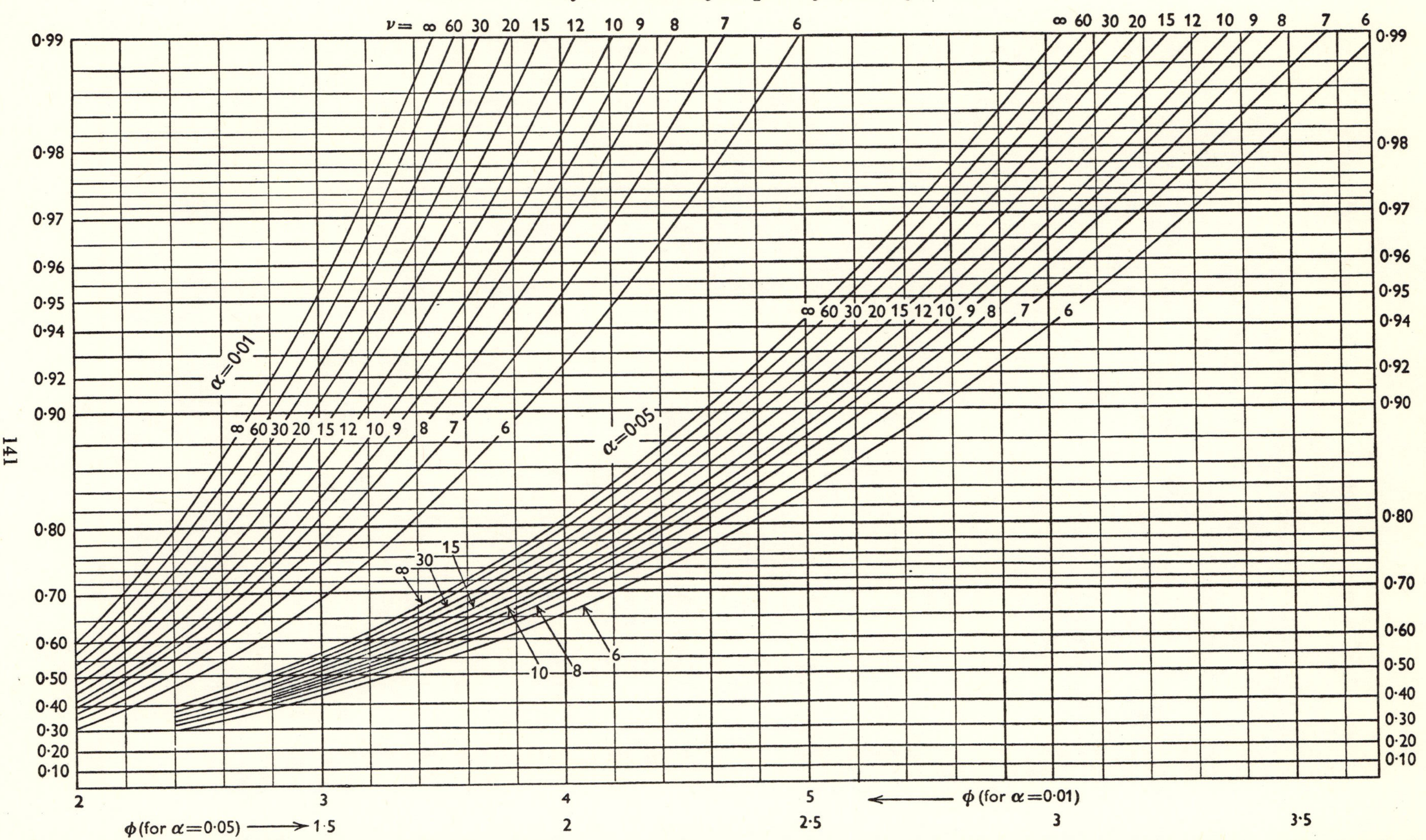

If y' is a normal variable with $\mathscr{E}(y')=\Delta$ and if s_y^2 is an independent mean-square estimate of its variance, σ_y^2, based on ν degrees of freedom, then $t'=y'/s_y$ is distributed as a non-central t. The chart shows on a logarithmic scale the chance of establishing significance according to the value of $\phi=\Delta/\sigma_y \times 1/\sqrt{2}$, using a two-tail t-test, at either the 5 or 1% levels ($\alpha=0{\cdot}05$ or $0{\cdot}01$). If a single-tail t-test is appropriate, then the curves marked $\alpha=0{\cdot}05$ and $0{\cdot}01$ give, respectively, the chance of establishing significance at the 2·5 and 0·5% levels.

Table 11. *Test for comparisons involving two variances which must be separately estimated*

Upper 5 % critical values of $v = \dfrac{(y-\eta)}{\sqrt{(\lambda_1 s_1^2 + \lambda_2 s_2^2)}}$

(i.e. upper 10 % critical values of $|v|$*)**

ν_2	$\dfrac{\lambda_1 s_1^2}{\lambda_1 s_1^2+\lambda_2 s_2^2}$ ν_1	0·0	0·1	0·2	0·3	0·4	0·5	0·6	0·7	0·8	0·9	1·0
6	6	1·94	1·90	1·85	1·80	1·76	1·74	1·76	1·80	1·85	1·90	1·94
	8	1·94	1·90	1·85	1·80	1·76	1·73	1·74	1·76	1·79	1·82	1·86
	10	1·94	1·90	1·85	1·80	1·76	1·73	1·73	1·74	1·76	1·78	1·81
	15	1·94	1·90	1·85	1·80	1·76	1·73	1·71	1·71	1·72	1·73	1·75
	20	1·94	1·90	1·85	1·80	1·76	1·73	1·71	1·70	1·70	1·71	1·72
	∞	1·94	1·90	1·85	1·80	1·76	1·72	1·69	1·67	1·66	1·65	1·64
8	6	1·86	1·82	1·79	1·76	1·74	1·73	1·76	1·80	1·85	1·90	1·94
	8	1·86	1·82	1·79	1·76	1·73	1·73	1·73	1·76	1·79	1·82	1·86
	10	1·86	1·82	1·79	1·76	1·73	1·72	1·72	1·74	1·76	1·78	1·81
	15	1·86	1·82	1·79	1·76	1·73	1·71	1·71	1·71	1·72	1·73	1·75
	20	1·86	1·82	1·79	1·76	1·73	1·71	1·70	1·70	1·70	1·71	1·72
	∞	1·86	1·82	1·79	1·75	1·72	1·70	1·68	1·66	1·65	1·65	1·64
10	6	1·81	1·78	1·76	1·74	1·73	1·73	1·76	1·80	1·85	1·90	1·94
	8	1·81	1·78	1·76	1·74	1·72	1·72	1·73	1·76	1·79	1·82	1·86
	10	1·81	1·78	1·76	1·73	1·72	1·71	1·72	1·73	1·76	1·78	1·81
	15	1·81	1·78	1·76	1·73	1·72	1·70	1·70	1·71	1·72	1·73	1·75
	20	1·81	1·78	1·76	1·73	1·71	1·70	1·69	1·69	1·70	1·71	1·72
	∞	1·81	1·78	1·76	1·73	1·71	1·69	1·67	1·66	1·65	1·65	1·64
15	6	1·75	1·73	1·72	1·71	1·71	1·73	1·76	1·80	1·85	1·90	1·94
	8	1·75	1·73	1·72	1·71	1·71	1·71	1·73	1·76	1·79	1·82	1·86
	10	1·75	1·73	1·72	1·71	1·70	1·70	1·72	1·73	1·76	1·78	1·81
	15	1·75	1·73	1·72	1·70	1·70	1·69	1·70	1·70	1·72	1·73	1·75
	20	1·75	1·73	1·72	1·70	1·69	1·69	1·69	1·69	1·70	1·71	1·72
	∞	1·75	1·73	1·72	1·70	1·68	1·67	1·66	1·65	1·65	1·65	1·64
20	6	1·72	1·71	1·70	1·70	1·71	1·73	1·76	1·80	1·85	1·90	1·94
	8	1·72	1·71	1·70	1·70	1·70	1·71	1·73	1·76	1·79	1·82	1·86
	10	1·72	1·71	1·70	1·69	1·69	1·70	1·71	1·73	1·76	1·78	1·81
	15	1·72	1·71	1·70	1·69	1·69	1·69	1·69	1·70	1·72	1·73	1·75
	20	1·72	1·71	1·70	1·69	1·68	1·68	1·68	1·69	1·70	1·71	1·72
	∞	1·72	1·71	1·70	1·68	1·67	1·66	1·66	1·65	1·65	1·65	1·64
∞	6	1·64	1·65	1·66	1·67	1·69	1·72	1·76	1·80	1·85	1·90	1·94
	8	1·64	1·65	1·65	1·66	1·68	1·70	1·72	1·75	1·79	1·82	1·86
	10	1·64	1·65	1·65	1·66	1·67	1·69	1·71	1·73	1·76	1·78	1·81
	15	1·64	1·65	1·65	1·65	1·66	1·67	1·68	1·70	1·72	1·73	1·75
	20	1·64	1·65	1·65	1·65	1·66	1·66	1·67	1·68	1·70	1·71	1·72
	∞	1·64	1·64	1·64	1·64	1·64	1·64	1·64	1·64	1·64	1·64	1·64

* y is normally distributed about η with variance $\lambda_1\sigma_1^2+\lambda_2\sigma_2^2$, and s_1^2 and s_2^2 are independent estimates of σ_1^2 and σ_2^2, based on ν_1 and ν_2 degrees of freedom, respectively. λ_1 and λ_2 are known constants.

In the problem of comparing the means of samples taken from two normal populations, put $y=(\bar{x}_1-\bar{x}_2)$, $\nu_1=(n_1-1)$, $\nu_2=(n_2-1)$, $\lambda_1=1/n_1$ and $\lambda_2=1/n_2$, where n_1 and n_2 are the sample sizes.

Table 11 (*continued*)

Upper 2·5 % critical values of $v = \dfrac{(y-\eta)}{\sqrt{(\lambda_1 s_1^2 + \lambda_2 s_2^2)}}$

(*i.e. upper* 5 % *critical values of* $|v|$)

ν_2	$\dfrac{\lambda_1 s_1^2}{(\lambda_1 s_1^2+\lambda_2 s_2^2)}$ ν_1	0·0	0·1	0·2	0·3	0·4	0·5	0·6	0·7	0·8	0·9	1·0
8	8	2·31	2·25	2·20	2·14	2·10	2·08	2·10	2·14	2·20	2·25	2·31
	10	2·31	2·25	2·20	2·15	2·10	2·08	2·08	2·11	2·14	2·19	2·23
	12	2·31	2·25	2·20	2·15	2·10	2·07	2·07	2·08	2·11	2·14	2·18
	15	2·31	2·25	2·20	2·15	2·10	2·07	2·05	2·06	2·08	2·10	2·13
	20	2·31	2·25	2·20	2·15	2·10	2·06	2·04	2·04	2·05	2·07	2·09
	∞	2·31	2·25	2·20	2·14	2·09	2·05	2·01	1·99	1·97	1·96	1·96
10	8	2·23	2·19	2·14	2·11	2·08	2·08	2·10	2·15	2·20	2·25	2·31
	10	2·23	2·18	2·14	2·11	2·08	2·06	2·08	2·11	2·14	2·18	2·23
	12	2·23	2·18	2·14	2·10	2·07	2·06	2·06	2·08	2·11	2·14	2·18
	15	2·23	2·18	2·14	2·10	2·07	2·05	2·05	2·06	2·08	2·10	2·13
	20	2·23	2·18	2·14	2·10	2·07	2·05	2·04	2·04	2·05	2·06	2·09
	∞	2·23	2·18	2·14	2·10	2·06	2·03	2·00	1·98	1·97	1·96	1·96
12	8	2·18	2·14	2·11	2·08	2·07	2·07	2·10	2·15	2·20	2·25	2·31
	10	2·18	2·14	2·11	2·08	2·06	2·06	2·07	2·10	2·14	2·18	2·23
	12	2·18	2·14	2·11	2·08	2·06	2·05	2·06	2·08	2·11	2·14	2·18
	15	2·18	2·14	2·11	2·08	2·06	2·04	2·04	2·06	2·08	2·10	2·13
	20	2·18	2·14	2·11	2·08	2·05	2·04	2·03	2·03	2·05	2·06	2·09
	∞	2·18	2·14	2·11	2·07	2·04	2·02	1·99	1·98	1·97	1·96	1·96
15	8	2·13	2·10	2·08	2·06	2·05	2·07	2·10	2·15	2·20	2·25	2·31
	10	2·13	2·10	2·08	2·06	2·05	2·05	2·07	2·10	2·14	2·18	2·23
	12	2·13	2·10	2·08	2·06	2·04	2·04	2·06	2·08	2·11	2·14	2·18
	15	2·13	2·10	2·08	2·05	2·04	2·03	2·04	2·05	2·08	2·10	2·13
	20	2·13	2·10	2·08	2·05	2·04	2·03	2·03	2·03	2·05	2·06	2·09
	∞	2·13	2·10	2·07	2·05	2·02	2·00	1·99	1·97	1·97	1·96	1·96
20	8	2·09	2·07	2·05	2·04	2·04	2·06	2·10	2·15	2·20	2·25	2·31
	10	2·09	2·06	2·05	2·04	2·04	2·05	2·07	2·10	2·14	2·18	2·23
	12	2·09	2·06	2·05	2·03	2·03	2·04	2·05	2·08	2·11	2·14	2·18
	15	2·09	2·06	2·05	2·03	2·03	2·03	2·04	2·05	2·08	2·10	2·13
	20	2·09	2·06	2·05	2·03	2·02	2·02	2·02	2·03	2·05	2·06	2·09
	∞	2·09	2·06	2·04	2·02	2·01	1·99	1·98	1·97	1·96	1·96	1·96
∞	8	1·96	1·96	1·97	1·99	2·01	2·05	2·09	2·14	2·20	2·25	2·31
	10	1·96	1·96	1·97	1·98	2·00	2·03	2·06	2·10	2·14	2·18	2·23
	12	1·96	1·96	1·97	1·98	1·99	2·02	2·04	2·07	2·11	2·14	2·18
	15	1·96	1·96	1·97	1·97	1·99	2·00	2·02	2·05	2·07	2·10	2·13
	20	1·96	1·96	1·96	1·97	1·98	1·99	2·01	2·02	2·04	2·06	2·09
	∞	1·96	1·96	1·96	1·96	1·96	1·96	1·96	1·96	1·96	1·96	1·96

Table 11 (*continued*). *Test for comparisons involving two variances which must be separately estimated*

Upper 1% *critical values of* $v = \dfrac{(y-\eta)}{\sqrt{(\lambda_1 s_1^2 + \lambda_2 s_2^2)}}$

(*i.e. upper* 2% *critical values of* $|v|$)*

ν_2	ν_1 \ $\dfrac{\lambda_1 s_1^2}{\lambda_1 s_1^2+\lambda_2 s_2^2}$	0·0	0·1	0·2	0·3	0·4	0·5	0·6	0·7	0·8	0·9	1·0
10	10	2·76	2·70	2·63	2·56	2·51	2·50	2·51	2·56	2·63	2·70	2·76
	12	2·76	2·70	2·63	2·56	2·51	2·49	2·49	2·52	2·57	2·62	2·68
	15	2·76	2·70	2·63	2·56	2·51	2·48	2·47	2·48	2·52	2·56	2·60
	20	2·76	2·70	2·63	2·56	2·51	2·47	2·45	2·45	2·47	2·49	2·53
	30	2·76	2·70	2·63	2·56	2·50	2·46	2·43	2·42	2·42	2·44	2·46
	∞	2·76	2·70	2·63	2·56	2·50	2·44	2·40	2·36	2·34	2·33	2·33
12	10	2·68	2·62	2·57	2·52	2·49	2·49	2·51	2·56	2·63	2·70	2·76
	12	2·68	2·62	2·57	2·52	2·48	2·47	2·48	2·52	2·57	2·62	2·68
	15	2·68	2·62	2·57	2·52	2·48	2·46	2·46	2·48	2·52	2·56	2·60
	20	2·68	2·62	2·57	2·52	2·48	2·45	2·44	2·45	2·47	2·49	2·53
	30	2·68	2·62	2·57	2·52	2·47	2·44	2·42	2·41	2·42	2·44	2·46
	∞	2·68	2·62	2·57	2·51	2·46	2·42	2·38	2·36	2·34	2·33	2·33
15	10	2·60	2·56	2·52	2·48	2·47	2·48	2·51	2·56	2·63	2·70	2·76
	12	2·60	2·56	2·52	2·48	2·46	2·46	2·48	2·52	2·57	2·62	2·68
	15	2·60	2·56	2·51	2·48	2·45	2·45	2·45	2·48	2·51	2·56	2·60
	20	2·60	2·56	2·51	2·48	2·45	2·43	2·43	2·44	2·46	2·49	2·53
	30	2·60	2·56	2·51	2·47	2·44	2·42	2·41	2·41	2·42	2·44	2·46
	∞	2·60	2·56	2·51	2·47	2·43	2·40	2·37	2·35	2·34	2·33	2·33
20	10	2·53	2·49	2·47	2·45	2·45	2·47	2·51	2·56	2·63	2·70	2·76
	12	2·53	2·49	2·47	2·45	2·44	2·45	2·48	2·52	2·57	2·62	2·68
	15	2·53	2·49	2·46	2·44	2·43	2·43	2·45	2·48	2·51	2·56	2·60
	20	2·53	2·49	2·46	2·44	2·42	2·42	2·42	2·44	2·46	2·49	2·53
	30	2·53	2·49	2·46	2·44	2·42	2·40	2·40	2·40	2·42	2·43	2·46
	∞	2·53	2·49	2·46	2·43	2·40	2·38	2·36	2·34	2·33	2·33	2·33
30	10	2·46	2·44	2·42	2·42	2·43	2·46	2·50	2·56	2·63	2·70	2·76
	12	2·46	2·44	2·42	2·41	2·42	2·44	2·47	2·52	2·57	2·62	2·68
	15	2·46	2·44	2·42	2·41	2·41	2·42	2·44	2·47	2·51	2·56	2·60
	20	2·46	2·43	2·42	2·40	2·40	2·40	2·42	2·44	2·46	2·49	2·53
	30	2·46	2·43	2·42	2·40	2·39	2·39	2·39	2·40	2·42	2·43	2·46
	∞	2·46	2·43	2·41	2·39	2·37	2·36	2·35	2·34	2·33	2·33	2·33
∞	10	2·33	2·33	2·34	2·36	2·40	2·44	2·50	2·56	2·63	2·70	2·76
	12	2·33	2·33	2·34	2·36	2·38	2·42	2·46	2·51	2·57	2·62	2·68
	15	2·33	2·33	2·34	2·35	2·37	2·40	2·43	2·47	2·51	2·56	2·60
	20	2·33	2·33	2·33	2·34	2·36	2·38	2·40	2·43	2·46	2·49	2·53
	30	2·33	2·33	2·33	2·34	2·35	2·36	2·37	2·39	2·41	2·43	2·46
	∞	2·33	2·33	2·33	2·33	2·33	2·33	2·33	2·33	2·33	2·33	2·33

* y is normally distributed about η with variance $\lambda_1\sigma_1^2+\lambda_2\sigma_2^2$, and s_1^2 and s_2^2 are independent estimates of σ_1^2 and σ_2^2, based on ν_1 and ν_2 degrees of freedom, respectively. λ_1 and λ_2 are known constants.

In the problem of comparing the means of samples taken from two normal populations, put $y=(\bar{x}_1-\bar{x}_2)$, $\nu_1=(n_1-1)$, $\nu_2=(n_2-1)$, $\lambda_1=1/n_1$ and $\lambda_2=1/n_2$, where n_1 and n_2 are the sample sizes.

Table 11 (*continued*)

Upper 0·5 % *critical values of* $v = \dfrac{y-\eta}{\sqrt{(\lambda_1 s_1^2 + \lambda_2 s_2^2)}}$

(*i.e. upper* 1 % *critical values of* $|\,v\,|$)

ν_2	$\dfrac{\lambda_1 s_1^2}{(\lambda_1 s_1^2+\lambda_2 s_2^2)}$ ν_1	0·0	0·1	0·2	0·3	0·4	0·5	0·6	0·7	0·8	0·9	1·0
10	10	3·17	3·08	3·00	2·90	2·82	2·79	2·82	2·90	3·00	3·08	3·17
	12	3·17	3·08	3·00	2·91	2·82	2·78	2·79	2·84	2·91	2·98	3·05
	15	3·17	3·08	3·00	2·91	2·82	2·77	2·76	2·78	2·83	2·89	2·95
	20	3·17	3·08	3·00	2·91	2·82	2·76	2·73	2·74	2·76	2·80	2·85
	30	3·17	3·08	3·00	2·91	2·82	2·75	2·71	2·69	2·70	2·72	2·75
	∞	3·17	3·08	2·99	2·91	2·82	2·74	2·67	2·63	2·60	2·58	2·58
12	10	3·05	2·98	2·91	2·84	2·79	2·78	2·82	2·91	3·00	3·08	3·17
	12	3·05	2·98	2·91	2·84	2·78	2·76	2·78	2·84	2·91	2·98	3·05
	15	3·05	2·98	2·91	2·84	2·78	2·75	2·75	2·78	2·83	2·89	2·95
	20	3·05	2·98	2·91	2·84	2·78	2·74	2·72	2·73	2·76	2·80	2·85
	30	3·05	2·98	2·91	2·84	2·77	2·73	2·70	2·69	2·70	2·72	2·75
	∞	3·05	2·98	2·91	2·84	2·77	2·71	2·65	2·62	2·59	2·58	2·58
15	10	2·95	2·89	2·83	2·78	2·76	2·77	2·82	2·91	3·00	3·08	3·17
	12	2·95	2·89	2·83	2·78	2·75	2·75	2·78	2·84	2·91	2·98	3·05
	15	2·95	2·89	2·83	2·78	2·74	2·73	2·74	2·78	2·83	2·89	2·95
	20	2·95	2·89	2·83	2·78	2·74	2·71	2·71	2·73	2·76	2·80	2·85
	30	2·95	2·89	2·83	2·78	2·73	2·70	2·68	2·68	2·70	2·72	2·75
	∞	2·95	2·89	2·83	2·77	2·72	2·67	2·64	2·61	2·59	2·58	2·58
20	10	2·85	2·80	2·76	2·74	2·73	2·76	2·82	2·91	3·00	3·08	3·17
	12	2·85	2·80	2·76	2·73	2·72	2·74	2·78	2·84	2·91	2·98	3·05
	15	2·85	2·80	2·76	2·73	2·71	2·71	2·74	2·78	2·83	2·89	2·95
	20	2·85	2·80	2·76	2·73	2·70	2·70	2·70	2·73	2·76	2·80	2·85
	30	2·85	2·80	2·76	2·72	2·69	2·68	2·67	2·68	2·70	2·72	2·75
	∞	2·85	2·80	2·76	2·72	2·68	2·65	2·62	2·60	2·59	2·58	2·58
30	10	2·75	2·72	2·70	2·69	2·71	2·75	2·82	2·91	3·00	3·08	3·17
	12	2·75	2·72	2·70	2·69	2·70	2·73	2·77	2·84	2·91	2·98	3·05
	15	2·75	2·72	2·70	2·68	2·68	2·70	2·73	2·78	2·83	2·89	2·95
	20	2·75	2·72	2·70	2·68	2·67	2·68	2·69	2·72	2·76	2·80	2·85
	30	2·75	2·72	2·69	2·67	2·66	2·66	2·66	2·67	2·69	2·72	2·75
	∞	2·75	2·72	2·69	2·66	2·64	2·62	2·60	2·59	2·58	2·58	2·58
∞	10	2·58	2·58	2·60	2·63	2·67	2·74	2·82	2·91	2·99	3·08	3·17
	12	2·58	2·58	2·59	2·62	2·65	2·71	2·77	2·84	2·91	2·98	3·05
	15	2·58	2·58	2·59	2·61	2·64	2·67	2·72	2·77	2·83	2·89	2·95
	20	2·58	2·58	2·59	2·60	2·62	2·65	2·68	2·72	2·76	2·80	2·85
	30	2·58	2·58	2·58	2·59	2·60	2·62	2·64	2·66	2·69	2·72	2·75
	∞	2·58	2·58	2·58	2·58	2·58	2·58	2·58	2·58	2·58	2·58	2·58

Table 12. *Percentage points of the t-distribution*

ν	$Q=0{\cdot}4$ $2Q=0{\cdot}8$	0·25 0·5	0·1 0·2	0·05 0·1	0·025 0·05	0·01 0·02	0·005 0·01	0·0025 0·005	0·001 0·002	0·0005 0·001
1	0·325	1·000	3·078	6·314	12·706	31·821	63·657	127·32	318·31	636·62
2	·289	0·816	1·886	2·920	4·303	6·965	9·925	14·089	22·327	31·598
3	·277	·765	1·638	2·353	3·182	4·541	5·841	7·453	10·214	12·924
4	·271	·741	1·533	2·132	2·776	3·747	4·604	5·598	7·173	8·610
5	0·267	0·727	1·476	2·015	2·571	3·365	4·032	4·773	5·893	6·869
6	·265	·718	1·440	1·943	2·447	3·143	3·707	4·317	5·208	5·959
7	·263	·711	1·415	1·895	2·365	2·998	3·499	4·029	4·785	5·408
8	·262	·706	1·397	1·860	2·306	2·896	3·355	3·833	4·501	5·041
9	·261	·703	1·383	1·833	2·262	2·821	3·250	3·690	4·297	4·781
10	0·260	0·700	1·372	1·812	2·228	2·764	3·169	3·581	4·144	4·587
11	·260	·697	1·363	1·796	2·201	2·718	3·106	3·497	4·025	4·437
12	·259	·695	1·356	1·782	2·179	2·681	3·055	3·428	3·930	4·318
13	·259	·694	1·350	1·771	2·160	2·650	3·012	3·372	3·852	4·221
14	·258	·692	1·345	1·761	2·145	2·624	2·977	3·326	3·787	4·140
15	0·258	0·691	1·341	1·753	2·131	2·602	2·947	3·286	3·733	4·073
16	·258	·690	1·337	1·746	2·120	2·583	2·921	3·252	3·686	4·015
17	·257	·689	1·333	1·740	2·110	2·567	2·898	3·222	3·646	3·965
18	·257	·688	1·330	1·734	2·101	2·552	2·878	3·197	3·610	3·922
19	·257	·688	1·328	1·729	2·093	2·539	2·861	3·174	3·579	3·883
20	0·257	0·687	1·325	1·725	2·086	2·528	2·845	3·153	3·552	3·850
21	·257	·686	1·323	1·721	2·080	2·518	2·831	3·135	3·527	3·819
22	·256	·686	1·321	1·717	2·074	2·508	2·819	3·119	3·505	3·792
23	·256	·685	1·319	1·714	2·069	2·500	2·807	3·104	3·485	3·767
24	·256	·685	1·318	1·711	2·064	2·492	2·797	3·091	3·467	3·745
25	0·256	0·684	1·316	1·708	2·060	2·485	2·787	3·078	3·450	3·725
26	·256	·684	1·315	1·706	2·056	2·479	2·779	3·067	3·435	3·707
27	·256	·684	1·314	1·703	2·052	2·473	2·771	3·057	3·421	3·690
28	·256	·683	1·313	1·701	2·048	2·467	2·763	3·047	3·408	3·674
29	·256	·683	1·311	1·699	2·045	2·462	2·756	3·038	3·396	3·659
30	0·256	0·683	1·310	1·697	2·042	2·457	2·750	3·030	3·385	3·646
40	·255	·681	1·303	1·684	2·021	2·423	2·704	2·971	3·307	3·551
60	·254	·679	1·296	1·671	2·000	2·390	2·660	2·915	3·232	3·460
120	·254	·677	1·289	1·658	1·980	2·358	2·617	2·860	3·160	3·373
∞	·253	·674	1·282	1·645	1·960	2·326	2·576	2·807	3·090	3·291

$Q=1-P(t|\nu)$ is the upper-tail area of the distribution for ν degrees of freedom, appropriate for use in a single-tail test. For a two-tail test, $2Q$ must be used.

Table 13. *Percentage points for the distribution of the correlation coefficient, r, when* $\rho=0$

ν	$Q=0{\cdot}05$ $2Q=0{\cdot}1$	0·025 0·05	0·01 0·02	0·005 0·01	0·0025 0·005	0·0005 0·001	ν	$Q=0{\cdot}05$ $2Q=0{\cdot}1$	0·025 0·05	0·01 0·02	0·005 0·01	0·0025 0·005	0·0005 0·001
1	0·9877	$0{\cdot}9^2692$	$0{\cdot}9^3507$	$0{\cdot}9^3877$	$0{\cdot}9^4692$	$0{\cdot}9^5877$	16	0·400	0·468	0·543	0·590	0·631	0·708
2	·9000	·9500	·9800	$\cdot 9^2000$	$\cdot 9^2500$	$\cdot 9^3000$	17	·389	·456	·529	·575	·616	·693
3	·805	·878	·9343	·9587	·9740	$\cdot 9^2114$	18	·378	·444	·516	·561	·602	·679
4	·729	·811	·882	·9172	·9417	·9741	19	·369	·433	·503	·549	·589	·665
5	·669	·754	·833	·875	·9056	·9509	20	·360	·423	·492	·537	·576	·652
6	0·621	0·707	0·789	0·834	0·870	0·9249	25	0·323	0·381	0·445	0·487	0·524	0·597
7	·582	·666	·750	·798	·836	·898	30	·296	·349	·409	·449	·484	·554
8	·549	·632	·715	·765	·805	·872	35	·275	·325	·381	·418	·452	·519
9	·521	·602	·685	·735	·776	·847	40	·257	·304	·358	·393	·425	·490
10	·497	·576	·658	·708	·750	·823	45	·243	·288	·338	·372	·403	·465
11	0·476	0·553	0·634	0·684	0·726	0·801	50	0·231	0·273	0·322	0·354	0·384	0·443
12	·457	·532	·612	·661	·703	·780	60	·211	·250	·295	·325	·352	·408
13	·441	·514	·592	·641	·683	·760	70	·195	·232	·274	·302	·327	·380
14	·426	·497	·574	·623	·664	·742	80	·183	·217	·257	·283	·307	·357
15	·412	·482	·558	·606	·647	·725	90	·173	·205	·242	·267	·290	·338
							100	·164	·195	·230	·254	·276	·321

$Q=1-P(r|\nu, \rho=0)$ is the upper-tail area of the distribution of r appropriate for use in a single-tail test. For a two-tail test, $2Q$ must be used. If r is calculated from n paired observations, enter the table with $\nu=n-2$. For partial correlations enter with $\nu=n-k-2$, where k is the number of variables held constant.

The columns for $2Q=0{\cdot}2$, 0·02 and 0·001 in Table 12 and $2Q=0{\cdot}02$ and 0·001 in Table 13 have been taken from *Statistical Tables for Biological, Agricultural and Medical Research*, Tables III and VI (Fisher & Yates, 1963) by permission of the authors and the publishers, Messrs Oliver and Boyd.

Table 14. *The z-transformation of the correlation coefficient*, $z = \tanh^{-1} r$

	r (3rd decimal)					Proportional parts, for right side→										*r* (3rd decimal)					
r	**·000**	**·002**	**·004**	**·006**	**·008**	**1**	**2**	**3**	**4**	**5**	**6**	**7**	**8**	**9**	**10**	**·000**	**·002**	**·004**	**·006**	**·008**	***r***
·00	·0000	·0020	·0040	·0060	·0080	1	3	4	5	7	8	9	11	12	13	·5493	·5520	·5547	·5573	·5600	**·50**
1	·0100	·0120	·0140	·0160	·0180	1	3	4	5	7	8	10	11	12	14	·5627	·5654	·5682	·5709	·5736	**1**
2	·0200	·0220	·0240	·0260	·0280	1	3	4	6	7	8	10	11	13	14	·5763	·5791	·5818	·5846	·5874	**2**
3	·0300	·0320	·0340	·0360	·0380	1	3	4	6	7	8	10	11	13	14	·5901	·5929	·5957	·5985	·6013	**3**
4	·0400	·0420	·0440	·0460	·0480	1	3	4	6	7	9	10	11	13	14	·6042	·6070	·6098	·6127	·6155	**4**
·05	·0500	·0520	·0541	·0561	·0581	1	3	4	6	7	9	10	12	13	14	·6184	·6213	·6241	·6270	·6299	**·55**
6	·0601	·0621	·0641	·0661	·0681	1	3	4	6	7	9	10	12	13	15	·6328	·6358	·6387	·6416	·6446	**6**
7	·0701	·0721	·0741	·0761	·0782	1	3	4	6	7	9	10	12	14	15	·6475	·6505	·6535	·6565	·6595	**7**
8	·0802	·0822	·0842	·0862	·0882	2	3	5	6	8	9	11	12	14	15	·6625	·6655	·6685	·6716	·6746	**8**
9	·0902	·0923	·0943	·0963	·0983	2	3	5	6	8	9	11	12	14	15	·6777	·6807	·6838	·6869	·6900	**9**
·10	·1003	·1024	·1044	·1064	·1084	2	3	5	6	8	9	11	13	14	16	·6931	·6963	·6994	·7026	·7057	**·60**
1	·1104	·1125	·1145	·1165	·1186	2	3	5	6	8	10	11	13	14	16	·7089	·7121	·7153	·7185	·7218	**1**
2	·1206	·1226	·1246	·1267	·1287	2	3	5	7	8	10	11	13	15	16	·7250	·7283	·7315	·7348	·7381	**2**
3	·1307	·1328	·1348	·1368	·1389	2	3	5	7	8	10	12	13	15	17	·7414	·7447	·7481	·7514	·7548	**3**
4	·1409	·1430	·1450	·1471	·1491	2	3	5	7	9	10	12	14	15	17	·7582	·7616	·7650	·7684	·7718	**4**
·15	·1511	·1532	·1552	·1573	·1593	2	4	5	7	9	11	12	14	16	18	·7753	·7788	·7823	·7858	·7893	**·65**
6	·1614	·1634	·1655	·1676	·1696	2	4	5	7	9	11	13	14	16	18	·7928	·7964	·7999	·8035	·8071	**6**
7	·1717	·1737	·1758	·1779	·1799	2	4	6	7	9	11	13	15	17	18	·8107	·8144	·8180	·8217	·8254	**7**
8	·1820	·1841	·1861	·1882	·1903	2	4	6	8	9	11	13	15	17	19	·8291	·8328	·8366	·8404	·8441	**8**
9	·1923	·1944	·1965	·1986	·2007	2	4	6	8	10	12	14	15	17	19	·8480	·8518	·8556	·8595	·8634	**9**
·20	·2027	·2048	·2069	·2090	·2111	2	4	6	8	10	12	14	16	18	20	·8673	·8712	·8752	·8792	·8832	**·70**
1	·2132	·2153	·2174	·2195	·2216	2	4	6	8	10	12	15	16	18	20	·8872	·8912	·8953	·8994	·9035	**1**
2	·2237	·2258	·2279	·2300	·2321	2	4	6	8	11	13	15	17	19	21	·9076	·9118	·9160	·9202	·9245	**2**
3	·2342	·2363	·2384	·2405	·2427	2	4	7	9	11	13	15	17	20	22	·9287	·9330	·9373	·9417	·9461	**3**
4	·2448	·2469	·2490	·2512	·2533	2	4	7	9	11	13	16	18	20	22	·9505	·9549	·9594	·9639	·9684	**4**
·25	·2554	·2575	·2597	·2618	·2640	1	2	3	4	5	6	7	9	10	11	0·973	0·978	0·982	0·987	0·991	**·75**
6	·2661	·2683	·2704	·2726	·2747	1	2	3	4	5	6	8	9	10	11	0·996	1·001	1·006	1·011	1·015	**6**
7	·2769	·2790	·2812	·2833	·2855	1	2	3	4	5	6	8	9	10	11	1·020	1·025	1·030	1·035	1·040	**7**
8	·2877	·2899	·2920	·2942	·2964	1	2	3	4	5	7	8	9	10	11	1·045	1·050	1·056	1·061	1·066	**8**
9	·2986	·3008	·3029	·3051	·3073	1	2	3	4	5	7	8	9	10	11	1·071	1·077	1·082	1·088	1·093	**9**
·30	·3095	·3117	·3139	·3161	·3183	1	2	3	4	6	7	8	9	10	11	1·099	1·104	1·110	1·116	1·121	**·80**
1	·3205	·3228	·3250	·3272	·3294	1	2	3	4	6	7	8	9	10	11	1·127	1·133	1·139	1·145	1·151	**1**
2	·3316	·3339	·3361	·3383	·3406	1	2	3	4	6	7	8	9	10	11	1·157	1·163	1·169	1·175	1·182	**2**
3	·3428	·3451	·3473	·3496	·3518	1	2	3	5	6	7	8	9	10	11	1·188	1·195	1·201	1·208	1·214	**3**
4	·3541	·3564	·3586	·3609	·3632	1	2	3	5	6	7	8	9	10	11	1·221	1·228	1·235	1·242	1·249	**4**
·35	·3654	·3677	·3700	·3723	·3746	1	2	3	5	6	7	8	9	10	11	1·256	1·263	1·271	1·278	1·286	**·85**
6	·3769	·3792	·3815	·3838	·3861	1	2	3	5	6	7	8	9	10	12	1·293	1·301	1·309	1·317	1·325	**6**
7	·3884	·3907	·3931	·3954	·3977	1	2	3	5	6	7	8	9	10	12	1 333	1·341	1·350	1·358	1·367	**7**
8	·4001	·4024	·4047	·4071	·4094	1	2	4	5	6	7	8	9	11	12	1·376	1·385	1·394	1·403	1·412	**8**
9	·4118	·4142	·4165	·4189	·4213	1	2	4	5	6	7	8	9	11	12	1·422	1·432	1·442	1·452	1·462	**9**
·40	·4236	·4260	·4284	·4308	·4332	1	2	4	5	6	7	8	10	11	12	1·472	1·483	1·494	1·505	1·516	**·90**
1	·4356	·4380	·4404	·4428	·4453	1	2	4	5	6	7	8	10	11	12	1·528	1·539	1·551	1·564	1·576	**1**
2	·4477	·4501	·4526	·4550	·4574	1	2	4	5	6	7	9	10	11	12	1·589	1·602	1·616	1·630	1·644	**2**
3	·4599	·4624	·4648	·4673	·4698	1	2	4	5	6	7	9	10	11	12	1·658	1·673	1·689	1·705	1·721	**3**
4	·4722	·4747	·4772	·4797	·4822	1	2	4	5	6	7	9	10	11	12	1·738	1·756	1·774	1·792	1·812	**4**
·45	·4847	·4872	·4897	·4922	·4948	1	3	4	5	6	8	9	10	11	13	1·832	1·853	1·874	1·897	1·921	**·95**
6	·4973	·4999	·5024	·5049	·5075	1	3	4	5	6	8	9	10	11	13	1·946	1·972	2·000	2·029	2·060	**6**
7	·5101	·5126	·5152	·5178	·5204	1	3	4	5	6	8	9	10	12	13	2·092	2·127	2·165	2·205	2·249	**7**
8	·5230	·5256	·5282	·5308	·5334	1	3	4	5	7	8	9	10	12	13	2·298	2·351	2·410	2·477	2·555	**8**
9	·5361	·5387	·5413	·5440	·5466	1	3	4	5	7	8	9	11	12	13	2·647	2·759	2·903	3·106	3·453	**9**
r	**·000**	**·002**	**·004**	**·006**	**·008**	**1**	**2**	**3**	**4**	**5**	**6**	**7**	**8**	**9**	**10**	**·000**	**·002**	**·004**	**·006**	**·008**	***r***
	r (3rd decimal)					←Proportional parts, for left side										*r* (3rd decimal)					

Interpolation

(1) $0 \leqslant r \leqslant 0{\cdot}25$: find argument r_0 nearest to r and form $z = z(r_0) + \Delta r$ (where $\Delta r = r - r_0$), e.g. for $r = 0{\cdot}2042$, $z = 0{\cdot}2069 + 0{\cdot}0002 = 0{\cdot}2071$.

(2) $0{\cdot}25 \leqslant r \leqslant 0{\cdot}75$: find argument r_0 nearest to r and form $z = z(r_0) \pm P$, where P is the proportional part for $\Delta r = r - r_0$, e.g. for $r = 0{\cdot}5146$, $z = 0{\cdot}5682 + 0{\cdot}0008 = 0{\cdot}5690$; for $r = 0{\cdot}5372$, $z = 0{\cdot}6013 - 0{\cdot}0011 = 0{\cdot}6002$.

(3) $0{\cdot}75 \leqslant r \leqslant 0{\cdot}98$: use linear interpolation to get 3-decimal place accuracy.

(4) $0{\cdot}98 \leqslant r < 1$: form $z = -\frac{1}{2} \log_e (1 - r) + 0{\cdot}097 + \frac{1}{4} r$, with the help of Table 53.

Table 15. *Chart giving confidence limits for the population correlation coefficient, ρ, given the sample coefficient, r. Confidence coefficient,* $1-2\alpha = 0{\cdot}95$

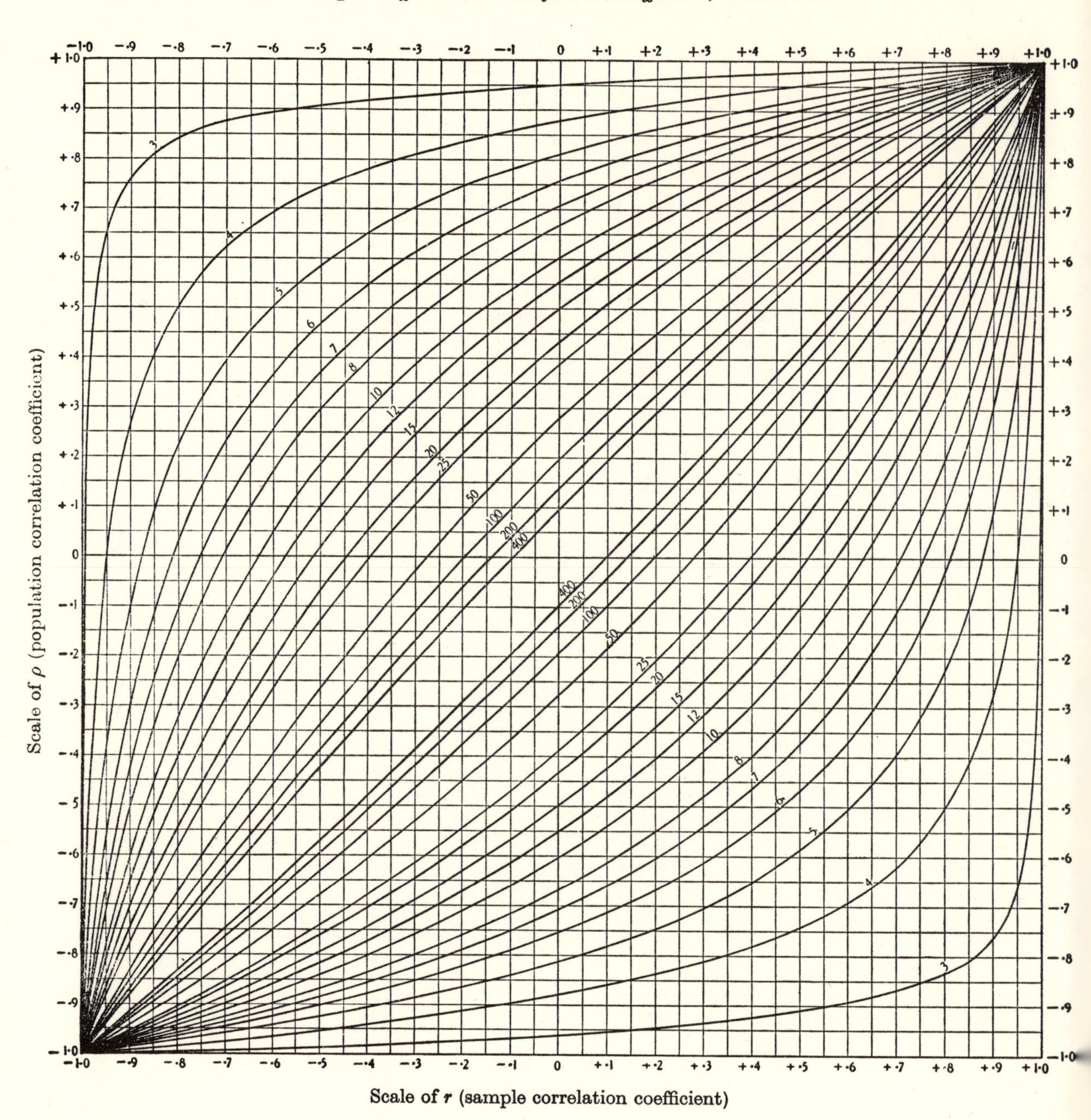

The numbers on the curves indicate sample size. The chart can also be used to determine upper and lower 2·5 % significance points for r, given ρ.

Table 15 (*continued*). *Confidence coefficient*, $1-2\alpha=0{\cdot}99$

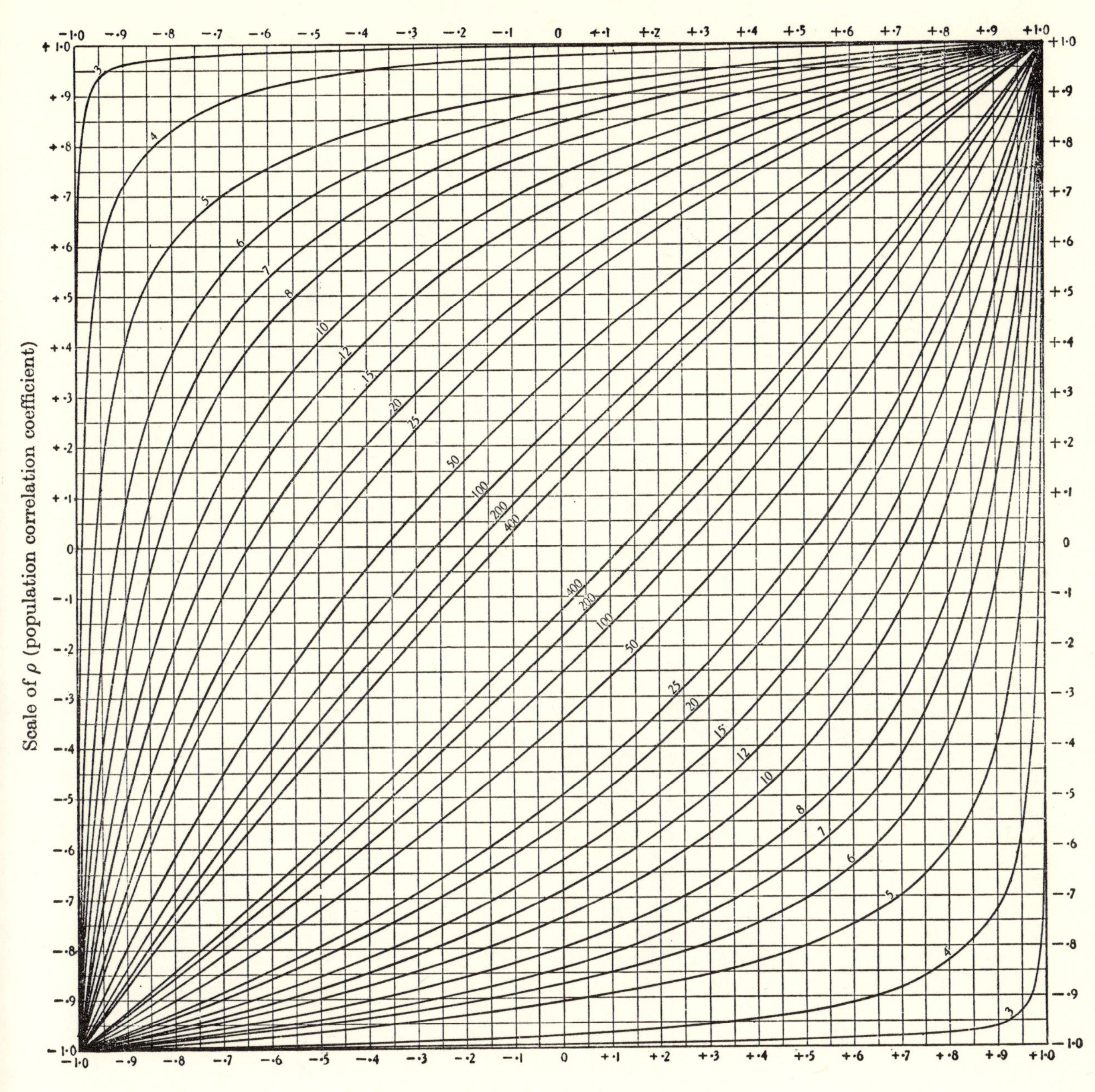

The numbers on the curves indicate sample size. The chart can also be used to determine upper and lower 0·5 % significance points for r, given ρ.

Table 16. *Percentage points of the* B*-distribution*

50% *points for x*

$\nu_1 = 2b, \quad \nu_2 = 2a$

ν_2 \ ν_1	1	2	3	4	5	6	7	8	9
1	0·50000	0·25000	0·16319	0·12061	0·095526	0·079033	0·067378	0·058711	0·052015
2	·75000	·50000	·37004	·29289	·24214	·20630	·17966	·15910	·14276
3	·83681	·62996	·50000	·41363	·35245	·30695	·27181	·24386	·22112
4	·87939	·70711	·58637	·50000	·43555	·38573	·34609	·31381	·28703
5	0·90447	0·75786	0·64755	0·56445	0·50000	0·44867	0·40684	0·37213	0·34286
6	·92097	·79370	·69305	·61427	·55133	·50000	·45737	·42141	·39068
7	·93262	·82034	·72819	·65391	·59316	·54263	·50000	·46355	·43205
8	·94129	·84090	·75614	·68619	·62787	·57859	·53645	·50000	·46818
9	·94799	·85724	·77888	·71297	·65714	·60932	·56795	·53182	·50000
10	0·95331	0·87055	0·79775	0·73555	0·68214	0·63588	0·59546	0·55984	0·52824
11	·95765	·88159	·81366	·75484	·70376	·65907	·61968	·58471	·55346
12	·96125	·89090	·82725	·77151	·72262	·67948	·64116	·60692	·57613
13	·96429	·89885	·83899	·78606	·73923	·69759	·66035	·62687	·59661
14	·96689	·90572	·84924	·79887	·75396	·71376	·67760	·64490	·61520
15	0·96913	0·91172	0·85827	0·81023	0·76712	0·72830	0·69318	0·66127	0·63216
16	·97109	·91700	·86627	·82038	·77894	·74143	·70732	·67620	·64768
17	·97282	·92169	·87342	·82950	·78963	·75334	·72022	·68986	·66195
18	·97435	·92587	·87985	·83774	·79932	·76421	·73203	·70242	·67511
19	·97572	·92964	·88565	·84522	·80817	·77417	·74288	·71401	·68729
20	0·97695	0·93303	0·89092	0·85204	0·81626	0·78331	0·75289	0·72472	0·69858
21	·97806	·93612	·89573	·85828	·82371	·79175	·76215	·73467	·70909
22	·97907	·93893	·90013	·86402	·83057	·79955	·77074	·74392	·71889
23	·97999	·94151	·90417	·86931	·83692	·80679	·77873	·75254	·72805
24	·98083	·94387	·90790	·87421	·84281	·81353	·78619	·76061	·73663
25	0·98161	0·94606	0·91135	0·87875	0·84828	0·81981	0·79315	0·76816	0·74469
26	·98232	·94808	·91455	·88298	·85340	·82568	·79968	·77526	·75227
27	·98298	·94995	·91753	·88692	·85817	·83118	·80581	·78193	·75941
28	·98360	·95170	·92031	·89060	·86265	·83635	·81157	·78821	·76615
29	·98417	·95332	·92290	·89406	·86685	·84120	·81701	·79415	·77253
30	0·98470	0·95484	0·92534	0·89730	0·87080	0·84578	0·82214	0·79976	0·77857
40	·98855	·96594	·94324	·92136	·90038	·88030	·86107	·84266	·82501
60	·99238	·97716	·96164	·94645	·93166	·91731	·90338	·88985	·87672
120	·99620	·98851	·98056	·97264	·96482	·95710	·94951	·94202	·93465
∞	1·00000	1·00000	1·00000	1·00000	1·00000	1·00000	1·00000	1·00000	1·00000

This table gives the values of x for which

$$I_x(a, b) = \int_0^x u^{a-1}(1-u)^{b-1}\,du \Big/ \int_0^1 u^{a-1}(1-u)^{b-1}\,du = 0{\cdot}50,$$

where $a = \frac{1}{2}\nu_2$, $b = \frac{1}{2}\nu_1$.

Table 16 (*continued*)

50% points for x

$\nu_1 = 2b, \quad \nu_2 = 2a$

ν_2 \ ν_1	10	12	15	20	24	30	40	60	120
1	0·046687	0·038746	0·030867	0·023051	0·019168	0·015301	0·011450	0·0076165	0·0037997
2	·12945	·10910	·088278	·066967	·056126	·045158	·034064	·022840	·011486
3	·20225	·17275	·14173	·10908	·092099	·074664	·056756	·038355	·019444
4	·26445	·22849	·18977	·14796	·12579	·10270	·078644	·053552	·027361
5	0·31786	0·27738	0·23288	0·18374	0·15719	0·12920	0·099622	0·068335	0·035184
6	·36412	·32052	·27170	·21669	·18647	·15422	·11970	·082690	·042896
7	·40454	·35884	·30682	·24711	·21381	·17786	·13893	·096624	·050494
8	·44016	·39308	·33873	·27528	·23939	·20024	·15734	·11015	·057977
9	·47176	·42387	·36784	·30142	·26337	·22143	·17499	·12328	·065345
10	0·50000	0·45169	0·39451	0·32575	0·28589	0·24154	0·19192	0·13603	0·072602
11	·52538	·47696	·41902	·34845	·30707	·26064	·20818	·14842	·079747
12	·54831	·50000	·44162	·36967	·32704	·27880	·22379	·16046	·086785
13	·56912	·52110	·46254	·38956	·34589	·29610	·23880	·17217	·093716
14	·58811	·54049	·48194	·40823	·36371	·31258	·25325	·18355	·10054
15	0·60549	0·55838	0·50000	0·42579	0·38059	0·32832	0·26715	0·19463	0·10727
16	·62147	·57492	·51684	·44234	·39660	·34334	·28055	·20541	·11390
17	·63621	·59027	·53258	·45797	·41181	·35772	·29347	·21591	·12043
18	·64984	·60456	·54733	·47274	·42626	·37147	·30593	·22613	·12686
19	·66248	·61788	·56118	·48673	·44002	·38465	·31796	·23609	·13320
20	0·67425	0·63033	0·57421	0·50000	0·45314	0·39729	0·32958	0·24580	0·13945
21	·68522	·64200	·58649	·51260	·46566	·40942	·34082	·25527	·14561
22	·69548	·65295	·59807	·52458	·47762	·42108	·35168	·26450	·15168
23	·70509	·66325	·60903	·53599	·48905	·43228	·36219	·27350	·15767
24	·71411	·67296	·61941	·54686	·50000	·44305	·37237	·28229	·16357
25	0·72260	0·68213	0·62924	0·55723	0·51049	0·45343	0·38222	0·29086	0·16939
26	·73060	·69079	·63859	·56714	·52054	·46342	·39177	·29924	·17513
27	·73815	·69900	·64747	·57662	·53019	·47306	·40103	·30741	·18080
28	·74529	·70678	·65593	·58569	·53946	·48236	·41001	·31540	·18638
29	·75205	·71417	·66399	·59438	·54837	·49133	·41873	·32321	·19189
30	0·75846	0·72120	0·67168	0·60271	0·55695	0·50000	0·42720	0·33084	0·19732
40	·80808	·77621	·73285	·67042	·62763	·57280	·50000	·39866	·24791
60	·86397	·83954	·80537	·75420	·71771	·66916	·60134	·50000	·33209
120	·92740	·91321	·89273	·86055	·83643	·80268	·75209	·66791	·50000
∞	1·00000	1·00000	1·00000	1·00000	1·00000	1·00000	1·00000	1·00000	1·00000

For $\nu_1 = \infty$, $x = 0$

Table 16. *Percentage points of the* B-*distribution* (*continued*)

Lower 25 % *points for x*

$\nu_1 = 2b, \quad \nu_2 = 2a$

ν_2 \ ν_1	1	2	3	4	5	6	7	8	9
1	0·14645	0·062500	0·039063	0·028310	0·022173	0·018215	0·015452	0·013416	0·011853
2	·43750	·25000	·17452	·13397	·10870	·091440	·078908	·069395	·061929
3	·59715	·39685	·29801	·23885	·19937	·17113	·14991	·13339	·12015
4	·68878	·50000	·39448	·32635	·27852	·24302	·21560	·19376	·17596
5	0·74711	0·57435	0·46936	0·39775	0·34546	0·30550	0·27390	0·24828	0·22707
6	·78726	·62996	·52848	·45632	·40198	·35944	·32516	·29692	·27323
7	·81650	·67295	·57609	·50494	·45001	·40614	·37021	·34022	·31478
8	·83872	·70711	·61516	·54582	·49117	·44680	·40996	·37885	·35219
9	·85616	·73487	·64773	·58060	·52678	·48245	·44521	·41343	·38597
10	0·87021	0·75786	0·67529	0·61052	0·55783	0·51390	0·47662	0·44451	0·41655
11	·88177	·77720	·69888	·63651	·58513	·54184	·50475	·47257	·44435
12	·89144	·79370	·71931	·65929	·60930	·56679	·53009	·49801	·46970
13	·89966	·80793	·73716	·67941	·63085	·58921	·55300	·52116	·49289
14	·90672	·82034	·75288	·69730	·65017	·60946	·57382	·54230	·51419
15	0·91285	0·83124	0·76684	0·71332	0·66758	0·62782	0·59282	0·56169	0·53380
16	·91823	·84090	·77932	·72773	·68336	·64456	·61021	·57953	·55192
17	·92298	·84951	·79053	·74077	·69772	·65986	·62619	·59599	·56870
18	·92721	·85724	·80066	·75263	·71084	·67391	·64093	·61122	·58428
19	·93100	·86422	·80986	·76345	·72287	·68686	·65456	·62536	·59879
20	0·93442	0·87055	0·81825	0·77337	0·73395	0·69882	0·66720	0·63852	0·61234
21	·93751	·87632	·82593	·78250	·74418	·70991	·67895	·65079	·62500
22	·94033	·88159	·83299	·79092	·75366	·72021	·68991	·66226	·63688
23	·94290	·88644	·83950	·79871	·76246	·72981	·70015	·67300	·64803
24	·94526	·89090	·84553	·80595	·77066	·73878	·70973	·68309	·65852
25	0·94744	0·89503	0·85112	0·81268	0·77831	0·74717	0·71873	0·69258	0·66841
26	·94944	·89885	·85632	·81896	·78547	·75505	·72719	·70152	·67774
27	·95130	·90241	·86116	·82484	·79218	·76244	·73515	·70995	·68657
28	·95303	·90572	·86570	·83035	·79848	·76941	·74267	·71793	·69492
29	·95464	·90882	·86994	·83552	·80442	·77598	·74977	·72548	·70285
30	0·95614	0·91172	0·87393	0·84039	0·81002	0·78219	0·75649	0·73263	0·71038
40	·96706	·93303	·90351	·87685	·85230	·82947	·80809	·78797	·76896
60	·97801	·95484	·93434	·91548	·89782	·88113	·86525	·85009	·83557
120	·98899	·97716	·96648	·95647	·94692	·93774	·92887	·92025	·91187
∞	1·00000	1·00000	1·00000	1·00000	1·00000	1·00000	1·00000	1·00000	1·00000

This table gives the values of x for which

$$I_x(a, b) = \int_0^x u^{a-1}(1-u)^{b-1}\,du \bigg/ \int_0^1 u^{a-1}(1-u)^{b-1}\,du = 0{\cdot}25,$$

where $a = \frac{1}{2}\nu_2$, $b = \frac{1}{2}\nu_1$.

Table 16 (*continued*)

Lower 25 % points for x

$\nu_1 = 2b, \quad \nu_2 = 2a$

ν_2 \ ν_1	10	12	15	20	24	30	40	60	120
1	0·010616	0·0087814	0·0069734	0·0051914	0·0043101	0·0034353	0·0025669	0·0017049	$0 \cdot 0^3 84926$
2	·055912	·046816	·037631	·028358	·023688	·018996	·014281	·0095436	·0047832
3	·10930	·092592	·075324	·057467	·048307	·038986	·029500	·019844	·010012
4	·16116	·13797	·11350	·087610	·074095	·060174	·045827	·031031	·015764
5	0·20922	0·18082	0·15026	0·11726	0·099749	0·081498	0·062458	0·042571	0·021774
6	·25307	·22058	·18500	·14585	·12475	·10251	·079043	·054222	·027923
7	·29291	·25724	·21758	·17316	·14887	·12301	·095405	·065857	·034143
8	·32908	·29099	·24802	·19913	·17203	·14289	·11145	·077403	·040396
9	·36198	·32205	·27644	·22376	·19420	·16211	·12712	·088814	·046656
10	0·39196	0·35068	0·30297	0·24710	0·21538	0·18064	0·14240	0·10006	0·052904
11	·41938	·37712	·32776	·26921	·23562	·19850	·15726	·11113	·059127
12	·44451	·40158	·35094	·29017	·25493	·21570	·17171	·12200	·065317
13	·46762	·42426	·37265	·31004	·27338	·23226	·18575	·13267	·071466
14	·48893	·44534	·39302	·32889	·29100	·24819	·19938	·14314	·077570
15	0·50863	0·46497	0·41215	0·34679	0·30784	0·26353	0·21261	0·15341	0·083624
16	·52691	·48330	·43016	·36380	·32395	·27831	·22546	·16347	·089625
17	·54389	·50043	·44712	·37998	·33936	·29254	·23793	·17332	·095571
18	·55972	·51649	·46312	·39539	·35411	·30625	·25004	·18298	·10146
19	·57449	·53156	·47825	·41008	·36824	·31946	·26179	·19244	·10729
20	0·58832	0·54574	0·49256	0·42409	0·38179	0·33221	0·27321	0·20170	0·11307
21	·60129	·55909	·50613	·43746	·39478	·34450	·28430	·21078	·11878
22	·61348	·57169	·51900	·45025	·40726	·35637	·29507	·21966	·12443
23	·62495	·58360	·53122	·46247	·41925	·36783	·30554	·22837	·13003
24	·63576	·59487	·54285	·47418	·43078	·37891	·31572	·23690	·13556
25	0·64597	0·60556	0·55392	0·48539	0·44186	0·38961	0·32561	0·24525	0·14103
26	·65563	·61570	·56447	·49615	·45253	·39996	·33524	·25343	·14645
27	·66478	·62533	·57455	·50647	·46281	·40997	·34460	·26145	·15180
28	·67346	·63450	·58417	·51639	·47272	·41967	·35371	·26931	·15710
29	·68170	·64323	·59337	·52591	·48228	·42906	·36259	·27701	·16234
30	0·68954	0·65156	0·60217	0·53508	0·49150	0·43815	0·37122	0·28455	0·16752
40	·75095	·71758	·67308	·61054	·56853	·51555	·44650	·35245	·21626
60	·82163	·79529	·75915	·70620	·66914	·62056	·55390	·45636	·29913
120	·90370	·88794	·86554	·83103	·80557	·77041	·71852	·63381	·46918
∞	1·00000	1·00000	1·00000	1·00000	1·00000	1·00000	1·00000	1·00000	1·00000

For $\nu_1 = \infty$, $x = 0$

Table 16. *Percentage points of the* B-*distribution* (*continued*)

Lower 10 % *points for* x

$\nu_1 = 2b, \quad \nu_2 = 2a$

$\nu_2 \backslash \nu_1$	1	2	3	4	5	6	7	8	9
1	0·024472	0·010000	0·0061812	0·0044577	0·0034819	0·0028553	0·0024193	0·0020986	0·0018528
2	·19000	·10000	·067830	·051317	·041268	·034511	·029654	·025996	·023141
3	·35136	·21544	·15648	·12310	·10154	·086434	·075257	·066647	·059809
4	·46812	·31623	·24136	·19580	·16493	·14256	·12558	·11223	·10147
5	0·55185	0·39811	0·31529	0·26204	0·22457	0·19664	0·17498	0·15766	0·14349
6	·61375	·46416	·37816	·32046	·27858	·24664	·22139	·20091	·18394
7	·66104	·51795	·43151	·37151	·32685	·29210	·26421	·24127	·22207
8	·69821	·56234	·47700	·41611	·36982	·33319	·30339	·27860	·25764
9	·72814	·59948	·51610	·45522	·40811	·37029	·33915	·31299	·29067
10	0·75273	0·63096	0·54996	0·48968	0·44232	0·40382	0·37178	0·34462	0·32128
11	·77328	·65793	·57954	·52022	·47300	·43419	·40159	·37374	·34963
12	·79069	·68129	·60555	·54744	·50062	·46178	·42889	·40058	·37592
13	·80564	·70170	·62860	·57181	·52560	·48693	·45393	·42535	·40032
14	·81861	·71969	·64915	·59375	·54827	·50992	·47697	·44827	·42299
15	0·82996	0·73564	0·66758	0·61360	0·56893	0·53100	0·49822	0·46951	0·44410
16	·83998	·74989	·68419	·63164	·58783	·55040	·51787	·48924	·46380
17	·84889	·76270	·69923	·64809	·60517	·56829	·53608	·50760	·48219
18	·85686	·77426	·71293	·66315	·62114	·58484	·55300	·52473	·49942
19	·86403	·78476	·72544	·67699	·63588	·60020	·56876	·54074	·51557
20	0·87052	0·79433	0·73691	0·68976	0·64954	0·61448	0·58347	0·55574	0·53073
21	·87643	·80309	·74747	·70156	·66222	·62779	·59722	·56980	·54500
22	·88181	·81113	·75722	·71250	·67403	·64022	·61011	·58302	·55845
23	·88675	·81855	·76625	·72268	·68504	·65187	·62222	·59547	·57115
24	·89129	·82540	·77464	·73216	·69535	·66279	·63361	·60721	·58314
25	0·89549	0·83176	0·78245	0·74103	0·70500	0·67305	0·64434	0·61829	0·59450
26	·89937	·83768	·78973	·74933	·71407	·68271	·65446	·62878	·60526
27	·90297	·84319	·79655	·75711	·72260	·69183	·66403	·63871	·61548
28	·90633	·84834	·80294	·76443	·73064	·70044	·67310	·64813	·62518
29	·90946	·85317	·80894	·77132	·73823	·70858	·68168	·65708	·63442
30	0·91239	0·85770	0·81459	0·77783	0·74541	0·71630	0·68984	0·66559	0·64322
40	·93381	·89125	·85693	·82706	·80025	·77578	·75319	·73219	·71255
60	·95555	·92612	·90182	·88023	·86048	·84213	·82490	·80864	·79321
120	·97761	·96235	·94944	·93773	·92679	·91643	·90653	·89702	·88785
∞	1·00000	1·00000	1·00000	1·00000	1·00000	1·00000	1·00000	1·00000	1·00000

This table gives the values of x for which

$$I_x(a, b) = \int_0^x u^{a-1}(1-u)^{b-1}\,du \Big/ \int_0^1 u^{a-1}(1-u)^{b-1}\,du = 0{\cdot}10,$$

where $a = \frac{1}{2}\nu_2$, $b = \frac{1}{2}\nu_1$.

Table 16 (*continued*)

Lower 10% *points for* x

$\nu_1 = 2b, \quad \nu_2 = 2a$

$\nu_2 \backslash \nu_1$	10	12	15	20	24	30	40	60	120
1	0·0016585	0·0013709	0·0010878	$0{\cdot}0^{3}80919$	$0{\cdot}0^{3}67157$	$0{\cdot}0^{3}53506$	$0{\cdot}0^{3}39965$	$0{\cdot}0^{3}26535$	$0{\cdot}0^{3}13213$
2	·020852	·017407	·013950	·010481	·0087416	·0069994	·0052542	·0035059	·0017545
3	·054246	·045740	·037035	·028119	·023579	·018982	·014327	·0096132	·0048379
4	·092595	·078823	·064459	·049452	·041691	·033749	·025617	·017288	·0087521
5	0·13167	0·11307	0·093336	0·072324	0·061295	0·049889	0·038083	0·025851	0·013167
6	·16964	·14685	·12228	·095653	·081477	·066668	·051174	·034941	·017906
7	·20573	·17941	·15059	·11886	·10173	·083668	·064573	·044345	·022866
8	·23966	·21040	·17792	·14161	·12177	·10064	·078083	·053928	·027978
9	·27139	·23970	·20411	·16374	·14141	·11743	·091577	·063600	·033197
10	0·30097	0·26732	0·22908	0·18513	0·16056	0·13394	0·10497	0·073298	0·038489
11	·32853	·29330	·25284	·20576	·17915	·15010	·11820	·082977	·043832
12	·35422	·31772	·27540	·22559	·19716	·16587	·13123	·092604	·049206
13	·37817	·34068	·29682	·24464	·21457	·18124	·14403	·10216	·054597
14	·40053	·36228	·31715	·26292	·23139	·19619	·15659	·11161	·059993
15	0·42143	0·38261	0·33645	0·28045	0·24762	0·21072	0·16889	0·12096	0·065386
16	·44100	·40176	·35478	·29726	·26327	·22483	·18093	·13019	·070768
17	·45934	·41983	·37219	·31338	·27837	·23853	·19270	·13930	·076134
18	·47657	·43689	·38875	·32885	·29293	·25182	·20420	·14828	·081478
19	·49277	·45302	·40451	·34369	·30697	·26471	·21544	·15712	·086796
20	0·50803	0·46829	0·41952	0·35793	0·32051	0·27721	0·22642	0·16583	0·092085
21	·52243	·48276	·43382	·37161	·33358	·28934	·23713	·17440	·097342
22	·53603	·49649	·44747	·38475	·34619	·30111	·24759	·18283	·10257
23	·54889	·50953	·46049	·39738	·35836	·31253	·25781	·19112	·10775
24	·56108	·52193	·47294	·40954	·37012	·32361	·26778	·19928	·11290
25	0·57264	0·53373	0·48485	0·42123	0·38147	0·33437	0·27751	0·20730	0·11801
26	·58361	·54498	·49624	·43248	·39245	·34481	·28701	·21518	·12308
27	·59405	·55571	·50716	·44333	·40306	·35495	·29629	·22293	·12811
28	·60398	·56595	·51763	·45378	·41332	·36479	·30534	·23054	·13310
29	·61344	·57574	·52767	·46386	·42325	·37436	·31419	·23803	·13804
30	0·62247	0·58511	0·53731	0·47359	0·43286	0·38366	0·32283	0·24539	0·14295
40	·69412	·66034	·61599	·55476	·51428	·46386	·39910	·31243	·18960
60	·77851	·75104	·71386	·66029	·62333	·57545	·51067	·41750	·27063
120	·87897	·86198	·83814	·80192	·77553	·73946	·68688	·60235	·44158
∞	1·00000	1·00000	1·00000	1·00000	1·00000	1·00000	1·00000	1·00000	1·00000

For $\nu_1 = \infty$, $x = 0$

Table 16. *Percentage points of the* B*-distribution* (*continued*)

Lower 5 % *points for* x

$\nu_1 = 2b, \quad \nu_2 = 2a$

ν_2 \ ν_1	1	2	3	4	5	6	7	8	9
1	0·0061558	0·0025000	0·0015429	0·0011119	0·0^{3}86820	0·0^{3}71179	0·0^{3}60300	0·0^{3}52300	0·0^{3}46170
2	·097500	·050000	·033617	·025321	·020308	·016952	·014548	·012741	·011334
3	·22852	·13572	·097308	·076010	·062413	·052962	·046007	·040671	·036447
4	·34163	·22361	·16825	·13535	·11338	·097611	·085727	·076440	·068979
5	0·43074	0·30171	0·23553	0·19403	0·16528	0·14409	0·12778	0·11482	0·10427
6	·50053	·36840	·29599	·24860	·21477	·18926	·16927	·15316	·13989
7	·55593	·42489	·34929	·29811	·26063	·23182	·20890	·19019	·17461
8	·60071	·47287	·39607	·34259	·30260	·27134	·24613	·22532	·20783
9	·63751	·51390	·43716	·38245	·34080	·30777	·28082	·25835	·23930
10	0·66824	0·54928	0·47338	0·41820	0·37553	0·34126	0·31301	0·28924	0·26894
11	·69425	·58003	·50546	·45033	·40712	·37203	·34283	·31807	·29677
12	·71654	·60696	·53402	·47930	·43590	·40031	·37044	·34494	·32286
13	·73583	·63073	·55958	·50551	·46219	·42635	·39604	·37000	·34732
14	·75268	·65184	·58256	·52932	·48626	·45036	·41980	·39338	·37025
15	0·76754	0·67070	0·60333	0·55102	0·50836	0·47255	0·44187	0·41521	0·39176
16	·78072	·68766	·62217	·57086	·52872	·49310	·46242	·43563	·41196
17	·79249	·70297	·63933	·58907	·54750	·51217	·48159	·45474	·43094
18	·80307	·71687	·65503	·60584	·56490	·52991	·49949	·47267	·44880
19	·81263	·72954	·66944	·62131	·58103	·54645	·51624	·48951	·46564
20	0·82131	0·74113	0·68271	0·63564	0·59605	0·56189	0·53194	0·50535	0·48152
21	·82923	·75178	·69496	·64894	·61004	·57635	·54669	·52027	·49652
22	·83647	·76160	·70632	·66132	·62312	·58990	·56056	·53434	·51071
23	·84313	·77067	·71687	·67287	·63536	·60263	·57363	·54764	·52415
24	·84927	·77908	·72669	·68366	·64684	·61461	·58596	·56022	·53689
25	0·85494	0·78690	0·73586	0·69377	0·65764	0·62590	0·59761	0·57213	0·54898
26	·86021	·79418	·74444	·70327	·66780	·63656	·60864	·58343	·56048
27	·86511	·80099	·75249	·71219	·67738	·64663	·61909	·59416	·57141
28	·86967	·80736	·76004	·72060	·68643	·65617	·62900	·60436	·58183
29	·87394	·81334	·76715	·72854	·69499	·66522	·63842	·61407	·59177
30	0·87794	0·81896	0·77386	0·73604	0·70311	0·67381	0·64738	0·62332	0·60125
40	·90734	·86089	·82447	·79327	·76559	·74053	·71758	·69636	·67663
60	·93748	·90497	·87881	·85591	·83517	·81606	·79824	·78150	·76569
120	·96837	·95130	·93720	·92458	·91290	·90192	·89148	·88150	·87191
∞	1·00000	1·00000	1·00000	1·00000	1·00000	1·00000	1·00000	1·00000	1·00000

This table gives the values of x for which

$$I_x(a, b) = \int_0^x u^{a-1}(1-u)^{b-1}\,du \bigg/ \int_0^1 u^{a-1}(1-u)^{b-1}\,du = 0{\cdot}05,$$

where $a = \frac{1}{2}\nu_2$, $b = \frac{1}{2}\nu_1$.

Table 16 (*continued*)

Lower 5 % *points for* x

$\nu_1 = 2b, \quad \nu_2 = 2a$

ν_2 \ ν_1	10	12	15	20	24	30	40	60	120
1	0·0^{3}41325	0·0^{3}34154	0·0^{3}27098	0·0^{3}20156	0·0^{3}16727	0·0^{3}13326	0·0^{4}99535	0·0^{4}66082	0·0^{4}32904
2	·010206	·0085124	·0068158	·0051162	·0042653	·0034137	·0025614	·0017083	·0^{3}85452
3	·033020	·027794	·022465	·017026	·014264	·011472	·0086511	·0057991	·0029157
4	·062850	·053376	·043541	·033319	·028053	·022679	·017191	·011585	·0058568
5	0·095510	0·081790	0·067312	0·051995	0·043994	0·035747	0·027240	0·018458	0·0093841
6	·12876	·11111	·092207	·071870	·061103	·049898	·038224	·026043	·013317
7	·16142	·14029	·11733	·092238	·078783	·064651	·049781	·034103	·017540
8	·19290	·16875	·14216	·11267	·096658	·079695	·061676	·042481	·021976
9	·22292	·19618	·16638	·13288	·11449	·094827	·073748	·051068	·026572
10	0·25137	0·22244	0·18984	0·15272	0·13211	0·10991	0·085885	0·059786	0·031288
11	·27823	·24746	·21244	·17207	·14943	·12484	·098008	·068575	·036094
12	·30354	·27125	·23413	·19086	·16636	·13955	·11006	·077394	·040967
13	·32737	·29383	·25492	·20908	·18288	·15401	·12199	·086209	·045889
14	·34981	·31524	·27481	·22669	·19895	·16818	·13377	·094994	·050847
15	0·37095	0·33554	0·29382	0·24370	0·21457	0·18203	0·14539	0·10373	0·055827
16	·39086	·35480	·31199	·26011	·22972	·19556	·15682	·11240	·060821
17	·40965	·37307	·32936	·27594	·24441	·20877	·16805	·12099	·065820
18	·42738	·39041	·34596	·29120	·25865	·22164	·17908	·12950	·070818
19	·44414	·40689	·36183	·30591	·27244	·23418	·18989	·13791	·075809
20	0·45999	0·42256	0·37701	0·32009	0·28580	0·24639	0·20050	0·14622	0·080789
21	·47501	·43746	·39154	·33375	·29874	·25828	·21088	·15442	·085753
22	·48925	·45165	·40544	·34693	·31126	·26985	·22106	·16252	·090698
23	·50276	·46518	·41877	·35964	·32340	·28112	·23102	·17051	·095622
24	·51560	·47808	·43154	·37190	·33515	·29208	·24077	·17838	·10052
25	0·52782	0·49040	0·44379	0·38373	0·34653	0·30275	0·25032	0·18615	0·10539
26	·53945	·50217	·45554	·39516	·35756	·31314	·25966	·19379	·11024
27	·55054	·51343	·46683	·40619	·36826	·32325	·26880	·20133	·11505
28	·56112	·52420	·47768	·41685	·37862	·33309	·27775	·20875	·11983
29	·57122	·53452	·48812	·42715	·38867	·34267	·28650	·21606	·12458
30	0·58088	0·54442	0·49816	0·43711	0·39842	0·35200	0·29507	0·22326	0·12930
40	·65819	·62459	·58083	·52099	·48175	·43321	·37136	·28936	·17453
60	·75070	·72282	·68535	·63185	·59522	·54807	·48477	·39458	·25416
120	·86266	·84504	·82047	·78342	·75661	·72016	·66738	·58326	·42519
∞	1·00000	1·00000	1·00000	1·00000	1·00000	1·00000	1·00000	1·00000	1·00000

For $\nu_1 = \infty$, $x = 0$

Table 16. *Percentage points of the* B-*distribution* (*continued*)

Lower 2·5 % *points for* x

$\nu_1 = 2b, \quad \nu_2 = 2a$

ν_2 \ ν_1	1	2	3	4	5	6	7	8	9
1	0·0015413	0·0³62500	0·0³38558	0·0³27783	0·0³21691	0·0³17782	0·0³15064	0·0³13065	0·0³11533
2	·049375	·025000	·016737	·012579	·010076	·0084038	·0072076	·0063095	·0056104
3	·14675	·085499	·060830	·047316	·038748	·032820	·028471	·025143	·022513
4	·24664	·15811	·11786	·094299	·078706	·067586	·059243	·052745	·047539
5	0·33318	0·22865	0·17674	0·14471	0·12275	0·10669	0·094390	0·084663	0·076770
6	·40505	·29240	·23259	·19412	·16696	·14663	·13081	·11812	·10770
7	·46442	·34855	·28375	·24063	·20942	·18562	·16681	·15153	·13886
8	·51378	·39764	·32993	·28358	·24933	·22278	·20151	·18405	·16944
9	·55524	·44054	·37137	·32290	·28642	·25774	·23450	·21523	·19897
10	0·59043	0·47818	0·40855	0·35877	0·32071	0·29042	0·26561	0·24486	0·22722
11	·62062	·51135	·44194	·39146	·35234	·32085	·29482	·27288	·25409
12	·64677	·54074	·47202	·42128	·38149	·34914	·32219	·29930	·27957
13	·66961	·56693	·49920	·44853	·40838	·37545	·34779	·32416	·30368
14	·68973	·59038	·52385	·47349	·43321	·39991	·37175	·34755	·32646
15	0·70756	0·61149	0·54628	0·49641	0·45618	0·42268	0·39418	0·36955	0·34799
16	·72349	·63058	·56676	·51750	·47746	·44390	·41520	·39026	·36833
17	·73778	·64792	·58553	·53697	·49723	·46372	·43490	·40976	·38756
18	·75069	·66373	·60278	·55498	·51561	·48224	·45341	·42814	·40575
19	·76239	·67821	·61869	·57169	·53276	·49959	·47081	·44549	·42297
20	0·77305	0·69150	0·63339	0·58722	0·54877	0·51586	0·48719	0·46187	0·43928
21	·78280	·70376	·64702	·60169	·56375	·53115	·50263	·47736	·45475
22	·79176	·71509	·65970	·61520	·57780	·54553	·51720	·49202	·46943
23	·80001	·72559	·67150	·62785	·59099	·55908	·53098	·50592	·48338
24	·80763	·73535	·68253	·63970	·60341	·57187	·54401	·51911	·49664
25	0·81469	0·74445	0·69285	0·65084	0·61511	0·58396	0·55636	0·53163	0·50927
26	·82126	·75295	·70253	·66132	·62615	·59540	·56808	·54354	·52130
27	·82738	·76090	·71162	·67119	·63658	·60624	·57921	·55488	·53278
28	·83310	·76836	·72018	·68052	·64646	·61652	·58980	·56568	·54373
29	·83845	·77538	·72825	·68933	·65582	·62630	·59988	·57599	·55420
30	0·84347	0·78198	0·73587	0·69768	0·66471	0·63559	0·60948	0·58582	0·56421
40	·88059	·83157	·79381	·76184	·73369	·70839	·68533	·66411	·64446
60	·91904	·88430	·85681	·83298	·81156	·79193	·77372	·75668	·74065
120	·95883	·94037	·92535	·91201	·89975	·88828	·87743	·86708	·85717
∞	1·00000	1·00000	1·00000	1·00000	1·00000	1·00000	1·00000	1·00000	1·00000

This table gives the values of x for which

$$I_x(a, b) = \int_0^x u^{a-1}(1-u)^{b-1}\,du \Big/ \int_0^1 u^{a-1}(1-u)^{b-1}\,du = 0{\cdot}025,$$

where $a = \frac{1}{2}\nu_2$, $b = \frac{1}{2}\nu_1$.

Table 16 (*continued*)

Lower 2·5 % points for x

$\nu_1 = 2b, \quad \nu_2 = 2a$

ν_2 \ ν_1	10	12	15	20	24	30	40	60	120
1	0·0^{3}10323	0·0^{4}85313	0·0^{4}67686	0·0^{4}50345	0·0^{4}41780	0·0^{4}33285	0·0^{4}24860	0·0^{4}16505	0·0^{5}82180
2	·0050508	·0042107	·0033700	·0025286	·0021076	·0016864	·0012651	·0^{3}84357	·0^{3}42187
3	·020382	·017139	·013838	·010477	·0087725	·0070519	·0053148	·0035607	·0017893
4	·043272	·036693	·029885	·022831	·019207	·015514	·011749	·0079110	·0039956
5	0·070233	0·060028	0·049302	0·038002	0·032119	0·026068	0·019841	0·013428	0·0068184
6	·098988	·085233	·070563	·054861	·046579	·037985	·029056	·019767	·010092
7	·12818	·11113	·092695	·072663	·061969	·050772	·039029	·026691	·013702
8	·15701	·13700	·11508	·090920	·077872	·064092	·049508	·034033	·017569
9	·18504	·16240	·13732	·10931	·094004	·077712	·060314	·041675	·021634
10	0·21201	0·18709	0·15917	0·12760	0·11017	0·091466	0·071319	0·049528	0·025854
11	·23780	·21091	·18048	·14565	·12623	·10523	·082426	·057528	·030196
12	·26238	·23379	·20115	·16336	·14210	·11893	·093564	·065622	·034634
13	·28574	·25571	·22112	·18067	·15770	·13249	·10468	·073771	·039147
14	·30790	·27667	·24039	·19753	·17299	·14588	·11573	·081944	·043718
15	0·32893	0·29668	0·25893	0·21392	0·18793	0·15905	0·12669	0·090115	0·048335
16	·34888	·31578	·27676	·22983	·20252	·17198	·13753	·098266	·052985
17	·36779	·33400	·29389	·24525	·21674	·18466	·14823	·10638	·057659
18	·38574	·35138	·31034	·26019	·23058	·19708	·15878	·11444	·062348
19	·40278	·36797	·32614	·27465	·24404	·20922	·16916	·12244	·067047
20	0·41896	0·38380	0·34132	0·28864	0·25713	0·22110	0·17938	0·13038	0·071749
21	·43435	·39893	·35589	·30218	·26985	·23270	·18943	·13823	·076449
22	·44900	·41338	·36990	·31528	·28221	·24402	·19930	·14601	·081144
23	·46294	·42720	·38335	·32795	·29422	·25508	·20899	·15370	·085828
24	·47623	·44042	·39629	·34021	·30588	·26587	·21850	·16130	·090500
25	0·48891	0·45307	0·40874	0·35207	0·31721	0·27640	0·22783	0·16881	0·095156
26	·50101	·46520	·42071	·36355	·32821	·28667	·23698	·17622	·099794
27	·51257	·47682	·43223	·37466	·33890	·29669	·24596	·18354	·10441
28	·52363	·48797	·44333	·38542	·34928	·30647	·25476	·19076	·10901
29	·53421	·49867	·45403	·39584	·35937	·31601	·26339	·19789	·11358
30	0·54435	0·50895	0·46434	0·40594	0·36918	0·32531	0·27185	0·20492	0·11812
40	·62616	·59296	·54999	·49168	·45370	·40697	·34780	·26997	·16201
60	·72550	·69743	·65992	·60674	·57056	·52422	·46239	·37498	·24027
120	·84764	·82954	·80442	·76678	·73968	·70299	·65017	·56658	·41107
∞	1·00000	1·00000	1·00000	1·00000	1·00000	1·00000	1·00000	1·00000	1·00000

For $\nu_1 = \infty$, $x = 0$

Table 16. *Percentage points of the* B-*distribution (continued)*

Lower 1 % *points for x*

$\nu_1 = 2b, \quad \nu_2 = 2a$

ν_2 \ ν_1	1	2	3	4	5	6	7	8	9
1	$0\cdot0^{3}24672$	$0\cdot0^{3}10000$	$0\cdot0^{4}61686$	$0\cdot0^{4}44446$	$0\cdot0^{4}34699$	$0\cdot0^{4}28446$	$0\cdot0^{4}24097$	$0\cdot0^{4}20899$	$0\cdot0^{4}18449$
2	·019900	·010000	·0066778	·0050126	·0040121	·0033445	·0028674	·0025094	·0022309
3	·080827	·046416	·032834	·025458	·020807	·017599	·015252	·013458	·012043
4	·15874	·10000	·073960	·058903	·049014	·041999	·036754	·032682	·029426
5	0·23520	0·15849	0·12142	0·098877	0·083563	0·072429	0·063948	0·057264	0·051857
6	·30387	·21544	·16979	·14087	·12065	·10564	·094014	·084730	·077136
7	·36370	·26827	·21636	·18236	·15801	·13959	·12511	·11341	·10375
8	·41540	·31623	·25997	·22207	·19437	·17307	·15612	·14227	·13073
9	·46009	·35938	·30024	·25945	·22910	·20543	·18637	·17066	·15745
10	0·49890	0·39811	0·33719	0·29431	0·26191	0·23632	0·21551	0·19820	0·18355
11	·53279	·43288	·37099	·32667	·29271	·26560	·24335	·22469	·20879
12	·56258	·46416	·40191	·35664	·32153	·29323	·26981	·25003	·23307
13	·58893	·49239	·43021	·38437	·34845	·31924	·29487	·27417	·25631
14	·61238	·51795	·45615	·41006	·37358	·34369	·31858	·29712	·27851
15	0·63336	0·54117	0·47999	0·43387	0·39706	0·36666	0·34098	0·31891	0·29968
16	·65224	·56234	·50193	·45597	·41899	·38826	·36214	·33958	·31985
17	·66930	·58171	·52219	·47651	·43951	·40857	·38213	·35920	·33905
18	·68479	·59948	·54094	·49565	·45872	·42768	·40103	·37781	·35733
19	·69892	·61585	·55832	·51350	·47674	·44568	·41890	·39547	·37473
20	0·71185	0·63096	0·57447	0·53018	0·49366	0·46266	0·43581	0·41224	0·39131
21	·72372	·64495	·58952	·54581	·50958	·47868	·45184	·42818	·40711
22	·73467	·65793	·60357	·56046	·52456	·49383	·46703	·44333	·42217
23	·74479	·67002	·61671	·57422	·53869	·50816	·48144	·45775	·43653
24	·75417	·68129	·62903	·58717	·55204	·52174	·49514	·47149	·45025
25	0·76290	0·69183	0·64059	0·59938	0·56466	0·53461	0·50816	0·48458	0·46335
26	·77103	·70170	·65147	·61090	·57660	·54683	·52055	·49706	·47587
27	·77862	·71097	·66172	·62180	·58793	·55845	·53236	·50899	·48785
28	·78573	·71969	·67139	·63211	·59868	·56951	·54362	·52038	·49932
29	·79240	·72790	·68054	·64188	·60890	·58004	·55437	·53127	·51031
30	0·79867	0·73564	0·68919	0·65116	0·61862	0·59008	0·56464	0·54170	0·52085
40	·84541	·79433	·75561	·72316	·69482	·66950	·64656	·62555	·60617
60	·89449	·85770	·82898	·80433	·78233	·76227	·74376	·72651	·71034
120	·94599	·92612	·91014	·89607	·88321	·87124	·85995	·84924	·83900
∞	1·00000	1·00000	1·00000	1·00000	1·00000	1·00000	1·00000	1·00000	1·00000

This table gives the values of x for which

$$I_x(a, b) = \int_0^x u^{a-1}(1-u)^{b-1}\,du \bigg/ \int_0^1 u^{a-1}(1-u)^{b-1}\,du = 0\cdot01,$$

where $a = \frac{1}{2}\nu_2$, $b = \frac{1}{2}\nu_1$.

Table 16 (*continued*)

Lower 1 % points for x

$\nu_1 = 2b, \quad \nu_2 = 2a$

ν_2 \ ν_1	10	12	15	20	24	30	40	60	120
1	$0{\cdot}0^{4}16513$	$0{\cdot}0^{4}13647$	$0{\cdot}0^{4}10827$	$0{\cdot}0^{5}80531$	$0{\cdot}0^{5}66831$	$0{\cdot}0^{5}53242$	$0{\cdot}0^{5}39766$	$0{\cdot}0^{5}26400$	$0{\cdot}0^{5}13145$
2	·0020080	·0016737	·0013391	·0010045	$\cdot 0^{3}83718$	$\cdot 0^{3}66980$	$\cdot 0^{3}50239$	$\cdot 0^{3}33496$	$\cdot 0^{3}16749$
3	·010898	·0091569	·0073877	·0055887	·0046777	·0037588	·0028317	·0018964	$\cdot 0^{3}95252$
4	·026763	·022665	·018435	·014065	·011824	·0095436	·0072226	·0048595	·0024525
5	0·047389	0·040434	0·033149	0·025503	0·021534	0·017459	0·013275	0·0089747	0·0045520
6	·070804	·060840	·050258	·038982	·033057	·026923	·020567	·013973	·0071235
7	·095628	·082714	·068820	·053801	·045816	·037481	·028767	·019640	·010065
8	·12095	·10526	·088177	·069455	·059390	·048797	·037625	·025815	·013300
9	·14619	·12796	·10787	·085584	·073472	·060623	·046957	·032376	·016768
10	0·17097	0·15044	0·12760	0·10193	0·087838	0·072776	0·056621	0·039229	0·020426
11	·19506	·17250	·14713	·11830	·10232	·085117	·066512	·046303	·024237
12	·21834	·19398	·16633	·13458	·11681	·097542	·076547	·053541	·028173
13	·24073	·21479	·18510	·15065	·13120	·10997	·086660	·060896	·032212
14	·26220	·23489	·20338	·16646	·14544	·12235	·096802	·068334	·036335
15	0·28276	0·25426	0·22113	0·18196	0·15948	0·13462	0·10693	0·075823	0·040526
16	·30240	·27289	·23833	·19711	·17327	·14676	·11702	·083341	·044772
17	·32117	·29079	·25497	·21189	·18680	·15873	·12704	·090866	·049062
18	·33910	·30797	·27105	·22630	·20005	·17053	·13697	·098383	·053386
19	·35622	·32446	·28658	·24032	·21301	·18212	·14680	·10588	·057738
20	0·37257	0·34029	0·30157	0·25395	0·22567	0·19351	0·15651	0·11334	0·062109
21	·38818	·35548	·31603	·26721	·23803	·20468	·16609	·12076	·066494
22	·40311	·37005	·32999	·28008	·25008	·21563	·17554	·12812	·070888
23	·41738	·38405	·34345	·29258	·26184	·22637	·18486	·13543	·075285
24	·43103	·39749	·35645	·30472	·27329	·23687	·19403	·14268	·079683
25	0·44410	0·41040	0·36899	0·31651	0·28446	0·24716	0·20305	0·14986	0·084077
26	·45661	·42280	·38109	·32795	·29534	·25723	·21193	·15697	·088464
27	·46861	·43473	·39278	·33906	·30594	·26707	·22066	·16401	·092842
28	·48011	·44621	·40407	·34985	·31626	·27670	·22925	·17096	·097209
29	·49115	·45726	·41498	·36032	·32632	·28612	·23768	·17785	·10156
30	0·50175	0·46789	0·42552	0·37049	0·33612	0·29534	0·24597	0·18465	0·10590
40	·58819	·55573	·51398	·45778	·42144	·37700	·32111	·24819	·14811
60	·69511	·66701	·62969	·57717	·54167	·49647	·43655	·35258	·22459
120	·82918	·81062	·78497	·74677	·71942	·68259	·62988	·54709	·39479
∞	1·00000	1·00000	1·00000	1·00000	1·00000	1·00000	1·00000	1·00000	1·00000

For $\nu_1 = \infty$, $x = 0$

Table 16. *Percentage points of the* B-*distribution* (*continued*)

Lower 0·5 % *points for* x

$\nu_1 = 2b, \quad \nu_2 = 2a$

ν_2 \ ν_1	1	2	3	4	5	6	7	8	9
1	$0{\cdot}0^{4}61684$	$0{\cdot}0^{4}25000$	$0{\cdot}0^{4}15421$	$0{\cdot}0^{4}11111$	$0{\cdot}0^{5}86745$	$0{\cdot}0^{5}71112$	$0{\cdot}0^{5}60240$	$0{\cdot}0^{5}52245$	$0{\cdot}0^{5}46121$
2	·0099750	·0050000	·0033361	·0025031	·0020030	·0016695	·0014311	·0012524	·0011133
3	·051237	·029240	·020632	·015976	·013046	·011028	·0095530	·0084269	·0075388
4	·11321	·070711	·052099	·041400	·034399	·029445	·025748	·022881	·020592
5	0·17995	0·12011	0·091593	0·074378	0·062737	0·054301	0·047891	0·042849	0·038776
6	·24356	·17100	·13408	·11088	·094759	·082829	·073618	·066279	·060287
7	·30126	·22007	·17656	·14830	·12818	·11303	·10116	·091593	·083708
8	·35261	·26591	·21745	·18510	·16159	·14360	·12933	·11770	·10804
9	·39799	·30808	·25604	·22046	·19415	·17373	·15736	·14389	·13261
10	0·43809	0·34657	0·29204	0·25399	0·22542	0·20297	0·18478	0·16970	0·15697
11	·47360	·38162	·32543	·28554	·25517	·23105	·21132	·19484	·18083
12	·50517	·41352	·35632	·31509	·28332	·25783	·23682	·21914	·20401
13	·53337	·44258	·38487	·34270	·30986	·28328	·26120	·24250	·22642
14	·55865	·46912	·41127	·36848	·33484	·30739	·28444	·26489	·24798
15	0·58144	0·49340	0·43569	0·39255	0·35833	0·33022	0·30656	0·28629	0·26869
16	·60206	·51567	·45832	·41503	·38042	·35180	·32757	·30672	·28852
17	·62080	·53616	·47932	·43605	·40120	·37221	·34754	·32620	·30752
18	·63789	·55505	·49885	·45571	·42076	·39151	·36650	·34478	·32568
19	·65354	·57251	·51703	·47414	·43917	·40976	·38451	·36248	·34305
20	0·66791	0·58870	0·53400	0·49144	0·45654	0·42705	0·40162	0·37936	0·35966
21	·68117	·60375	·54987	·50768	·47293	·44343	·41789	·39546	·37554
22	·69341	·61775	·56472	·52297	·48841	·45896	·43337	·41082	·39073
23	·70477	·63083	·57866	·53738	·50306	·47369	·44810	·42547	·40527
24	·71532	·64305	·59176	·55098	·51693	·48769	·46213	·43947	·41918
25	0·72516	0·65451	0·60409	0·56382	0·53007	0·50100	0·47551	0·45285	0·43251
26	·73434	·66527	·61572	·57597	·54255	·51367	·48827	·46564	·44528
27	·74294	·67539	·62669	·58749	·55440	·52574	·50045	·47788	·45752
28	·75100	·68492	·63707	·59841	·56568	·53724	·51210	·48960	·46927
29	·75857	·69392	·64690	·60878	·57642	·54822	·52324	·50083	·48054
30	0·76570	0·70242	0·65622	0·61864	0·58665	0·55871	0·53389	0·51159	0·49137
40	·81920	·76727	·72823	·69571	·66744	·64229	·61956	·59882	·57973
60	·87598	·83811	·80877	·78370	·76142	·74119	·72256	·70526	·68907
120	·93619	·91548	·89893	·88442	·87120	·85892	·84739	·83645	·82602
∞	1·00000	1·00000	1·00000	1·00000	1·00000	1·00000	1·00000	1·00000	1·00000

This table gives the values of x for which

$$I_x(a, b) = \int_0^x u^{a-1}(1-u)^{b-1}\,du \Big/ \int_0^1 u^{a-1}(1-u)^{b-1}\,du = 0{\cdot}005,$$

where $a = \frac{1}{2}\nu_2$, $b = \frac{1}{2}\nu_1$.

Table 16 (*continued*)

Lower 0·5 % *points for x*

$\nu_1 = 2b, \quad \nu_2 = 2a$

ν_2 \ ν_1	10	12	15	20	24	30	40	60	120
1	0·0^{5}41280	0·0^{5}34116	0·0^{5}27067	0·0^{5}20132	0·0^{5}16707	0·0^{5}13310	0·0^{6}99411	0·0^{6}65998	0·0^{6}32862
2	·0010020	·0^{3}83507	·0^{3}66812	·0^{3}50113	·0^{3}41762	·0^{3}33411	·0^{3}25060	·0^{3}16707	·0^{4}83539
3	·0068204	·0057290	·0046206	·0034943	·0029242	·0023493	·0017696	·0011848	·0^{3}59503
4	·018721	·015844	·012879	·0098197	·0082522	·0066584	·0050373	·0033880	·0017093
5	0·035415	0·030191	0·024729	0·019006	0·016040	0·012998	0·0098777	0·0066743	0·0033833
6	·055299	·047464	·039162	·030337	·025709	·020924	·015973	·010844	·0055240
7	·077090	·066592	·055329	·043189	·036749	·030038	·023034	·015712	·0080441
8	·099867	·086788	·072586	·057076	·048759	·040023	·030828	·021129	·010873
9	·12300	·10749	·090464	·071635	·061433	·050635	·039174	·026976	·013953
10	0·14606	0·12831	0·10862	0·086595	0·074540	0·061684	0·047930	0·033162	0·017241
11	·16876	·14898	·12683	·10175	·087903	·073027	·056985	·039611	·020700
12	·19092	·16931	·14489	·11696	·10139	·084550	·066252	·046265	·024302
13	·21242	·18919	·16270	·13211	·11490	·096167	·075662	·053076	·028022
14	·23320	·20853	·18017	·14710	·12835	·10781	·085158	·060005	·031841
15	0·25323	0·22728	0·19725	0·16190	0·14170	0·11942	0·094697	0·067019	0·035743
16	·27248	·24543	·21388	·17644	·15488	·13097	·10424	·074093	·039714
17	·29098	·26295	·23006	·19070	·16787	·14241	·11377	·081204	·043741
18	·30872	·27986	·24576	·20465	·18065	·15373	·12324	·088333	·047815
19	·32574	·29615	·26099	·21829	·19319	·16489	·13265	·095466	·051927
20	0·34206	0·31184	0·27575	0·23160	0·20549	0·17590	0·14198	0·10259	0·056070
21	·35770	·32696	·29004	·24458	·21753	·18673	·15122	·10969	·060237
22	·37269	·34151	·30387	·25723	·22932	·19738	·16036	·11676	·064421
23	·38707	·35552	·31725	·26954	·24084	·20784	·16938	·12380	·068619
24	·40087	·36901	·33020	·28153	·25210	·21811	·17829	·13078	·072825
25	0·41411	0·38200	0·34272	0·29320	0·26309	0·22818	0·18707	0·13772	0·077035
26	·42682	·39452	·35484	·30456	·27383	·23806	·19573	·14461	·081247
27	·43903	·40658	·36657	·31560	·28432	·24775	·20426	·15143	·085457
28	·45076	·41821	·37792	·32635	·29455	·25724	·21266	·15819	·089662
29	·46204	·42942	·38891	·33681	·30454	·26654	·22093	·16489	·093860
30	0·47289	0·44024	0·39954	0·34698	0·31429	0·27565	0·22907	0·17152	0·098048
40	·56205	·53024	·48950	·43493	·39980	·35700	·30341	·23388	·13907
60	·67384	·64584	·60879	·55688	·52194	·47762	·41913	·33759	·21421
120	·81604	·79720	·77125	·73277	·70531	·66845	·61590	·53378	·38380
∞	1·00000	1·00000	1·00000	1·00000	1·00000	1·00000	1·00000	1·00000	1·00000

For $\nu_1 = \infty$, $x = 0$

Table 16. *Percentage points of the* B-*distribution (continued)*

Lower 0·25 % *points for x*

$\nu_1 = 2b, \quad \nu_2 = 2a$

ν_2 \ ν_1	1	2	3	4	5	6	7	8	9
1	$0{\cdot}0^{4}15421$	$0{\cdot}0^{5}62500$	$0{\cdot}0^{5}38553$	$0{\cdot}0^{5}27778$	$0{\cdot}0^{5}21686$	$0{\cdot}0^{5}17778$	$0{\cdot}0^{5}15060$	$0{\cdot}0^{5}13061$	$0{\cdot}0^{5}11530$
2	·0049938	·0025000	·0016674	·0012508	·0010008	$\cdot 0^{3}83403$	$\cdot 0^{3}71492$	$\cdot 0^{3}62559$	$\cdot 0^{3}55610$
3	·032403	·018420	·012978	·010040	·0081943	·0069244	·0059966	·0052886	·0047305
4	·080525	·050000	·036742	·029152	·024198	·020698	·018089	·016069	·014456
5	0·13734	0·091028	0·069178	0·056059	0·047218	0·040827	0·035979	0·032171	0·029099
6	·19476	·13572	·10601	·087454	·074607	·065131	·057831	·052024	·047291
7	·24902	·18053	·14424	·12083	·10424	·091778	·082047	·074220	·067778
8	·29873	·22361	·18208	·15455	·13464	·11946	·10745	·097696	·089597
9	·34367	·26410	·21856	·18764	·16489	·14730	·13324	·12170	·11205
10	0·38408	0·30171	0·25316	0·21954	0·19441	0·17475	0·15887	0·14573	0·13466
11	·42037	·33643	·28571	·24995	·22287	·20145	·18398	·16943	·15708
12	·45301	·36840	·31616	·27876	·25010	·22719	·20837	·19258	·17910
13	·48243	·39782	·34458	·30594	·27601	·25188	·23191	·21504	·20056
14	·50904	·42489	·37107	·33153	·30060	·27547	·25452	·23673	·22138
15	0·53319	0·44984	0·39575	0·35558	0·32388	0·29794	0·27619	0·25760	0·24150
16	·55517	·47287	·41876	·37819	·34591	·31932	·29690	·27765	·26089
17	·57525	·49417	·44023	·39943	·36674	·33965	·31668	·29687	·27956
18	·59365	·51390	·46029	·41941	·38644	·35897	·33555	·31528	·29749
19	·61057	·53223	·47906	·43822	·40508	·37732	·35356	·33290	·31471
20	0·62617	0·54928	0·49664	0·45593	0·42272	0·39476	0·37074	0·34976	0·33124
21	·64060	·56518	·51313	·47264	·43943	·41135	·38712	·36590	·34711
22	·65398	·58003	·52863	·48841	·45526	·42713	·40276	·38135	·36233
23	·66641	·59393	·54321	·50331	·47029	·44214	·41769	·39614	·37693
24	·67800	·60696	·55695	·51741	·48455	·45645	·43195	·41029	·39095
25	0·68882	0·61921	0·56991	0·53076	0·49811	0·47008	0·44558	0·42386	0·40441
26	·69895	·63073	·58216	·54343	·51100	·48309	·45861	·43686	·41734
27	·70845	·64159	·59375	·55545	·52328	·49550	·47108	·44933	·42976
28	·71737	·65184	·60473	·56688	·53498	·50736	·48302	·46129	·44170
29	·72577	·66153	·61515	·57776	·54615	·51870	·49446	·47277	·45319
30	0·73369	0·67070	0·62505	0·58811	0·55680	0·52956	0·50543	0·48380	0·46423
40	·79348	·74113	·70204	·66963	·64157	·61667	·59425	·57382	·55507
60	·85761	·81896	·78919	·76385	·74141	·72108	·70241	·68511	·66895
120	·92636	·90497	·88794	·87306	·85954	·84702	·83527	·82416	·81357
∞	1·00000	1·00000	1·00000	1·00000	1·00000	1·00000	1·00000	1·00000	1·00000

This tables gives the values of x for which

$$I_x(a, b) = \int_0^x u^{a-1}(1-u)^{b-1}\,du \bigg/ \int_0^1 u^{a-1}(1-u)^{b-1}\,du = 0{\cdot}0025,$$

where $a = \frac{1}{2}\nu_2$, $b = \frac{1}{2}\nu_1$.

Table 16 (*continued*). *Lower* 0·25 % *points for* x

$\nu_1 = 2b, \quad \nu_2 = 2a$

$\nu_2 \backslash \nu_1$	10	12	15	20	24	30	40	60	120
1	0·0^{5}10320	0·0^{6}85289	0·0^{6}67667	0·0^{6}50330	0·0^{6}41767	0·0^{6}33275	0·0^{6}24852	0·0^{6}16499	0·0^{7}82154
2	·0^{3}50050	·0^{3}41710	·0^{3}33370	·0^{3}25028	·0^{3}20857	·0^{3}16686	·0^{3}12515	·0^{4}83434	·0^{4}41718
3	·0042791	·0035937	·0028978	·0021910	·0018333	·0014727	·0011092	·0^{3}74260	·0^{3}37289
4	·013139	·011115	·0090313	·0068827	·0057826	·0046647	·0035282	·0023724	·0011966
5	0·026565	0·022632	0·018524	0·014228	0·012002	0·0097228	0·0073857	0·0049884	0·0025277
6	·043354	·037180	·030649	·023720	·020091	·016343	·012469	·0084602	·0043072
7	·062379	·053830	·044676	·034832	·029620	·024195	·018541	·012638	·0064653
8	·082760	·071838	·060007	·047120	·040224	·032992	·025391	·017387	·0089391
9	·10385	·090643	·076179	·060230	·051609	·042499	·032849	·022598	·011676
10	0·12520	0·10983	0·092843	0·073889	0·063544	0·052533	0·040776	0·028181	0·014633
11	·14647	·12912	·10974	·087888	·075849	·062945	·049062	·034061	·017776
12	·16745	·14828	·12668	·10206	·088383	·073619	·057615	·040179	·021074
13	·18799	·16717	·14352	·11630	·10104	·084463	·066365	·046487	·024504
14	·20799	·18569	·16016	·13049	·11373	·095402	·075252	·052943	·028046
15	0·22739	0·20377	0·17652	0·14458	0·12638	0·10638	0·084228	0·059513	0·031682
16	·24616	·22136	·19255	·15849	·13896	·11734	·093255	·066170	·035399
17	·26427	·23843	·20822	·17221	·15140	·12826	·10230	·072890	·039183
18	·28173	·25498	·22349	·18569	·16370	·13910	·11133	·079652	·043025
19	·29855	·27099	·23837	·19891	·17581	·14984	·12034	·086441	·046915
20	0·31473	0·28647	0·25283	0·21187	0·18773	0·16045	0·12930	0·093240	0·050844
21	·33030	·30143	·26689	·22454	·19944	·17093	·13819	·10004	·054807
22	·34527	·31588	·28053	·23692	·21093	·18127	·14701	·10682	·058795
23	·35968	·32983	·29377	·24901	·22219	·19145	·15574	·11359	·062805
24	·37353	·34330	·30661	·26082	·23322	·20146	·16438	·12033	·066831
25	0·38686	0·35631	0·31907	0·27233	0·24403	0·21131	0·17292	0·12703	0·070869
26	·39968	·36886	·33115	·28356	·25460	·22099	·18136	·13369	·074915
27	·41203	·38099	·34286	·29451	·26494	·23050	·18968	·14031	·078965
28	·42391	·39270	·35422	·30518	·27506	·23983	·19790	·14687	·083017
29	·43536	·40402	·36523	·31558	·28495	·24899	·20600	·15339	·087068
30	0·44640	0·41496	0·37591	0·32571	0·29462	0·25798	0·21398	0·15985	0·091115
40	·53774	·50663	·46693	·41399	·38004	·33882	·28740	·22099	·13099
60	·65378	·62595	·58923	·53801	·50367	·46023	·40313	·32392	·20483
120	·80346	·78440	·75823	·71954	·69203	·65519	·60285	·52143	·37370
∞	1·00000	1·00000	1·00000	1·00000	1·00000	1·00000	1·00000	1·00000	1·00000

For $\nu_1 = \infty$, $x = 0$.

Table 16. *Percentage points of the* B-*distribution* (*continued*)

Lower 0·1 % *points for* x

$\nu_1 = 2b, \quad \nu_2 = 2a$

ν_2 \ ν_1	1	2	3	4	5	6	7	8	9
1	$0{\cdot}0^{5}24674$	$0{\cdot}0^{5}10000$	$0{\cdot}0^{6}61685$	$0{\cdot}0^{6}44444$	$0{\cdot}0^{6}34698$	$0{\cdot}0^{6}28444$	$0{\cdot}0^{6}24096$	$0{\cdot}0^{6}20898$	$0{\cdot}0^{6}18448$
2	·0019990	·0010000	$\cdot 0^{3}66678$	$\cdot 0^{3}50013$	$\cdot 0^{3}40012$	$\cdot 0^{3}33344$	$\cdot 0^{3}28582$	$\cdot 0^{3}25009$	$\cdot 0^{3}22231$
3	·017644	·010000	·0070369	·0054407	·0044385	·0037496	·0032465	·0028627	·0025603
4	·051192	·031623	·023184	·018370	·015235	·013023	·011376	·010102	·0090853
5	0·095821	0·063096	0·047798	0·038658	0·032518	0·028089	0·024736	0·022105	0·019984
6	·14455	·10000	·077811	·064038	·054539	·047552	·042181	·037916	·034444
7	·19313	·13895	·11056	·092358	·079519	·069913	·062428	·056419	·051482
8	·23942	·17783	·14417	·12201	·10607	·093954	·084400	·076655	·070238
9	·28251	·21544	·17750	·15192	·13320	·11878	·10729	·097886	·090036
10	0·32217	0·25119	0·20983	0·18139	0·16025	0·14377	0·13051	0·11957	0·11037
11	·35846	·28480	·24079	·20999	·18678	·16851	·15366	·14132	·13088
12	·39160	·31623	·27019	·23748	·21254	·19270	·17645	·16286	·15129
13	·42187	·34551	·29798	·26374	·23735	·21618	·19872	·18401	·17142
14	·44953	·37276	·32416	·28872	·26115	·23885	·22033	·20464	·19114
15	0·47487	0·39811	0·34879	0·31243	0·28389	0·26064	0·24121	0·22467	0·21037
16	·49813	·42170	·37194	·33489	·30558	·28154	·26134	·24405	·22905
17	·51952	·44367	·39370	·35615	·32624	·30155	·28070	·26277	·24714
18	·53926	·46416	·41415	·37628	·34590	·32069	·29928	·28080	·26464
19	·55750	·48329	·43340	·39532	·36460	·33897	·31711	·29816	·28153
20	0·57440	0·50119	0·45152	0·41336	0·38239	0·35643	0·33420	0·31486	0·29783
21	·59011	·51795	·46860	·43044	·39932	·37311	·35058	·33091	·31354
22	·60472	·53367	·48471	·44664	·41543	·38905	·36628	·34634	·32867
23	·61836	·54844	·49993	·46201	·43078	·40428	·38133	·36117	·34325
24	·63111	·56234	·51432	·47660	·44540	·41883	·39575	·37541	·35729
25	0·64306	0·57544	0·52795	0·49046	0·45934	0·43275	0·40958	0·38910	0·37082
26	·65427	·58780	·54086	·50365	·47265	·44607	·42284	·40227	·38385
27	·66481	·59948	·55311	·51621	·48535	·45882	·43557	·41493	·39641
28	·67474	·61054	·56475	·52817	·49749	·47103	·44779	·42710	·40851
29	·68410	·62102	·57582	·53958	·50910	·48273	·45952	·43882	·42017
30	0·69295	0·63096	0·58636	0·55048	0·52021	0·49396	0·47080	0·45010	0·43142
40	·76032	·70795	·66908	·63702	·60937	·58494	·56299	·54306	·52481
60	·83365	·79433	·76420	·73867	·71614	·69579	·67716	·65993	·64388
120	·91338	·89125	·87372	·85844	·84460	·83182	·81985	·80854	·79780
∞	1·00000	1·00000	1·00000	1·00000	1·00000	1·00000	1·00000	1·00000	1·00000

This table gives the values of x for which

$$I_x(a,b) = \int_0^x u^{a-1}(1-u)^{b-1}\,du \Big/ \int_0^1 u^{a-1}(1-u)^{b-1}\,du = 0{\cdot}001,$$

where $a = \frac{1}{2}\nu_2$, $b = \frac{1}{2}\nu_1$.

Table 16 (*continued*). *Lower* 0·1 % *points for x*

$\nu_1 = 2b, \quad \nu_2 = 2a$

ν_2 \ ν_1	10	12	15	20	24	30	40	60	120
1	$0{\cdot}0^{6}16512$	$0{\cdot}0^{6}13646$	$0{\cdot}0^{6}10827$	$0{\cdot}0^{7}80527$	$0{\cdot}0^{7}66827$	$0{\cdot}0^{7}53240$	$0{\cdot}0^{7}39764$	$0{\cdot}0^{7}26399$	$0{\cdot}0^{7}13145$
2	$\cdot0^{3}20008$	$\cdot0^{3}16674$	$\cdot0^{3}13339$	$\cdot0^{3}10005$	$\cdot0^{4}83372$	$\cdot0^{4}66698$	$\cdot0^{4}50024$	$\cdot0^{4}33349$	$\cdot0^{4}16675$
3	·0023158	·0019445	·0015677	·0011851	$\cdot0^{3}99159$	$\cdot0^{3}79650$	$\cdot0^{3}59984$	$\cdot0^{3}40156$	$\cdot0^{3}20162$
4	·0082555	·0069815	·0056704	·0043196	·0036285	·0029264	·0022129	·0014877	$\cdot0^{3}75019$
5	0·018237	0·015527	0·012701	0·0097486	0·0082211	0·0066572	0·0050552	0·0034131	0·0017288
6	·031561	·027043	·022273	·017221	·014579	·011854	·0090391	·0061296	·0031189
7	·047351	·040819	·033840	·026352	·022395	·018281	·013999	·0095351	·0048744
8	·064830	·056206	·046888	·036766	·031361	·025702	·019764	·013522	·0069454
9	·083377	·072674	·060988	·048141	·041214	·033908	·026183	·017993	·0092863
10	0·10252	0·089813	0·075797	0·060214	0·051732	0·042723	0·033125	0·022865	0·011858
11	·12192	·10731	·091048	·072774	·062738	·052005	·040484	·028068	·014627
12	·14131	·12493	·10654	·085655	·074088	·061635	·048171	·033544	·017565
13	·16051	·14250	·12210	·098725	·085665	·071517	·056112	·039242	·020649
14	·17939	·15989	·13763	·11188	·097379	·081573	·064245	·045122	·023858
15	0·19787	0·17700	0·15302	0·12503	0·10915	0·091740	0·072521	0·051149	0·027174
16	·21588	·19378	·16821	·13813	·12093	·10196	·080896	·057292	·030584
17	·23338	·21018	·18316	·15111	·13266	·11221	·089336	·063528	·034073
18	·25035	·22616	·19781	·16394	·14431	·12243	·097811	·069833	·037632
19	·26678	·24171	·21216	·17659	·15585	·13260	·10630	·076190	·041249
20	0·28268	0·25682	0·22618	0·18904	0·16725	0·14270	0·11477	0·082583	0·044917
21	·29803	·27149	·23985	·20127	·17849	·15272	·12322	·088998	·048629
22	·31287	·28571	·25319	·21326	·18957	·16263	·13163	·095423	·052376
23	·32719	·29949	·26617	·22502	·20047	·17242	·13998	·10185	·056154
24	·34101	·31285	·27881	·23654	·21118	·18209	·14826	·10827	·059957
25	0·35435	0·32578	0·29110	0·24780	0·22169	0·19162	0·15648	0·11467	0·063780
26	·36723	·33831	·30306	·25882	·23202	·20102	·16461	·12104	·067620
27	·37966	·35044	·31469	·26959	·24214	·21027	·17266	·12739	·071472
28	·39166	·36218	·32599	·28011	·25206	·21937	·18062	·13370	·075333
29	·40324	·37356	·33698	·29039	·26179	·22833	·18848	·13998	·079199
30	0·41443	0·38458	0·34766	0·30043	0·27131	0·23713	0·19625	0·14621	0·083069
40	·50798	·47784	·43954	·38873	·35629	·31705	·26833	·20574	·12150
60	·62883	·60130	·56512	·51488	·48135	·43908	·38377	·30748	·19364
120	·78755	·76828	·74189	·70304	·67550	·63876	·58676	·50630	·36144
∞	1·00000	1·00000	1·00000	1·00000	1·00000	1·00000	1·00000	1·00000	1·00000

For $\nu_1 = \infty$, $x = 0$.

Table 17. *Chart for determining the probability levels of the incomplete* B-*function*, $I_x(a, b)$

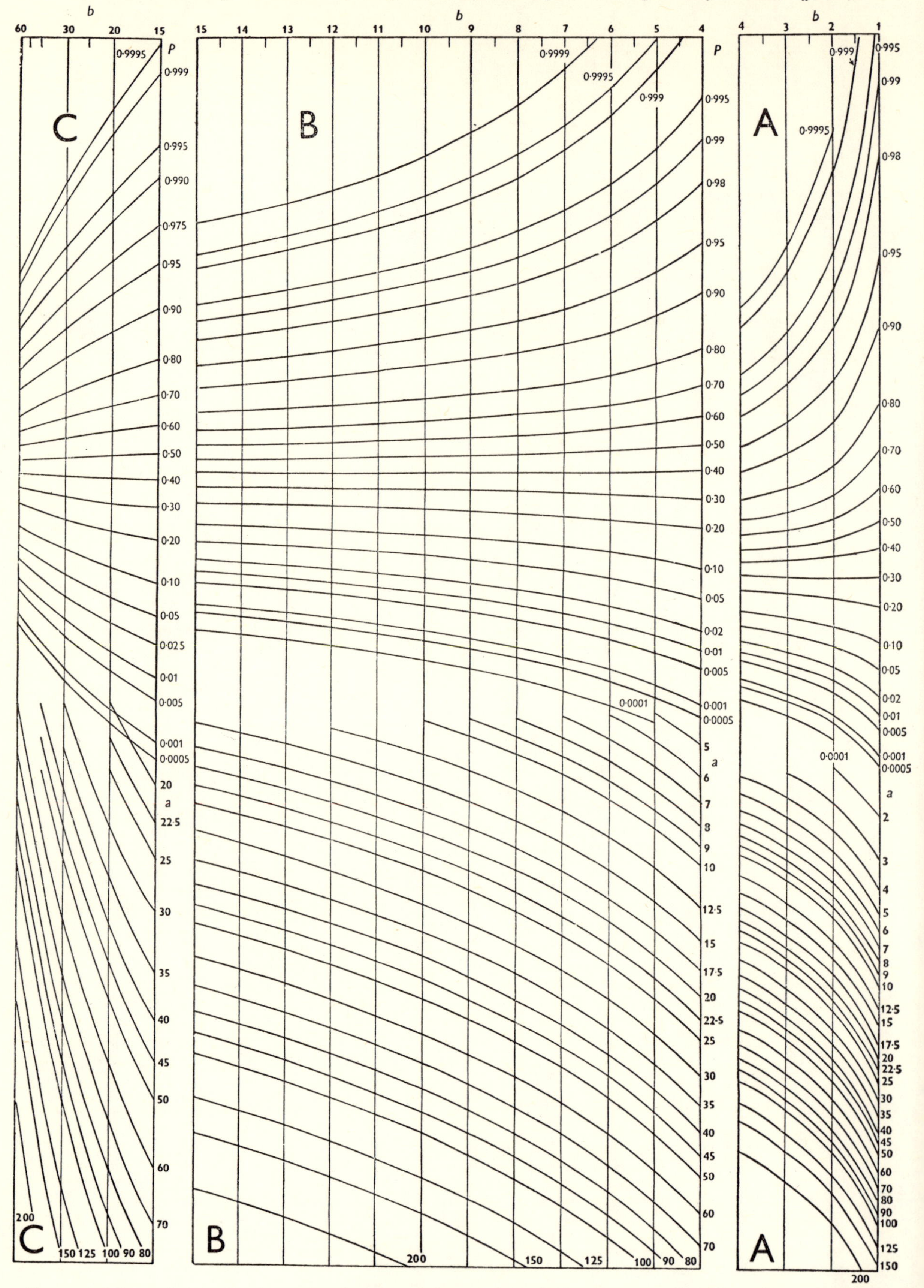

For use with x-scale contained in pocket at end of volume. In each section, each member of the lower family of curves corresponds to a constant value of a, and each member of the upper family to a constant value of $P = I_x(a, b)$. For directions as to use, see Introduction, pp. 33–4.

Table 18. *Percentage points of the F-distribution (variance ratio)*

Upper 25 % points

ν_2 \ ν_1	1	2	3	4	5	6	7	8	9	10	12	15	20	24	30	40	60	120	∞
1	5·83	7·50	8·20	8·58	8·82	8·98	9·10	9·19	9·26	9·32	9·41	9·49	9·58	9·63	9·67	9·71	9·76	9·80	9·85
2	2·57	3·00	3·15	3·23	3·28	3·31	3·34	3·35	3·37	3·38	3·39	3·41	3·43	3·43	3·44	3·45	3·46	3·47	3·48
3	2·02	2·28	2·36	2·39	2·41	2·42	2·43	2·44	2·44	2·44	2·45	2·46	2·46	2·46	2·47	2·47	2·47	2·47	2·47
4	1·81	2·00	2·05	2·06	2·07	2·08	2·08	2·08	2·08	2·08	2·08	2·08	2·08	2·08	2·08	2·08	2·08	2·08	2·08
5	1·69	1·85	1·88	1·89	1·89	1·89	1·89	1·89	1·89	1·89	1·89	1·89	1·88	1·88	1·88	1·88	1·87	1·87	1·87
6	1·62	1·76	1·78	1·79	1·79	1·78	1·78	1·78	1·77	1·77	1·77	1·76	1·76	1·75	1·75	1·75	1·74	1·74	1·74
7	1·57	1·70	1·72	1·72	1·71	1·71	1·70	1·70	1·69	1·69	1·68	1·68	1·67	1·67	1·66	1·66	1·65	1·65	1·65
8	1·54	1·66	1·67	1·66	1·66	1·65	1·64	1·64	1·63	1·63	1·62	1·62	1·61	1·60	1·60	1·59	1·59	1·58	1·58
9	1·51	1·62	1·63	1·63	1·62	1·61	1·60	1·60	1·59	1·59	1·58	1·57	1·56	1·56	1·55	1·54	1·54	1·53	1·53
10	1·49	1·60	1·60	1·59	1·59	1·58	1·57	1·56	1·56	1·55	1·54	1·53	1·52	1·52	1·51	1·51	1·50	1·49	1·48
11	1·47	1·58	1·58	1·57	1·56	1·55	1·54	1·53	1·53	1·52	1·51	1·50	1·49	1·49	1·48	1·47	1·47	1·46	1·45
12	1·46	1·56	1·56	1·55	1·54	1·53	1·52	1·51	1·51	1·50	1·49	1·48	1·47	1·46	1·45	1·45	1·44	1·43	1·42
13	1·45	1·55	1·55	1·53	1·52	1·51	1·50	1·49	1·49	1·48	1·47	1·46	1·45	1·44	1·43	1·42	1·42	1·41	1·40
14	1·44	1·53	1·53	1·52	1·51	1·50	1·49	1·48	1·47	1·46	1·45	1·44	1·43	1·42	1·41	1·41	1·40	1·39	1·38
15	1·43	1·52	1·52	1·51	1·49	1·48	1·47	1·46	1·46	1·45	1·44	1·43	1·41	1·41	1·40	1·39	1·38	1·37	1·36
16	1·42	1·51	1·51	1·50	1·48	1·47	1·46	1·45	1·44	1·44	1·43	1·41	1·40	1·39	1·38	1·37	1·36	1·35	1·34
17	1·42	1·51	1·50	1·49	1·47	1·46	1·45	1·44	1·43	1·43	1·41	1·40	1·39	1·38	1·37	1·36	1·35	1·34	1·33
18	1·41	1·50	1·49	1·48	1·46	1·45	1·44	1·43	1·42	1·42	1·40	1·39	1·38	1·37	1·36	1·35	1·34	1·33	1·32
19	1·41	1·49	1·49	1·47	1·46	1·44	1·43	1·42	1·41	1·41	1·40	1·38	1·37	1·36	1·35	1·34	1·33	1·32	1·30
20	1·40	1·49	1·48	1·47	1·45	1·44	1·43	1·42	1·41	1·40	1·39	1·37	1·36	1·35	1·34	1·33	1·32	1·31	1·29
21	1·40	1·48	1·48	1·46	1·44	1·43	1·42	1·41	1·40	1·39	1·38	1·37	1·35	1·34	1·33	1·32	1·31	1·30	1·28
22	1·40	1·48	1·47	1·45	1·44	1·42	1·41	1·40	1·39	1·39	1·37	1·36	1·34	1·33	1·32	1·31	1·30	1·29	1·28
23	1·39	1·47	1·47	1·45	1·43	1·42	1·41	1·40	1·39	1·38	1·37	1·35	1·34	1·33	1·32	1·31	1·30	1·28	1·27
24	1·39	1·47	1·46	1·44	1·43	1·41	1·40	1·39	1·38	1·38	1·36	1·35	1·33	1·32	1·31	1·30	1·29	1·28	1·26
25	1·39	1·47	1·46	1·44	1·42	1·41	1·40	1·39	1·38	1·37	1·36	1·34	1·33	1·32	1·31	1·29	1·28	1·27	1·25
26	1·38	1·46	1·45	1·44	1·42	1·41	1·39	1·38	1·37	1·37	1·35	1·34	1·32	1·31	1·30	1·29	1·28	1·26	1·25
27	1·38	1·46	1·45	1·43	1·42	1·40	1·39	1·38	1·37	1·36	1·35	1·33	1·32	1·31	1·30	1·28	1·27	1·26	1·24
28	1·38	1·46	1·45	1·43	1·41	1·40	1·39	1·38	1·37	1·36	1·34	1·33	1·31	1·30	1·29	1·28	1·27	1·25	1·24
29	1·38	1·45	1·45	1·43	1·41	1·40	1·38	1·37	1·36	1·35	1·34	1·32	1·31	1·30	1·29	1·27	1·26	1·25	1·23
30	1·38	1·45	1·44	1·42	1·41	1·39	1·38	1·37	1·36	1·35	1·34	1·32	1·30	1·29	1·28	1·27	1·26	1·24	1·23
40	1·36	1·44	1·42	1·40	1·39	1·37	1·36	1·35	1·34	1·33	1·31	1·30	1·28	1·26	1·25	1·24	1·22	1·21	1·19
60	1·35	1·42	1·41	1·38	1·37	1·35	1·33	1·32	1·31	1·30	1·29	1·27	1·25	1·24	1·22	1·21	1·19	1·17	1·15
120	1·34	1·40	1·39	1·37	1·35	1·33	1·31	1·30	1·29	1·28	1·26	1·24	1·22	1·21	1·19	1·18	1·16	1·13	1·10
∞	1·32	1·39	1·37	1·35	1·33	1·31	1·29	1·28	1·27	1·25	1·24	1·22	1·19	1·18	1·16	1·14	1·12	1·08	1·00

$F = \frac{s_1^2}{s_2^2} = \frac{S_1}{\nu_1} \Big/ \frac{S_2}{\nu_2}$, where $s_1^2 = S_1/\nu_1$ and $s_2^2 = S_2/\nu_2$ are independent mean squares estimating a common variance σ^2 and based on ν_1 and ν_2 degrees of freedom, respectively.

Table 18. *Percentage points of the F-distribution (continued)*

Upper 10 % *points*

ν_2 \ ν_1	1	2	3	4	5	6	7	8	9	10	12	15	20	24	30	40	60	120	∞
1	39·86	49·50	53·59	55·83	57·24	58·20	58·91	59·44	59·86	60·19	60·71	61·22	61·74	62·00	62·26	62·53	62·79	63·06	63·33
2	8·53	9·00	9·16	9·24	9·29	9·33	9·35	9·37	9·38	9·39	9·41	9·42	9·44	9·45	9·46	9·47	9·47	9·48	9·49
3	5·54	5·46	5·39	5·34	5·31	5·28	5·27	5·25	5·24	5·23	5·22	5·20	5·18	5·18	5·17	5·16	5·15	5·14	5·13
4	4·54	4·32	4·19	4·11	4·05	4·01	3·98	3·95	3·94	3·92	3·90	3·87	3·84	3·83	3·82	3·80	3·79	3·78	3·76
5	4·06	3·78	3·62	3·52	3·45	3·40	3·37	3·34	3·32	3·30	3·27	3·24	3·21	3·19	3·17	3·16	3·14	3·12	3·10
6	3·78	3·46	3·29	3·18	3·11	3·05	3·01	2·98	2·96	2·94	2·90	2·87	2·84	2·82	2·80	2·78	2·76	2·74	2·72
7	3·59	3·26	3·07	2·96	2·88	2·83	2·78	2·75	2·72	2·70	2·67	2·63	2·59	2·58	2·56	2·54	2·51	2·49	2·47
8	3·46	3·11	2·92	2·81	2·73	2·67	2·62	2·59	2·56	2·54	2·50	2·46	2·42	2·40	2·38	2·36	2·34	2·32	2·29
9	3·36	3·01	2·81	2·69	2·61	2·55	2·51	2·47	2·44	2·42	2·38	2·34	2·30	2·28	2·25	2·23	2·21	2·18	2·16
10	3·29	2·92	2·73	2·61	2·52	2·46	2·41	2·38	2·35	2·32	2·28	2·24	2·20	2·18	2·16	2·13	2·11	2·08	2·06
11	3·23	2·86	2·66	2·54	2·45	2·39	2·34	2·30	2·27	2·25	2·21	2·17	2·12	2·10	2·08	2·05	2·03	2·00	1·97
12	3·18	2·81	2·61	2·48	2·39	2·33	2·28	2·24	2·21	2·19	2·15	2·10	2·06	2·04	2·01	1·99	1·96	1·93	1·90
13	3·14	2·76	2·56	2·43	2·35	2·28	2·23	2·20	2·16	2·14	2·10	2·05	2·01	1·98	1·96	1·93	1·90	1·88	1·85
14	3·10	2·73	2·52	2·39	2·31	2·24	2·19	2·15	2·12	2·10	2·05	2·01	1·96	1·94	1·91	1·89	1·86	1·83	1·80
15	3·07	2·70	2·49	2·36	2·27	2·21	2·16	2·12	2·09	2·06	2·02	1·97	1·92	1·90	1·87	1·85	1·82	1·79	1·76
16	3·05	2·67	2·46	2·33	2·24	2·18	2·13	2·09	2·06	2·03	1·99	1·94	1·89	1·87	1·84	1·81	1·78	1·75	1·72
17	3·03	2·64	2·44	2·31	2·22	2·15	2·10	2·06	2·03	2·00	1·96	1·91	1·86	1·84	1·81	1·78	1·75	1·72	1·69
18	3·01	2·62	2·42	2·29	2·20	2·13	2·08	2·04	2·00	1·98	1·93	1·89	1·84	1·81	1·78	1·75	1·72	1·69	1·66
19	2·99	2·61	2·40	2·27	2·18	2·11	2·06	2·02	1·98	1·96	1·91	1·86	1·81	1·79	1·76	1·73	1·70	1·67	1·63
20	2·97	2·59	2·38	2·25	2·16	2·09	2·04	2·00	1·96	1·94	1·89	1·84	1·79	1·77	1·74	1·71	1·68	1·64	1·61
21	2·96	2·57	2·36	2·23	2·14	2·08	2·02	1·98	1·95	1·92	1·87	1·83	1·78	1·75	1·72	1·69	1·66	1·62	1·59
22	2·95	2·56	2·35	2·22	2·13	2·06	2·01	1·97	1·93	1·90	1·86	1·81	1·76	1·73	1·70	1·67	1·64	1·60	1·57
23	2·94	2·55	2·34	2·21	2·11	2·05	1·99	1·95	1·92	1·89	1·84	1·80	1·74	1·72	1·69	1·66	1·62	1·59	1·55
24	2·93	2·54	2·33	2·19	2·10	2·04	1·98	1·94	1·91	1·88	1·83	1·78	1·73	1·70	1·67	1·64	1·61	1·57	1·53
25	2·92	2·53	2·32	2·18	2·09	2·02	1·97	1·93	1·89	1·87	1·82	1·77	1·72	1·69	1·66	1·63	1·59	1·56	1·52
26	2·91	2·52	2·31	2·17	2·08	2·01	1·96	1·92	1·88	1·86	1·81	1·76	1·71	1·68	1·65	1·61	1·58	1·54	1·50
27	2·90	2·51	2·30	2·17	2·07	2·00	1·95	1·91	1·87	1·85	1·80	1·75	1·70	1·67	1·64	1·60	1·57	1·53	1·49
28	2·89	2·50	2·29	2·16	2·06	2·00	1·94	1·90	1·87	1·84	1·79	1·74	1·69	1·66	1·63	1·59	1·56	1·52	1·48
29	2·89	2·50	2·28	2·15	2·06	1·99	1·93	1·89	1·86	1·83	1·78	1·73	1·68	1·65	1·62	1·58	1·55	1·51	1·47
30	2·88	2·49	2·28	2·14	2·05	1·98	1·93	1·88	1·85	1·82	1·77	1·72	1·67	1·64	1·61	1·57	1·54	1·50	1·46
40	2·84	2·44	2·23	2·09	2·00	1·93	1·87	1·83	1·79	1·76	1·71	1·66	1·61	1·57	1·54	1·51	1·47	1·42	1·38
60	2·79	2·39	2·18	2·04	1·95	1·87	1·82	1·77	1·74	1·71	1·66	1·60	1·54	1·51	1·48	1·44	1·40	1·35	1·29
120	2·75	2·35	2·13	1·99	1·90	1·82	1·77	1·72	1·68	1·65	1·60	1·55	1·48	1·45	1·41	1·37	1·32	1·26	1·19
∞	2·71	2·30	2·08	1·94	1·85	1·77	1·72	1·67	1·63	1·60	1·55	1·49	1·42	1·38	1·34	1·30	1·24	1·17	1·00

$F=\frac{s_1^2}{s_2^2}=\frac{S_1}{\nu_1}\Big/\frac{S_2}{\nu_2}$, where $s_1^2=S_1/\nu_1$ and $s_2^2=S_2/\nu_2$ are independent mean squares estimating a common variance σ^2 and based on ν_1 and ν_2 degrees of freedom, respectively.

Table 18 (*continued*)

Upper 5% points

ν_2 \ ν_1	1	2	3	4	5	6	7	8	9	10	12	15	20	24	30	40	60	120	∞
1	161·4	199·5	215·7	224·6	230·2	234·0	236·8	238·9	240·5	241·9	243·9	245·9	248·0	249·1	250·1	251·1	252·2	253·3	254·3
2	18·51	19·00	19·16	19·25	19·30	19·33	19·35	19·37	19·38	19·40	19·41	19·43	19·45	19·45	19·46	19·47	19·48	19·49	19·50
3	10·13	9·55	9·28	9·12	9·01	8·94	8·89	8·85	8·81	8·79	8·74	8·70	8·66	8·64	8·62	8·59	8·57	8·55	8·53
4	7·71	6·94	6·59	6·39	6·26	6·16	6·09	6·04	6·00	5·96	5·91	5·86	5·80	5·77	5·75	5·72	5·69	5·66	5·63
5	6·61	5·79	5·41	5·19	5·05	4·95	4·88	4·82	4·77	4·74	4·68	4·62	4·56	4·53	4·50	4·46	4·43	4·40	4·36
6	5·99	5·14	4·76	4·53	4·39	4·28	4·21	4·15	4·10	4·06	4·00	3·94	3·87	3·84	3·81	3·77	3·74	3·70	3·67
7	5·59	4·74	4·35	4·12	3·97	3·87	3·79	3·73	3·68	3·64	3·57	3·51	3·44	3·41	3·38	3·34	3·30	3·27	3·23
8	5·32	4·46	4·07	3·84	3·69	3·58	3·50	3·44	3·39	3·35	3·28	3·22	3·15	3·12	3·08	3·04	3·01	2·97	2·93
9	5·12	4·26	3·86	3·63	3·48	3·37	3·29	3·23	3·18	3·14	3·07	3·01	2·94	2·90	2·86	2·83	2·79	2·75	2·71
10	4·96	4·10	3·71	3·48	3·33	3·22	3·14	3·07	3·02	2·98	2·91	2·85	2·77	2·74	2·70	2·66	2·62	2·58	2·54
11	4·84	3·98	3·59	3·36	3·20	3·09	3·01	2·95	2·90	2·85	2·79	2·72	2·65	2·61	2·57	2·53	2·49	2·45	2·40
12	4·75	3·89	3·49	3·26	3·11	3·00	2·91	2·85	2·80	2·75	2·69	2·62	2·54	2·51	2·47	2·43	2·38	2·34	2·30
13	4·67	3·81	3·41	3·18	3·03	2·92	2·83	2·77	2·71	2·67	2·60	2·53	2·46	2·42	2·38	2·34	2·30	2·25	2·21
14	4·60	3·74	3·34	3·11	2·96	2·85	2·76	2·70	2·65	2·60	2·53	2·46	2·39	2·35	2·31	2·27	2·22	2·18	2·13
15	4·54	3·68	3·29	3·06	2·90	2·79	2·71	2·64	2·59	2·54	2·48	2·40	2·33	2·29	2·25	2·20	2·16	2·11	2·07
16	4·49	3·63	3·24	3·01	2·85	2·74	2·66	2·59	2·54	2·49	2·42	2·35	2·28	2·24	2·19	2·15	2·11	2·06	2·01
17	4·45	3·59	3·20	2·96	2·81	2·70	2·61	2·55	2·49	2·45	2·38	2·31	2·23	2·19	2·15	2·10	2·06	2·01	1·96
18	4·41	3·55	3·16	2·93	2·77	2·66	2·58	2·51	2·46	2·41	2·34	2·27	2·19	2·15	2·11	2·06	2·02	1·97	1·92
19	4·38	3·52	3·13	2·90	2·74	2·63	2·54	2·48	2·42	2·38	2·31	2·23	2·16	2·11	2·07	2·03	1·98	1·93	1·88
20	4·35	3·49	3·10	2·87	2·71	2·60	2·51	2·45	2·39	2·35	2·28	2·20	2·12	2·08	2·04	1·99	1·95	1·90	1·84
21	4·32	3·47	3·07	2·84	2·68	2·57	2·49	2·42	2·37	2·32	2·25	2·18	2·10	2·05	2·01	1·96	1·92	1·87	1·81
22	4·30	3·44	3·05	2·82	2·66	2·55	2·46	2·40	2·34	2·30	2·23	2·15	2·07	2·03	1·98	1·94	1·89	1·84	1·78
23	4·28	3·42	3·03	2·80	2·64	2·53	2·44	2·37	2·32	2·27	2·20	2·13	2·05	2·01	1·96	1·91	1·86	1·81	1·76
24	4·26	3·40	3·01	2·78	2·62	2·51	2·42	2·36	2·30	2·25	2·18	2·11	2·03	1·98	1·94	1·89	1·84	1·79	1·73
25	4·24	3·39	2·99	2·76	2·60	2·49	2·40	2·34	2·28	2·24	2·16	2·09	2·01	1·96	1·92	1·87	1·82	1·77	1·71
26	4·23	3·37	2·98	2·74	2·59	2·47	2·39	2·32	2·27	2·22	2·15	2·07	1·99	1·95	1·90	1·85	1·80	1·75	1·69
27	4·21	3·35	2·96	2·73	2·57	2·46	2·37	2·31	2·25	2·20	2·13	2·06	1·97	1·93	1·88	1·84	1·79	1·73	1·67
28	4·20	3·34	2·95	2·71	2·56	2·45	2·36	2·29	2·24	2·19	2·12	2·04	1·96	1·91	1·87	1·82	1·77	1·71	1·65
29	4·18	3·33	2·93	2·70	2·55	2·43	2·35	2·28	2·22	2·18	2·10	2·03	1·94	1·90	1·85	1·81	1·75	1·70	1·64
30	4·17	3·32	2·92	2·69	2·53	2·42	2·33	2·27	2·21	2·16	2·09	2·01	1·93	1·89	1·84	1·79	1·74	1·68	1·62
40	4·08	3·23	2·84	2·61	2·45	2·34	2·25	2·18	2·12	2·08	2·00	1·92	1·84	1·79	1·74	1·69	1·64	1·58	1·51
60	4·00	3·15	2·76	2·53	2·37	2·25	2·17	2·10	2·04	1·99	1·92	1·84	1·75	1·70	1·65	1·59	1·53	1·47	1·39
120	3·92	3·07	2·68	2·45	2·29	2·17	2·09	2·02	1·96	1·91	1·83	1·75	1·66	1·61	1·55	1·50	1·43	1·35	1·25
∞	3·84	3·00	2·60	2·37	2·21	2·10	2·01	1·94	1·88	1·83	1·75	1·67	1·57	1·52	1·46	1·39	1·32	1·22	1·00

$F = \dfrac{s_1^2}{s_2^2} = \dfrac{S_1}{\nu_1}\Big/\dfrac{S_2}{\nu_2}$, where $s_1^2 = S_1/\nu_1$ and $s_2^2 = S_2/\nu_2$ are independent mean squares estimating a common variance σ^2 and based on ν_1 and ν_2 degrees of freedom, respectively.

Table 18. *Percentage points of the F-distribution (continued)*

Upper 2·5 % points

ν_2 \ ν_1	1	2	3	4	5	6	7	8	9	10	12	15	20	24	30	40	60	120	∞
1	647·8	799·5	864·2	899·6	921·8	937·1	948·2	956·7	963·3	968·6	976·7	984·9	993·1	997·2	1001	1006	1010	1014	1018
2	38·51	39·00	39·17	39·25	39·30	39·33	39·36	39·37	39·39	39·40	39·41	39·43	39·45	39·46	39·46	39·47	39·48	39·49	39·50
3	17·44	16·04	15·44	15·10	14·88	14·73	14·62	14·54	14·47	14·42	14·34	14·25	14·17	14·12	14·08	14·04	13·99	13·95	13·90
4	12·22	10·65	9·98	9·60	9·36	9·20	9·07	8·98	8·90	8·84	8·75	8·66	8·56	8·51	8·46	8·41	8·36	8·31	8·26
5	10·01	8·43	7·76	7·39	7·15	6·98	6·85	6·76	6·68	6·62	6·52	6·43	6·33	6·28	6·23	6·18	6·12	6·07	6·02
6	8·81	7·26	6·60	6·23	5·99	5·82	5·70	5·60	5·52	5·46	5·37	5·27	5·17	5·12	5·07	5·01	4·96	4·90	4·85
7	8·07	6·54	5·89	5·52	5·29	5·12	4·99	4·90	4·82	4·76	4·67	4·57	4·47	4·42	4·36	4·31	4·25	4·20	4·14
8	7·57	6·06	5·42	5·05	4·82	4·65	4·53	4·43	4·36	4·30	4·20	4·10	4·00	3·95	3·89	3·84	3·78	3·73	3·67
9	7·21	5·71	5·08	4·72	4·48	4·32	4·20	4·10	4·03	3·96	3·87	3·77	3·67	3·61	3·56	3·51	3·45	3·39	3·33
10	6·94	5·46	4·83	4·47	4·24	4·07	3·95	3·85	3·78	3·72	3·62	3·52	3·42	3·37	3·31	3·26	3·20	3·14	3·08
11	6·72	5·26	4·63	4·28	4·04	3·88	3·76	3·66	3·59	3·53	3·43	3·33	3·23	3·17	3·12	3·06	3·00	2·94	2·88
12	6·55	5·10	4·47	4·12	3·89	3·73	3·61	3·51	3·44	3·37	3·28	3·18	3·07	3·02	2·96	2·91	2·85	2·79	2·72
13	6·41	4·97	4·35	4·00	3·77	3·60	3·48	3·39	3·31	3·25	3·15	3·05	2·95	2·89	2·84	2·78	2·72	2·66	2·60
14	6·30	4·86	4·24	3·89	3·66	3·50	3·38	3·29	3·21	3·15	3·05	2·95	2·84	2·79	2·73	2·67	2·61	2·55	2·49
15	6·20	4·77	4·15	3·80	3·58	3·41	3·29	3·20	3·12	3·06	2·96	2·86	2·76	2·70	2·64	2·59	2·52	2·46	2·40
16	6·12	4·69	4·08	3·73	3·50	3·34	3·22	3·12	3·05	2·99	2·89	2·79	2·68	2·63	2·57	2·51	2·45	2·38	2·32
17	6·04	4·62	4·01	3·66	3·44	3·28	3·16	3·06	2·98	2·92	2·82	2·72	2·62	2·56	2·50	2·44	2·38	2·32	2·25
18	5·98	4·56	3·95	3·61	3·38	3·22	3·10	3·01	2·93	2·87	2·77	2·67	2·56	2·50	2·44	2·38	2·32	2·26	2·19
19	5·92	4·51	3·90	3·56	3·33	3·17	3·05	2·96	2·88	2·82	2·72	2·62	2·51	2·45	2·39	2·33	2·27	2·20	2·13
20	5·87	4·46	3·86	3·51	3·29	3·13	3·01	2·91	2·84	2·77	2·68	2·57	2·46	2·41	2·35	2·29	2·22	2·16	2·09
21	5·83	4·42	3·82	3·48	3·25	3·09	2·97	2·87	2·80	2·73	2·64	2·53	2·42	2·37	2·31	2·25	2·18	2·11	2·04
22	5·79	4·38	3·78	3·44	3·22	3·05	2·93	2·84	2·76	2·70	2·60	2·50	2·39	2·33	2·27	2·21	2·14	2·08	2·00
23	5·75	4·35	3·75	3·41	3·18	3·02	2·90	2·81	2·73	2·67	2·57	2·47	2·36	2·30	2·24	2·18	2·11	2·04	1·97
24	5·72	4·32	3·72	3·38	3·15	2·99	2·87	2·78	2·70	2·64	2·54	2·44	2·33	2·27	2·21	2·15	2·08	2·01	1·94
25	5·69	4·29	3·69	3·35	3·13	2·97	2·85	2·75	2·68	2·61	2·51	2·41	2·30	2·24	2·18	2·12	2·05	1·98	1·91
26	5·66	4·27	3·67	3·33	3·10	2·94	2·82	2·73	2·65	2·59	2·49	2·39	2·28	2·22	2·16	2·09	2·03	1·95	1·88
27	5·63	4·24	3·65	3·31	3·08	2·92	2·80	2·71	2·63	2·57	2·47	2·36	2·25	2·19	2·13	2·07	2·00	1·93	1·85
28	5·61	4·22	3·63	3·29	3·06	2·90	2·78	2·69	2·61	2·55	2·45	2·34	2·23	2·17	2·11	2·05	1·98	1·91	1·83
29	5·59	4·20	3·61	3·27	3·04	2·88	2·76	2·67	2·59	2·53	2·43	2·32	2·21	2·15	2·09	2·03	1·96	1·89	1·81
30	5·57	4·18	3·59	3·25	3·03	2·87	2·75	2·65	2·57	2·51	2·41	2·31	2·20	2·14	2·07	2·01	1·94	1·87	1·79
40	5·42	4·05	3·46	3·13	2·90	2·74	2·62	2·53	2·45	2·39	2·29	2·18	2·07	2·01	1·94	1·88	1·80	1·72	1·64
60	5·29	3·93	3·34	3·01	2·79	2·63	2·51	2·41	2·33	2·27	2·17	2·06	1·94	1·88	1·82	1·74	1·67	1·58	1·48
120	5·15	3·80	3·23	2·89	2·67	2·52	2·39	2·30	2·22	2·16	2·05	1·94	1·82	1·76	1·69	1·61	1·53	1·43	1·31
∞	5·02	3·69	3·12	2·79	2·57	2·41	2·29	2·19	2·11	2·05	1·94	1·83	1·71	1·64	1·57	1·48	1·39	1·27	1·00

$F=\frac{s_1^2}{s_2^2}=\frac{S_1}{\nu_1}\Big/\frac{S_2}{\nu_2}$, where $s_1^2=S_1/\nu_1$ and $s_2^2=S_2/\nu_2$ are independent mean squares estimating a common variance σ^2 and based on ν_1 and ν_2 degrees of freedom, respectively.

Table 18 (*continued*)

Upper 1 % points

ν_2 \ ν_1	1	2	3	4	5	6	7	8	9	10	12	15	20	24	30	40	60	120	∞
1	4052	4999·5	5403	5625	5764	5859	5928	5981	6022	6056	6106	6157	6209	6235	6261	6287	6313	6339	6366
2	98·50	99·00	99·17	99·25	99·30	99·33	99·36	99·37	99·39	99·40	99·42	99·43	99·45	99·46	99·47	99·47	99·48	99·49	99·50
3	34·12	30·82	29·46	28·71	28·24	27·91	27·67	27·49	27·35	27·23	27·05	26·87	26·69	26·60	26·50	26·41	26·32	26·22	26·13
4	21·20	18·00	16·69	15·98	15·52	15·21	14·98	14·80	14·66	14·55	14·37	14·20	14·02	13·93	13·84	13·75	13·65	13·56	13·46
5	16·26	13·27	12·06	11·39	10·97	10·67	10·46	10·29	10·16	10·05	9·89	9·72	9·55	9·47	9·38	9·29	9·20	9·11	9·02
6	13·75	10·92	9·78	9·15	8·75	8·47	8·26	8·10	7·98	7·87	7·72	7·56	7·40	7·31	7·23	7·14	7·06	6·97	6·88
7	12·25	9·55	8·45	7·85	7·46	7·19	6·99	6·84	6·72	6·62	6·47	6·31	6·16	6·07	5·99	5·91	5·82	5·74	5·65
8	11·26	8·65	7·59	7·01	6·63	6·37	6·18	6·03	5·91	5·81	5·67	5·52	5·36	5·28	5·20	5·12	5·03	4·95	4·86
9	10·56	8·02	6·99	6·42	6·06	5·80	5·61	5·47	5·35	5·26	5·11	4·96	4·81	4·73	4·65	4·57	4·48	4·40	4·31
10	10·04	7·56	6·55	5·99	5·64	5·39	5·20	5·06	4·94	4·85	4·71	4·56	4·41	4·33	4·25	4·17	4·08	4·00	3·91
11	9·65	7·21	6·22	5·67	5·32	5·07	4·89	4·74	4·63	4·54	4·40	4·25	4·10	4·02	3·94	3·86	3·78	3·69	3·60
12	9·33	6·93	5·95	5·41	5·06	4·82	4·64	4·50	4·39	4·30	4·16	4·01	3·86	3·78	3·70	3·62	3·54	3·45	3·36
13	9·07	6·70	5·74	5·21	4·86	4·62	4·44	4·30	4·19	4·10	3·96	3·82	3·66	3·59	3·51	3·43	3·34	3·25	3·17
14	8·86	6·51	5·56	5·04	4·69	4·46	4·28	4·14	4·03	3·94	3·80	3·66	3·51	3·43	3·35	3·27	3·18	3·09	3·00
15	8·68	6·36	5·42	4·89	4·56	4·32	4·14	4·00	3·89	3·80	3·67	3·52	3·37	3·29	3·21	3·13	3·05	2·96	2·87
16	8·53	6·23	5·29	4·77	4·44	4·20	4·03	3·89	3·78	3·69	3·55	3·41	3·26	3·18	3·10	3·02	2·93	2·84	2·75
17	8·40	6·11	5·18	4·67	4·34	4·10	3·93	3·79	3·68	3·59	3·46	3·31	3·16	3·08	3·00	2·92	2·83	2·75	2·65
18	8·29	6·01	5·09	4·58	4·25	4·01	3·84	3·71	3·60	3·51	3·37	3·23	3·08	3·00	2·92	2·84	2·75	2·66	2·57
19	8·18	5·93	5·01	4·50	4·17	3·94	3·77	3·63	3·52	3·43	3·30	3·15	3·00	2·92	2·84	2·76	2·67	2·58	2·49
20	8·10	5·85	4·94	4·43	4·10	3·87	3·70	3·56	3·46	3·37	3·23	3·09	2·94	2·86	2·78	2·69	2·61	2·52	2·42
21	8·02	5·78	4·87	4·37	4·04	3·81	3·64	3·51	3·40	3·31	3·17	3·03	2·88	2·80	2·72	2·64	2·55	2·46	2·36
22	7·95	5·72	4·82	4·31	3·99	3·76	3·59	3·45	3·35	3·26	3·12	2·98	2·83	2·75	2·67	2·58	2·50	2·40	2·31
23	7·88	5·66	4·76	4·26	3·94	3·71	3·54	3·41	3·30	3·21	3·07	2·93	2·78	2·70	2·62	2·54	2·45	2·35	2·26
24	7·82	5·61	4·72	4·22	3·90	3·67	3·50	3·36	3·26	3·17	3·03	2·89	2·74	2·66	2·58	2·49	2·40	2·31	2·21
25	7·77	5·57	4·68	4·18	3·85	3·63	3·46	3·32	3·22	3·13	2·99	2·85	2·70	2·62	2·54	2·45	2·36	2·27	2·17
26	7·72	5·53	4·64	4·14	3·82	3·59	3·42	3·29	3·18	3·09	2·96	2·81	2·66	2·58	2·50	2·42	2·33	2·23	2·13
27	7·68	5·49	4·60	4·11	3·78	3·56	3·39	3·26	3·15	3·06	2·93	2·78	2·63	2·55	2·47	2·38	2·29	2·20	2·10
28	7·64	5·45	4·57	4·07	3·75	3·53	3·36	3·23	3·12	3·03	2·90	2·75	2·60	2·52	2·44	2·35	2·26	2·17	2·06
29	7·60	5·42	4·54	4·04	3·73	3·50	3·33	3·20	3·09	3·00	2·87	2·73	2·57	2·49	2·41	2·33	2·23	2·14	2·03
30	7·56	5·39	4·51	4·02	3·70	3·47	3·30	3·17	3·07	2·98	2·84	2·70	2·55	2·47	2·39	2·30	2·21	2·11	2·01
40	7·31	5·18	4·31	3·83	3·51	3·29	3·12	2·99	2·89	2·80	2·66	2·52	2·37	2·29	2·20	2·11	2·02	1·92	1·80
60	7·08	4·98	4·13	3·65	3·34	3·12	2·95	2·82	2·72	2·63	2·50	2·35	2·20	2·12	2·03	1·94	1·84	1·73	1·60
120	6·85	4·79	3·95	3·48	3·17	2·96	2·79	2·66	2·56	2·47	2·34	2·19	2·03	1·95	1·86	1·76	1·66	1·53	1·38
∞	6·63	4·61	3·78	3·32	3·02	2·80	2·64	2·51	2·41	2·32	2·18	2·04	1·88	1·79	1·70	1·59	1·47	1·32	1·00

$F=\frac{s_1^2}{s_2^2}=\frac{S_1}{\nu_1}\Big/\frac{S_2}{\nu_2}$, where $s_1^2=S_1/\nu_1$ and $s_2^2=S_2/\nu_2$ are independent mean squares estimating a common variance σ^2 and based on ν_1 and ν_2 degrees of freedom, respectively.

Table 18. *Percentage points of the F-distribution (continued)*

Upper 0·5 % points

ν_2 \ ν_1	1	2	3	4	5	6	7	8	9	10	12	15	20	24	30	40	60	120	∞
1	16211	20000	21615	22500	23056	23437	23715	23925	24091	24224	24426	24630	24836	24940	25044	25148	25253	25359	25465
2	198·5	199·0	199·2	199·2	199·3	199·3	199·4	199·4	199·4	199·4	199·4	199·4	199·4	199·5	199·5	199·5	199·5	199·5	199·5
3	55·55	49·80	47·47	46·19	45·39	44·84	44·43	44·13	43·88	43·69	43·39	43·08	42·78	42·62	42·47	42·31	42·15	41·99	41·83
4	31·33	26·28	24·26	23·15	22·46	21·97	21·62	21·35	21·14	20·97	20·70	20·44	20·17	20·03	19·89	19·75	19·61	19·47	19·32
5	22·78	18·31	16·53	15·56	14·94	14·51	14·20	13·96	13·77	13·62	13·38	13·15	12·90	12·78	12·66	12·53	12·40	12·27	12·14
6	18·63	14·54	12·92	12·03	11·46	11·07	10·79	10·57	10·39	10·25	10·03	9·81	9·59	9·47	9·36	9·24	9·12	9·00	8·88
7	16·24	12·40	10·88	10·05	9·52	9·16	8·89	8·68	8·51	8·38	8·18	7·97	7·75	7·65	7·53	7·42	7·31	7·19	7·08
8	14·69	11·04	9·60	8·81	8·30	7·95	7·69	7·50	7·34	7·21	7·01	6·81	6·61	6·50	6·40	6·29	6·18	6·06	5·95
9	13·61	10·11	8·72	7·96	7·47	7·13	6·88	6·69	6·54	6·42	6·23	6·03	5·83	5·73	5·62	5·52	5·41	5·30	5·19
10	12·83	9·43	8·08	7·34	6·87	6·54	6·30	6·12	5·97	5·85	5·66	5·47	5·27	5·17	5·07	4·97	4·86	4·75	4·64
11	12·23	8·91	7·60	6·88	6·42	6·10	5·86	5·68	5·54	5·42	5·24	5·05	4·86	4·76	4·65	4·55	4·44	4·34	4·23
12	11·75	8·51	7·23	6·52	6·07	5·76	5·52	5·35	5·20	5·09	4·91	4·72	4·53	4·43	4·33	4·23	4·12	4·01	3·90
13	11·37	8·19	6·93	6·23	5·79	5·48	5·25	5·08	4·94	4·82	4·64	4·46	4·27	4·17	4·07	3·97	3·87	3·76	3·65
14	11·06	7·92	6·68	6·00	5·56	5·26	5·03	4·86	4·72	4·60	4·43	4·25	4·06	3·96	3·86	3·76	3·66	3·55	3·44
15	10·80	7·70	6·48	5·80	5·37	5·07	4·85	4·67	4·54	4·42	4·25	4·07	3·88	3·79	3·69	3·58	3·48	3·37	3·26
16	10·58	7·51	6·30	5·64	5·21	4·91	4·69	4·52	4·38	4·27	4·10	3·92	3·73	3·64	3·54	3·44	3·33	3·22	3·11
17	10·38	7·35	6·16	5·50	5·07	4·78	4·56	4·39	4·25	4·14	3·97	3·79	3·61	3·51	3·41	3·31	3·21	3·10	2·98
18	10·22	7·21	6·03	5·37	4·96	4·66	4·44	4·28	4·14	4·03	3·86	3·68	3·50	3·40	3·30	3·20	3·10	2·99	2·87
19	10·07	7·09	5·92	5·27	4·85	4·56	4·34	4·18	4·04	3·93	3·76	3·59	3·40	3·31	3·21	3·11	3·00	2·89	2·78
20	9·94	6·99	5·82	5·17	4·76	4·47	4·26	4·09	3·96	3·85	3·68	3·50	3·32	3·22	3·12	3·02	2·92	2·81	2·69
21	9·83	6·89	5·73	5·09	4·68	4·39	4·18	4·01	3·88	3·77	3·60	3·43	3·24	3·15	3·05	2·95	2·84	2·73	2·61
22	9·73	6·81	5·65	5·02	4·61	4·32	4·11	3·94	3·81	3·70	3·54	3·36	3·18	3·08	2·98	2·88	2·77	2·66	2·55
23	9·63	6·73	5·58	4·95	4·54	4·26	4·05	3·88	3·75	3·64	3·47	3·30	3·12	3·02	2·92	2·82	2·71	2·60	2·48
24	9·55	6·66	5·52	4·89	4·49	4·20	3·99	3·83	3·69	3·59	3·42	3·25	3·06	2·97	2·87	2·77	2·66	2·55	2·43
25	9·48	6·60	5·46	4·84	4·43	4·15	3·94	3·78	3·64	3·54	3·37	3·20	3·01	2·92	2·82	2·72	2·61	2·50	2·38
26	9·41	6·54	5·41	4·79	4·38	4·10	3·89	3·73	3·60	3·49	3·33	3·15	2·97	2·87	2·77	2·67	2·56	2·45	2·33
27	9·34	6·49	5·36	4·74	4·34	4·06	3·85	3·69	3·56	3·45	3·28	3·11	2·93	2·83	2·73	2·63	2·52	2·41	2·29
28	9·28	6·44	5·32	4·70	4·30	4·02	3·81	3·65	3·52	3·41	3·25	3·07	2·89	2·79	2·69	2·59	2·48	2·37	2·25
29	9·23	6·40	5·28	4·66	4·26	3·98	3·77	3·61	3·48	3·38	3·21	3·04	2·86	2·76	2·66	2·56	2·45	2·33	2·21
30	9·18	6·35	5·24	4·62	4·23	3·95	3·74	3·58	3·45	3·34	3·18	3·01	2·82	2·73	2·63	2·52	2·42	2·30	2·18
40	8·83	6·07	4·98	4·37	3·99	3·71	3·51	3·35	3·22	3·12	2·95	2·78	2·60	2·50	2·40	2·30	2·18	2·06	1·93
60	8·49	5·79	4·73	4·14	3·76	3·49	3·29	3·13	3·01	2·90	2·74	2·57	2·39	2·29	2·19	2·08	1·96	1·83	1·69
120	8·18	5·54	4·50	3·92	3·55	3·28	3·09	2·93	2·81	2·71	2·54	2·37	2·19	2·09	1·98	1·87	1·75	1·61	1·43
∞	7·88	5·30	4·28	3·72	3·35	3·09	2·90	2·74	2·62	2·52	2·36	2·19	2·00	1·90	1·79	1·67	1·53	1·36	1·00

$F=\frac{s_1^2}{s_2^2}=\frac{S_1}{\nu_1}\Big/\frac{S_2}{\nu_2}$, where $s_1^2=S_1/\nu_1$ and $s_2^2=S_2/\nu_2$ are independent mean squares estimating a common variance σ^2 and based on ν_1 and ν_2 degrees of freedom, respectively

Table 18 (*continued*)

Upper 0·1 % *points*

ν_2 \ ν_1	1	2	3	4	5	6	7	8	9	10	12	15	20	24	30	40	60	120	∞
1	4053*	5000*	5404*	5625*	5764*	5859*	5929*	5981*	6023*	6056*	6107*	6158*	6209*	6235*	6261*	6287*	6313*	6340*	6366*
2	998·5	999·0	999·2	999·2	999·3	999·3	999·4	999·4	999·4	999·4	999·4	999·4	999·4	999·5	999·5	999·5	999·5	999·5	999·5
3	167·0	148·5	141·1	137·1	134·6	132·8	131·6	130·6	129·9	129·2	128·3	127·4	126·4	125·9	125·4	125·0	124·5	124·0	123·5
4	74·14	61·25	56·18	53·44	51·71	50·53	49·66	49·00	48·47	48·05	47·41	46·76	46·10	45·77	45·43	45·09	44·75	44·40	44·05
5	47·18	37·12	33·20	31·09	29·75	28·84	28·16	27·64	27·24	26·92	26·42	25·91	25·39	25·14	24·87	24·60	24·33	24·06	23·79
6	35·51	27·00	23·70	21·92	20·81	20·03	19·46	19·03	18·69	18·41	17·99	17·56	17·12	16·89	16·67	16·44	16·21	15·99	15·75
7	29·25	21·69	18·77	17·19	16·21	15·52	15·02	14·63	14·33	14·08	13·71	13·32	12·93	12·73	12·53	12·33	12·12	11·91	11·70
8	25·42	18·49	15·83	14·39	13·49	12·86	12·40	12·04	11·77	11·54	11·19	10·84	10·48	10·30	10·11	9·92	9·73	9·53	9·33
9	22·86	16·39	13·90	12·56	11·71	11·13	10·70	10·37	10·11	9·89	9·57	9·24	8·90	8·72	8·55	8·37	8·19	8·00	7·81
10	21·04	14·91	12·55	11·28	10·48	9·92	9·52	9·20	8·96	8·75	8·45	8·13	7·80	7·64	7·47	7·30	7·12	6·94	6·76
11	19·69	13·81	11·56	10·35	9·58	9·05	8·66	8·35	8·12	7·92	7·63	7·32	7·01	6·85	6·68	6·52	6·35	6·17	6·00
12	18·64	12·97	10·80	9·63	8·89	8·38	8·00	7·71	7·48	7·29	7·00	6·71	6·40	6·25	6·09	5·93	5·76	5·59	5·42
13	17·81	12·31	10·21	9·07	8·35	7·86	7·49	7·21	6·98	6·80	6·52	6·23	5·93	5·78	5·63	5·47	5·30	5·14	4·97
14	17·14	11·78	9·73	8·62	7·92	7·43	7·08	6·80	6·58	6·40	6·13	5·85	5·56	5·41	5·25	5·10	4·94	4·77	4·60
15	16·59	11·34	9·34	8·25	7·57	7·09	6·74	6·47	6·26	6·08	5·81	5·54	5·25	5·10	4·95	4·80	4·64	4·47	4·31
16	16·12	10·97	9·00	7·94	7·27	6·81	6·46	6·19	5·98	5·81	5·55	5·27	4·99	4·85	4·70	4·54	4·39	4·23	4·06
17	15·72	10·66	8·73	7·68	7·02	6·56	6·22	5·96	5·75	5·58	5·32	5·05	4·78	4·63	4·48	4·33	4·18	4·02	3·85
18	15·38	10·39	8·49	7·46	6·81	6·35	6·02	5·76	5·56	5·39	5·13	4·87	4·59	4·45	4·30	4·15	4·00	3·84	3·67
19	15·08	10·16	8·28	7·26	6·62	6·18	5·85	5·59	5·39	5·22	4·97	4·70	4·43	4·29	4·14	3·99	3·84	3·68	3·51
20	14·82	9·95	8·10	7·10	6·46	6·02	5·69	5·44	5·24	5·08	4·82	4·56	4·29	4·15	4·00	3·86	3·70	3·54	3·38
21	14·59	9·77	7·94	6·95	6·32	5·88	5·56	5·31	5·11	4·95	4·70	4·44	4·17	4·03	3·88	3·74	3·58	3·42	3·26
22	14·38	9·61	7·80	6·81	6·19	5·76	5·44	5·19	4·99	4·83	4·58	4·33	4·06	3·92	3·78	3·63	3·48	3·32	3·15
23	14·19	9·47	7·67	6·69	6·08	5·65	5·33	5·09	4·89	4·73	4·48	4·23	3·96	3·82	3·68	3·53	3·38	3·22	3·05
24	14·03	9·34	7·55	6·59	5·98	5·55	5·23	4·99	4·80	4·64	4·39	4·14	3·87	3·74	3·59	3·45	3·29	3·14	2·97
25	13·88	9·22	7·45	6·49	5·88	5·46	5·15	4·91	4·71	4·56	4·31	4·06	3·79	3·66	3·52	3·37	3·22	3·06	2·89
26	13·74	9·12	7·36	6·41	5·80	5·38	5·07	4·83	4·64	4·48	4·24	3·99	3·72	3·59	3·44	3·30	3·15	2·99	2·82
27	13·61	9·02	7·27	6·33	5·73	5·31	5·00	4·76	4·57	4·41	4·17	3·92	3·66	3·52	3·38	3·23	3·08	2·92	2·75
28	13·50	8·93	7·19	6·25	5·66	5·24	4·93	4·69	4·50	4·35	4·11	3·86	3·60	3·46	3·32	3·18	3·02	2·86	2·69
29	13·39	8·85	7·12	6·19	5·59	5·18	4·87	4·64	4·45	4·29	4·05	3·80	3·54	3·41	3·27	3·12	2·97	2·81	2·64
30	13·29	8·77	7·05	6·12	5·53	5·12	4·82	4·58	4·39	4·24	4·00	3·75	3·49	3·36	3·22	3·07	2·92	2·76	2·59
40	12·61	8·25	6·60	5·70	5·13	4·73	4·44	4·21	4·02	3·87	3·64	3·40	3·15	3·01	2·87	2·73	2·57	2·41	2·23
60	11·97	7·76	6·17	5·31	4·76	4·37	4·09	3·87	3·69	3·54	3·31	3·08	2·83	2·69	2·55	2·41	2·25	2·08	1·89
120	11·38	7·32	5·79	4·95	4·42	4·04	3·77	3·55	3·38	3·24	3·02	2·78	2·53	2·40	2·26	2·11	1·95	1·76	1·54
∞	10·83	6·91	5·42	4·62	4·10	3·74	3·47	3·27	3·10	2·96	2·74	2·51	2·27	2·13	1·99	1·84	1·66	1·45	1·00

* Multiply these entries by 100.

This 0·1 % table is based on the following sources: Colcord & Deming (1935); Fisher & Yates (1953, Table V) used with the permission of the authors and of Messrs Oliver and Boyd; Norton (1952).

Table 19. *Percentage points of the largest variance ratio,* $s^2_{max.}/s^2_0$

Upper 5 % *points*

ν \ k	1	2	3	4	5	6	7	8	9	10
10	4·96	6·79	8·00	8·96	9·78	10·52	11·18	11·79	12·36	12·87
12	4·75	6·44	7·53	8·37	9·06	9·68	10·20	10·68	11·12	11·53
15	4·54	6·12	7·11	7·86	8·47	8·98	9·43	9·82	10·19	10·52
20	4·35	5·81	6·72	7·40	7·94	8·39	8·79	9·13	9·44	9·71
30	4·17	5·52	6·36	6·97	7·46	7·87	8·21	8·51	8·79	9·03
60	4·00	5·25	6·02	6·58	7·02	7·38	7·68	7·96	8·20	8·41
∞	3·84	5·00	5·70	6·21	6·60	6·92	7·20	7·44	7·65	7·84

Upper 1 % *points*

ν \ k	1	2	3	4	5	6	7	8	9	10
10	10·04	13·17	15·08	16·43	17·43	18·25	18·91	19·48	19·97	20·41
12	9·33	11·88	13·52	14·73	15·69	16·47	17·12	17·68	18·16	18·60
15	8·68	10·82	12·18	13·21	14·03	14·72	15·30	15·81	16·26	16·66
20	8·10	9·93	11·08	11·93	12·61	13·19	13·67	14·09	14·49	14·83
30	7·56	9·16	10·14	10·86	11·43	11·90	12·31	12·66	12·97	13·26
60	7·08	8·49	9·34	9·95	10·43	10·82	11·15	11·45	11·72	11·95
∞	6·63	7·88	8·61	9·15	9·54	9·87	10·16	10·41	10·62	10·82

k is the number of independent variance estimates, each based on 1 degree of freedom, of which $s^2_{max.}$ is the largest. ν denotes the degrees of freedom of the independent 'error' mean square, s^2_0.

Table 20. *Moment constants of the mean deviation and of the range*

n	Mean deviation, m/σ					Range, $W=w/\sigma$						
	Expectation	S.D.	Variance	β_1	β_2	$\mathscr{E}(W_n)=d_n$	S.D.	V_n = variance	β_1	β_2	d_n/V_n	d_n^2/V_n
2	0·564 190	0·4263	0·18169	0·991	3·869	1·12838	0·8525	0·72676	0·9906	3·869	1·55	1·75
3	·651 470	·3419	·11692	·417	3·286	1·69257	·8884	·78920	·4176	3·286	2·14	3·63
4	·690 988	·2970	·08822	·298	3·252	2·05875	·8798	·77406	·2735	3·188	2·66	5·48
5	0·713 650	0·2663	0·07094	0·230	3·197	2·32593	0·8641	0·74664	0·2167	3·169	3·12	7·25
6	·728 366	·2436	·05934	·187	3·161	2·53441	·8480	·71917	·1892	3·170	3·52	8·93
7	·738 698	·2258	·05101	·157	3·136	2·70436	·8332	·69423	·1743	3·176	3·90	10·53
8	·746 353	·2115	·04473	·136	3·118	2·84720	·8198	·67212	·1659	3·184	4·24	12·06
9	·752 253	·1996	·03982	·119	3·104	2·97003	·8078	·65260	·1609	3·192	4·55	13·52
10	0·756 940	0·1894	0·03589	0·106	3·0927	3·07751	0·7971	0·63529	0·1581	3·200	4·84	14·91
11	·760 753	·1807	·03266	·0961	3·0838	3·17287	·7873	·61986	·1566	3·207	5·12	16·2
12	·763 916	·1731	·02997	·0876	3·0765	3·25846	·7785	·60603	·1559	3·214	5·38	17·5
13	·766 583	·1664	·02769	·0805	3·0703	3·33598	·7704	·59354	·1559	3·221	5·62	18·8
14	·768 861	·1604	·02573	·0744	3·0650	3·40676	·7630	·58220	·1563	3·227	5·85	19·9
15	0·770 830	0·1550	0·02403	0·0692	3·0605	3·47183	0·7562	0·57186	0·1569	3·233	6·07	21·1
16	·772 548	·1501	·02254	·0647	3·0566	3·53198	·7499	·56236	·1578	3·238	6·28	22·2
17	·774 062	·1457	·02122	·0607	3·0531	3·58788	·7441	·55361	·1588	3·243	6·48	23·3
18	·775 404	·1416	·02005	·0572	3·0501	3·64006	·7386	·54552	·1598	3·248	6·67	24·3
19	·776 604	·1378	·01900	·0541	3·0473	3·68896	·7335	·53800	·1610	3·253	6·86	25·3
20	0·777 682	0·1344	0·01806	0·0513	3·0449	3·73495	0·7287	0·53098	0·1622	3·257	7·03	26·3
30	·784 474	·1098	·01206	·0338	3·0296							
60	·791 208	·0777	·00604	·0167	3·0146							

The unit is the population standard deviation.

Table 21. *Percentage points of the distribution of the mean deviation*

Size of sample n	0·1	0·5	1·0	2·5	5·0	10·0
Lower percentage points						
2	0·001	0·004	0·009	0·022	0·044	0·089
3	0·022	0·052	0·073	0·116	0·166	0·238
4	0·066	0·114	0·145	0·199	0·254	0·328
5	0·112	0·170	0·203	0·260	0·315	0·386
6	0·153	0·215	0·250	0·306	0·360	0·428
7	0·190	0·252	0·287	0·342	0·394	0·459
8	0·220	0·283	0·318	0·372	0·422	0·484
9	0·247	0·310	0·344	0·396	0·445	0·504
10	0·271	0·333	0·366	0·417	0·464	0·521
Normal approximation						
10	0·171	0·269	0·316	0·386	0·445	0·514

Size of sample n	10·0	5·0	2·5	1·0	0·5	0·1
Upper percentage points						
2	1·163	1·386	1·585	1·821	1·985	2·327
3	1·117	1·276	1·417	1·586	1·703	1·949
4	1·089	1·224	1·344	1·489	1·590	1·806
5	1·069	1·187	1·292	1·419	1·507	1·693
6	1·052	1·158	1·253	1·366	1·445	1·613
7	1·038	1·135	1·222	1·325	1·397	1·550
8	1·026	1·116	1·196	1·292	1·358	1·499
9	1·016	1·100	1·175	1·264	1·326	1·457
10	1·007	1·086	1·156	1·240	1·299	1·422
Normal approximation						
10	1·000	1·069	1·128	1·198	1·245	1·342

The unit is the population standard deviation. For $n > 10$, see Introduction, pp. 41–2, 83.

Table 22. *Percentage points of the distribution of the range*

Size of sample n	Factor $1/d_n$	Lower percentage points						Upper percentage points					
		0·1	0·5	1·0	2·5	5·0	10·0	10·0	5·0	2·5	1·0	0·5	0·1
2	0·8862	0·00	0·01	0·02	0·04	0·09	0·18	2·33	2·77	3·17	3·64	3·97	4·65
3	·5908	0·06	0·13	0·19	0·30	0·43	0·62	2·90	3·31	3·68	4·12	4·42	5·06
4	·4857	0·20	0·34	0·43	0·59	0·76	0·98	3·24	3·63	3·98	4·40	4·69	5·31
5	·4299	0·37	0·55	0·67	0·85	1·03	1·26	3·48	3·86	4·20	4·60	4·89	5·48
6	0·3946	0·53	0·75	0·87	1·07	1·25	1·49	3·66	4·03	4·36	4·76	5·03	5·62
7	·3698	0·69	0·92	1·05	1·25	1·44	1·68	3·81	4·17	4·49	4·88	5·15	5·73
8	·3512	0·83	1·08	1·20	1·41	1·60	1·84	3·93	4·29	4·60	4·99	5·25	5·82
9	·3367	0·97	1·21	1·34	1·55	1·74	1·97	4·04	4·39	4·70	5·08	5·34	5·90
10	0·3249	1·08	1·33	1·47	1·67	1·86	2·09	4·13	4·47	4·78	5·16	5·42	5·97
11	·3152	1·19	1·45	1·58	1·78	1·97	2·20	4·21	4·55	4·86	5·23	5·49	6·04
12	·3069	1·29	1·55	1·68	1·88	2·07	2·30	4·28	4·62	4·92	5·29	5·55	6·09
13	·2998	1·39	1·64	1·77	1·98	2·16	2·39	4·35	4·68	4·99	5·35	5·60	6·14
14	·2935	1·47	1·72	1·86	2·06	2·24	2·47	4·41	4·74	5·04	5·40	5·65	6·19
15	0·2880	1·55	1·80	1·93	2·14	2·32	2·54	4·47	4·80	5·09	5·45	5·70	6·23
16	·2831	1·62	1·88	2·01	2·21	2·39	2·61	4·52	4·85	5·14	5·49	5·74	6·27
17	·2787	1·69	1·94	2·07	2·27	2·45	2·67	4·57	4·89	5·18	5·54	5·78	6·31
18	·2747	1·76	2·01	2·14	2·34	2·52	2·73	4·61	4·93	5·22	5·57	5·82	6·35
19	·2711	1·82	2·07	2·20	2·39	2·57	2·79	4·65	4·97	5·26	5·61	5·86	6·38
20	0·2677	1·88	2·12	2·25	2·45	2·63	2·84	4·69	5·01	5·30	5·65	5·89	6·41

The unit is the population standard deviation.
Estimate of σ = range (or mean range) in a sample of n observations × $1/d_n$.

Table 23. *Probability integral of the range, W, in normal samples of size n*

W \ n	2	3	4	5	6	7	8	9	10
0·00	0·0000	0·0000							
0·05	·0282	·0007	0·0000						
0·10	·0564	·0028	·0001						
0·15	·0845	·0062	·0004	0·0000					
0·20	·1125	·0110	·0010	·0001					
0·25	0·1403	0·0171	0·0020	0·0002	0·0000				
0·30	·1680	·0245	·0034	·0004	·0001				
0·35	·1955	·0332	·0053	·0008	·0001				
0·40	·2227	·0431	·0079	·0014	·0002	0·0000			
0·45	·2497	·0543	·0111	·0022	·0004	·0001			
0·50	0·2763	0·0666	0·0152	0·0033	0·0007	0·0002	0·0000		
0·55	·3027	·0800	·0200	·0048	·0011	·0003	·0001		
0·60	·3286	·0944	·0257	·0068	·0017	·0004	·0001	0·0000	
0·65	·3542	·1099	·0322	·0092	·0026	·0007	·0002	·0001	
0·70	·3794	·1263	·0398	·0121	·0036	·0011	·0003	·0001	
0·75	0·4041	0·1436	0·0483	0·0157	0·0050	0·0016	0·0005	0·0002	0·0000
0·80	·4284	·1616	·0578	·0200	·0068	·0023	·0008	·0002	·0001
0·85	·4522	·1805	·0682	·0250	·0090	·0032	·0011	·0004	·0001
0·90	·4755	·2000	·0797	·0308	·0117	·0044	·0016	·0006	·0002
0·95	·4983	·2201	·0922	·0375	·0150	·0059	·0023	·0009	·0003
1·00	0·5205	0·2407	0·1057	0·0450	0·0188	0·0078	0·0032	0·0013	0·0005
1·05	·5422	·2618	·1201	·0535	·0234	·0101	·0043	·0018	·0008
1·10	·5633	·2833	·1355	·0629	·0287	·0129	·0058	·0025	·0011
1·15	·5839	·3052	·1517	·0733	·0348	·0163	·0075	·0035	·0016
1·20	·6039	·3272	·1688	·0847	·0417	·0203	·0098	·0047	·0022
1·25	0·6232	0·3495	0·1867	0·0970	0·0495	0·0249	0·0125	0·0062	0·0030
1·30	·6420	·3719	·2054	·1104	·0583	·0304	·0157	·0080	·0041
1·35	·6602	·3943	·2248	·1247	·0680	·0366	·0195	·0103	·0054
1·40	·6778	·4168	·2448	·1400	·0787	·0437	·0240	·0131	·0071
1·45	·6948	·4392	·2654	·1562	·0904	·0516	·0292	·0164	·0092
1·50	0·7112	0·4614	0·2865	0·1733	0·1031	0·0606	0·0353	0·0204	0·0117
1·55	·7269	·4835	·3080	·1913	·1168	·0705	·0421	·0250	·0148
1·60	·7421	·5053	·3299	·2101	·1315	·0814	·0499	·0304	·0184
1·65	·7567	·5269	·3521	·2296	·1473	·0934	·0587	·0366	·0227
1·70	·7707	·5481	·3745	·2498	·1639	·1064	·0684	·0437	·0278
1·75	0·7841	0·5690	0·3970	0·2706	0·1815	0·1204	0·0792	0·0517	0·0336
1·80	·7969	·5894	·4197	·2920	·2000	·1355	·0910	·0607	·0403
1·85	·8092	·6094	·4423	·3138	·2193	·1516	·1039	·0707	·0479
1·90	·8209	·6290	·4649	·3361	·2394	·1686	·1178	·0818	·0565
1·95	·8321	·6480	·4874	·3587	·2602	·1867	·1329	·0939	·0661
2·00	0·8427	0·6665	0·5096	0·3816	0·2816	0·2056	0·1489	0·1072	0·0768
2·05	·8528	·6845	·5317	·4046	·3035	·2254	·1661	·1216	·0886
2·10	·8624	·7019	·5534	·4277	·3260	·2460	·1842	·1371	·1015
2·15	·8716	·7187	·5748	·4508	·3489	·2673	·2032	·1536	·1155
2·20	·8802	·7349	·5957	·4739	·3720	·2893	·2232	·1712	·1307
2·25	0·8884	0·7505	0·6163	0·4969	0·3955	0·3118	0·2440	0·1899	0·1470
2·30	·8961	·7655	·6363	·5196	·4190	·3348	·2656	·2095	·1645
2·35	·9034	·7799	·6559	·5421	·4427	·3582	·2878	·2300	·1829
2·40	·9103	·7937	·6748	·5643	·4663	·3820	·3107	·2514	·2025
2·45	·9168	·8069	·6932	·5861	·4899	·4059	·3341	·2735	·2229
2·50	0·9229	0·8195	0·7110	0·6075	0·5132	0·4300	0·3579	0·2963	0·2443

Table 23 (*continued*)

W \ n	11	12	13	14	15	16	17	18	19	20
0·85	0·0000									
0·90	·0001									
0·95	·0001	0·0000								
1·00	0·0002	0·0001	0·0000							
1·05	·0003	·0001	·0001							
1·10	·0005	·0002	·0001	0·0000						
1·15	·0007	·0003	·0001	·0001	0·0000					
1·20	·0010	·0005	·0002	·0001	·0001					
1·25	0·0015	0·0007	0·0004	0·0002	0·0001	0·0000				
1·30	·0021	·0010	·0005	·0003	·0001	·0001	0·0000			
1·35	·0028	·0015	·0008	·0004	·0002	·0001	·0001			
1·40	·0038	·0021	·0011	·0006	·0003	·0002	·0001	0·0000		
1·45	·0051	·0028	·0016	·0009	·0005	·0003	·0001	·0001	0·0000	
1·50	0·0067	0·0038	0·0022	0·0012	0·0007	0·0004	0·0002	0·0001	0·0001	0·0000
1·55	·0087	·0051	·0030	·0017	·0010	·0006	·0003	·0002	·0001	·0001
1·60	·0111	·0067	·0040	·0024	·0014	·0008	·0005	·0003	·0002	·0001
1·65	·0140	·0086	·0053	·0032	·0020	·0012	·0007	·0004	·0003	·0002
1·70	·0176	·0111	·0070	·0044	·0027	·0017	·0011	·0007	·0004	·0003
1·75	0·0217	0·0140	0·0090	0·0058	0·0037	0·0023	0·0015	0·0010	0·0006	0·0004
1·80	·0266	·0175	·0115	·0075	·0049	·0032	·0021	·0014	·0009	·0006
1·85	·0323	·0217	·0145	·0097	·0065	·0043	·0029	·0019	·0013	·0008
1·90	·0388	·0266	·0182	·0124	·0084	·0057	·0039	·0026	·0018	·0012
1·95	·0463	·0323	·0225	·0156	·0108	·0075	·0052	·0036	·0024	·0017
2·00	0·0548	0·0389	0·0276	0·0195	0·0137	0·0097	0·0068	0·0048	0·0033	0·0023
2·05	·0643	·0465	·0335	·0241	·0173	·0124	·0088	·0063	·0045	·0032
2·10	·0748	·0550	·0403	·0295	·0215	·0156	·0114	·0082	·0060	·0043
2·15	·0866	·0646	·0481	·0357	·0265	·0196	·0144	·0106	·0078	·0058
2·20	·0994	·0753	·0569	·0429	·0323	·0242	·0182	·0136	·0102	·0076
2·25	0·1134	0·0872	0·0669	0·0511	0·0390	0·0297	0·0226	0·0172	0·0130	0·0099
2·30	·1286	·1003	·0779	·0604	·0468	·0361	·0279	·0214	·0165	·0127
2·35	·1450	·1145	·0902	·0709	·0556	·0435	·0340	·0265	·0207	·0161
2·40	·1624	·1299	·1036	·0825	·0655	·0519	·0411	·0325	·0256	·0202
2·45	·1810	·1466	·1183	·0953	·0766	·0615	·0493	·0394	·0315	·0251
2·50	0·2007	0·1643	0·1342	0·1094	0·0890	0·0722	0·0585	0·0474	0·0383	0·0309

Table 23. *Probability integral of the range* (*continued*)

W \ n	2	3	4	5	6	7	8	9	10
2·50	0·9229	0·8195	0·7110	0·6075	0·5132	0·4300	0·3579	0·2963	0·2443
2·55	·9286	·8315	·7282	·6283	·5364	·4541	·3820	·3198	·2665
2·60	·9340	·8429	·7448	·6487	·5592	·4782	·4064	·3437	·2894
2·65	·9390	·8537	·7607	·6685	·5816	·5022	·4310	·3680	·3130
2·70	·9438	·8640	·7759	·6877	·6036	·5259	·4555	·3927	·3372
2·75	0·9482	0·8737	0·7905	0·7063	0·6252	0·5494	0·4801	0·4175	0·3617
2·80	·9523	·8828	·8045	·7242	·6461	·5725	·5045	·4425	·3867
2·85	·9561	·8915	·8177	·7415	·6665	·5952	·5286	·4675	·4119
2·90	·9597	·8996	·8304	·7581	·6863	·6174	·5525	·4923	·4372
2·95	·9630	·9073	·8424	·7739	·7055	·6391	·5760	·5171	·4625
3·00	0·9661	0·9145	0·8537	0·7891	0·7239	0·6601	0·5991	0·5415	0·4878
3·05	·9690	·9212	·8645	·8036	·7416	·6806	·6216	·5656	·5129
3·10	·9716	·9275	·8746	·8174	·7587	·7003	·6436	·5892	·5378
3·15	·9741	·9334	·8842	·8305	·7750	·7194	·6649	·6124	·5623
3·20	·9763	·9388	·8931	·8429	·7905	·7377	·6856	·6350	·5864
3·25	0·9784	0·9439	0·9016	0·8546	0·8053	0·7553	0·7055	0·6569	0·6099
3·30	·9804	·9487	·9095	·8657	·8194	·7721	·7248	·6782	·6329
3·35	·9822	·9531	·9168	·8761	·8327	·7881	·7432	·6988	·6553
3·40	·9838	·9572	·9237	·8859	·8454	·8034	·7609	·7186	·6769
3·45	·9853	·9610	·9302	·8951	·8573	·8179	·7778	·7376	·6978
3·50	0·9867	0·9644	0·9361	0·9037	0·8685	0·8316	0·7938	0·7558	0·7180
3·55	·9879	·9677	·9417	·9117	·8790	·8446	·8091	·7732	·7373
3·60	·9891	·9706	·9468	·9192	·8889	·8568	·8236	·7898	·7558
3·65	·9901	·9734	·9516	·9261	·8981	·8683	·8372	·8055	·7735
3·70	·9911	·9759	·9560	·9326	·9067	·8790	·8501	·8204	·7903
3·75	0·9920	0·9782	0·9600	0·9386	0·9147	0·8891	0·8622	0·8345	0·8062
3·80	·9928	·9803	·9637	·9441	·9222	·8985	·8736	·8477	·8212
3·85	·9935	·9822	·9672	·9493	·9291	·9073	·8842	·8602	·8355
3·90	·9942	·9840	·9703	·9540	·9355	·9155	·8941	·8718	·8488
3·95	·9948	·9856	·9732	·9583	·9415	·9230	·9034	·8827	·8614
4·00	0·9953	0·9870	0·9758	0·9623	0·9469	0·9300	0·9120	0·8929	0·8731
4·05	·9958	·9883	·9782	·9660	·9520	·9365	·9199	·9024	·8841
4·10	·9963	·9895	·9804	·9693	·9566	·9425	·9273	·9112	·8943
4·15	·9967	·9906	·9824	·9724	·9608	·9480	·9341	·9193	·9038
4·20	·9970	·9916	·9842	·9752	·9647	·9530	·9404	·9268	·9126
4·25	0·9973	0·9925	0·9859	0·9777	0·9682	0·9576	0·9461	0·9338	0·9208
4·30	·9976	·9933	·9874	·9800	·9715	·9619	·9514	·9402	·9283
4·35	·9979	·9941	·9887	·9821	·9744	·9657	·9562	·9460	·9352
4·40	·9981	·9947	·9899	·9840	·9771	·9692	·9607	·9514	·9416
4·45	·9983	·9953	·9910	·9857	·9795	·9724	·9647	·9563	·9474
4·50	0·9985	0·9958	0·9920	0·9873	0·9817	0·9754	0·9684	0·9608	0·9527
4·55	·9987	·9963	·9929	·9887	·9837	·9780	·9717	·9649	·9576
4·60	·9989	·9967	·9937	·9899	·9855	·9804	·9747	·9686	·9620
4·65	·9990	·9971	·9944	·9911	·9871	·9825	·9775	·9719	·9660
4·70	·9991	·9974	·9951	·9921	·9885	·9845	·9799	·9750	·9696
4·75	0·9992	0·9977	0·9956	0·9930	0·9898	0·9862	0·9822	0·9777	0·9729
4·80	·9993	·9980	·9962	·9938	·9910	·9878	·9842	·9802	·9759
4·85	·9994	·9982	·9966	·9945	·9920	·9892	·9860	·9824	·9786
4·90	·9995	·9985	·9970	·9952	·9930	·9904	·9876	·9844	·9810
4·95	·9995	·9986	·9974	·9958	·9938	·9916	·9890	·9862	·9832
5·00	0·9996	0·9988	0·9977	0·9963	0·9945	0·9926	0·9903	0·9878	0·9851

Table 23 (*continued*)

W \ n	11	12	13	14	15	16	17	18	19	20
2·50	0·2007	0·1643	0·1342	0·1094	0·0890	0·0722	0·0585	0·0474	0·0383	0·0309
2·55	·2213	·1833	·1513	·1247	·1025	·0842	·0690	·0565	·0462	·0377
2·60	·2429	·2032	·1696	·1413	·1174	·0974	·0807	·0668	·0552	·0455
2·65	·2653	·2243	·1891	·1590	·1335	·1119	·0937	·0783	·0654	·0545
2·70	·2885	·2462	·2096	·1780	·1509	·1278	·1080	·0911	·0768	·0647
2·75	0·3124	0·2690	0·2311	0·1981	0·1696	0·1449	0·1236	0·1053	0·0896	0·0761
2·80	·3368	·2926	·2536	·2194	·1894	·1632	·1405	·1208	·1037	·0889
2·85	·3618	·3169	·2770	·2416	·2103	·1828	·1587	·1376	·1191	·1031
2·90	·3870	·3417	·3011	·2647	·2323	·2036	·1782	·1557	·1360	·1186
2·95	·4125	·3670	·3258	·2887	·2553	·2255	·1989	·1752	·1541	·1355
3·00	0·4382	0·3927	0·3511	0·3134	0·2792	0·2484	0·2207	0·1959	0·1736	0·1537
3·05	·4639	·4186	·3769	·3387	·3039	·2723	·2436	·2177	·1944	·1733
3·10	·4895	·4446	·4029	·3645	·3292	·2969	·2675	·2407	·2163	·1942
3·15	·5150	·4706	·4291	·3907	·3551	·3223	·2922	·2646	·2394	·2163
3·20	·5401	·4965	·4554	·4171	·3814	·3483	·3177	·2894	·2634	·2395
3·25	0·5649	0·5222	0·4817	0·4437	0·4080	0·3748	0·3438	0·3151	0·2884	0·2638
3·30	·5893	·5475	·5078	·4703	·4348	·4016	·3704	·3413	·3142	·2890
3·35	·6131	·5725	·5337	·4967	·4617	·4286	·3974	·3681	·3407	·3150
3·40	·6363	·5970	·5592	·5230	·4885	·4557	·4246	·3953	·3676	·3416
3·45	·6589	·6209	·5842	·5489	·5150	·4827	·4519	·4227	·3950	·3688
3·50	0·6807	0·6442	0·6087	0·5744	0·5413	0·5096	0·4792	0·4502	0·4226	0·3964
3·55	·7017	·6668	·6326	·5994	·5672	·5362	·5063	·4777	·4504	·4242
3·60	·7220	·6886	·6558	·6237	·5926	·5624	·5332	·5051	·4781	·4522
3·65	·7414	·7096	·6782	·6474	·6173	·5881	·5597	·5322	·5056	·4801
3·70	·7600	·7298	·6999	·6704	·6414	·6132	·5856	·5588	·5329	·5078
3·75	0·7776	0·7491	0·7206	0·6925	0·6648	0·6376	0·6110	0·5850	0·5598	0·5352
3·80	·7944	·7675	·7406	·7138	·6874	·6613	·6357	·6106	·5861	·5622
3·85	·8103	·7850	·7596	·7342	·7090	·6842	·6596	·6355	·6118	·5887
3·90	·8254	·8016	·7777	·7537	·7298	·7062	·6827	·6596	·6369	·6145
3·95	·8395	·8173	·7948	·7723	·7497	·7273	·7050	·6829	·6611	·6397
4·00	0·8528	0·8321	0·8111	0·7899	0·7686	0·7474	0·7263	0·7053	0·6845	0·6640
4·05	·8653	·8460	·8264	·8066	·7866	·7666	·7466	·7268	·7070	·6874
4·10	·8769	·8590	·8408	·8223	·8036	·7848	·7660	·7472	·7285	·7099
4·15	·8878	·8712	·8543	·8371	·8196	·8021	·7844	·7667	·7491	·7315
4·20	·8978	·8826	·8669	·8509	·8347	·8183	·8018	·7852	·7686	·7520
4·25	0·9072	0·8931	0·8787	0·8639	0·8488	0·8336	0·8182	0·8027	0·7871	0·7715
4·30	·9158	·9029	·8896	·8760	·8620	·8479	·8336	·8191	·8046	·7899
4·35	·9238	·9120	·8998	·8872	·8744	·8613	·8480	·8346	·8210	·8074
4·40	·9312	·9204	·9092	·8976	·8858	·8737	·8615	·8490	·8364	·8237
4·45	·9379	·9281	·9178	·9073	·8964	·8853	·8740	·8625	·8508	·8391
4·50	0·9441	0·9352	0·9258	0·9162	0·9062	0·8960	0·8856	0·8750	0·8643	0·8534
4·55	·9498	·9417	·9332	·9244	·9153	·9060	·8964	·8867	·8768	·8667
4·60	·9550	·9476	·9399	·9319	·9236	·9151	·9064	·8975	·8884	·8791
4·65	·9597	·9530	·9460	·9388	·9313	·9235	·9155	·9074	·8991	·8906
4·70	·9639	·9579	·9516	·9451	·9382	·9312	·9240	·9165	·9089	·9012
4·75	0·9678	0·9624	0·9567	0·9508	0·9446	0·9383	0·9317	0·9249	0·9180	0·9110
4·80	·9713	·9665	·9614	·9560	·9505	·9447	·9387	·9326	·9263	·9199
4·85	·9745	·9702	·9656	·9608	·9557	·9505	·9452	·9396	·9339	·9281
4·90	·9774	·9735	·9694	·9650	·9605	·9559	·9510	·9460	·9409	·9356
4·95	·9799	·9765	·9728	·9689	·9649	·9607	·9563	·9518	·9472	·9424
5·00	0·9822	0·9791	0·9759	0·9724	0·9688	0·9650	0·9611	0·9571	0·9529	0·9486

Table 23. *Probability integral of the range* (*continued*)

W \ n	2	3	4	5	6	7	8	9	10
5·00	0·9996	0·9988	0·9977	0·9963	0·9945	0·9926	0·9903	0·9878	0·9851
5·05	·9996	·9990	·9980	·9967	·9952	·9935	·9915	·9893	·9869
5·10	·9997	·9991	·9982	·9971	·9958	·9942	·9925	·9906	·9884
5·15	·9997	·9992	·9985	·9975	·9963	·9950	·9934	·9917	·9898
5·20	·9998	·9993	·9987	·9978	·9968	·9956	·9942	·9927	·9911
5·25	0·9998	0·9994	0·9988	0·9981	0·9972	0·9961	0·9949	·09936	0·9922
5·30	·9998	·9995	·9990	·9983	·9975	·9966	·9956	·9944	·9931
5·35	·9998	·9995	·9991	·9985	·9979	·9971	·9961	·9951	·9940
5·40	·9999	·9996	·9992	·9987	·9981	·9974	·9966	·9957	·9948
5·45	·9999	·9997	·9993	·9989	·9984	·9978	·9971	·9963	·9954
5·50	0·9999	0·9997	0·9994	0·9990	0·9986	0·9981	0·9974	0·9968	0·9960
5·55	·9999	·9997	·9995	·9992	·9988	·9983	·9978	·9972	·9965
5·60	·9999	·9998	·9996	·9993	·9989	·9985	·9981	·9976	·9970
5·65	·9999	·9998	·9996	·9994	·9991	·9987	·9983	·9979	·9974
5·70	0·9999	·9998	·9997	·9995	·9992	·9989	·9986	·9982	·9977
5·75	1·0000	0·9999	0·9997	0·9995	0·9993	0·9991	0·9988	0·9984	0·9980
5·80		·9999	·9998	·9996	·9994	·9992	·9989	·9986	·9983
5·85		·9999	·9998	·9997	·9995	·9993	·9991	·9988	·9985
5·90		·9999	·9998	·9997	·9996	·9994	·9992	·9990	·9988
5·95		·9999	·9998	·9997	·9996	·9995	·9993	·9991	·9989
6·00		0·9999	0·9999	0·9998	0·9997	0·9996	0·9994	0·9993	0·9991
6·05		0·9999	·9999	·9998	·9997	·9996	·9995	·9994	·9992
6·10		1·0000	·9999	·9998	·9998	·9997	·9996	·9995	·9993
6·15			·9999	·9999	·9998	·9997	·9996	·9995	·9994
6·20			·9999	·9999	·9998	·9998	·9997	·9996	·9995
6·25			0·9999	0·9999	0·9999	0·9998	0·9997	0·9997	0·9996
6·30			1·0000	·9999	·9999	·9998	·9998	·9997	·9996
6·35				·9999	·9999	·9999	·9998	·9998	·9997
6·40				·9999	·9999	·9999	·9998	·9998	·9997
6·45				0·9999	·9999	·9999	·9999	·9998	·9998
6·50				1·0000	0·9999	0·9999	0·9999	0·9999	0·9998
6·55					0·9999	·9999	·9999	·9999	·9998
6·60					1·0000	·9999	·9999	·9999	·9999
6·65						0·9999	·9999	·9999	·9999
6·70						1·0000	0·9999	·9999	·9999
6·75							1·0000	0·9999	0·9999
6·80								0·9999	·9999
6·85								1·0000	0·9999
6·90									1·0000
6·95									
7·00									
7·05									
7·10									
7·15									
7·20									
7·25									

Table 23 (*continued*)

W \ n	11	12	13	14	15	16	17	18	19	20
5·00	0·9822	0·9791	0·9759	0·9724	0·9688	0·9650	0·9611	0·9571	0·9529	0·9486
5·05	·9843	·9816	·9786	·9756	·9723	·9690	·9655	·9618	·9581	·9543
5·10	·9862	·9837	·9811	·9784	·9755	·9725	·9694	·9661	·9628	·9593
5·15	·9878	·9856	·9833	·9809	·9783	·9757	·9729	·9700	·9670	·9639
5·20	·9893	·9874	·9853	·9832	·9809	·9785	·9760	·9735	·9708	·9681
5·25	0·9906	0·9889	0·9871	0·9852	0·9832	0·9811	0·9789	0·9766	0·9742	0·9718
5·30	·9917	·9903	·9887	·9870	·9852	·9833	·9814	·9794	·9773	·9751
5·35	·9928	·9915	·9901	·9886	·9870	·9854	·9836	·9819	·9800	·9781
5·40	·9937	·9925	·9913	·9900	·9886	·9872	·9856	·9841	·9824	·9807
5·45	·9945	·9935	·9924	·9913	·9900	·9888	·9874	·9860	·9846	·9831
5·50	0·9952	0·9943	0·9934	0·9924	0·9913	0·9902	0·9890	0·9878	0·9865	0·9852
5·55	·9958	·9951	·9942	·9933	·9924	·9914	·9904	·9893	·9882	·9870
5·60	·9964	·9957	·9950	·9942	·9934	·9925	·9916	·9907	·9897	·9887
5·65	·9969	·9963	·9956	·9950	·9943	·9935	·9927	·9919	·9910	·9901
5·70	·9973	·9968	·9962	·9956	·9950	·9944	·9937	·9930	·9922	·9914
5·75	0·9976	0·9972	0·9967	0·9962	0·9957	0·9951	0·9945	0·9939	0·9932	0·9925
5·80	·9980	·9976	·9972	·9967	·9963	·9958	·9952	·9947	·9941	·9935
5·85	·9982	·9979	·9976	·9972	·9968	·9963	·9959	·9954	·9949	·9944
5·90	·9985	·9982	·9979	·9976	·9972	·9968	·9964	·9960	·9956	·9952
5·95	·9987	·9985	·9982	·9979	·9976	·9973	·9969	·9966	·9962	·9958
6·00	0·9989	0·9987	0·9984	0·9982	0·9979	0·9977	0·9974	0·9971	0·9967	0·9964
6·05	·9990	·9989	·9987	·9985	·9982	·9980	·9977	·9975	·9972	·9969
6·10	·9992	·9990	·9989	·9987	·9985	·9983	·9981	·9978	·9976	·9973
6·15	·9993	·9992	·9990	·9989	·9987	·9985	·9983	·9981	·9979	·9977
6·20	·9994	·9993	·9992	·9990	·9989	·9987	·9986	·9984	·9982	·9980
6·25	0·9995	0·9994	0·9993	0·9992	0·9990	0·9989	0·9988	0·9986	0·9985	0·9983
6·30	·9996	·9995	·9994	·9993	·9992	·9991	·9990	·9988	·9987	·9986
6·35	·9996	·9996	·9995	·9994	·9993	·9992	·9991	·9990	·9989	·9988
6·40	·9997	·9996	·9996	·9995	·9994	·9993	·9992	·9992	·9991	·9990
6·45	·9997	·9997	·9996	·9996	·9995	·9994	·9994	·9993	·9992	·9991
6·50	0·9998	0·9997	0·9997	0·9996	0·9996	0·9995	0·9995	0·9994	0·9993	0·9993
6·55	·9998	·9998	·9997	·9997	·9996	·9996	·9995	·9995	·9994	·9994
6·60	·9998	·9998	·9998	·9997	·9997	·9997	·9996	·9996	·9995	·9995
6·65	·9999	·9998	·9998	·9998	·9997	·9997	·9997	·9996	·9996	·9995
6·70	·9999	·9999	·9998	·9998	·9998	·9998	·9997	·9997	·9997	·9996
6·75	0·9999	0·9999	0·9999	0·9998	0·9998	0·9998	0·9998	0·9997	0·9997	0·9997
6·80	·9999	·9999	·9999	·9999	·9998	·9998	·9998	·9998	·9998	·9997
6·85	·9999	·9999	·9999	·9999	·9999	·9999	·9998	·9998	·9998	·9998
6·90	0·9999	·9999	·9999	·9999	·9999	·9999	·9999	·9998	·9998	·9998
6·95	1·0000	0·9999	·9999	·9999	·9999	·9999	·9999	·9999	·9999	·9998
7·00		1·0000	0·9999	0·9999	0·9999	0·9999	0·9999	0·9999	0·9999	0·9999
7·05			1·0000	0·9999	·9999	·9999	·9999	·9999	·9999	·9999
7·10				1·0000	0·9999	0·9999	·9999	·9999	·9999	·9999
7·15					1·0000	1·0000	0·9999	·9999	·9999	·9999
7·20							1·0000	0·9999	0·9999	·9999
7·25								1·0000	1·0000	0·9999
7·30										1·0000

Table 24. *Percentage points of the extreme standardized deviate from population mean,* $(x_n - \mu)/\sigma$ *or* $(\mu - x_1)/\sigma$

n	Lower percentage points						Upper percentage points					
	0·1	**0·5**	**1·0**	**2·5**	**5·0**	**10·0**	**10·0**	**5·0**	**2·5**	**1·0**	**0·5**	**0·1**
1	−3·090	−2·576	−2·326	−1·960	−1·645	−1·282	1·282	1·645	1·960	2·326	2·576	3·090
2	−1·858	−1·471	−1·282	−1·002	−0·760	−0·478	1·632	1·955	2·239	2·575	2·807	3·290
3	−1·282	−0·950	−0·788	−0·546	−0·336	−0·090	1·818	2·121	2·391	2·712	2·935	3·403
4	−0·924	−0·625	−0·478	−0·259	−0·068	0·157	1·943	2·234	2·494	2·806	3·023	3·481
5	−0·671	−0·395	−0·258	−0·055	0·124	0·334	2·036	2·319	2·572	2·877	3·090	3·540
6	−0·478	−0·218	−0·090	0·102	0·271	0·471	2·111	2·386	2·635	2·934	3·143	3·588
7	−0·325	−0·077	0·045	·229	·390	·582	2·172	2·442	2·687	2·981	3·188	3·628
8	−0·198	0·039	·157	·333	·489	·674	2·224	2·490	2·731	3·022	3·227	3·662
9	−0·090	·138	·252	·423	·574	·753	2·269	2·531	2·769	3·057	3·260	3·692
10	0·003	·224	·334	·500	·647	·822	2·309	2·568	2·803	3·089	3·290	3·719
11	0·084	0·300	0·407	0·568	0·711	0·882	2·344	2·601	2·834	3·117	3·317	3·743
12	·157	·367	·471	·629	·769	·936	2·376	2·630	2·862	3·143	3·341	3·765
13	·222	·427	·529	·684	·821	·985	2·406	2·657	2·887	3·166	3·363	3·785
14	·281	·482	·582	·733	·868	1·029	2·432	2·682	2·910	3·187	3·383	3·803
15	·334	·531	·630	·779	·911	1·070	2·457	2·705	2·932	3·207	3·402	3·820
16	0·384	0·577	0·674	0·821	0·951	1·107	2·480	2·726	2·952	3·226	3·420	3·836
17	·429	·620	·715	·859	·988	1·142	2·502	2·746	2·970	3·243	3·436	3·851
18	·471	·659	·753	·895	1·022	1·175	2·522	2·765	2·988	3·259	3·452	3·865
19	·511	·696	·788	·929	1·054	1·205	2·541	2·783	3·004	3·275	3·466	3·878
20	·547	·730	·822	·960	1·084	1·233	2·559	2·799	3·020	3·289	3·480	3·890
21	0·582	0·762	0·853	0·990	1·113	1·260	2·576	2·815	3·034	3·303	3·493	3·902
22	·614	·793	·882	1·018	1·139	1·285	2·592	2·830	3·049	3·316	3·506	3·914
23	·645	·821	·910	1·044	1·164	1·309	2·607	2·844	3·062	3·328	3·517	3·924
24	·674	·848	·936	1·069	1·188	1·332	2·621	2·857	3·075	3·340	3·529	3·934
25	·702	·874	·961	1·093	1·211	1·353	2·635	2·870	3·087	3·351	3·539	3·944
26	0·728	0·899	0·985	1·116	1·233	1·374	2·648	2·883	3·098	3·362	3·550	3·954
27	·753	·922	1·008	1·137	1·253	1·393	2·661	2·895	3·110	3·373	3·560	3·963
28	·777	·945	1·029	1·158	1·273	1·412	2·673	2·906	3·120	3·383	3·569	3·971
29	·800	·966	1·050	1·178	1·292	1·430	2·685	2·917	3·130	3·392	3·578	3·980
30	·822	·987	1·070	1·197	1·310	1·447	2·696	2·928	3·141	3·402	3·587	3·988

Table 25. *Percentage points of the extreme standardized deviate from sample mean,* $(x_n - \bar{x})/\sigma$ *or* $(\bar{x} - x_1)/\sigma$

Size of sample n	Lower percentage points						Upper percentage points					
	0·1	0·5	1·0	2·5	5·0	10·0	10·0	5·0	2·5	1·0	0·5	0·1
3	0·03	0·06	0·09	0·14	0·20	0·29	1·50	1·74	1·95	2·22	2·40	2·78
4	0·09	0·16	0·20	0·27	0·35	0·45	1·70	1·94	2·16	2·43	2·62	3·01
5	0·16	0·25	0·30	0·38	0·47	0·58	1·83	2·08	2·30	2·57	2·76	3·17
6	0·23	0·33	0·38	0·48	0·56	0·68	1·94	2·18	2·41	2·68	2·87	3·28
7	0·30	0·40	0·46	0·56	0·65	0·76	2·02	2·27	2·49	2·76	2·95	3·36
8	0·36	0·47	0·53	0·62	0·72	0·84	2·09	2·33	2·56	2·83	3·02	3·43
9	0·41	0·53	0·59	0·69	0·78	0·90	2·15	2·39	2·61	2·88	3·07	3·48

Table 26. *Upper percentage points of the extreme studentized deviate from the sample mean* $(x_n - \bar{x})/s_\nu$ *or* $(\bar{x} - x_1)/s_\nu$

	ν \ n	3	4	5	6	7	8	9	10	12
10% points	10	1·68	1·92	2·09	2·23	2·33	2·42	2·50	2·56	2·68
	11	1·66	1·90	2·07	2·20	2·30	2·39	2·46	2·53	2·64
	12	1·65	1·88	2·05	2·17	2·28	2·36	2·44	2·50	2·61
	13	1·63	1·86	2·03	2·16	2·26	2·34	2·41	2·47	2·58
	14	1·62	1·85	2·01	2·14	2·24	2·32	2·39	2·45	2·56
	15	1·61	1·84	2·00	2·12	2·22	2·31	2·38	2·44	2·54
	16	1·61	1·83	1·99	2·11	2·21	2·29	2·36	2·42	2·52
	17	1·60	1·82	1·98	2·10	2·20	2·28	2·35	2·41	2·51
	18	1·59	1·82	1·97	2·09	2·19	2·27	2·34	2·39	2·49
	19	1·59	1·81	1·96	2·08	2·18	2·26	2·33	2·38	2·48
	20	1·58	1·80	1·96	2·08	2·17	2·25	2·32	2·37	2·47
	24	1·57	1·78	1·94	2·05	2·15	2·22	2·29	2·34	2·44
	30	1·55	1·77	1·92	2·03	2·12	2·20	2·26	2·32	2·41
	40	1·54	1·75	1·90	2·01	2·10	2·17	2·23	2·29	2·38
	60	1·52	1·73	1·87	1·98	2·07	2·14	2·20	2·26	2·35
	120	1·51	1·71	1·85	1·96	2·05	2·12	2·18	2·23	2·32
	∞	1·50	1·70	1·83	1·94	2·02	2·09	2·15	2·20	2·28
5% points	5	2·37	2·71	2·95	3·15	3·30	3·43	3·54	3·64	3·80
	6	2·24	2·55	2·78	2·95	3·09	3·21	3·31	3·39	3·54
	7	2·15	2·45	2·66	2·82	2·95	3·06	3·15	3·23	3·37
	8	2·09	2·37	2·57	2·72	2·85	2·95	3·04	3·12	3·25
	9	2·04	2·32	2·51	2·65	2·78	2·87	2·96	3·03	3·15
	10	2·01	2·27	2·46	2·60	2·72	2·81	2·89	2·96	3·08
	11	1·98	2·24	2·42	2·56	2·67	2·76	2·84	2·91	3·03
	12	1·96	2·21	2·39	2·52	2·63	2·72	2·80	2·87	2·98
	13	1·94	2·19	2·36	2·50	2·60	2·69	2·76	2·83	2·94
	14	1·93	2·17	2·34	2·47	2·57	2·66	2·74	2·80	2·91
	15	1·91	2·15	2·32	2·45	2·55	2·64	2·71	2·77	2·88
	16	1·90	2·14	2·31	2·43	2·53	2·62	2·69	2·75	2·86
	17	1·89	2·13	2·29	2·42	2·52	2·60	2·67	2·73	2·84
	18	1·88	2·11	2·28	2·40	2·50	2·58	2·65	2·71	2·82
	19	1·87	2·11	2·27	2·39	2·49	2·57	2·64	2·70	2·80
	20	1·87	2·10	2·26	2·38	2·47	2·56	2·63	2·68	2·78
	24	1·84	2·07	2·23	2·34	2·44	2·52	2·58	2·64	2·74
	30	1·82	2·04	2·20	2·31	2·40	2·48	2·54	2·60	2·69
	40	1·80	2·02	2·17	2·28	2·37	2·44	2·50	2·56	2·65
	60	1·78	1·99	2·14	2·25	2·33	2·41	2·47	2·52	2·61
	120	1·76	1·96	2·11	2·22	2·30	2·37	2·43	2·48	2·57
	∞	1·74	1·94	2·08	2·18	2·27	2·33	2·39	2·44	2·52
2·5% points	10	2·34	2·63	2·83	2·98	3·10	3·20	3·29	3·36	3·49
	11	2·30	2·58	2·77	2·92	3·03	3·13	3·22	3·29	3·41
	12	2·27	2·54	2·73	2·87	2·98	3·08	3·16	3·23	3·35
	13	2·24	2·51	2·69	2·83	2·94	3·03	3·11	3·18	3·29
	14	2·22	2·48	2·66	2·79	2·90	2·99	3·07	3·14	3·25
	15	2·20	2·45	2·63	2·76	2·87	2·96	3·04	3·11	3·21
	16	2·18	2·43	2·61	2·74	2·84	2·93	3·01	3·08	3·18
	17	2·17	2·42	2·59	2·72	2·82	2·91	2·98	3·05	3·15
	18	2·15	2·40	2·57	2·70	2·80	2·89	2·96	3·02	3·12
	19	2·14	2·39	2·56	2·68	2·78	2·87	2·94	3·00	3·10
	20	2·13	2·37	2·54	2·67	2·77	2·85	2·92	2·98	3·08
	24	2·10	2·34	2·50	2·62	2·72	2·80	2·87	2·93	3·02
	30	2·07	2·30	2·46	2·58	2·67	2·75	2·81	2·87	2·96
	40	2·04	2·27	2·42	2·53	2·62	2·70	2·76	2·82	2·91
	60	2·01	2·23	2·38	2·49	2·58	2·65	2·71	2·76	2·85
	120	1·98	2·20	2·34	2·45	2·53	2·60	2·66	2·71	2·79
	∞	1·95	2·16	2·30	2·41	2·49	2·56	2·61	2·66	2·74

Table 26 (*continued*). *Upper percentage points of the extreme studentized deviate from the sample mean* $(x_n - \bar{x})/s_\nu$ *or* $(\bar{x} - x_1)/s_\nu$

	ν \ n	3	4	5	6	7	8	9	10	12
1% points	5	3·65	4·11	4·45	4·70	4·93	5·11	5·26	5·39	5·62
	6	3·32	3·72	4·02	4·24	4·43	4·58	4·71	4·82	5·01
	7	3·11	3·48	3·74	3·94	4·11	4·25	4·37	4·46	4·63
	8	2·96	3·31	3·56	3·74	3·89	4·02	4·13	4·22	4·38
	9	2·86	3·19	3·41	3·59	3·73	3·86	3·95	4·04	4·19
	10	2·78	3·10	3·32	3·48	3·62	3·73	3·82	3·90	4·04
	11	2·72	3·02	3·24	3·39	3·52	3·63	3·72	3·79	3·93
	12	2·67	2·96	3·17	3·32	3·45	3·55	3·64	3·71	3·84
	13	2·63	2·92	3·12	3·27	3·38	3·48	3·57	3·64	3·76
	14	2·60	2·88	3·07	3·22	3·33	3·43	3·51	3·58	3·70
	15	2·57	2·84	3·03	3·17	3·29	3·38	3·46	3·53	3·65
	16	2·54	2·81	3·00	3·14	3·25	3·34	3·42	3·49	3·60
	17	2·52	2·79	2·97	3·11	3·22	3·31	3·38	3·45	3·56
	18	2·50	2·77	2·95	3·08	3·19	3·28	3·35	3·42	3·53
	19	2·49	2·75	2·93	3·06	3·16	3·25	3·33	3·39	3·50
	20	2·47	2·73	2·91	3·04	3·14	3·23	3·30	3·37	3·47
	24	2·42	2·68	2·84	2·97	3·07	3·16	3·23	3·29	3·38
	30	2·38	2·62	2·79	2·91	3·01	3·08	3·15	3·21	3·30
	40	2·34	2·57	2·73	2·85	2·94	3·02	3·08	3·13	3·22
	60	2·29	2·52	2·68	2·79	2·88	2·95	3·01	3·06	3·15
	120	2·25	2·48	2·62	2·73	2·82	2·89	2·95	3·00	3·08
	∞	2·22	2·43	2·57	2·68	2·76	2·83	2·88	2·93	3·01
0·5% points	10	3·12	3·46	3·70	3·87	4·02	4·14	4·24	4·33	4·47
	11	3·04	3·37	3·59	3·76	3·90	4·01	4·11	4·19	4·33
	12	2·98	3·29	3·51	3·67	3·80	3·91	4·00	4·08	4·21
	13	2·93	3·23	3·44	3·60	3·72	3·83	3·92	3·99	4·12
	14	2·88	3·18	3·38	3·54	3·66	3·76	3·85	3·92	4·04
	15	2·84	3·13	3·33	3·48	3·60	3·70	3·78	3·86	3·98
	16	2·81	3·10	3·29	3·44	3·56	3·65	3·73	3·80	3·92
	17	2·78	3·07	3·26	3·40	3·52	3·61	3·68	3·75	3·86
	18	2·76	3·04	3·23	3·37	3·48	3·57	3·64	3·71	3·82
	19	2·74	3·01	3·20	3·34	3·45	3·54	3·61	3·68	3·79
	20	2·72	2·99	3·17	3·31	3·42	3·51	3·58	3·65	3·75
	24	2·66	2·92	3·10	3·23	3·33	3·42	3·49	3·55	3·65
	30	2·60	2·86	3·03	3·15	3·25	3·33	3·40	3·46	3·55
	40	2·55	2·79	2·96	3·08	3·17	3·25	3·31	3·37	3·46
	60	2·50	2·73	2·89	3·01	3·10	3·17	3·23	3·28	3·37
	120	2·45	2·67	2·83	2·94	3·02	3·09	3·15	3·20	3·28
	∞	2·40	2·62	2·76	2·87	2·95	3·02	3·07	3·12	3·20
0·1% points	10	4·0	4·3	4·6	4·8	5·0	5·2	5·3	5·4	5·6
	11	3·8	4·2	4·5	4·7	4·8	5·0	5·1	5·2	5·3
	12	3·7	4·1	4·3	4·5	4·7	4·8	4·9	5·0	5·1
	13	3·6	4·0	4·2	4·4	4·5	4·6	4·7	4·8	5·0
	14	3·5	3·9	4·1	4·3	4·4	4·5	4·6	4·7	4·9
	15	3·5	3·8	4·0	4·2	4·3	4·4	4·5	4·6	4·8
	16	3·4	3·7	4·0	4·1	4·3	4·4	4·5	4·5	4·7
	17	3·4	3·7	3·9	4·1	4·2	4·3	4·4	4·5	4·6
	18	3·3	3·6	3·9	4·0	4·1	4·2	4·3	4·4	4·5
	19	3·3	3·6	3·8	4·0	4·1	4·2	4·3	4·4	4·5
	20	3·3	3·6	3·8	3·9	4·0	4·1	4·2	4·3	4·4
	24	3·2	3·5	3·7	3·8	3·9	4·0	4·1	4·2	4·3
	30	3·1	3·4	3·6	3·7	3·8	3·9	4·0	4·0	4·1
	40	3·0	3·3	3·5	3·6	3·7	3·7	3·8	3·9	4·0
	60	2·9	3·2	3·4	3·5	3·6	3·6	3·7	3·8	3·8
	120	2·9	3·1	3·3	3·4	3·5	3·5	3·6	3·6	3·7
	∞	2·8	3·0	3·2	3·3	3·4	3·4	3·5	3·5	3·6

Table 26a. *Percentage points of* $(x_n - \bar{x})/S$ *or* $(\bar{x} - x_1)/S$. *Upper* 5% *points*

ν \ n	3	4	5	6	7	8	9	10	12	15	20
0	0·8154	0·844	0·836	0·815	0·791	0·768	0·746	0·725	0·689	0·644	0·586
1	0·789	0·800	0·789	0·771	0·752	0·733	0·714	0·697	0·666	0·626	0·574
2	·741	·752	·745	·732	·717	·701	·686	·672	·644	·609	·562
3	·692	·707	·705	·697	·686	·673	·661	·648	·625	·594	·550
4	·648	·668	·671	·666	·658	·648	·638	·627	·607	·579	·540
5	·610	·634	·640	·638	·633	·625	·617	·608	·591	·566	·530
6	0·577	0·604	0·613	0·614	0·610	0·604	0·598	0·591	0·575	0·553	0·520
7	·549	·578	·589	·591	·590	·586	·580	·574	·561	·541	·511
8	·524	·554	·567	·571	·571	·569	·564	·559	·548	·530	·502
9	·502	·533	·547	·553	·554	·553	·550	·546	·536	·520	·494
10	·483	·515	·530	·536	·539	·538	·536	·533	·524	·510	·486
12	0·450	0·482	0·499	0·507	0·511	0·512	0·511	0·509	0·503	0·492	0·472
15	·411	·443	·461	·471	·476	·479	·480	·479	·476	·468	·452
20	·363	·395	·413	·425	·431	·436	·438	·439	·439	·435	·424
24	·335	·366	·384	·396	·403	·408	·411	·413	·415	·413	·405
30	·303	·332	·350	·362	·370	·375	·379	·382	·385	·385	·381
40	·266	·292	·309	·321	·329	·335	·339	·342	·347	·349	·349
50	·239	·264	·280	·291	·299	·305	·310	·313	·318	·222	·323

Upper 1% *points*

ν \ n	3	4	5	6	7	8	9	10	12	15	20
0	0·8165	0·8617	0·8739	0·8695	0·8566	0·8394	0·8211	0·8032	0·7687	0·7228	0·6614
1	0·8111	0·8431	0·8478	0·8400	0·8263	0·8104	0·7942	0·7780	0·7465	0·7048	0·6483
2	·7904	·8155	·8176	·8094	·7971	·7833	·7688	·7541	·7260	·6879	·6356
3	·7614	·7844	·7865	·7800	·7698	·7579	·7450	·7320	·7070	·6723	·6239
4	·7299	·7532	·7570	·7527	·7444	·7341	·7229	·7116	·6890	·6576	·6127
5	·6990	·7238	·7297	·7274	·7207	·7120	·7026	·6926	·6724	·6438	·6020
6	0·6703	0·6968	0·7045	0·7037	0·6987	0·6918	0·6837	0·6748	0·6548	0·6306	0·5917
7	·6442	·6720	·6812	·6819	·6786	·6729	·6659	·6581	·6422	·6182	·5822
8	·6204	·6491	·6597	·6620	·6599	·6554	·6495	·6428	·6286	·6066	·5727
9	·5986	·6282	·6401	·6436	·6424	·6389	·6341	·6284	·6156	·5956	·5638
10	·5788	·6091	·6219	·6263	·6263	·6237	·6198	·6148	·6031	·5851	·5556
12	0·5441	0·5723	0·5895	0·5956	0·5971	0·5962	0·5935	0·5899	0·5808	0·5654	0·5398
15	·5017	·5333	·5489	·5566	·5600	·5607	·5597	·5576	·5513	·5393	·5183
20	·4482	·4795	·4962	·5055	·5106	·5132	·5140	·5136	·5104	·5031	·4880
24	·4158	·4463	·4633	·4732	·4792	·4826	·4844	·4850	·4837	·4785	·4668
30	·3779	·4073	·4240	·4346	·4412	·4455	·4480	·4496	·4501	·4479	·4393
40	·3328	·3600	·3763	·3869	·3940	·3990	·4023	·4047	·4071	·4074	·4038
50	·3006	·3162	·3416	·3519	·3591	·3642	·3681	·3711	·3744	·3766	·3751

Note: $S^2 = \sum_1^n (x_i - \bar{x})^2 + \sum_1^{\nu+1} (y_i - \bar{y})^2$, with y_i independent of x_i.

Table 26*b*. *Percentage points of max* $|x_i - \bar{x}|/S$. *Upper* 5 % *points*

ν \ n	3	4	5	6	7	8	9	10	12	15	20
0	0·8162	0·855	0·857	0·844	0·825	0·804	0·783	0·763	0·727	0·681	0·621
1	0·803	0·824	0·820	0·807	0·789	0·771	0·753	0·736	0·704	0·663	0·608
2	·769	·786	·782	·771	·757	·741	·726	·711	·683	·646	·596
3	·729	·747	·746	·738	·727	·714	·701	·688	·664	·630	·584
4	·690	·711	·713	·708	·670	·689	·678	·667	·646	·616	·573
5	·654	·678	·684	·681	·675	·667	·658	·648	·629	·602	·563
6	0·620	0·649	0·657	0·657	0·652	0·646	0·638	0·630	0·613	0·589	0·553
7	·590	·622	·633	·635	·632	·627	·621	·614	·599	·577	·544
8	·563	·598	·610	·614	·613	·609	·604	·598	·586	·566	·535
9	·541	·576	·590	·595	·595	·593	·589	·584	·573	·555	·526
10	·520	·556	·572	·578	·579	·578	·575	·571	·561	·545	·518
12	0·484	0·522	0·539	0·547	0·550	0·550	0·549	0·546	0·539	0·525	0·502
15	·442	·480	·499	·509	·514	·516	·516	·514	·510	·500	·482
20	·390	·427	·447	·459	·466	·469	·471	·472	·471	·465	·452
24	·360	·395	·416	·428	·436	·440	·443	·444	·445	·442	·432
30	·325	·359	·379	·391	·399	·405	·408	·411	·413	·413	·407
40	·284	·315	·334	·346	·355	·361	·365	·368	·372	·375	·372
50	·255	·284	·302	·314	·323	·328	·333	·336	·341	·345	·345

Upper 1 % *points*

ν \ n	3	4	5	6	7	8	9	10	12	15	20
0	0·8165	0·864	0·881	0·882	0·874	0·860	0·843	0·827	0·794	0·750	0·688
1	0·814	0·851	0·862	0·858	0·847	0·833	0·819	0·804	0·773	0·732	0·674
2	·800	·830	·837	·831	·821	·808	·795	·781	·753	·715	·662
3	·778	·805	·809	·805	·796	·785	·772	·759	·735	·700	·651
4	·751	·777	·782	·779	·772	·762	·751	·740	·717	·686	·639
5	·724	·750	·757	·755	·749	·741	·731	·721	·701	·671	·628
6	0·698	0·725	0·733	0·733	0·728	0·721	0·713	0·704	0·685	0·658	0·617
7	·673	·702	·711	·712	·708	·703	·695	·687	·671	·646	·608
8	·651	·680	·690	·692	·690	·685	·679	·672	·657	·634	·599
9	·629	·659	·671	·674	·673	·669	·664	·657	·644	·623	·591
10	·609	·640	·653	·657	·657	·654	·649	·644	·632	·612	·581
12	0·573	0·607	0·621	0·626	0·628	0·626	0·623	0·619	0·609	0·593	0·565
15	·529	·564	·580	·587	·590	·590	·588	·586	·579	·566	·544
20	·473	·508	·525	·534	·539	·541	·542	·541	·537	·528	·512
24	·439	·473	·491	·501	·507	·510	·511	·511	·509	·503	·490
30	·399	·432	·450	·460	·467	·470	·473	·474	·474	·471	·462
40	·351	·382	·399	·410	·417	·422	·426	·427	·429	·429	·424
50	·317	·346	·363	·373	·381	·386	·389	·392	·395	·397	·394

Note: $S^2 = \sum_1^n (x_i - \bar{x})^2 + \sum_1^{\nu+1} (y_i - \bar{y})^2$, with y_i independent of x_i.

Table 27. *Mean range in normal samples of size n*

n	0	1	2	3	4	5	6	7	8	9	n
0	—	—	1·12838	1·69257	2·05875+	2·32593	2·53441	2·70436	2·84720	2·97003	0
10	3·07751	3·17287	3·25846	3·33598	3·40676	3·47183	3·53198	3·58788	3·64006	3·68896	10
20	3·73495+	3·77834	3·81938	3·85832	3·89535−	3·93063	3·96432	3·99654	4·02741	4·05704	20
30	4·08552	4·11293	4·13934	4·16482	4·18943	4·21322	4·23625−	4·25855+	4·28018	4·30117	30
40	4·32155	4·34135	4·36063	4·37938	4·39764	4·41544	4·43279	4·44972	4·46624	4·48238	40
50	4·49815−	4·51356	4·52864	4·54339	4·55783	4·57197	4·58582	4·59939	4·61270	4·62575+	50
60	4·63856	4·65112	4·66346	4·67557	4·68747	4·69916	4·71065−	4·72194	4·73305−	4·74397	60
70	4·75472	4·76529	4·77570	4·78595+	4·79604	4·80598	4·81578	4·82543	4·83493	4·84431	70
80	4·85355−	4·86266	4·87165−	4·88051	4·88926	4·89789	4·90641	4·91481	4·92311	4·93131	80
90	4·93940	4·94739	4·95529	4·96309	4·97079	4·97841	4·98593	4·99337	5·00073	5·00800	90
100	5·01519	5·02230	5·02933	5·03628	5·04316	5·04997	5·05670	5·06337	5·06996	5·07649	100
110	5·08295−	5·08934	5·09568	5·10195−	5·10815+	5·11430	5·12039	5·12642	5·13239	5·13831	110
120	5·14417	5·14998	5·15573	5·16144	5·16709	5·17269	5·17824	5·18374	5·18919	5·19460	120
130	5·19996	5·20528	5·21055−	5·21578	5·22096	5·22610	5·23120	5·23625+	5·24127	5·24624	130
140	5·25118	5·25608	5·26094	5·26576	5·27054	5·27529	5·28000	5·28468	5·28932	5·29392	140
150	5·29849	5·30303	5·30754	5·31201	5·31645−	5·32086	5·32523	5·32958	5·33389	5·33818	150
160	5·34244	5·34666	5·35086	5·35503	5·35917	5·36328	5·36737	5·37142	5·37545+	5·37946	160
170	5·38344	5·38739	5·39132	5·39522	5·39910	5·40295+	5·40678	5·41059	5·41437	5·41812	170
180	5·42186	5·42557	5·42926	5·43293	5·43657	5·44019	5·44380	5·44738	5·45093	5·45447	180
190	5·45799	5·46149	5·46497	5·46842	5·47186	5·47528	5·47868	5·48206	5·48542	5·48876	190
200	5·49209	5·49539	5·49868	5·50195−	5·50520	5·50843	5·51165+	5·51485+	5·51803	5·52120	200
210	5·52435−	5·52748	5·53060	5·53370	5·53678	5·53985+	5·54291	5·54594	5·54897	5·55197	210
220	5·55497	5·55794	5·56091	5·56385+	5·56679	5·56971	5·57261	5·57550	5·57838	5·58124	220
230	5·58409	5·58692	5·58975−	5·59255+	5·59535−	5·59813	5·60090	5·60366	5·60640	5·60913	230
240	5·61185−	5·61456	5·61725−	5·61993	5·62260	5·62526	5·62790	5·63054	5·63316	5·63577	240
250	5·63837	5·64096	5·64353	5·64610	5·64865+	5·65119	5·65373	5·65625−	5·65876	5·66126	250
260	5·66375−	5·66623	5·66869	5·67115+	5·67360	5·67604	5·67847	5·68088	5·68329	5·68569	260
270	5·68808	5·69046	5·69282	5·69518	5·69753	5·69987	5·70221	5·70453	5·70684	5·70914	270
280	5·71144	5·71372	5·71600	5·71827	5·72053	5·72278	5·72502	5·72725+	5·72948	5·73170	280
290	5·73390	5·73610	5·73829	5·74048	5·74265+	5·74482	5·74698	5·74913	5·75127	5·75341	290
300	5·75553	5·75765+	5·75977	5·76187	5·76397	5·76605+	5·76814	5·77021	5·77228	5·77434	300
310	5·77639	5·77843	5·78047	5·78250	5·78453	5·78654	5·78855+	5·79055+	5·79255+	5·79454	310
320	5·79652	5·79850	5·80046	5·80243	5·80438	5·80633	5·80827	5·81021	5·81214	5·81406	320
330	5·81598	5·81789	5·81979	5·82169	5·82358	5·82546	5·82734	5·82922	5·83108	5·83294	330
340	5·83480	5·83665−	5·83849	5·84033	5·84216	5·84398	5·84580	5·84762	5·84942	5·85123	340
350	5·85302	5·85482	5·85660	5·85838	5·86016	5·86192	5·86369	5·86545−	5·86720	5·86895−	350
360	5·87069	5·87243	5·87416	5·87588	5·87761	5·87932	5·88103	5·88274	5·88444	5·88614	360
370	5·88783	5·88951	5·89119	5·89287	5·89454	5·89621	5·89787	5·89952	5·90118	5·90282	370
380	5·90447	5·90610	5·90774	5·90936	5·91099	5·91261	5·91422	5·91583	5·91744	5·91904	380
390	5·92063	5·92223	5·92381	5·92540	5·92697	5·92855−	5·93012	5·93168	5·93325−	5·93480	390
400	5·93636	5·93790	5·93945−	5·94099	5·94253	5·94406	5·94558	5·94711	5·94863	5·95014	400
410	5·95166	5·95316	5·95467	5·95617	5·95766	5·95915+	5·96064	5·96212	5·96360	5·96508	410
420	5·96655+	5·96802	5·96949	5·97095−	5·97240	5·97386	5·97531	5·97675+	5·97820	5·97963	420
430	5·98107	5·98250	5·98393	5·98535+	5·98677	5·98819	5·98960	5·99101	5·99242	5·99382	430
440	5·99522	5·99662	5·99801	5·99940	6·00079	6·00217	6·00355−	6·00492	6·00630	6·00766	440
450	6·00903	6·01039	6·01175+	6·01311	6·01446	6·01581	6·01716	6·01850	6·01984	6·02117	450
460	6·02251	6·02384	6·02516	6·02649	6·02781	6·02913	6·03044	6·03175+	6·03306	6·03437	460
470	6·03567	6·03697	6·03826	6·03956	6·04085−	6·04214	6·04342	6·04470	6·04598	6·04726	470
480	6·04853	6·04980	6·05107	6·05233	6·05359	6·05485+	6·05611	6·05736	6·05861	6·05986	480
490	6·06110	6·06234	6·06358	6·06482	6·06605+	6·06728	6·06851	6·06974	6·07096	6·07218	490

n	0	10	20	30	40	50	60	70	80	90	n
500	6·07340	6·08543	6·09721	6·10874	6·12004	6·13112	6·14198	6·15263	6·16308	6·17333	500
600	6·18340	6·19329	6·20301	6·21255+	6·22194	6·23116	6·24024	6·24916	6·25794	6·26659	600
700	6·27510	6·28347	6·29173	6·29985+	6·30786	6·31576	6·32353	6·33120	6·33877	6·34623	700
800	6·35358	6·36084	6·36800	6·37507	6·38205−	6·38894	6·39574	6·40245+	6·40909	6·41564	800
900	6·42211	6·42851	6·43483	6·44108	6·44725+	6·45335+	6·45939	6·46536	6·47126	6·47710	900
1000	6·48287										1000

The unit is the population standard deviation.

Table 28. *Mean positions of ranked normal deviates (normal order statistics)*

i \ n	2	3	4	5	6	7	8	9	10	11	12
1	0·564	0·846	1·029	1·163	1·267	1·352	1·424	1·485	1·539	1·586	1·629
2		·000	0·297	0·495	0·642	0·757	0·852	0·932	1·001	1·062	1·116
3				·000	·202	·353	·473	·572	0·656	0·729	0·793
4						·000	·153	·275	·376	·462	·537
5								0·000	0·123	0·225	0·312
6										·000	·103

i \ n	13	14	15	16	17	18	19	20	21	22	23	24	25
1	1·668	1·703	1·736	1·766	1·794	1·820	1·844	1·867	1·89	1·91	1·93	1·95	1·97
2	1·164	1·208	1·248	1·285	1·319	1·350	1·380	1·408	1·43	1·46	1·48	1·50	1·52
3	0·850	0·901	0·948	0·990	1·029	1·066	1·099	1·131	1·16	1·19	1·21	1·24	1·26
4	·603	·662	·715	·763	0·807	0·848	0·886	0·921	0·95	0·98	1·01	1·04	1·07
5	0·388	0·456	0·516	0·570	0·619	0·665	0·707	0·745	0·78	0·82	0·85	0·88	0·91
6	·190	·267	·335	·396	·451	·502	·548	·590	·63	·67	·70	·73	·76
7	·000	·088	·165	·234	·295	·351	·402	·448	·49	·53	·57	·60	·64
8			·000	·077	·146	·208	·264	·315	·36	·41	·45	·48	·52
9					·000	·069	·131	·187	·24	·29	·33	·37	·41
10							0·000	0·062	0·12	0·17	0·22	0·26	0·30
11									·00	·06	·11	·16	·20
12											·00	·05	·10
13													·00

i \ n	26	28	30	32	34	36	38	40	42	44	46	48	50
1	1·98	2·01	2·04	2·07	2·09	2·12	2·14	2·16	2·18	2·20	2·22	2·23	2·25
2	1·54	1·58	1·62	1·65	1·68	1·70	1·73	1·75	1·78	1·80	1·82	1·84	1·85
3	1·29	1·33	1·36	1·40	1·43	1·46	1·49	1·52	1·54	1·57	1·59	1·61	1·63
4	1·09	1·14	1·18	1·22	1·25	1·28	1·32	1·34	1·37	1·40	1·42	1·44	1·46
5	0·93	0·98	1·03	1·07	1·11	1·14	1·17	1·20	1·23	1·26	1·28	1·31	1·33
6	·79	·85	0·89	0·94	0·98	1·02	1·05	1·08	1·11	1·14	1·17	1·19	1·22
7	·67	·73	·78	·82	·87	0·91	0·94	0·98	1·01	1·04	1·07	1·09	1·12
8	·55	·61	·67	·72	·76	·81	·85	·88	0·91	0·95	0·98	1·00	1·03
9	·44	·51	·57	·62	·67	·71	·75	·79	·83	·86	·89	0·92	0·95
10	0·34	0·41	0·47	0·53	0·58	0·63	0·67	0·71	0·75	0·78	0·81	0·84	0·87
11	·24	·32	·38	·44	·50	·54	·59	·63	·67	·71	·74	·77	·80
12	·14	·22	·29	·36	·41	·47	·51	·56	·60	·64	·67	·70	·74
13	·05	·13	·21	·28	·34	·39	·44	·49	·53	·57	·60	·64	·67
14		·04	·12	·20	·26	·32	·37	·42	·46	·50	·54	·58	·61
15			0·04	0·12	0·18	0·24	0·30	0·35	0·40	0·44	0·48	0·52	0·55
16				·04	·11	·17	·23	·28	·33	·38	·42	·46	·49
17					·04	·10	·16	·22	·27	·32	·36	·40	·44
18						·03	·10	·16	·21	·26	·30	·34	·38
19							·03	·09	·15	·20	·25	·29	·33
20								0·03	0·09	0·14	0·19	0·24	0·28
21									·03	·09	·14	·18	·23
22										·03	·08	·13	·18
23											·03	·08	·13
24												·03	·07
25													0·03

The table gives the expectation $\xi(i|n)$ of the ith largest observation in a sample of n normal deviates, the unit being the population standard deviation.

The columns for n from 21 to 50 are taken from *Statistical Tables for Biological, Agricultural and Medical Research* (Fisher & Yates, 1963, Table XX) by permission of the authors and the publishers, Messrs Oliver and Boyd.

Table 29. *Percentage points of the studentized range,* $q = (x_n - x_1)/s_\nu$. *Upper* 10% *points*

ν \ n	2	3	4	5	6	7	8	9	10
1	8·93	13·44	16·36	18·49	20·15	21·51	22·64	23·62	24·48
2	4·13	5·73	6·77	7·54	8·14	8·63	9·05	9·41	9·72
3	3·33	4·47	5·20	5·74	6·16	6·51	6·81	7·06	7·29
4	3·01	3·98	4·59	5·03	5·39	5·68	5·93	6·14	6·33
5	2·85	3·72	4·26	4·66	4·98	5·24	5·46	5·65	5·82
6	2·75	3·56	4·07	4·44	4·73	4·97	5·17	5·34	5·50
7	2·68	3·45	3·93	4·28	4·55	4·78	4·97	5·14	5·28
8	2·63	3·37	3·83	4·17	4·43	4·65	4·83	4·99	5·13
9	2·59	3·32	3·76	4·08	4·34	4·54	4·72	4·87	5·01
10	2·56	3·27	3·70	4·02	4·26	4·47	4·64	4·78	4·91
11	2·54	3·23	3·66	3·96	4·20	4·40	4·57	4·71	4·84
12	2·52	3·20	3·62	3·92	4·16	4·35	4·51	4·65	4·78
13	2·50	3·18	3·59	3·88	4·12	4·30	4·46	4·60	4·72
14	2·49	3·16	3·56	3·85	4·08	4·27	4·42	4·56	4·68
15	2·48	3·14	3·54	3·83	4·05	4·23	4·39	4·52	4·64
16	2·47	3·12	3·52	3·80	4·03	4·21	4·36	4·49	4·61
17	2·46	3·11	3·50	3·78	4·00	4·18	4·33	4·46	4·58
18	2·45	3·10	3·49	3·77	3·98	4·16	4·31	4·44	4·55
19	2·45	3·09	3·47	3·75	3·97	4·14	4·29	4·42	4·53
20	2·44	3·08	3·46	3·74	3·95	4·12	4·27	4·40	4·51
24	2·42	3·05	3·42	3·69	3·90	4·07	4·21	4·34	4·44
30	2·40	3·02	3·39	3·65	3·85	4·02	4·16	4·28	4·38
40	2·38	2·99	3·35	3·60	3·80	3·96	4·10	4·21	4·32
60	2·36	2·96	3·31	3·56	3·75	3·91	4·04	4·16	4·25
120	2·34	2·93	3·28	3·52	3·71	3·86	3·99	4·10	4·19
∞	2·33	2·90	3·24	3·48	3·66	3·81	3·93	4·04	4·13

ν \ n	11	12	13	14	15	16	17	18	19	20
1	25·24	25·92	26·54	27·10	27·62	28·10	28·54	28·96	29·35	29·71
2	10·01	10·26	10·49	10·70	10·89	11·07	11·24	11·39	11·54	11·68
3	7·49	7·67	7·83	7·98	8·12	8·25	8·37	8·48	8·58	8·68
4	6·49	6·65	6·78	6·91	7·02	7·13	7·23	7·33	7·41	7·50
5	5·97	6·10	6·22	6·34	6·44	6·54	6·63	6·71	6·79	6·86
6	5·64	5·76	5·87	5·98	6·07	6·16	6·25	6·32	6·40	6·47
7	5·41	5·53	5·64	5·74	5·83	5·91	5·99	6·06	6·13	6·19
8	5·25	5·36	5·46	5·56	5·64	5·72	5·80	5·87	5·93	6·00
9	5·13	5·23	5·33	5·42	5·51	5·58	5·66	5·72	5·79	5·85
10	5·03	5·13	5·23	5·32	5·40	5·47	5·54	5·61	5·67	5·73
11	4·95	5·05	5·15	5·23	5·31	5·38	5·45	5·51	5·57	5·63
12	4·89	4·99	5·08	5·16	5·24	5·31	5·37	5·44	5·49	5·55
13	4·83	4·93	5·02	5·10	5·18	5·25	5·31	5·37	5·43	5·48
14	4·79	4·88	4·97	5·05	5·12	5·19	5·26	5·32	5·37	5·43
15	4·75	4·84	4·93	5·01	5·08	5·15	5·21	5·27	5·32	5·38
16	4·71	4·81	4·89	4·97	5·04	5·11	5·17	5·23	5·28	5·33
17	4·68	4·77	4·86	4·93	5·01	5·07	5·13	5·19	5·24	5·30
18	4·65	4·75	4·83	4·90	4·98	5·04	5·10	5·16	5·21	5·26
19	4·63	4·72	4·80	4·88	4·95	5·01	5·07	5·13	5·18	5·23
20	4·61	4·70	4·78	4·85	4·92	4·99	5·05	5·10	5·16	5·20
24	4·54	4·63	4·71	4·78	4·85	4·91	4·97	5·02	5·07	5·12
30	4·47	4·56	4·64	4·71	4·77	4·83	4·89	4·94	4·99	5·03
40	4·41	4·49	4·56	4·63	4·69	4·75	4·81	4·86	4·90	4·95
60	4·34	4·42	4·49	4·56	4·62	4·67	4·73	4·78	4·82	4·86
120	4·28	4·35	4·42	4·48	4·54	4·60	4·65	4·69	4·74	4·78
∞	4·21	4·28	4·35	4·41	4·47	4·52	4·57	4·61	4·65	4·69

n: size of sample from which range obtained. ν: degrees of freedom of independent s_ν.

Table 29 (*continued*). *Percentage points of the studentized range,* $q=(x_n-x_1)/s_\nu$.
Upper 5 % points

ν \ n	2	3	4	5	6	7	8	9	10
1	17·97	26·98	32·82	37·08	40·41	43·12	45·40	47·36	49·07
2	6·08	8·33	9·80	10·88	11·74	12·44	13·03	13·54	13·99
3	4·50	5·91	6·82	7·50	8·04	8·48	8·85	9·18	9·46
4	3·93	5·04	5·76	6·29	6·71	7·05	7·35	7·60	7·83
5	3·64	4·60	5·22	5·67	6·03	6·33	6·58	6·80	6·99
6	3·46	4·34	4·90	5·30	5·63	5·90	6·12	6·32	6·49
7	3·34	4·16	4·68	5·06	5·36	5·61	5·82	6·00	6·16
8	3·26	4·04	4·53	4·89	5·17	5·40	5·60	5·77	5·92
9	3·20	3·95	4·41	4·76	5·02	5·24	5·43	5·59	5·74
10	3·15	3·88	4·33	4·65	4·91	5·12	5·30	5·46	5·60
11	3·11	3·82	4·26	4·57	4·82	5·03	5·20	5·35	5·49
12	3·08	3·77	4·20	4·51	4·75	4·95	5·12	5·27	5·39
13	3·06	3·73	4·15	4·45	4·69	4·88	5·05	5·19	5·32
14	3·03	3·70	4·11	4·41	4·64	4·83	4·99	5·13	5·25
15	3·01	3·67	4·08	4·37	4·59	4·78	4·94	5·08	5·20
16	3·00	3·65	4·05	4·33	4·56	4·74	4·90	5·03	5·15
17	2·98	3·63	4·02	4·30	4·52	4·70	4·86	4·99	5·11
18	2·97	3·61	4·00	4·28	4·49	4·67	4·82	4·96	5·07
19	2·96	3·59	3·98	4·25	4·47	4·65	4·79	4·92	5·04
20	2·95	3·58	3·96	4·23	4·45	4·62	4·77	4·90	5·01
24	2·92	3·53	3·90	4·17	4·37	4·54	4·68	4·81	4·92
30	2·89	3·49	3·85	4·10	4·30	4·46	4·60	4·72	4·82
40	2·86	3·44	3·79	4·04	4·23	4·39	4·52	4·63	4·73
60	2·83	3·40	3·74	3·98	4·16	4·31	4·44	4·55	4·65
120	2·80	3·36	3·68	3·92	4·10	4·24	4·36	4·47	4·56
∞	2·77	3·31	3·63	3·86	4·03	4·17	4·29	4·39	4·47

ν \ n	11	12	13	14	15	16	17	18	19	20
1	50·59	51·96	53·20	54·33	55·36	56·32	57·22	58·04	58·83	59·56
2	14·39	14·75	15·08	15·38	15·65	15·91	16·14	16·37	16·57	16·77
3	9·72	9·95	10·15	10·35	10·52	10·69	10·84	10·98	11·11	11·24
4	8·03	8·21	8·37	8·52	8·66	8·79	8·91	9·03	9·13	9·23
5	7·17	7·32	7·47	7·60	7·72	7·83	7·93	8·03	8·12	8·21
6	6·65	6·79	6·92	7·03	7·14	7·24	7·34	7·43	7·51	7·59
7	6·30	6·43	6·55	6·66	6·76	6·85	6·94	7·02	7·10	7·17
8	6·05	6·18	6·29	6·39	6·48	6·57	6·65	6·73	6·80	6·87
9	5·87	5·98	6·09	6·19	6·28	6·36	6·44	6·51	6·58	6·64
10	5·72	5·83	5·93	6·03	6·11	6·19	6·27	6·34	6·40	6·47
11	5·61	5·71	5·81	5·90	5·98	6·06	6·13	6·20	6·27	6·33
12	5·51	5·61	5·71	5·80	5·88	5·95	6·02	6·09	6·15	6·21
13	5·43	5·53	5·63	5·71	5·79	5·86	5·93	5·99	6·05	6·11
14	5·36	5·46	5·55	5·64	5·71	5·79	5·85	5·91	5·97	6·03
15	5·31	5·40	5·49	5·57	5·65	5·72	5·78	5·85	5·90	5·96
16	5·26	5·35	5·44	5·52	5·59	5·66	5·73	5·79	5·84	5·90
17	5·21	5·31	5·39	5·47	5·54	5·61	5·67	5·73	5·79	5·84
18	5·17	5·27	5·35	5·43	5·50	5·57	5·63	5·69	5·74	5·79
19	5·14	5·23	5·31	5·39	5·46	5·53	5·59	5·65	5·70	5·75
20	5·11	5·20	5·28	5·36	5·43	5·49	5·55	5·61	5·66	5·71
24	5·01	5·10	5·18	5·25	5·32	5·38	5·44	5·49	5·55	5·59
30	4·92	5·00	5·08	5·15	5·21	5·27	5·33	5·38	5·43	5·47
40	4·82	4·90	4·98	5·04	5·11	5·16	5·22	5·27	5·31	5·36
60	4·73	4·81	4·88	4·94	5·00	5·06	5·11	5·15	5·20	5·24
120	4·64	4·71	4·78	4·84	4·90	4·95	5·00	5·04	5·09	5·13
∞	4·55	4·62	4·68	4·74	4·80	4·85	4·89	4·93	4·97	5·01

n: size of sample from which range obtained. ν: degrees of freedom of independent s_ν.

Table 29 (*continued*). *Upper* 1 % *points*

ν \ n	2	3	4	5	6	7	8	9	10
1	90·03	135·0	164·3	185·6	202·2	215·8	227·2	237·0	245·6
2	14·04	19·02	22·29	24·72	26·63	28·20	29·53	30·68	31·69
3	8·26	10·62	12·17	13·33	14·24	15·00	15·64	16·20	16·69
4	6·51	8·12	9·17	9·96	10·58	11·10	11·55	11·93	12·27
5	5·70	6·98	7·80	8·42	8·91	9·32	9·67	9·97	10·24
6	5·24	6·33	7·03	7·56	7·97	8·32	8·61	8·87	9·10
7	4·95	5·92	6·54	7·01	7·37	7·68	7·94	8·17	8·37
8	4·75	5·64	6·20	6·62	6·96	7·24	7·47	7·68	7·86
9	4·60	5·43	5·96	6·35	6·66	6·91	7·13	7·33	7·49
10	4·48	5·27	5·77	6·14	6·43	6·67	6·87	7·05	7·21
11	4·39	5·15	5·62	5·97	6·25	6·48	6·67	6·84	6·99
12	4·32	5·05	5·50	5·84	6·10	6·32	6·51	6·67	6·81
13	4·26	4·96	5·40	5·73	5·98	6·19	6·37	6·53	6·67
14	4·21	4·89	5·32	5·63	5·88	6·08	6·26	6·41	6·54
15	4·17	4·84	5·25	5·56	5·80	5·99	6·16	6·31	6·44
16	4·13	4·79	5·19	5·49	5·72	5·92	6·08	6·22	6·35
17	4·10	4·74	5·14	5·43	5·66	5·85	6·01	6·15	6·27
18	4·07	4·70	5·09	5·38	5·60	5·79	5·94	6·08	6·20
19	4·05	4·67	5·05	5·33	5·55	5·73	5·89	6·02	6·14
20	4·02	4·64	5·02	5·29	5·51	5·69	5·84	5·97	6·09
24	3·96	4·55	4·91	5·17	5·37	5·54	5·69	5·81	5·92
30	3·89	4·45	4·80	5·05	5·24	5·40	5·54	5·65	5·76
40	3·82	4·37	4·70	4·93	5·11	5·26	5·39	5·50	5·60
60	3·76	4·28	4·59	4·82	4·99	5·13	5·25	5·36	5·45
120	3·70	4·20	4·50	4·71	4·87	5·01	5·12	5·21	5·30
∞	3·64	4·12	4·40	4·60	4·76	4·88	4·99	5·08	5·16

ν \ n	11	12	13	14	15	16	17	18	19	20
1	253·2	260·0	266·2	271·8	277·0	281·8	286·3	290·4	294·3	298·0
2	32·59	33·40	34·13	34·81	35·43	36·00	36·53	37·03	37·50	37·95
3	17·13	17·53	17·89	18·22	18·52	18·81	19·07	19·32	19·55	19·77
4	12·57	12·84	13·09	13·32	13·53	13·73	13·91	14·08	14·24	14·40
5	10·48	10·70	10·89	11·08	11·24	11·40	11·55	11·68	11·81	11·93
6	9·30	9·48	9·65	9·81	9·95	10·08	10·21	10·32	10·43	10·54
7	8·55	8·71	8·86	9·00	9·12	9·24	9·35	9·46	9·55	9·65
8	8·03	8·18	8·31	8·44	8·55	8·66	8·76	8·85	8·94	9·03
9	7·65	7·78	7·91	8·03	8·13	8·23	8·33	8·41	8·49	8·57
10	7·36	7·49	7·60	7·71	7·81	7·91	7·99	8·08	8·15	8·23
11	7·13	7·25	7·36	7·46	7·56	7·65	7·73	7·81	7·88	7·95
12	6·94	7·06	7·17	7·26	7·36	7·44	7·52	7·59	7·66	7·73
13	6·79	6·90	7·01	7·10	7·19	7·27	7·35	7·42	7·48	7·55
14	6·66	6·77	6·87	6·96	7·05	7·13	7·20	7·27	7·33	7·39
15	6·55	6·66	6·76	6·84	6·93	7·00	7·07	7·14	7·20	7·26
16	6·46	6·56	6·66	6·74	6·82	6·90	6·97	7·03	7·09	7·15
17	6·38	6·48	6·57	6·66	6·73	6·81	6·87	6·94	7·00	7·05
18	6·31	6·41	6·50	6·58	6·65	6·73	6·79	6·85	6·91	6·97
19	6·25	6·34	6·43	6·51	6·58	6·65	6·72	6·78	6·84	6·89
20	6·19	6·28	6·37	6·45	6·52	6·59	6·65	6·71	6·77	6·82
24	6·02	6·11	6·19	6·26	6·33	6·39	6·45	6·51	6·56	6·61
30	5·85	5·93	6·01	6·08	6·14	6·20	6·26	6·31	6·36	6·41
40	5·69	5·76	5·83	5·90	5·96	6·02	6·07	6·12	6·16	6·21
60	5·53	5·60	5·67	5·73	5·78	5·84	5·89	5·93	5·97	6·01
120	5·37	5·44	5·50	5·56	5·61	5·66	5·71	5·75	5·79	5·83
∞	5·23	5·29	5·35	5·40	5·45	5·49	5·54	5·57	5·61	5·65

Table 29*a*. *Two-sample analogue of Student's test. Values of* $u = |\bar{x}_1 - \bar{x}_2|/(w_1 + w_2)$ *exceeded with probability* α (*for a single-tailed test* α *must be halved*)

n_1	n_2	Probability (α) 0·10	0·05	0·02	0·01
2	2	1·161	1·714	2·776	3·958
	3	0·693	0·915	1·255	1·557
	4	0·556	0·732	1·002	1·242
	5	0·478	0·619	0·827	1·008
	6	0·429	0·549	0·721	0·865
	7	0·396	0·502	0·652	0·776
	8	0·372	0·469	0·603	0·713
	9	0·353	0·443	0·567	0·666
	10	0·338	0·423	0·538	0·630
	11	0·326	0·407	0·515	0·601
	12	0·316	0·393	0·496	0·557
	13	0·307	0·382	0·480	0·557
	14	0·300	0·372	0·467	0·541
	15	0·294	0·363	0·455	0·526
	16	0·287	0·356	0·445	0·513
	17	0·282	0·349	0·436	0·502
	18	0·278	0·343	0·428	0·492
	19	0·274	0·338	0·420	0·483
	20	0·270	0·333	0·414	0·475
3	3	0·487	0·635	0·860	1·050
	4	0·398	0·511	0·663	0·814
	5	0·339	0·429	0·556	0·660
	6	0·311	0·391	0·501	0·590
	7	0·288	0·360	0·458	0·536
	8	0·271	0·338	0·427	0·498
	9	0·258	0·321	0·404	0·469
	10	0·248	0·307	0·385	0·446
	11	0·240	0·296	0·370	0·427
	12	0·232	0·287	0·358	0·412
	13	0·226	0·279	0·347	0·399
	14	0·221	0·272	0·338	0·388
	15	0·216	0·266	0·330	0·378
	16	0·212	0·261	0·323	0·370
	17	0·209	0·256	0·317	0·362
	18	0·205	0·252	0·311	0·356
	19	0·202	0·248	0·306	0·350
	20	0·200	0·245	0·302	0·344
4	4	0·322	0·407	0·526	0·620
	5	0·282	0·353	0·450	0·528
	6	0·256	0·319	0·403	0·469
	7	0·237	0·294	0·370	0·429
	8	0·224	0·276	0·346	0·399
	9	0·213	0·263	0·327	0·377
	10	0·204	0·252	0·313	0·359
	11	0·197	0·242	0·301	0·345
	12	0·191	0·235	0·291	0·333
	13	0·186	0·228	0·282	0·322
	14	0·182	0·223	0·275	0·314
	15	0·178	0·218	0·268	0·306
4	16	0·175	0·213	0·263	0·299
	17	0·172	0·210	0·258	0·293
	18	0·169	0·206	0·253	0·288
	19	0·166	0·203	0·249	0·283
	20	0·164	0·200	0·246	0·279
5	5	0·247	0·307	0·387	0·450
	6	0·224	0·277	0·347	0·402
	7	0·208	0·256	0·319	0·368
	8	0·195	0·240	0·299	0·343
	9	0·186	0·228	0·282	0·323
	10	0·178	0·218	0·270	0·309
	11	0·172	0·210	0·260	0·296
	12	0·167	0·204	0·251	0·286
	13	0·162	0·198	0·244	0·277
	14	0·158	0·193	0·237	0·270
	15	0·155	0·189	0·232	0·263
	16	0·152	0·185	0·227	0·257
	17	0·149	0·182	0·222	0·252
	18	0·147	0·179	0·218	0·248
	19	0·144	0·176	0·215	0·244
	20	0·142	0·173	0·212	0·240
6	6	0·203	0·250	0·312	0·359
	7	0·188	0·240	0·287	0·329
	8	0·177	0·217	0·268	0·307
	9	0·168	0·206	0·254	0·289
	10	0·161	0·197	0·242	0·276
	11	0·155	0·189	0·233	0·265
	12	0·150	0·183	0·225	0·255
	13	0·146	0·178	0·218	0·247
	14	0·142	0·173	0·212	0·241
	15	0·139	0·169	0·207	0·235
	16	0·136	0·166	0·203	0·229
	17	0·134	0·163	0·199	0·225
	18	0·131	0·160	0·195	0·221
	19	0·129	0·157	0·192	0·217
	20	0·128	0·155	0·189	0·214
7	7	0·174	0·213	0·263	0·301
	8	0·163	0·200	0·246	0·281
	9	0·155	0·189	0·233	0·265
	10	0·148	0·181	0·222	0·252
	11	0·143	0·174	0·213	0·242
	12	0·138	0·168	0·206	0·233
	13	0·134	0·163	0·199	0·226
	14	0·131	0·159	0·194	0·220
	15	0·128	0·155	0·189	0·214
	16	0·125	0·152	0·185	0·209
	17	0·123	0·149	0·181	0·205
	18	0·121	0·146	0·178	0·201
	19	0·119	0·144	0·175	0·198
	20	0·117	0·142	0·172	0·195

Table 29*a* (*continued*)

n_1	n_2	Probability (α)			
		0·10	0·05	0·02	0·01
8	8	0·153	0·187	0·231	0·262
	9	0·145	0·177	0·217	0·247
	10	0·139	0·169	0·207	0·235
	11	0·133	0·162	0·199	0·225
	12	0·129	0·157	0·192	0·217
	13	0·125	0·152	0·186	0·210
	14	0·122	0·148	0·180	0·204
	15	0·119	0·144	0·176	0·199
	16	0·116	0·141	0·172	0·194
	17	0·114	0·138	0·168	0·190
	18	0·112	0·136	0·165	0·186
	19	0·110	0·134	0·162	0·183
	20	0·109	0·132	0·160	0·180
9	9	0·137	0·167	0·205	0·233
	10	0·131	0·160	0·195	0·221
	11	0·126	0·153	0·187	0·212
	12	0·122	0·148	0·180	0·204
	13	0·118	0·143	0·175	0·197
	14	0·115	0·139	0·170	0·192
	15	0·112	0·136	0·165	0·187
	16	0·110	0·133	0·162	0·182
	17	0·107	0·130	0·158	0·178
	18	0·106	0·128	0·155	0·175
	19	0·104	0·126	0·152	0·172
	20	0·102	0·124	0·150	0·169
10	10	0·125	0·152	0·186	0·210
	11	0·120	0·146	0·178	0·201
	12	0·116	0·141	0·171	0·194
	13	0·112	0·136	0·166	0·187
	14	0·109	0·133	0·161	0·182
	15	0·107	0·129	0·157	0·177
	16	0·104	0·126	0·153	0·173
	17	0·102	0·124	0·150	0·169
	18	0·100	0·121	0·147	0·165
	19	0·098	0·119	0·144	0·162
	20	0·097	0·117	0·142	0·160
11	11	0·115	0·140	0·170	0·193
	12	0·111	0·135	0·164	0·185
	13	0·108	0·131	0·159	0·179
	14	0·105	0·127	0·154	0·174
	15	0·102	0·123	0·150	0·169
	16	0·100	0·121	0·146	0·165
	17	0·098	0·118	0·143	0·161
	18	0·096	0·116	0·140	0·158
	19	0·094	0·114	0·138	0·155
	20	0·092	0·112	0·135	0·152
12	12	0·107	0·130	0·158	0·178
	13	0·104	0·126	0·153	0·172
	14	0·101	0·122	0·148	0·167
	15	0·098	0·119	0·144	0·162
	16	0·096	0·116	0·140	0·158
	17	0·094	0·113	0·137	0·154
	18	0·092	0·111	0·134	0·151
	19	0·090	0·109	0·132	0·149
	20	0·089	0·107	0·130	0·146
13	13	0·100	0·121	0·147	0·166
	14	0·097	0·118	0·143	0·161
	15	0·095	0·115	0·139	0·156
	16	0·092	0·112	0·135	0·152
	17	0·090	0·109	0·132	0·149
	18	0·089	0·107	0·130	0·146
	19	0·087	0·105	0·127	0·143
	20	0·086	0·103	0·125	0·140
14	14	0·094	0·114	0·138	0·156
	15	0·092	0·111	0·135	0·151
	16	0·090	0·108	0·131	0·147
	17	0·088	0·106	0·128	0·144
	18	0·086	0·104	0·125	0·141
	19	0·084	0·102	0·123	0·138
	20	0·083	0·101	0·121	0·135
15	15	0·089	0·108	0·131	0·147
	16	0·087	0·105	0·127	0·143
	17	0·085	0·103	0·124	0·140
	18	0·083	0·101	0·122	0·137
	19	0·082	0·099	0·119	0·134
	20	0·080	0·097	0·117	0·131
16	16	0·085	0·103	0·124	0·139
	17	0·083	0·100	0·121	0·136
	18	0·081	0·098	0·118	0·133
	19	0·080	0·096	0·116	0·130
	20	0·078	0·094	0·114	0·128
17	17	0·081	0·098	0·118	0·132
	18	0·079	0·096	0·115	0·130
	19	0·078	0·094	0·113	0·127
	20	0·076	0·092	0·111	0·124
18	18	0·077	0·093	0·113	0·126
	19	0·076	0·092	0·110	0·124
	20	0·074	0·090	0·108	0·121
19	19	0·074	0·090	0·108	0·121
	20	0·073	0·088	0·106	0·119
20	20	0·071	0·086	0·104	0·116

$\bar{x}_1$, $\bar{x}_2$ are the means and w_1, w_2 the ranges in two independent samples containing n_1, n_2 observations, respectively.

Table 29*b*. *Upper percentage points of the ratio of two independent ranges,* $F' = w_1/w_2$

Values of α

n_2	n_1	0·500	0·250	0·100	0·050	0·025	0·010	0·005	0·001
2	2	1·000	2·414	6·314	12·71	25·45	63·66	127·3	636·6
	3	1·653	3·698	9·501	19·07	38·19	95·49	191·0	955·0
	4	2·063	4·523	11·57	23·21	46·45	116·1	232·3	1162
	5	2·358	5·123	13·07	26·22	52·48	131·2	262·5	1312
	6	2·587	5·591	14·25	28·57	57·18	143·0	286·0	1430
	7	2·772	5·971	15·21	30·49	61·02	152·6	305·1	1526
	8	2·927	6·291	16·01	32·10	64·24	160·6	321·3	1607
2	9	3·060	6·565	16·70	33·49	67·01	167·6	335·1	1676
	10	3·176	6·805	17·31	34·70	69·44	173·6	347·3	1736
	11	3·279	7·018	17·85	35·77	71·59	179·0	358·0	1790
	12	3·371	7·209	18·33	36·74	73·52	183·8	367·7	1838
	13	3·454	7·382	18·76	37·61	75·27	188·2	376·4	1882
	14	3·530	7·540	19·16	38·41	76·87	192·2	384·4	1922
	15	3·600	7·685	19·53	39·15	78·34	195·9	391·8	1959
3	2	0·6050	1·189	2·167	3·194	4·607	7·370	10·46	23·49
	3	1·000	1·735	3·009	4·373	6·267	9·986	14·16	31·76
	4	1·246	2·082	3·555	5·144	7·355	11·71	16·59	37·19
	5	1·422	2·335	3·956	5·712	8·159	12·98	18·39	41·22
	6	1·559	2·533	4·272	6·160	8·794	13·98	19·81	44·39
	7	1·669	2·695	4·531	6·528	9·315	14·81	20·98	47·00
	8	1·762	2·831	4·750	6·839	9·756	15·50	21·97	49·21
3	9	1·841	2·948	4·939	7·108	10·14	16·11	22·82	51·12
	10	1·911	3·050	5·105	7·344	10·47	16·64	23·57	52·80
	11	1·972	3·142	5·252	7·554	10·77	17·11	24·24	54·31
	12	2·028	3·224	5·385	7·743	11·04	17·54	24·84	55·66
	13	2·077	3·298	5·505	7·914	11·28	17·92	25·39	56·88
	14	2·123	3·366	5·616	8·072	11·50	18·28	25·89	58·01
	15	2·165	3·428	5·717	8·216	11·71	18·60	26·35	59·04
4	2	0·4847	0·9032	1·497	2·027	2·662	3·725	4·755	8·250
	3	·8028	1·296	2·011	2·663	3·453	4·789	6·090	10·52
	4	1·000	1·542	2·341	3·075	3·971	5·489	6·971	12·03
	5	1·141	1·721	2·584	3·381	4·356	6·010	7·626	13·15
	6	1·251	1·862	2·775	3·623	4·660	6·424	8·147	14·04
	7	1·340	1·976	2·933	3·822	4·911	6·765	8·578	14·77
	8	1·414	2·073	3·066	3·990	5·125	7·056	8·944	15·40
4	9	1·478	2·156	3·181	4·137	5·310	7·307	9·262	15·95
	10	1·533	2·229	3·283	4·265	5·473	7·529	9·542	16·43
	11	1·583	2·294	3·373	4·380	5·618	7·728	9·792	16·85
	12	1·627	2·353	3·454	4·484	5·749	7·906	10·02	17·24
	13	1·667	2·406	3·528	4·578	5·869	8·069	10·22	17·60
	14	1·703	2·454	3·596	4·664	5·978	8·218	10·41	17·92
	15	1·737	2·499	3·658	4·744	6·079	8·356	10·59	18·22
5	2	0·4240	0·7717	1·228	1·602	2·020	2·664	3·242	4·993
	3	·7032	1·098	1·621	2·059	2·553	3·324	4·020	6·145
	4	·8761	1·301	1·871	2·353	2·900	3·757	4·532	6·905
	5	1·000	1·448	2·055	2·570	3·157	4·079	4·914	7·474
	6	1·096	1·563	2·200	2·742	3·361	4·335	5·219	7·929
	7	1·174	1·657	2·319	2·884	3·530	4·547	5·471	8·307
	8	1·239	1·736	2·420	3·004	3·674	4·728	5·686	8·628

w_1, w_2 are the ranges in two independent samples containing n_1, n_2 observations, respectively.

Table 29*b* (*continued*)

Values of α

n_2	n_1	0·500	0·250	0·100	0·050	0·025	0·010	0·005	0·001
5	**9**	1·295	1·805	2·507	3·109	3·798	4·885	5·874	8·907
	10	1·343	1·865	2·584	3·201	3·908	5·024	6·039	9·156
	11	1·387	1·918	2·653	3·283	4·007	5·149	6·187	9·378
	12	1·425	1·966	2·714	3·357	4·096	5·261	6·321	9·580
	13	1·460	2·010	2·771	3·425	4·176	5·363	6·443	9·764
	14	1·492	2·050	2·822	3·487	4·251	5·457	6·555	9·932
	15	1·522	2·086	2·870	3·544	4·319	5·544	6·659	10·09
6	**2**	0·3866	0·6944	1·080	1·381	1·704	2·176	2·580	3·720
	3	·6416	·9827	1·411	1·752	2·120	2·665	3·135	4·472
	4	·7995	1·161	1·620	1·988	2·388	2·983	3·498	4·967
	5	·9125	1·290	1·773	2·162	2·586	3·220	3·768	5·339
	6	1·000	1·391	1·894	2·300	2·744	3·409	3·985	5·636
	7	1·071	1·474	1·993	2·414	2·875	3·565	4·164	5·883
	8	1·130	1·543	2·077	2·510	2·986	3·699	4·317	6·093
6	**9**	1·181	1·603	2·150	2·594	3·082	3·815	4·451	6·277
	10	1·226	1·655	2·214	2·668	3·168	3·918	4·569	6·441
	11	1·265	1·702	2·271	2·734	3·244	4·010	4·675	6·588
	12	1·301	1·744	2·323	2·794	3·313	4·093	4·770	6·721
	13	1·333	1·783	2·370	2·849	3·376	4·169	4·858	6·842
	14	1·362	1·818	2·413	2·899	3·433	4·238	4·938	6·954
	15	1·389	1·850	2·452	2·945	3·487	4·303	5·013	7·058
7	**2**	0·3608	0·6426	0·9853	1·245	1·514	1·897	2·213	3·064
	3	·5991	·9063	1·279	1·564	1·864	2·294	2·652	3·624
	4	·7465	1·069	1·463	1·767	2·088	2·551	2·937	3·992
	5	·8521	1·186	1·597	1·916	2·254	2·742	3·150	4·267
	6	·9338	1·278	1·703	2·034	2·385	2·894	3·320	4·487
	7	1·000	1·353	1·790	2·131	2·494	3·020	3·461	4·671
	8	1·056	1·416	1·863	2·213	2·586	3·127	3·581	4·827
7	**9**	1·103	1·470	1·927	2·285	2·667	3·221	3·686	4·964
	10	1·145	1·518	1·983	2·348	2·738	3·304	3·779	5·086
	11	1·182	1·560	2·033	2·405	2·802	3·378	3·863	5·196
	12	1·215	1·598	2·079	2·456	2·859	3·445	3·939	5·295
	13	1·245	1·633	2·120	2·503	2·912	3·507	4·008	5·386
	14	1·272	1·665	2·157	2·545	2·960	3·563	4·071	5·469
	15	1·297	1·694	2·192	2·585	3·004	3·615	4·130	5·547
8	**2**	0·3417	0·6050	0·9188	1·151	1·387	1·715	1·979	2·667
	3	·5675	·8513	1·187	1·437	1·695	2·056	2·349	3·119
	4	·7073	1·002	1·354	1·618	1·891	2·275	2·588	3·413
	5	·8073	1·112	1·475	1·750	2·035	2·437	2·765	3·633
	6	·8847	1·197	1·571	1·855	2·150	2·566	2·907	3·810
	7	·9474	1·266	1·649	1·941	2·244	2·673	3·025	3·957
	8	1·000	1·324	1·716	2·014	2·325	2·765	3·125	4·082
8	**9**	1·045	1·375	1·773	2·078	2·395	2·844	3·213	4·192
	10	1·085	1·419	1·824	2·134	2·457	2·915	3·291	4·290
	11	1·119	1·458	1·870	2·184	2·512	2·978	3·361	4·377
	12	1·151	1·494	1·911	2·229	2·563	3·035	3·424	4·457
	13	1·179	1·526	1·948	2·271	2·608	3·088	3·482	4·530
	14	1·205	1·555	1·982	2·309	2·650	3·136	3·535	4·598
	15	1·229	1·582	2·013	2·344	2·689	3·180	3·584	4·660

Table 29*b* (*continued*). *Upper percentage points of the ratio of two independent ranges,* $F' = w_1/w_2$

Values of α

n_2	n_1	0·500	0·250	0·100	0·050	0·025	0·010	0·005	0·001
9	2	0·3268	0·5763	0·8690	1·082	1·296	1·587	1·817	2·402
	3	·5430	·8093	1·118	1·345	1·574	1·889	2·141	2·784
	4	·6767	·9521	1·273	1·509	1·750	2·083	2·349	3·032
	5	·7725	1·055	1·385	1·630	1·880	2·226	2·503	3·218
	6	·8465	1·135	1·474	1·725	1·983	2·339	2·626	3·366
	7	·9065	1·200	1·546	1·804	2·068	2·434	2·728	3·490
	8	·9568	1·255	1·607	1·870	2·140	2·514	2·816	3·595
9	9	1·000	1·303	1·661	1·928	2·203	2·584	2·892	3·687
	10	1·038	1·344	1·707	1·979	2·258	2·646	2·959	3·769
	11	1·071	1·381	1·749	2·025	2·308	2·702	3·020	3·843
	12	1·101	1·415	1·787	2·066	2·353	2·753	3·075	3·911
	13	1·128	1·445	1·821	2·104	2·394	2·799	3·125	3·972
	14	1·153	1·473	1·853	2·138	2·432	2·841	3·172	4·029
	15	1·176	1·498	1·882	2·170	2·467	2·880	3·215	4·082
10	2	0·3149	0·5534	0·8301	1·029	1·226	1·491	1·698	2·212
	3	·5233	·7761	1·065	1·274	1·483	1·766	1·989	2·547
	4	·6522	·9122	1·210	1·427	1·645	1·941	2·175	2·763
	5	·7444	1·010	1·316	1·539	1·764	2·070	2·313	2·925
	6	·8157	1·086	1·399	1·627	1·858	2·173	2·422	3·054
	7	·8736	1·148	1·466	1·700	1·935	2·258	2·514	3·161
	8	·9221	1·201	1·524	1·761	2·002	2·330	2·591	3·253
10	9	0·9637	1·246	1·574	1·815	2·059	2·394	2·659	3·333
	10	1·000	1·286	1·617	1·862	2·110	2·450	2·720	2·404
	11	1·032	1·321	1·657	1·904	2·155	2·500	2·774	3·469
	12	1·061	1·352	1·692	1·942	2·197	2·545	2·823	3·527
	13	1·087	1·381	1·724	1·977	2·234	2·587	2·868	3·581
	14	1·111	1·407	1·754	2·009	2·269	2·625	2·909	3·630
	15	1·133	1·432	1·781	2·039	2·301	2·661	2·947	3·676
11	2	0·3050	0·5347	0·7986	0·9863	1·171	1·416	1·605	2·069
	3	·5070	·7490	1·023	1·218	1·411	1·670	1·872	2·370
	4	·6318	·8797	1·160	1·362	1·562	1·831	2·042	2·563
	5	·7212	·9736	1·260	1·467	1·673	1·950	2·168	2·707
	6	·7903	1·047	1·339	1·549	1·760	2·045	2·268	2·822
	7	·8463	1·106	1·403	1·618	1·832	2·123	2·350	2·918
	8	·8933	1·156	1·457	1·675	1·894	2·189	2·421	2·999
11	9	0·9336	1·200	1·504	1·725	1·947	2·247	2·483	3·071
	10	·9688	1·238	1·546	1·770	1·994	2·299	2·538	3·134
	11	1·000	1·271	1·583	1·809	2·037	2·345	2·587	3·192
	12	1·028	1·302	1·616	1·845	2·075	2·387	2·632	3·244
	13	1·053	1·329	1·646	1·878	2·110	2·425	2·672	3·292
	14	1·076	1·355	1·674	1·908	2·142	2·460	2·710	3·336
	15	1·098	1·378	1·700	1·935	2·171	2·492	2·745	3·377
12	2	0·2967	0·5190	0·7726	0·9513	1·126	1·356	1·531	1·957
	3	·4932	·7264	·9872	1·172	1·353	1·594	1·779	2·232
	4	·6147	·8526	1·119	1·309	1·496	1·744	1·937	2·408
	5	·7016	·9432	1·215	1·408	1·600	1·855	2·053	2·539
	6	·7688	1·014	1·289	1·487	1·682	1·943	2·145	2·643
	7	·8233	1·071	1·351	1·551	1·749	2·015	2·222	2·730
	8	·8690	1·119	1·403	1·606	1·807	2·077	2·287	2·804

w_1, w_2 are the ranges in two independent samples containing n_1, n_2 observations, respectively.

Table 29*b* (*continued*)

Values of α

n_2	n_1	0·500	0·250	0·100	0·050	0·025	0·010	0·005	0·001
12	9	0·9082	1·161	1·447	1·653	1·857	2·131	2·344	2·868
	10	·9424	1·198	1·487	1·695	1·902	2·179	2·395	2·926
	11	·9728	1·230	1·522	1·732	1·941	2·222	2·440	2·978
	12	1·000	1·260	1·554	1·766	1·977	2·261	2·482	3·025
	13	1·025	1·286	1·583	1·797	2·010	2·296	2·519	3·069
	14	1·047	1·310	1·609	1·825	2·040	2·329	2·554	3·109
	15	1·068	1·333	1·634	1·851	2·068	2·359	2·586	3·146
13	2	0·2895	0·5056	0·7505	0·9220	1·088	1·306	1·471	1·866
	3	·4814	·7071	·9576	1·134	1·305	1·531	1·704	2·121
	4	·5999	·8295	1·084	1·264	1·440	1·673	1·851	2·284
	5	·6848	·9174	1·176	1·359	1·539	1·777	1·960	2·404
	6	·7503	·9856	1·248	1·434	1·617	1·860	2·047	2·501
	7	·8035	1·041	1·307	1·496	1·681	1·928	2·118	2·581
	8	·8481	1·088	1·357	1·547	1·736	1·986	2·179	2·648
13	9	0·8863	1·128	1·400	1·593	1·783	2·037	2·232	2·708
	10	·9198	1·164	1·438	1·633	1·825	2·082	2·280	2·761
	11	·9494	1·195	1·471	1·668	1·863	2·122	2·322	2·809
	12	·9760	1·224	1·502	1·701	1·897	2·158	2·360	2·852
	13	1·000	1·249	1·530	1·730	1·928	2·192	2·396	2·892
	14	1·022	1·273	1·555	1·757	1·956	2·222	2·428	2·929
	15	1·042	1·295	1·579	1·782	1·983	2·250	2·458	2·963
14	2	0·2833	0·4940	0·7316	0·8970	1·057	1·264	1·420	1·792
	3	·4710	·6904	·9322	1·101	1·265	1·479	1·641	2·030
	4	·5871	·8096	1·055	1·227	1·394	1·613	1·781	2·182
	5	·6701	·8950	1·144	1·318	1·488	1·712	1·883	2·295
	6	·7342	·9614	1·213	1·390	1·562	1·791	1·965	2·384
	7	·7863	1·016	1·270	1·449	1·624	1·855	2·032	2·458
	8	·8299	1·061	1·318	1·499	1·676	1·910	2·090	2·522
14	9	0·8673	1·100	1·359	1·542	1·721	1·958	2·139	2·577
	10	·9000	1·135	1·396	1·580	1·761	2·001	2·184	2·626
	11	·9290	1·165	1·428	1·614	1·797	2·038	2·224	2·671
	12	·9550	1·193	1·458	1·645	1·829	2·073	2·260	2·711
	13	·9785	1·218	1·484	1·673	1·859	2·105	2·293	2·748
	14	1·000	1·241	1·509	1·699	1·886	2·133	2·324	2·782
	15	1·020	1·262	1·532	1·723	1·911	2·160	2·352	2·814
15	2	0·2778	0·4838	0·7151	0·8753	1·029	1·228	1·377	1·729
	3	·4620	·6758	·9101	1·073	1·230	1·434	1·588	1·954
	4	·5757	·7922	1·029	1·194	1·354	1·562	1·721	2·098
	5	·6571	·8755	1·115	1·282	1·445	1·657	1·818	2·203
	6	·7201	·9403	1·182	1·352	1·516	1·731	1·896	2·287
	7	·7711	·9930	1·237	1·408	1·575	1·793	1·959	2·357
	8	·8138	1·037	1·284	1·456	1·625	1·845	2·014	2·416
15	9	0·8505	1·076	1·324	1·498	1·668	1·892	2·061	2·468
	10	·8826	1·109	1·359	1·535	1·707	1·932	2·104	2·514
	11	·9110	1·139	1·391	1·568	1·741	1·968	2·141	2·556
	12	·9365	1·166	1·419	1·598	1·772	2·001	2·176	2·594
	13	·9596	1·190	1·445	1·625	1·800	2·031	2·207	2·628
	14	·9807	1·212	1·469	1·649	1·826	2·058	2·236	2·660
	15	1·000	1·233	1·491	1·673	1·850	2·084	2·263	2·690

Table 29*c*. *Percentage points of the ratio of range to standard deviation, w/s, where w and s are derived from the same sample of n observations*

Size of sample n	Lower percentage points						Upper percentage points					
	0·0	**0·5**	**1·0**	**2·5**	**5·0**	**10·0**	**10·0**	**5·0**	**2·5**	**1·0**	**0·5**	**0·0**
3	1·732	1·735	1·737	1·745	1·758	1·782	1·997	1·999	2·000	2·000	2·000	2·000
4	1·732	1·83	1·87	1·93	1·98	2·04	2·409	2·429	2·439	2·445	2·447	2·449
5	1·826	1·98	2·02	2·09	2·15	2·22	2·712	2·753	2·782	2·803	2·813	2·828
6	1·826	2·11	2·15	2·22	2·28	2·37	2·949	3·012	3·056	3·095	3·115	3·162
7	1·871	2·22	2·26	2·33	2·40	2·49	3·143	3·222	3·282	3·338	3·369	3·464
8	1·871	2·31	2·35	2·43	2·50	2·59	3·308	3·399	3·471	3·543	3·585	3·742
9	1·897	2·39	2·44	2·51	2·59	2·68	3·449	3·552	3·634	3·720	3·772	4·000
10	1·897	2·46	2·51	2·59	2·67	2·76	3·57	3·685	3·777	3·875	3·935	4·243
11	1·915	2·53	2·58	2·66	2·74	2·84	3·68	3·80	3·903	4·012	4·079	4·472
12	1·915	2·59	2·64	2·72	2·80	2·90	3·78	3·91	4·02	4·134	4·208	4·690
13	1·927	2·64	2·70	2·78	2·86	2·96	3·87	4·00	4·12	4·244	4·325	4·899
14	1·927	2·70	2·75	2·83	2·92	3·02	3·95	4·09	4·21	4·34	4·431	5·099
15	1·936	2·74	2·80	2·88	2·97	3·07	4·02	4·17	4·29	4·44	4·53	5·292
16	1·936	2·79	2·84	2·93	3·01	3·12	4·09	4·24	4·37	4·52	4·62	5·477
17	1·944	2·83	2·88	2·97	3·06	3·17	4·15	4·31	4·44	4·60	4·70	5·657
18	1·944	2·87	2·92	3·01	3·10	3·21	4·21	4·37	4·51	4·67	4·78	5·831
19	1·949	2·90	2·96	3·05	3·14	3·25	4·27	4·43	4·57	4·74	4·85	6·000
20	1·949	2·94	2·99	3·09	3·18	3·29	4·32	4·49	4·63	4·80	4·91	6·164
25	1·961	3·09	3·15	3·24	3·34	3·45	4·53	4·71	4·87	5·06	5·19	6·93
30	1·966	3·21	3·27	3·37	3·47	3·59	4·70	4·89	5·06	5·26	5·40	7·62
35	1·972	3·32	3·38	3·48	3·58	3·70	4·84	5·04	5·21	5·42	5·57	8·25
40	1·975	3·41	3·47	3·57	3·67	3·79	4·96	5·16	5·34	5·56	5·71	8·83
45	1·978	3·49	3·55	3·66	3·75	3·88	5·06	5·26	5·45	5·67	5·83	9·38
50	1·980	3·56	3·62	3·73	3·83	3·95	5·14	5·35	5·54	5·77	5·93	9·90
55	1·982	3·62	3·69	3·80	3·90	4·02	5·22	5·43	5·63	5·86	6·02	10·39
60	1·983	3·68	3·75	3·86	3·96	4·08	5·29	5·51	5·70	5·94	6·10	10·86
65	1·985	3·74	3·80	3·91	4·01	4·14	5·35	5·57	5·77	6·01	6·17	11·31
70	1·986	3·79	3·85	3·96	4·06	4·19	5·41	5·63	5·83	6·07	6·24	11·75
75	1·987	3·83	3·90	4·01	4·11	4·24	5·46	5·68	5·88	6·13	6·30	12·17
80	1·987	3·88	3·94	4·05	4·16	4·28	5·51	5·73	5·93	6·18	6·35	12·57
85	1·988	3·92	3·99	4·09	4·20	4·33	5·56	5·78	5·98	6·23	6·40	12·96
90	1·989	3·96	4·02	4·13	4·24	4·36	5·60	5·82	6·03	6·27	6·45	13·34
95	1·990	3·99	4·06	4·17	4·27	4·40	5·64	5·86	6·07	6·32	6·49	13·71
100	1·990	4·03	4·10	4·21	4·31	4·44	5·68	5·90	6·11	6·36	6·53	14·07
150	1·993	4·32	4·38	4·48	4·59	4·72	5·96	6·18	6·39	6·64	6·82	17·26
200	1·995	4·53	4·59	4·68	4·78	4·90	6·15	6·39	6·60	6·84	7·01	19·95
500	1·998	5·06	5·13	5·25	5·37	5·49	6·72	6·94	7·15	7·42	7·60	31·59
1000	1·999	5·50	5·57	5·68	5·79	5·92	7·11	7·33	7·54	7·80	7·99	44·70

Table 30. *Tables for analysis of variance based on range*

A. *Scale factor, c, and equivalent degrees of freedom, ν, appropriate to a simple classification into k groups of n observations*

k \ n	2		3		4		5		6		7		8		9		10	
	ν	c	ν	c	ν	c	ν	c	ν	c	ν	c	ν	c	ν	c	ν	c
1	1·0	1·41	2·0	1·91	2·9	2·24	3·8	2·48	4·7	2·67	5·5	2·83	6·3	2·96	7·0	3·08	7·7	3·18
2	1·9	1·28	3·8	1·81	5·7	2·15	7·5	2·40	9·2	2·60	10·8	2·77	12·3	2·91	13·8	3·02	15·1	3·13
3	2·8	1·23	5·7	1·77	8·4	2·12	11·1	2·38	13·6	2·58	16·0	2·75	18·3	2·89	20·5	3·01	22·6	3·11
4	3·7	1·21	7·5	1·75	11·2	2·11	14·7	2·37	18·1	2·57	21·3	2·74	24·4	2·88	27·3	3·00	30·1	3·10
5	4·6	1·19	9·3	1·74	13·9	2·10	18·4	2·36	22·6	2·56	26·6	2·73	30·4	2·87	34·0	2·99	37·5	3·10
10	9·0	1·16	18·4	1·72	27·6	2·08	36·5	2·34	44·9	2·55	52·9	2·72	60·6	2·86	67·8	2·98	74·8	3·09
d_n		1·13		1·69		2·06		2·33		2·53		2·70		2·85		2·97		3·08
C.D.	0·88		1·82		2·74		3·62		4·47		5·27		6·03		6·76		7·45	

N.B. C.D. = constant difference.

B. *Scale factor, c, and equivalent degrees of freedom, ν, for analysis of double classification, with k blocks and n treatments*

k \ n	2		3		4		5		6		7		8		9	
	ν	c	ν	c	ν	c	ν	c	ν	c	ν	c	ν	c	ν	c
2	1·0	1·00	2·0	1·35	2·9	1·58	3·8	1·75	4·7	1·89	5·5	2·00	6·3	2·10	7·0	2·18
3	1·9	1·05	3·7	1·48	5·6	1·76	7·4	1·96	9·3	2·12	11·3	2·26	13·4	2·37	15·7	2·46
4	2·7	1·07	5·4	1·54	8·2	1·84	11·0	2·06	13·9	2·23	16·9	2·38	20·1	2·50	23·6	2·60
5	3·6	1·08	7·2	1·57	10·9	1·88	14·6	2·12	18·5	2·30	22·4	2·45	26·6	2·57	31·1	2·68
6	4·5	1·09	8·9	1·59	13·6	1·91	18·2	2·15	23·0	2·34	27·9	2·49	33·0	2·62	38·3	2·73
7	5·4	1·09	10·7	1·61	16·3	1·93	21·8	2·18	27·6	2·37	33·3	2·52	39·3	2·65	45·4	2·76
8	6·3	1·10	12·5	1·62	19·0	1·95	25·4	2·20	32·1	2·39	38·7	2·55	45·6	2·68	52·5	2·79
9	7·1	1·10	14·3	1·63	21·7	1·96	29·0	2·21	36·6	2·41	44·0	2·57	51·8	2·70	59·6	2·81
10	8·1	1·10	16·1	1·63	24·4	1·97	32·6	2·22	41·0	2·42	49·3	2·58	57·9	2·71	66·6	2·83
20	16·7	1·11	33·9	1·66	51·5	2·02	68·8	2·28	86·0	2·48	103	2·64	119	2·78	134	2·90
d_n		1·13		1·69		2·06		2·33		2·53		2·70		2·85		2·97
C.D.	0·87		1·80		2·71		3·62		4·50		5·33		6·10		6·79	

Table 31. *Percentage points of the ratio,* $s^2_{max.}/s^2_{min.}$

Upper 5 % points

ν \ k	2	3	4	5	6	7	8	9	10	11	12
2	39·0	87·5	142	202	266	333	403	475	550	626	704
3	15·4	27·8	39·2	50·7	62·0	72·9	83·5	93·9	104	114	124
4	9·60	15·5	20·6	25·2	29·5	33·6	37·5	41·1	44·6	48·0	51·4
5	7·15	10·8	13·7	16·3	18·7	20·8	22·9	24·7	26·5	28·2	29·9
6	5·82	8·38	10·4	12·1	13·7	15·0	16·3	17·5	18·6	19·7	20·7
7	4·99	6·94	8·44	9·70	10·8	11·8	12·7	13·5	14·3	15·1	15·8
8	4·43	6·00	7·18	8·12	9·03	9·78	10·5	11·1	11·7	12·2	12·7
9	4·03	5·34	6·31	7·11	7·80	8·41	8·95	9·45	9·91	10·3	10·7
10	3·72	4·85	5·67	6·34	6·92	7·42	7·87	8·28	8·66	9·01	9·34
12	3·28	4·16	4·79	5·30	5·72	6·09	6·42	6·72	7·00	7·25	7·48
15	2·86	3·54	4·01	4·37	4·68	4·95	5·19	5·40	5·59	5·77	5·93
20	2·46	2·95	3·29	3·54	3·76	3·94	4·10	4·24	4·37	4·49	4·59
30	2·07	2·40	2·61	2·78	2·91	3·02	3·12	3·21	3·29	3·36	3·39
60	1·67	1·85	1·96	2·04	2·11	2·17	2·22	2·26	2·30	2·33	2·36
∞	1·00	1·00	1·00	1·00	1·00	1·00	1·00	1·00	1·00	1·00	1·00

Upper 1 % points

ν \ k	2	3	4	5	6	7	8	9	10	11	12
2	199	448	729	1036	1362	1705	2063	2432	2813	3204	3605
3	47·5	85	120	151	184	21(6)	24(9)	28(1)	31(0)	33(7)	36(1)
4	23·2	37	49	59	69	79	89	97	106	113	120
5	14·9	22	28	33	38	42	46	50	54	57	60
6	11·1	15·5	19·1	22	25	27	30	32	34	36	37
7	8·89	12·1	14·5	16·5	18·4	20	22	23	24	26	27
8	7·50	9·9	11·7	13·2	14·5	15·8	16·9	17·9	18·9	19·8	21
9	6·54	8·5	9·9	11·1	12·1	13·1	13·9	14·7	15·3	16·0	16·6
10	5·85	7·4	8·6	9·6	10·4	11·1	11·8	12·4	12·9	13·4	13·9
12	4·91	6·1	6·9	7·6	8·2	8·7	9·1	9·5	9·9	10·2	10·6
15	4·07	4·9	5·5	6·0	6·4	6·7	7·1	7·3	7·5	7·8	8·0
20	3·32	3·8	4·3	4·6	4·9	5·1	5·3	5·5	5·6	5·8	5·9
30	2·63	3·0	3·3	3·4	3·6	3·7	3·8	3·9	4·0	4·1	4·2
60	1·96	2·2	2·3	2·4	2·4	2·5	2·5	2·6	2·6	2·7	2·7
∞	1·00	1·0	1·0	1·0	1·0	1·0	1·0	1·0	1·0	1·0	1·0

$s^2_{max.}$ is the largest and $s^2_{min.}$ the smallest in a set of k independent mean squares, each based on ν degrees of freedom.

Values in the column $k=2$ and in the rows $\nu=2$ and ∞ are exact. Elsewhere the third digit may be in error by a few units for the 5 % points and several units for the 1 % points. The third digit figures in brackets for $\nu=3$ are the most uncertain.

Note regarding Table 31c, pp. 264–5.

We have added to this 2nd impression of the 3rd edition of this volume a table due to R. T. Leslie and B. M. Brown (*Biometrika*, 1966, **53**, 226–7) giving the upper 5%, 2·5%, 1% and 0·5% points of the ratio w_{max}/w_{min}. Here w_{max} is the largest and w_{min} the smallest in a set of k independent ranges, each derived from a sample of n observations. On the null hypothesis these samples have been drawn from k normal populations having a common variance.

To avoid extensive repagination, Table 31c has been included on pp. 264–5 at the end of the volume.

Table 31*a*. *Percentage points of the ratio* $s^2_{\max.} \Big/ \sum_{t=1}^{k} s^2_t$. *Upper* 5% *points*

ν \ k	2	3	4	5	6	7	8	9	10	12	15	20
1	0·9985	0·9669	0·9065	0·8412	0·7808	0·7271	0·6798	0·6385	0·6020	0·5410	0·4709	0·3894
2	·9750	·8709	·7679	·6838	·6161	·5612	·5157	·4775	·4450	·3924	·3346	·2705
3	·9392	·7977	·6841	·5981	·5321	·4800	·4377	·4027	·3733	·3264	·2758	·2205
4	·9057	·7457	·6287	·5441	·4803	·4307	·3910	·3584	·3311	·2880	·2419	·1921
5	·8772	·7071	·5895	·5065	·4447	·3974	·3595	·3286	·3029	·2624	·2195	·1735
6	0·8534	0·6771	0·5598	0·4783	0·4184	0·3726	0·3362	0·3067	0·2823	0·2439	0·2034	0·1602
7	·8332	·6530	·5365	·4564	·3980	·3535	·3185	·2901	·2666	·2299	·1911	·1501
8	·8159	·6333	·5175	·4387	·3817	·3384	·3043	·2768	·2541	·2187	·1815	·1422
9	·8010	·6167	·5017	·4241	·3682	·3259	·2926	·2659	·2439	·2098	·1736	·1357
10	·7880	·6025	·4884	·4118	·3568	·3154	·2829	·2568	·2353	·2020	·1671	·1303
16	0·7341	0·5466	0·4366	0·3645	0·3135	0·2756	0·2462	0·2226	0·2032	0·1737	0·1429	0·1108
36	·6602	·4748	·3720	·3066	·2612	·2278	·2022	·1820	·1655	·1403	·1144	·0879
144	·5813	·4031	·3093	·2513	·2119	·1833	·1616	·1446	·1308	·1100	·0889	·0675
∞	·5000	·3333	·2500	·2000	·1667	·1429	·1250	·1111	·1000	·0833	·0667	·0500

Upper 1% *points*

ν \ k	2	3	4	5	6	7	8	9	10	12	15	20
1	0·9999	0·9933	0·9676	0·9279	0·8828	0·8376	0·7945	0·7544	0·7175	0·6528	0·5747	0·4799
2	·9950	·9423	·8643	·7885	·7218	·6644	·6152	·5727	·5358	·4751	·4069	·3297
3	·9794	·8831	·7814	·6957	·6258	·5685	·5209	·4810	·4469	·3919	·3317	·2654
4	·9586	·8335	·7212	·6329	·5635	·5080	·4627	·4251	·3934	·3428	·2882	·2288
5	·9373	·7933	·6761	·5875	·5195	·4659	·4226	·3870	·3572	·3099	·2593	·2048
6	0·9172	0·7606	0·6410	0·5531	0·4866	0·4347	0·3932	0·3592	0·3308	0·2861	0·2386	0·1877
7	·8988	·7335	·6129	·5259	·4608	·4105	·3704	·3378	·3106	·2680	·2228	·1748
8	·8823	·7107	·5897	·5037	·4401	·3911	·3522	·3207	·2945	·2535	·2104	·1646
9	·8674	·6912	·5702	·4854	·4229	·3751	·3373	·3067	·2813	·2419	·2002	·1567
10	·8539	·6743	·5536	·4697	·4084	·3616	·3248	·2950	·2704	·2320	·1918	·1501
16	0·7949	0·6059	0·4884	0·4094	0·3529	0·3105	0·2779	0·2514	0·2297	0·1961	0·1612	0·1248
36	·7067	·5153	·4057	·3351	·2858	·2494	·2214	·1992	·1811	·1535	·1251	·0960
144	·6062	·4230	·3251	·2644	·2229	·1929	·1700	·1521	·1376	·1157	·0934	·0709
∞	·5000	·3333	·2500	·2000	·1667	·1429	·1250	·1111	·1000	·0833	·0667	·0500

$s^2_{\max.}$ is the largest in a set of k independent mean squares, s^2_t, each based on ν degrees of freedom. This table is taken, with permission, from Eisenhart, Hastay & Wallis (1947).

Table 31*b*. *Percentage points of the ratio* $w_{\max.} \Big/ \sum_{t=1}^{k} w_t$. *Upper* 5% *points*

n \ k	2	3	4	5	6	7	8	9	10	12	15	20
2	0·962	0·813	0·681	0·581	0·508	0·451	0·407	0·369	0·339	0·290	0·239	0·188
3	·862	·667	·538	·451	·389	·342	·305	·276	·253	·216	·178	·138
4	·803	·601	·479	·398	·342	·300	·267	·241	·220	·188	·154	·119
5	·764	·563	·446	·369	·316	·278	·248	·224	·204	·173	·142	·110
6	0·736	0·539	0·425	0·351	0·300	0·263	0·234	0·211	0·193	0·163	0·134	0·104
7	·717	·521	·410	·338	·288	·253	·225	·203	·185	·157	·129	·099
8	·702	·507	·398	·328	·280	·245	·218	·197	·179	·152	·125	·096
9	·691	·498	·389	·320	·273	·239	·213	·192	·174	·148	·121	·094
10	·682	·489	·382	·314	·267	·234	·208	·188	·172	·146	·119	·091

$w_{\max.}$ is the largest in a set of k independent ranges, w_t, each derived from a sample of n observations. See Introduction for approximation when $10 < k \leqslant 50$.

Table 32. *Test for heterogeneity of variance: percentage points of M*

Upper 5 % points

k \ c_1	0·0	0·5	1·0	1·5	2·0	2·5	3·0	3·5	4·0	4·5	5·0	6·0	7·0	8·0	9·0	10·0	12·0	14·0
3 (a)	5·99	6·47	6·89	7·20	7·38	7·39	7·22	—	—	—	—	—	—	—	—	—	—	—
(b)	5·99	6·22	6·43	6·64	6·84	7·03	7·22	—	—	—	—	—	—	—	—	—	—	—
4 (a)	7·81	8·24	8·63	8·96	9·21	9·38	9·43	9·37	9·18	—	—	—	—	—	—	—	—	—
(b)	7·81	8·00	8·17	8·35	8·52	8·69	8·85	9·02	9·18	—	—	—	—	—	—	—	—	—
5 (a)	9·49	9·88	10·24	10·57	10·86	11·08	11·24	11·32	11·31	11·21	11·02	—	—	—	—	—	—	—
(b)	9·49	9·65	9·80	9·96	10·11	10·27	10·42	10·57	10·72	10·87	11·02	—	—	—	—	—	—	—
6 (a)	11·07	11·43	11·78	12·11	12·40	12·65	12·86	13·01	13·11	13·14	13·10	12·78	—	—	—	—	—	—
(b)	11·07	11·22	11·36	11·51	11·65	11·79	11·94	12·08	12·22	12·36	12·50	12·78	—	—	—	—	—	—
7 (a)	12·59	12·94	13·27	13·59	13·88	14·15	14·38	14·58	14·73	14·83	14·88	14·81	14·49	—	—	—	—	—
(b)	12·59	12·73	12·87	13·00	13·14	13·27	13·41	13·55	13·68	13·82	13·95	14·22	14·49	—	—	—	—	—
8 (a)	14·07	14·40	14·72	15·03	15·32	15·60	15·84	16·06	16·25	16·40	16·51	16·60	16·49	16·16	—	—	—	—
(b)	14·07	14·20	14·33	14·46	14·59	14·72	14·85	14·98	15·11	15·25	15·38	15·64	15·90	16·16	—	—	—	—
9 (a)	15·51	15·83	16·14	16·44	16·73	17·01	17·26	17·49	17·70	17·88	18·03	18·22	18·26	18·12	17·79	—	—	—
(b)	15·51	15·63	15·76	15·89	16·02	16·14	16·27	16·40	16·52	16·65	16·78	17·03	17·29	17·54	17·79	—	—	—
10 (a)	16·92	17·23	17·54	17·83	18·12	18·39	18·65	18·89	19·11	19·31	19·48	19·75	19·89	19·89	19·73	19·40	—	—
(b)	16·92	17·04	17·17	17·29	17·41	17·54	17·66	17·79	17·91	18·04	18·16	18·41	18·66	18·91	19·16	19·40	—	—
11 (a)	18·31	18·61	18·91	19·20	19·48	19·76	20·02	20·26	20·49	20·70	20·89	21·21	21·42	21·52	21·49	21·32	—	—
(b)	18·31	18·43	18·55	18·67	18·79	18·91	19·04	19·16	19·28	19·40	19·52	19·77	20·01	20·26	20·50	20·75	—	—
12 (a)	19·68	19·97	20·26	20·55	20·83	21·10	21·36	21·61	21·84	22·06	22·27	22·62	22·88	23·06	23·12	23·07	22·56	—
(b)	19·68	19·79	19·91	20·03	20·15	20·27	20·39	20·51	20·63	20·75	20·87	21·12	21·36	21·60	21·84	22·08	22·56	—
13 (a)	21·03	21·32	21·60	21·89	22·16	22·43	22·69	22·94	23·18	23·40	23·62	23·99	24·30	24·53	24·66	24·70	24·44	—
(b)	21·03	21·14	21·26	21·38	21·50	21·62	21·74	21·85	21·97	22·09	22·21	22·45	22·69	22·92	23·16	23·40	23·88	—
14 (a)	22·36	22·65	22·93	23·21	23·48	23·75	24·01	24·26	24·50	24·73	24·95	25·34	25·68	25·95	26·14	26·25	26·17	25·66
(b)	22·36	22·48	22·60	22·71	22·83	22·95	23·06	23·18	23·30	23·42	23·53	23·77	24·00	24·24	24·48	24·71	25·19	25·66
15 (a)	23·68	23·97	24·24	24·52	24·79	25·05	25·31	25·56	25·80	26·04	26·26	26·67	27·03	27·33	27·56	27·73	27·80	27·50
(b)	23·68	23·80	23·92	24·03	24·15	24·26	24·38	24·50	24·61	24·73	24·85	25·08	25·31	25·55	25·78	26·01	26·48	26·95

$M = \mathrm{N}\log_e\left\{\sum_{t=1}^{k}(\nu_t s_t^2)/\mathrm{N}\right\} - \sum_{t=1}^{k}(\nu_t \log_e s_t^2)$ $\mathrm{N} = \sum_{t=1}^{k}\nu_t$ $c_1 = \sum_{t=1}^{k}\frac{1}{\nu_t} - \frac{1}{\mathrm{N}}$ N.B. $\log_e x = 2{\cdot}3026 \log_{10} x$.

s_t^2 $(t = 1, 2, \ldots, k)$ are k independent mean square estimates of a variance, σ^2, based on ν_t degrees of freedom.

Table 32 (*continued*)

Upper 1 % *points*

k \ c_1	0·0	0·5	1·0	1·5	2·0	2·5	3·0	3·5	4·0	4·5	5·0	6·0	7·0	8·0	9·0	10·0	12·0	14·0
3 (*a*)	9·21	9·92	10·47	10·78	10·81	10·50	9·83	—	—	—	—	—	—	—	—	—	—	—
(*b*)	9·21	9·29	9·38	9·48	9·59	9·71	9·83	—	—	—	—	—	—	—	—	—	—	—
4 (*a*)	11·34	11·95	12·46	12·86	13·11	13·18	13·03	12·65	12·03	—	—	—	—	—	—	—	—	—
(*b*)	11·34	11·40	11·46	11·54	11·63	11·72	11·82	11·92	12·03	—	—	—	—	—	—	—	—	—
5 (*a*)	13·28	13·81	14·30	14·71	15·03	15·25	15·34	15·28	15·06	14·66	14·07	—	—	—	—	—	—	—
(*b*)	13·28	13·33	13·39	13·45	13·53	13·61	13·69	13·78	13·87	13·97	14·07	—	—	—	—	—	—	—
6 (*a*)	15·09	15·58	16·03	16·44	16·79	17·07	17·27	17·37	17·37	17·24	16·98	16·03	—	—	—	—	—	—
(*b*)	15·09	15·14	15·20	15·26	15·33	15·41	15·48	15·57	15·65	15·74	15·84	16·03	—	—	—	—	—	—
7 (*a*)	16·81	17·27	17·70	18·10	18·46	18·77	19·02	19·21	19·32	19·35	19·28	18·84	17·92	—	—	—	—	—
(*b*)	16·81	16·87	16·93	16·99	17·06	17·14	17·21	17·29	17·37	17·46	17·55	17·73	17·92	—	—	—	—	—
8 (*a*)	18·48	18·91	19·32	19·71	20·07	20·39	20·67	20·90	21·08	21·20	21·25	21·13	20·64	19·76	—	—	—	—
(*b*)	18·48	18·54	18·60	18·67	18·74	18·81	18·88	18·96	19·04	19·13	19·21	19·39	19·57	19·76	—	—	—	—
9 (*a*)	20·09	20·50	20·90	21·28	21·64	21·97	22·26	22·52	22·74	22·91	23·03	23·10	22·91	22·41	21·56	—	—	—
(*b*)	20·09	20·15	20·22	20·29	20·36	20·44	20·51	20·59	20·67	20·75	20·84	21·01	21·19	21·37	21·56	—	—	—
10 (*a*)	21·67	22·06	22·45	22·82	23·17	23·50	23·80	24·08	24·32	24·52	24·69	24·90	24·90	24·66	24·15	23·33	—	—
(*b*)	21·67	21·73	21·80	21·88	21·95	22·02	22·10	22·18	22·26	22·34	22·42	22·60	22·77	22·95	23·14	23·33	—	—
11 (*a*)	23·21	23·59	23·97	24·33	24·67	25·00	25·31	25·59	25·85	26·08	26·28	26·57	26·70	26·65	26·38	25·86	—	—
(*b*)	23·21	23·28	23·35	23·43	23·50	23·58	23·66	23·74	23·82	23·90	23·98	24·15	24·33	24·51	24·69	24·88	—	—
12 (*a*)	24·72	25·10	25·46	25·81	26·15	26·48	26·79	27·08	27·35	27·59	27·81	28·16	28·39	28·46	28·37	28·07	26·79	—
(*b*)	24·72	24·80	24·87	24·95	25·03	25·11	25·18	25·27	25·35	25·43	25·51	25·68	25·86	26·04	26·22	26·41	26·79	—
13 (*a*)	26·22	26·58	26·93	27·28	27·62	27·94	28·25	28·54	28·81	29·07	29·30	29·70	29·99	30·16	30·19	30·06	29·22	—
(*b*)	26·22	26·29	26·37	26·45	26·53	26·61	26·69	26·77	26·85	26·94	27·02	27·19	27·37	27·55	27·73	27·91	28·29	—
14 (*a*)	27·69	28·04	28·39	28·73	29·06	29·38	29·69	29·98	30·26	30·52	30·77	31·19	31·53	31·77	31·89	31·88	31·39	30·16
(*b*)	27·69	27·77	27·85	27·93	28·01	28·09	28·17	28·25	28·34	28·42	28·51	28·68	28·86	29·03	29·22	29·40	29·77	30·16
15 (*a*)	29·14	29·49	29·83	30·16	30·49	30·80	30·11	31·40	31·68	31·95	32·20	32·66	33·03	33·32	33·51	33·59	33·37	32·52
(*b*)	29·14	29·22	29·30	29·38	29·47	29·55	29·63	29·72	29·80	29·89	29·97	30·15	30·32	30·50	30·69	30·87	31·24	31·62

$M = \mathrm{N}\log_e\left\{\sum_{t=1}^{k}(\nu_t s_t^2)/\mathrm{N}\right\} - \sum_{t=1}^{k}(\nu_t \log_e s_t^2)$ $\quad \mathrm{N} = \sum_{t=1}^{k}\nu_t$ $\quad c_1 = \sum_{t=1}^{k}\frac{1}{\nu_t} - \frac{1}{\mathrm{N}}$ $\quad$ N.B. $\log_e x = 2{\cdot}3026 \log_{10} x$.

s_t^2 $(t = 1, 2, \ldots, k)$ are k independent mean square estimates of a variance, σ^2, based on ν_t degrees of freedom.

For Table 32*a* giving percentage points when all $\nu_t = \nu$ are the same see pp. 266–7.

Table 33. *Test for heterogeneity of variance: table to facilitate interpolation in Table 32*

k \ c_1		0·5	1·0	1·5	2·0	2·5	3·0	3·5	4·0	4·5	5·0	6·0	7·0	8·0	10·0	12·0	14·0
3	C	0·014	0·111	0·375	0·889	1·736	3·000	—	—	—	—	—	—	—	—	—	—
	ΔC	0·486	0·889	1·125	1·111	0·764	0·000	—	—	—	—	—	—	—	—	—	—
4	C	0·008	0·062	0·211	0·500	0·977	1·688	2·680	4·000	—	—	—	—	—	—	—	—
	ΔC	0·492	0·938	1·289	1·500	1·523	1·312	0·820	0·000	—	—	—	—	—	—	—	—
5	C	0·005	0·040	0·135	0·320	0·625	1·080	1·715	2·560	3·645	5·000	—	—	—	—	—	—
	ΔC	0·495	0·960	1·365	1·680	1·875	1·920	1·785	1·440	0·855	0·000	—	—	—	—	—	—
6	C	0·003	0·028	0·094	0·222	0·434	0·750	1·191	1·778	2·531	3·472	6·000	—	—	—	—	—
	ΔC	0·497	0·972	1·406	1·778	2·066	2·250	2·309	2·222	1·969	1·528	0·000	—	—	—	—	—
7	C	0·003	0·020	0·069	0·163	0·319	0·551	0·875	1·306	1·860	2·551	4·408	7·000	—	—	—	—
	ΔC	0·497	0·980	1·431	1·837	2·181	2·449	2·625	2·694	2·640	2·449	1·592	0·000	—	—	—	—
8	C	0·002	0·016	0·053	0·125	0·244	0·422	0·670	1·000	1·424	1·953	3·375	5·359	8·000	—	—	—
	ΔC	0·498	0·984	1·447	1·875	2·256	2·578	2·830	3·000	3·076	3·047	2·625	1·641	0·000	—	—	—
9	C	0·002	0·012	0·042	0·099	0·193	0·333	0·529	0·790	1·125	1·543	2·667	4·235	6·321	—	—	—
	ΔC	0·498	0·988	1·458	1·901	2·307	2·667	2·971	3·210	3·375	3·457	3·333	2·765	1·679	—	—	—
10	C	0·001	0·010	0·034	0·080	0·156	0·270	0·429	0·640	0·911	1·250	2·160	3·430	5·120	10·000	—	—
	ΔC	0·499	0·990	1·466	1·920	2·344	2·730	3·071	3·360	3·589	3·750	3·840	3·570	2·880	0·000	—	—
11	C	0·001	0·008	0·028	0·066	0·129	0·223	0·354	0·529	0·753	1·033	1·785	2·835	4·231	8·264	—	—
	ΔC	0·499	0·992	1·472	1·934	2·371	2·777	3·146	3·471	3·747	3·967	4·215	4·165	3·769	1·736	—	—
12	C	0·001	0·007	0·023	0·056	0·109	0·188	0·298	0·444	0·633	0·868	1·500	2·382	3·556	6·944	12·000	—
	ΔC	0·499	0·993	1·477	1·944	2·391	2·812	3·202	3·556	3·867	4·132	4·500	4·618	4·444	3·056	0·000	—
13	C	0·001	0·006	0·020	0·047	0·092	0·160	0·254	0·379	0·539	0·740	1·278	2·030	3·030	5·917	10·225	—
	ΔC	0·499	0·994	1·480	1·953	2·408	2·840	3·246	3·621	3·961	4·260	4·722	4·970	4·970	4·083	1·775	—
14	C	0·001	0·005	0·017	0·041	0·080	0·138	0·219	0·327	0·465	0·638	1·102	1·750	2·612	5·102	8·816	14·000
	ΔC	0·499	0·995	1·483	1·959	2·420	2·862	3·281	3·673	4·035	4·362	4·898	5·250	5·388	4·898	3·184	0·000
15	C	0·001	0·004	0·015	0·036	0·069	0·120	0·191	0·284	0·405	0·556	0·960	1·524	2·276	4·444	7·680	12·196
	ΔC	0·499	0·996	1·485	1·964	2·431	2·880	3·309	3·716	4·095	4·444	5·040	5·476	5·724	5·556	4·320	1·804

Table 34. *Tests for departure from normality*

A. *Percentage points of the distribution of a = (mean deviation)/(standard deviation)*

Size of sample n	$n-1$	Percentage points						Mean	Standard deviation
		Upper 1%	Upper 5%	Upper 10%	Lower 10%	Lower 5%	Lower 1%		
11	10	·9359	·9073	·8899	·7409	·7153	·6675	·81805	·05784
16	15	·9137	·8884	·8733	·7452	·7236	·6829	·81128	·04976
21	20	·9001	·8768	·8631	·7495	·7304	·6950	·80792	·04419
26	25	·8901	·8686	·8570	·7530	·7360	·7040	·80590	·04011
31	30	·8827	·8625	·8511	·7559	·7404	·7110	·80456	·03697
36	35	·8769	·8578	·8468	·7583	·7440	·7167	·80360	·03447
41	40	·8722	·8540	·8436	·7604	·7470	·7216	·80289	·03241
46	45	·8682	·8508	·8409	·7621	·7496	·7256	·80233	·03068
51	50	·8648	·8481	·8385	·7636	·7518	·7291	·80188	·02919
61	60	·8592	·8434	·8349	·7662	·7554	·7347	·80122	·02678
71	70	·8549	·8403	·8321	·7683	·7583	·7393	·80074	·02487
81	80	·8515	·8376	·8298	·7700	·7607	·7430	·80038	·02332
91	90	·8484	·8353	·8279	·7714	·7626	·7460	·80010	·02203
101	100	·8460	·8344	·8264	·7726	·7644	·7487	·79988	·02094
201	200	·8322	·8229	·8178	·7796	·7738	·7629	·79888	·01491
301	300	·8260	·8183	·8140	·7828	·7781	·7693	·79855	·01220
401	400	·8223	·8155	·8118	·7847	·7807	·7731	·79838	·01058
501	500	·8198	·8136	·8103	·7861	·7825	·7757	·79828	·00947
601	600	·8179	·8123	·8092	·7873	·7838	·7776	·79822	·00865
701	700	·8164	·8112	·8084	·7878	·7848	·7791	·79817	·00801
801	800	·8152	·8103	·8077	·7885	·7857	·7803	·79813	·00749
901	900	·8142	·8096	·8071	·7890	·7864	·7814	·79811	·00707
1001	1000	·8134	·8090	·8066	·7894	·7869	·7822	·79808	·00670

B. *Percentage points of the distribution of $\sqrt{b_1} = m_3/m_2^{\frac{3}{2}}$*

Size of sample n	Percentage points		Standard deviation	Size of sample n	Percentage points		Standard deviation	Size of sample n	Percentage points		Standard deviation
	5%	1%			5%	1%			5%	1%	
25	·711	1·061	·4354	200	·280	·403	·1706	1000	·127	·180	·0772
30	·662	·986	·4052	250	·251	·360	·1531	1200	·116	·165	·0705
35	·621	·923	·3804	300	·230	·329	·1400	1400	·107	·152	·0653
40	·587	·870	·3596	350	·213	·305	·1298	1600	·100	·142	·0611
45	·558	·825	·3418	400	·200	·285	·1216	1800	·095	·134	·0576
50	·534	·787	·3264	450	·188	·269	·1147	2000	·090	·127	·0547
				500	·179	·255	·1089				
60	·492	·723	·3009	550	·171	·243	·1039	2500	·080	·114	·0489
70	·459	·673	·2806	600	·163	·233	·0995	3000	·073	·104	·0447
80	·432	·631	·2638	650	·157	·224	·0956	3500	·068	·096	·0414
90	·409	·596	·2498	700	·151	·215	·0922	4000	·064	·090	·0387
100	·389	·567	·2377	750	·146	·208	·0891	4500	·060	·085	·0365
				800	·142	·202	·0863	5000	·057	·081	·0346
125	·350	·508	·2139	850	·138	·196	·0837				
150	·321	·464	·1961	900	·134	·190	·0814				
175	·298	·430	·1820	950	·130	·185	·0792				
200	·280	·403	·1706	1000	·127	·180	·0772				

N.B. As the sampling distribution of $\sqrt{b_1}$ is symmetrical about zero, the same values, with negative sign, correspond to the lower limits.

Table 34. *Tests for departure from normality (continued)*

C. *Percentage points of the distribution of* $b_2 = m_4/m_2^2$

Size of sample n	Percentage points Upper 1%	Upper 5%	Lower 5%	Lower 1%	Size of sample n	Percentage points Upper 1%	Upper 5%	Lower 5%	Lower 1%
50	4·88	3·99	2·15	1·95	700	3·50	3·31	2·72	2·62
75	4·59	3·87	2·27	2·08	800	3·46	3·29	2·74	2·65
100	4·39	3·77	2·35	2·18	900	3·43	3·28	2·75	2·66
125	4·24	3·71	2·40	2·24	1000	3·41	3·26	2·76	2·68
150	4·13	3·65	2·45	2·29					
200	3·98	3·57	2·51	2·37	1200	3·37	3·24	2·78	2·71
250	3·87	3·52	2·55	2·42	1400	3·34	3·22	2·80	2·72
300	3·79	3·47	2·59	2·46	1600	3·32	3·21	2·81	2·74
350	3·72	3·44	2·62	2·50	1800	3·30	3·20	2·82	2·76
400	3·67	3·41	2·64	2·52	2000	3·28	3·18	2·83	2·77
450	3·63	3·39	2·66	2·55					
500	3·60	3·37	2·67	2·57	2500	3·25	3·16	2·85	2·79
550	3·57	3·35	2·69	2·58	3000	3·22	3·15	2·86	2·81
600	3·54	3·34	2·70	2·60	3500	3·21	3·14	2·87	2·82
650	3·52	3·33	2·71	2·61	4000	3·19	3·13	2·88	2·83
					4500	3·18	3·12	2·88	2·84
					5000	3·17	3·12	2·89	2·85

Table 35. *Moments of* $s/\sigma = \chi/\sqrt{\nu}$ *and factors for determining confidence limits for* σ

Degrees of freedom ν	Moments of s/σ (or $\chi/\sqrt{\nu}$) Expectation	S.D.	S.D. $\times \sqrt{(2\nu)}$	β_1	β_2	Factors for confidence limits of σ: $1-2\alpha=0·95$ Lower	$1-2\alpha=0·95$ Upper	$1-2\alpha=0·99$ Lower	$1-2\alpha=0·99$ Upper
(1)	(2)	(3)	(4)	(5)	(6)	(7)	(8)	(9)	(10)
1	0·797 885	0·60281	0·8525	0·9906	3·8692	0·446	31·91	0·356	159·58
2	·886 227	·46325	·9265	·3983	3·2451	·521	6·28	·434	14·12
3	·921 318	·38881	·9524	·2359	3·1082	·566	3·73	·483	6·47
4	·939 986	·34121	·9651	·1646	3·0593	·599	2·87	·519	4·40
5	·951 533	·30755	·9725	·1255	3·0370	·624	2·45	·546	3·48
6	0·959 369	0·28216	0·9774	0·1011	3·0251	0·644	2·20	0·569	2·98
7	·965 030	·26214	·9808	·0845	3·0181	·661	2·04	·588	2·66
8	·969 311	·24584	·9834	·0725	3·0136	·675	1·92	·604	2·44
9	·972 659	·23224	·9853	·0634	3·0106	·688	1·83	·618	2·28
10	·975 350	·22066	·9868	·0564	3·00852	·699	1·75	·630	2·15
11	0·977 559	0·21066	0·9881	0·0507	3·00697	0·708	1·70	0·641	2·06
12	·979 406	·20190	·9891	·0461	3·00581	·717	1·65	·651	1·98
13	·980 971	·19415	·9900	·0422	3·00492	·725	1·61	·660	1·91
14	·982 316	·18723	·9907	·0390	3·00421	·732	1·58	·669	1·85
15	·983 484	·18100	·9914	·0362	3·00365	·739	1·55	·676	1·81
16	0·984 506	0·17535	0·9919	0·0337	3·00319	0·745	1·52	0·683	1·76
17	·985 410	·17020	·9924	·0316	3·00281	·750	1·50	·690	1·73
18	·986 214	·16547	·9928	·0297	3·00250	·756	1·48	·696	1·70
19	·986 934	·16112	·9932	·0281	3·00223	·760	1·46	·702	1·67
20	·987 583	·15710	·9936	·0266	3·00201	·765	1·44	·707	1·64
25	0·990 052	0·14070	0·9949	0·0210	3·00127	0·784	1·38	0·730	1·54
30	·991 703	·12855	·9958	·0174	3·00087	·799	1·34	·748	1·48
35	·992 884	·11909	·9964	·0148	3·00064	·811	1·30	·762	1·43
40	·993 770	·11145	·9968	·0129	3·00049	·821	1·28	·774	1·39
45	·994 460	·10511	·9972	·0114	3·00038	·829	1·26	·784	1·36
50	·995 013	·09975	·9975	·0103	3·00031	·837	1·24	·793	1·34
60	0·995 842	0·09110	0·9979	0·00851	3·00021	0·849	1·22	0·808	1·30
70	·996 435	·08436	·9982	·00727	3·00016	·858	1·20	·820	1·27
80	·996 880	·07893	·9984	·00635	3·00012	·866	1·18	·829	1·25
90	·997 226	·07443	·9986	·00563	3·00009	·873	1·17	·838	1·23
100	·997 503	·07062	·9987	·00506	3·00008	·879	1·16	·845	1·22

Table 36. *Test for the significance of the difference between two Poisson variables*

A. *Lower significance levels for $b < a$ in single-tail test*

$r=a+b$	Nominal significance level					$r=a+b$	Nominal significance level				
	0·10	**0·05**	**0·025**	**0·01**	**0·005**		**0·10**	**0·05**	**0·025**	**0·01**	**0·005**
1						**41**	15	14	13	12	11
2						**42**	16	15	14	13	12
3						**43**	16	15	14	13	12
4	0					**44**	17	16	15	13	13
5	0	0				**45**	17	16	15	14	13
6	0	0	0			**46**	18	16	15	14	13
7	1	0	0	0		**47**	18	17	16	15	14
8	1	1	0	0	0	**48**	19	17	16	15	14
9	2	1	1	0	0	**49**	19	18	17	15	15
10	2	1	1	0	0	**50**	19	18	17	16	15
11	2	2	1	1	0	**51**	20	19	18	16	15
12	3	2	2	1	1	**52**	20	19	18	17	16
13	3	3	2	1	1	**53**	21	20	18	17	16
14	4	3	2	2	1	**54**	21	20	19	18	17
15	4	3	3	2	2	**55**	22	20	19	18	17
16	4	4	3	2	2	**56**	22	21	20	18	17
17	5	4	4	3	2	**57**	23	21	20	19	18
18	5	5	4	3	3	**58**	23	22	21	19	18
19	6	5	4	4	3	**59**	24	22	21	20	19
20	6	5	5	4	3	**60**	24	23	21	20	19
21	7	6	5	4	4	**61**	24	23	22	20	20
22	7	6	5	5	4	**62**	25	24	22	21	20
23	7	7	6	5	4	**63**	25	24	23	21	20
24	8	7	6	5	5	**64**	26	24	23	22	21
25	8	7	7	6	5	**65**	26	25	24	22	21
26	9	8	7	6	6	**66**	27	25	24	23	22
27	9	8	7	7	6	**67**	27	26	25	23	22
28	10	9	8	7	6	**68**	28	26	25	23	22
29	10	9	8	7	7	**69**	28	27	25	24	23
30	10	10	9	8	7	**70**	29	27	26	24	23
31	11	10	9	8	7	**71**	29	28	26	25	24
32	11	10	9	8	8	**72**	30	28	27	25	24
33	12	11	10	9	8	**73**	30	28	27	26	25
34	12	11	10	9	9	**74**	30	29	28	26	25
35	13	12	11	10	9	**75**	31	29	28	26	25
36	13	12	11	10	9	**76**	31	30	28	27	26
37	14	13	12	10	10	**77**	32	30	29	27	26
38	14	13	12	11	10	**78**	32	31	29	28	27
39	15	13	12	11	11	**79**	33	31	30	28	27
40	15	14	13	12	11	**80**	33	32	30	29	28

B. *True significance levels resulting from long-run use of Table* 36 A *according to the nominal level and the value, m, of the common Poisson expectation*

m	Nominal significance level					m
	0·10	0·05	0·025	0·01	0·005	
2·5	0·032	0·012	0·004	0·0012	0·0003	2·5
5·0	·054	·023	·011	·0033	·0015	5·0
7·5	·061	·030	·014	·0044	·0021	7·5
10·0	·066	·032	·015	·0055	·0024	10·0
12·5	·068	·034	·015	·0059	·0027	12·5
15·0	0·071	0·035	0·015	0·0060	0·0029	15·0
17·5	·074	·036	·016	·0061	·0031	17·5
20·0	·076	·037	·017	·0065	·0032	20·0
22·5	·077	·037	·018	·0068	·0033	22·5
25·0	·077	·038	·018	·0070	·0034	25·0

Table 37. *Individual terms of certain binomial distributions*: $f(i \mid n, p) = \binom{n}{i} p^i (1-p)^{n-i}$

n	i \ p	0·01	0·02	0·04	0·06	0·08	0·1	0·2	0·3	0·4	0·5	i	n
5	0	0·95099	0·90392	0·81537	0·73390	0·65908	0·59049	0·32768	0·16807	0·07776	0·03125	5	5
	1	·04803	·09224	·16987	·23422˙	·28656	·32805	·40960	·36015	·25920	·15625	4	
	2	·00097	·00376	·01416	·02990	·04984	·07290	·20480	·30870	·34560	·31250	3	
	3	·00001	·00008	·00059	·00191	·00433	·00810	·05120	·13230	·23040	·31250	2	
	4			·00001	·00006	·00019	·00045	·00640	·02835	·07680	·15625	1	
	5						·00001	·00032	·00243	·01024	·03125	0	
10	0	0·90438˙	0·81707	0·66483	0·53862.	0·43439	0·34868	0·10737	0·02825	0·00605	0·00098	10	10
	1	·09135	·16675	·27701˙	·34380	·37773	·38742	·26844	·12106	·04031	·00977.	9	
	2	·00415	·01531˙	·05194	·09875	·14781	·19371	·30199	·23347	·12093	·04395.	8	
	3	·00011	·00083	·00577	·01681	·03427	·05740.	·20133	·26683	·21499	·11719	7	
	4		·00003	·00042	·00188	·00522	·01116	·08808	·20012	·25082	·20508	6	
	5			0·00002	0·00014	0·00054	0·00149	0·02642	0·10292	0·20066	0·24609	5	
	6				·00001	·00004	·00014	·00551	·03676	·11148	·20508	4	
	7						·00001	·00079	·00900	·04247	·11719	3	
	8							·00007	·00145	·01062	·04395.	2	
	9								·00014	·00157	·00977	1	
	10								·00001.	·00010	·00098	0	
15	0	0·86006	0·73857	0·54209	0·39529	0·28630	0·20589	0·03518˙	0·00475	0·00047	0·00003	15	15
	1	·13031	·22609	·33880˙	·37847	·37343	·34315	·13194	·03052	·00470	·00046	14	
	2	·00921˙	·03230	·09882	·16910˙	·22731	·26690	·23090	·09156	·02194	·00320	13	
	3	·00040	·00286	·01784	·04677	·08565	·12851	·25014	·17004	·06339	·01389	12	
	4	·00001	·00017	·00223	·00896	·02234	·04284.	·18760	·21862	·12678	·04166	11	
	5		0·00001	0·00020	0·00126	0·00427	0·01047	0·10318	0·20613	0·18594	0·09164	10	
	6			·00001	·00013	·00062	·00194	·04299	·14724	·20660	·15274	9	
	7				·00001	·00007	·00028	·01382	·08113	·17708	·19638	8	
	8					·00001	·00003	·00345˙	·03477	·11806	·19638	7	
	9							·00067	·01159	·06121	·15274	6	
	10							0·00010	0·00298	0·02449	0·09164	5	
	11							·00001	·00058	·00742	·04166	4	
	12								·00008	·00165	·01389	3	
	13								·00001	·00025	·00320	2	
	14									·00002	·00046	1	
	15										·00003	0	
20	0	0·81791	0·66761	0·44200	0·29011	0·18869	0·12158	0·01153	0·00080	0·00004		20	20
	1	·16523	·27249	·36834	·37035	·32816	·27017	·05765	·00684	·00049	0·00002	19	
	2	·01586	·05283	·14580	·22457	·27109	·28518	·13691	·02785.	·00309	·00018	18	
	3	·00096	·00647	·03645	·08601	·14144	·19012	·20536	·07160	·01235	·00109	17	
	4	·00004	·00056	·00645	·02333	·05227	·08978	·21820	·13042	·03499	·00462	16	
	5		0·00004	0·00086	0·00477.	0·01454	0·03192	0·17456	0·17886	0·07465	0·01479	15	
	6			·00009	·00076	·00316	·00887	·10910	·19164	·12441	·03696	14	
	7			·00001	·00010	·00055	·00197	·05455	·16426	·16588	·07393	13	
	8				·00001	·00008	·00036.	·02216	·11440	·17971	·12013	12	
	9					·00001	·00005	·00739	·06537	·15974	·16018	11	
	10						0·00001	0·00203	0·03082	0·11714	0·17620	10	
	11							·00046	·01201	·07099˙	·16018	9	
	12							·00009	·00386	·03550	·12013	8	
	13							·00001	·00102	·01456	·07393	7	
	14								·00022	·00485	·03696	6	
	15								0·00004	0·00129	0·01479	5	
	16								·00001.	·00027	·00462	4	
	17									·00004	·00109	3	
	18										·00018	2	
	19										·00002	1	
	20											0	
		0·99	0·98	0·96	0·94	0·92	0·9	0·8	0·7	0·6	0·5	p \ i	n

Where the totals of the individual terms differ from unity owing to the incidence of rounding off errors, an adjustment involving a minimum percentage error may be made as follows: entries marked with a high dot (e.g. 6˙) should be increased by a unit in the 5th decimal and those marked with a low dot (e.g. 4.) decreased by a unit.

Table 37 (*continued*)

n	i \ p	0·01	0·02	0·04	0·06	0·08	0·1	0·2	0·3	0·4	0·5	i	n
25	0	0·77782	0·60346˙	0·36040	0·21291	0·12436	0·07179	0·00378	0·00013			25	25
	1	·19642	·30789	·37541	·33975	·27036	·19942	·02361	·00144	0·00005		24	
	2	·02381	·07540	·18771	·26023	·28211	·26589	·07084	·00739	·00038	0·00001	23	
	3	·00184	·01180	·05996	·12735	·18807˙	·22650	·13577	·02428	·00194	·00007	22	
	4	·00010	·00132	·01374	·04471	·08995	·13842.	·18668	·05723	·00710	·00038	21	
	5	0·00000˙	0·00011	0·00240˙	0·01199	0·03285	0·06459	0·19602	0·10302	0·01989	0·00158	20	
	6		·00001	·00033	·00255	·00952	·02392	·16335	·14717	·04420	·00528	19	
	7			·00004	·00044	·00225	·00722	·11084	·17119	·07999	·01433	18	
	8				·00006	·00044	·00180	·06235	·16508	·11998	·03223	17	
	9				·00001	·00007	·00038	·02944	·13364	·15109.	·06089.	16	
	10					0·00001	0·00007	0·01178	0·09164	0·16116	0·09742	15	
	11						·00001	·00401	·05355	·14651	·13284	14	
	12							·00117	·02678	·11395	·15498	13	
	13							·00029	·01148	·07597	·15498	12	
	14							·00006	·00422	·04341	·13284	11	
	15							0·00001	0·00132	0·02122	0·09742	10	
	16								·00035	·00884	·06089.	9	
	17								·00008	·00312	·03223	8	
	18								·00002.	·00092	·01433	7	
	19									·00023	·00528	6	
	20									0·00005	0·00158	5	
	21									·00001	·00038	4	
	22										·00007	3	
	23										·00001	2	
30	0	0·73970	0·54548˙	0·29386	0·15626	0·08197	0·04239	0·00124	0·00002			30	30
	1	·22415	·33397	·36732	·29921	·21382˙	·14130	·00928˙	·00029			29	
	2	·03283	·09883	·22192˙	·27693	·26961	·22766	·03366	·00180	0·00004		28	
	3	·00310	·01882	·08630	·16498	·21881	·23609	·07853	·00720	·00027		27	
	4	·00021	·00259	·02427	·07108	·12843	·17707	·13252	·02084	·00120	0·00003	26	
	5	0·00001	0·00028	0·00526	0·02359	0·05807	0·10230˙	0·17228	0·04644	0·00415	0·00013	25	
	6		·00002	·00091	·00627˙	·02104	·04736	·17946	·08293	·01152	·00055	24	
	7			·00013	·00137	·00627	·01804	·15382	·12185	·02634	·00190	23	
	8			·00002	·00025	·00157	·00576	·11056	·15014	·05049	·00545	22	
	9				·00004	·00033	·00157	·06756	·15729	·08228	·01332	21	
	10				0·00001	0·00006	0·00037	0·03547	0·14156	0·11519	0·02798	20	
	11					·00001	·00007	·01612	·11031	·13962	·05088	19	
	12						·00001	·00638	·07485	·14738.	·08055	18	
	13							·00221	·04442	·13604	·11154	17	
	14							·00067	·02312.	·11013	·13544	16	
	15							0·00018	0·01057	0·07831	0·14446	15	
	16							·00004	·00425	·04895.	·13544	14	
	17							·00001	·00150	·02687	·11154	13	
	18								·00046	·01294	·08055	12	
	19								·00013	·00545	·05088	11	
	20								0·00003	0·00200	0·02798	10	
	21								·00001	·00063	·01332	9	
	22									·00017	·00545	8	
	23									·00004	·00190	7	
	24									·00001	·00055	6	
	25										0·00013	5	
	26										·00003	4	
		0·99	0·98	0·96	0·94	0·92	0·9	0·8	0·7	0·6	0·5	p \ i	n

27-2

Table 38. *Significance tests in a* 2 × 2 *contingency table*

	a	Probability			
		0·05	0·025	0·01	0·005
A=3 B=3	3	**0** ·050	—	—	—
A=4 B=4	4	**0** ·014	**0** ·014	—	—
3	4	**0** ·029	—	—	—
A=5 B=5	5	**1** ·024	**1** ·024	**0** ·004	**0** ·004
	4	**0** ·024	**0** ·024	—	—
4	5	**1** ·048	**0** ·008	**0** ·008	—
	4	**0** ·040	—	—	—
3	5	**0** ·018	**0** ·018	—	—
2	5	**0** ·048	—	—	—
A=6 B=6	6	**2** ·030	**1** ·008	**1** ·008	**0** ·001
	5	**1** ·040	**0** ·008	**0** ·008	—
	4	**0** ·030	—	—	—
5	6	**1** ·015+	**1** ·015+	**0** ·002	**0** ·002
	5	**0** ·013	**0** ·013	—	—
	4	**0** ·045+	—	—	—
4	6	**1** ·033	**0** ·005−	**0** ·005−	**0** ·005−
	5	**0** ·024	**0** ·024	—	—
3	6	**0** ·012	**0** ·012	—	—
	5	**0** ·048	—	—	—
2	6	**0** ·036	—	—	—
A=7 B=7	7	**3** ·035−	**2** ·010+	**1** ·002	**1** ·002
	6	**1** ·015−	**1** ·015−	**0** ·002	**0** ·002
	5	**0** ·010+	**0** ·010+	—	—
	4	**0** ·035−	—	—	—
6	7	**2** ·021	**2** ·021	**1** ·005−	**1** ·005−
	6	**1** ·025+	**0** ·004	**0** ·004	**0** ·004
	5	**0** ·016	**0** ·016	—	—
	4	**0** ·049	—	—	—
5	7	**2** ·045+	**1** ·010+	**0** ·001	**0** ·001
	6	**1** ·045+	**0** ·008	**0** ·008	—
	5	**0** ·027	—	—	—
4	7	**1** ·024	**1** ·024	**0** ·003	**0** ·003
	6	**0** ·015+	**0** ·015+	—	—
	5	**0** ·045+	—	—	—
3	7	**0** ·008	**0** ·008	**0** ·008	—
	6	**0** ·033	—	—	—
2	7	**0** ·028	—	—	—
A=8 B=8	8	**4** ·038	**3** ·013	**2** ·003	**2** ·003
	7	**2** ·020	**2** ·020	**1** ·005+	**0** ·001
	6	**1** ·020	**1** ·020	**0** ·003	**0** ·003
	5	**0** ·013	**0** ·013	—	—
	4	**0** ·038	—	—	—
7	8	**3** ·026	**2** ·007	**2** ·007	**1** ·001
	7	**2** ·035−	**1** ·009	**1** ·009	**0** ·001
	6	**1** ·032	**0** ·006	**0** ·006	—
	5	**0** ·019	**0** ·019	—	—
6	8	**2** ·015−	**2** ·015−	**1** ·003	**1** ·003
	7	**1** ·016	**1** ·016	**0** ·002	**0** ·002
	6	**0** ·009	**0** ·009	**0** ·009	—
	5	**0** ·028	—	—	—
5	8	**2** ·035−	**1** ·007	**1** ·007	**0** ·001
	7	**1** ·032	**0** ·005−	**0** ·005−	**0** ·005−
	6	**0** ·016	**0** ·016	—	—
	5	**0** ·044	—	—	—
4	8	**1** ·018	**1** ·018	**0** ·002	**0** ·002
	7	**0** ·010+	**0** ·010+	—	—
	6	**0** ·030	—	—	—
3	8	**0** ·006	**0** ·006	**0** ·006	—
	7	**0** ·024	**0** ·024	—	—
2	8	**0** ·022	**0** ·022	—	—
A=9 B=9	9	**5** ·041	**4** ·015−	**3** ·005−	**3** ·005−
	8	**3** ·025−	**3** ·025−	**2** ·008	**1** ·002
	7	**2** ·028	**1** ·008	**1** ·008	**0** ·001
	6	**1** ·025−	**1** ·025−	**0** ·005−	**0** ·005−
	5	**0** ·015−	**0** ·015−	—	—
	4	**0** ·041	—	—	—
8	9	**4** ·029	**3** ·009	**3** ·009	**2** ·002
	8	**3** ·043	**2** ·013	**1** ·003	**1** ·003
	7	**2** ·044	**1** ·012	**0** ·002	**0** ·002
	6	**1** ·036	**0** ·007	**0** ·007	—
	5	**0** ·020	**0** ·020	—	—
7	9	**3** ·019	**3** ·019	**2** ·005−	**2** ·005−
	8	**2** ·024	**2** ·024	**1** ·006	**0** ·001
	7	**1** ·020	**1** ·020	**0** ·003	**0** ·003
	6	**0** ·010+	**0** ·010+	—	—
	5	**0** ·029	—	—	—
6	9	**3** ·044	**2** ·011	**1** ·002	**1** ·002
	8	**2** ·047	**1** ·011	**0** ·001	**0** ·001
	7	**1** ·035−	**0** ·006	**0** ·006	—
	6	**0** ·017	**0** ·017	—	—
	5	**0** ·042	—	—	—

The table shows: (1) In bold type, for given a, A and B, the value of b ($<a$) which is just significant at the probability level quoted (single-tail test).

(2) In small type, for given A, B and $r=a+b$, the exact probability (if there is independence) that b is equal to or less than the integer shown in bold type.

Table 38 (*continued*)

	a	Probability 0·05	0·025	0·01	0·005
A=9 B=5	9	2 ·027	1 ·005-	1 ·005-	1 ·005-
	8	1 ·023	1 ·023	0 ·003	0 ·003
	7	0 ·010+	0 ·010+	—	—
	6	0 ·028	—	—	—
4	9	1 ·014	1 ·014	0 ·001	0 ·001
	8	0 ·007	0 ·007	0 ·007	—
	7	0 ·021	0 ·021	—	—
	6	0 ·049	—	—	—
3	9	1 ·045+	0 ·005-	0 ·005-	0 ·005-
	8	0 ·018	0 ·018	—	—
	7	0 ·045+	—	—	—
2	9	0 ·018	0 ·018	—	—
A=10 B=10	10	6 ·043	5 ·016	4 ·005+	3 ·002
	9	4 ·029	3 ·010-	3 ·010-	2 ·003
	8	3 ·035-	2 ·012	1 ·003	1 ·003
	7	2 ·035-	1 ·010-	1 ·010-	0 ·002
	6	1 ·029	0 ·005+	0 ·005+	—
	5	0 ·016	0 ·016	—	—
	4	0 ·043	—	—	—
9	10	5 ·033	4 ·011	3 ·003	3 ·003
	9	4 ·050-	3 ·017	2 ·005-	2 ·005-
	8	2 ·019	2 ·019	1 ·004	1 ·004
	7	1 ·015-	1 ·015-	0 ·002	0 ·002
	6	1 ·040	0 ·008	0 ·008	—
	5	0 ·022	0 ·022	—	—
8	10	4 ·023	4 ·023	3 ·007	2 ·002
	9	3 ·032	2 ·009	2 ·009	1 ·002
	8	2 ·031	1 ·008	1 ·008	0 ·001
	7	1 ·023	1 ·023	0 ·004	0 ·004
	6	0 ·011	0 ·011	—	—
	5	0 ·029	—	—	—
7	10	3 ·015-	3 ·015-	2 ·003	2 ·003
	9	2 ·018	2 ·018	1 ·004	1 ·004
	8	1 ·013	1 ·013	0 ·002	0 ·002
	7	1 ·036	0 ·006	0 ·006	—
	6	0 ·017	0 ·017	—	—
	5	0 ·041	—	—	—
6	10	3 ·036	2 ·008	2 ·008	1 ·001
	9	2 ·036	1 ·008	1 ·008	0 ·001
	8	1 ·024	1 ·024	0 ·003	0 ·003
	7	0 ·010+	0 ·010+	—	—
	6	0 ·026	—	—	—
5	10	2 ·022	2 ·022	1 ·004	1 ·004
	9	1 ·017	1 ·017	0 ·002	0 ·002
	8	1 ·047	0 ·007	0 ·007	—
	7	0 ·019	0 ·019	—	—
	6	0 ·042	—	—	—

	a	Probability 0·05	0·025	0·01	0·005
A=10 B=4	10	1 ·011	1 ·011	0 ·001	0 ·001
	9	1 ·041	0 ·005-	0 ·005-	0 ·005-
	8	0 ·015-	0 ·015-	—	—
	7	0 ·035-	—	—	—
3	10	1 ·038	0 ·003	0 ·003	0 ·003
	9	0 ·014	0 ·014	—	—
	8	0 ·035-	—	—	—
2	10	0 ·015+	0 ·015+	—	—
	9	0 ·045+	—	—	—
A=11 B=11	11	7 ·045+	6 ·018	5 ·006	4 ·002
	10	5 ·032	4 ·012	3 ·004	3 ·004
	9	4 ·040	3 ·015-	2 ·004	2 ·004
	8	3 ·043	2 ·015-	1 ·004	1 ·004
	7	2 ·040	1 ·012	0 ·002	0 ·002
	6	1 ·032	0 ·006	0 ·006	—
	5	0 ·018	0 ·018	—	—
	4	0 ·045+	—	—	—
10	11	6 ·035+	5 ·012	4 ·004	4 ·004
	10	4 ·021	4 ·021	3 ·007	2 ·002
	9	3 ·024	3 ·024	2 ·007	1 ·002
	8	2 ·023	2 ·023	1 ·006	0 ·001
	7	1 ·017	1 ·017	0 ·003	0 ·003
	6	1 ·043	0 ·009	0 ·009	—
	5	0 ·023	0 ·023	—	—
9	11	5 ·026	4 ·008	4 ·008	3 ·002
	10	4 ·038	3 ·012	2 ·003	2 ·003
	9	3 ·040	2 ·012	1 ·003	1 ·003
	8	2 ·035-	1 ·009	1 ·009	0 ·001
	7	1 ·025-	1 ·025-	0 ·004	0 ·004
	6	0 ·012	0 ·012	—	—
	5	0 ·030	—	—	—
8	11	4 ·018	4 ·018	3 ·005-	3 ·005-
	10	3 ·024	3 ·024	2 ·006	1 ·001
	9	2 ·022	2 ·022	1 ·005-	1 ·005-
	8	1 ·015-	1 ·015-	0 ·002	0 ·002
	7	1 ·037	0 ·007	0 ·007	—
	6	0 ·017	0 ·017	—	—
	5	0 ·040	—	—	—
7	11	4 ·043	3 ·011	2 ·002	2 ·002
	10	3 ·047	2 ·013	1 002	1 ·002
	9	2 ·039	1 ·009	1 ·009	0 ·001
	8	1 ·025-	1 ·025-	0 ·004	0 ·004
	7	0 ·010+	0 ·010+	—	—
	6	0 ·025-	0 ·025-	—	—
6	11	3 ·029	2 ·006	2 ·006	1 ·001
	10	2 ·028	1 ·005+	1 ·005+	0 ·001
	9	1 ·018	1 ·018	0 ·002	0 ·002

Table 38. *Significance tests in a 2 × 2 contingency table (continued)*

	a	Probability 0·05	0·025	0·01	0·005
A = 11 B = 6	8	**1** ·043	**0** ·007	**0** ·007	—
	7	**0** ·017	**0** ·017	—	—
	6	**0** ·037	—	—	—
5	11	**2** ·018	**2** ·018	**1** ·003	**1** ·003
	10	**1** ·013	**1** ·013	**0** ·001	**0** ·001
	9	**1** ·036	**0** ·005−	**0** ·005−	**0** ·005−
	8	**0** ·013	**0** ·013	—	—
	7	**0** ·029	—	—	—
4	11	**1** ·009	**1** ·009	**1** ·009	**0** ·001
	10	**1** ·033	**0** ·004	**0** ·004	**0** ·004
	9	**0** ·011	**0** ·011	—	—
	8	**0** ·026	—	—	—
3	11	**1** ·033	**0** ·003	**0** ·003	**0** ·003
	10	**0** ·011	**0** ·011	—	—
	9	**0** ·027	—	—	—
2	11	**0** ·013	**0** ·013	—	—
	10	**0** ·038	—	—	—
A = 12 B = 12	12	**8** ·047	**7** ·019	**6** ·007	**5** ·002
	11	**6** ·034	**5** ·014	**4** ·005−	**4** ·005−
	10	**5** ·045−	**4** ·018	**3** ·006	**2** ·002
	9	**4** ·050−	**3** ·020	**2** ·006	**1** ·001
	8	**3** ·050−	**2** ·018	**1** ·005−	**1** ·005−
	7	**2** ·045−	**1** ·014	**0** ·002	**0** ·002
	6	**1** ·034	**0** ·007	**0** ·007	—
	5	**0** ·019	**0** ·019	—	—
	4	**0** ·047	—	—	—
11	12	**7** ·037	**6** ·014	**5** ·005−	**5** ·005−
	11	**5** ·024	**5** ·024	**4** ·008	**3** ·002
	10	**4** ·029	**3** ·010+	**2** ·003	**2** ·003
	9	**3** ·030	**2** ·009	**2** ·009	**1** ·002
	8	**2** ·026	**1** ·007	**1** ·007	**0** ·001
	7	**1** ·019	**1** ·019	**0** ·003	**0** ·003
	6	**1** ·045−	**0** ·009	**0** ·009	—
	5	**0** ·024	**0** ·024	—	—
10	12	**6** ·029	**5** ·010−	**5** ·010−	**4** ·003
	11	**5** ·043	**4** ·015+	**3** ·005−	**3** ·005−
	10	**4** ·048	**3** ·017	**2** ·005−	**2** ·005−
	9	**3** ·046	**2** ·015−	**1** ·004	**1** ·004
	8	**2** ·038	**1** ·010+	**0** ·002	**0** ·002
	7	**1** ·026	**0** ·005−	**0** ·005−	**0** ·005−
	6	**0** ·012	**0** ·012	—	—
	5	**0** ·030	—	—	—
9	12	**5** ·021	**5** ·021	**4** ·006	**3** ·002
	11	**4** ·029	**3** ·009	**3** ·009	**2** ·002
	10	**3** ·029	**2** ·008	**2** ·008	**1** ·002
	9	**2** ·024	**2** ·024	**1** ·006	**0** ·001
	8	**1** ·016	**1** ·016	**0** ·002	**0** ·002

	a	Probability 0·05	0·025	0·01	0·005
A = 12 B = 9	7	**1** ·037	**0** ·007	**0** ·007	—
	6	**0** ·017	**0** ·017	—	—
	5	**0** ·039	—	—	—
8	12	**5** ·049	**4** ·014	**3** ·004	**3** ·004
	11	**3** ·018	**3** ·018	**2** ·004	**2** ·004
	10	**2** ·015+	**2** ·015+	**1** ·003	**1** ·003
	9	**2** ·040	**1** ·010−	**1** ·010−	**0** ·001
	8	**1** ·025−	**1** ·025−	**0** ·004	**0** ·004
	7	**0** ·010+	**0** ·010+	—	—
	6	**0** ·024	**0** ·024	—	—
7	12	**4** ·036	**3** ·009	**3** ·009	**2** ·002
	11	**3** ·038	**2** ·010−	**2** ·010−	**1** ·002
	10	**2** ·029	**1** ·006	**1** ·006	**0** ·001
	9	**1** ·017	**1** ·017	**0** ·002	**0** ·002
	8	**1** ·040	**0** ·007	**0** ·007	—
	7	**0** ·016	**0** ·016	—	—
	6	**0** ·034	—	—	—
6	12	**3** ·025−	**3** ·025−	**2** ·005−	**2** ·005−
	11	**2** ·022	**2** ·022	**1** ·004	**1** ·004
	10	**1** ·013	**1** ·013	**0** ·002	**0** ·002
	9	**1** ·032	**0** ·005−	**0** ·005−	**0** ·005−
	8	**0** ·011	**0** ·011	—	—
	7	**0** ·025−	**0** ·025−	—	—
	6	**0** ·050−	—	—	—
5	12	**2** ·015−	**2** ·015−	**1** ·002	**1** ·002
	11	**1** ·010−	**1** ·010−	**1** ·010−	**0** ·001
	10	**1** ·028	**0** ·003	**0** ·003	**0** ·003
	9	**0** ·009	**0** ·009	**0** ·009	—
	8	**0** ·020	**0** ·020	—	—
	7	**0** ·041	—	—	—
4	12	**2** ·050	**1** ·007	**1** ·007	**0** ·001
	11	**1** ·027	**0** ·003	**0** ·003	**0** ·003
	10	**0** ·008	**0** ·008	**0** ·008	—
	9	**0** ·019	**0** ·019	—	—
	8	**0** ·038	—	—	—
3	12	**1** ·029	**0** ·002	**0** ·002	**0** ·002
	11	**0** ·009	**0** ·009	**0** ·009	—
	10	**0** ·022	**0** ·022	—	—
	9	**0** ·044	—	—	—
2	12	**0** ·011	**0** ·011	—	—
	11	**0** ·033	—	—	—
A = 13 B = 13	13	**9** ·048	**8** ·020	**7** ·007	**6** ·003
	12	**7** ·037	**6** ·015+	**5** ·006	**4** ·002
	11	**6** ·048	**5** ·021	**4** ·008	**3** ·002
	10	**4** ·024	**4** ·024	**3** ·008	**2** ·002
	9	**3** ·024	**3** ·024	**2** ·008	**1** ·002
	8	**2** ·021	**2** ·021	**1** ·006	**0** ·001

The table shows: (1) In bold type, for given a, A and B, the value of b ($<a$) which is just significant at the probability level quoted (single-tail test).

(2) In small type, for given A, B and $r=a+b$, the exact probability (if there is independence) that b is equal to or less than the integer shown in bold type.

Table 38 (*continued*)

	a	Probability 0·05	0·025	0·01	0·005
A=13 B=13	7	2 ·048	1 ·015+	0 ·003	0 ·003
	6	1 ·037	0 ·007	0 ·007	—
	5	0 ·020	0 ·020	—	—
	4	0 ·048	—	—	—
12	13	8 ·039	7 ·015−	6 ·005+	5 ·002
	12	6 ·027	5 ·010−	5 ·010−	4 ·003
	11	5 ·033	4 ·013	3 ·004	3 ·004
	10	4 ·036	3 ·013	2 ·004	2 ·004
	9	3 ·034	2 ·011	1 ·003	1 ·003
	8	2 ·029	1 ·008	1 ·008	0 ·001
	7	1 ·020	1 ·020	0 ·004	0 ·004
	6	1 ·046	0 ·010−	0 ·010−	—
	5	0 ·024	0 ·024	—	—
11	13	7 ·031	6 ·011	5 ·003	5 ·003
	12	6 ·048	5 ·018	4 ·006	3 ·002
	11	4 ·021	4 ·021	3 ·007	2 ·002
	10	3 ·021	3 ·021	2 ·006	1 ·001
	9	3 ·050−	2 ·017	1 ·004	1 ·004
	8	2 ·040	1 ·011	0 ·002	0 ·002
	7	1 ·027	0 ·005−	0 ·005−	0 ·005−
	6	0 ·013	0 ·013	—	—
	5	0 ·030	—	—	—
10	13	6 ·024	6 ·024	5 ·007	4 ·002
	12	5 ·035−	4 ·012	3 ·003	3 ·003
	11	4 ·037	3 ·012	2 ·003	2 ·003
	10	3 ·033	2 ·010+	1 ·002	1 ·002
	9	2 ·026	1 ·006	1 ·006	0 ·001
	8	1 ·017	1 ·017	0 ·003	0 ·003
	7	1 ·038	0 ·007	0 ·007	—
	6	0 ·017	0 ·017	—	—
	5	0 ·038	—	—	—
9	13	5 ·017	5 ·017	4 ·005−	4 ·005−
	12	4 ·023	4 ·023	3 ·007	2 ·001
	11	3 ·022	3 ·022	2 ·006	1 ·001
	10	2 ·017	2 ·017	1 ·004	1 ·004
	9	2 ·040	1 ·010+	0 ·001	0 ·001
	8	1 ·025−	1 ·025−	0 ·004	0 ·004
	7	0 ·010+	0 ·010+	—	—
	6	0 ·023	0 ·023	—	—
	5	0 ·049	—	—	—
8	13	5 ·042	4 ·012	3 ·003	3 ·003
	12	4 ·047	3 ·014	2 ·003	2 ·003
	11	3 ·041	2 ·011	1 ·002	1 ·002
	10	2 ·029	1 ·007	1 ·007	0 ·001
	9	1 ·017	1 ·017	0 ·002	0 ·002
	8	1 ·037	0 ·006	0 ·006	—
	7	0 ·015−	0 ·015−	—	—
	6	0 ·032	—	—	—
7	13	4 ·031	3 ·007	3 ·007	2 ·001
	12	3 ·031	2 ·007	2 ·007	1 ·001

	a	Probability 0·05	0·025	0·01	0·005
A=13 B=7	11	2 ·022	2 ·022	1 ·004	1 ·004
	10	1 ·012	1 ·012	0 ·002	0 ·002
	9	1 ·029	0 ·004	0 ·004	0 ·004
	8	0 ·010+	0 ·010+	—	—
	7	0 ·022	0 ·022	—	—
	6	0 ·044	—	—	—
6	13	3 ·021	3 ·021	2 ·004	2 ·004
	12	2 ·017	2 ·017	1 ·003	1 ·003
	11	2 ·046	1 ·010−	1 ·010−	0 ·001
	10	1 ·024	1 ·024	0 ·003	0 ·003
	9	1 ·050−	0 ·008	0 ·008	—
	8	0 ·017	0 ·017	—	—
	7	0 ·034	—	—	—
5	13	2 ·012	2 ·012	1 ·002	1 ·002
	12	2 ·044	1 ·008	1 ·008	0 ·001
	11	1 ·022	1 ·022	0 ·002	0 ·002
	10	1 ·047	0 ·007	0 ·007	—
	9	0 ·015−	0 ·015−	—	—
	8	0 ·029	—	—	—
4	13	2 ·044	1 ·006	1 ·006	0 ·000
	12	1 ·022	1 ·022	0 ·002	0 ·002
	11	0 ·006	0 ·006	0 ·006	—
	10	0 ·015−	0 ·015−	—	—
	9	0 ·029	—	—	—
3	13	1 ·025	1 ·025	0 ·002	0 ·002
	12	0 ·007	0 ·007	0 ·007	—
	11	0 ·018	0 ·018	—	—
	10	0 ·036	—	—	—
2	13	0 ·010−	0 ·010−	0 ·010−	—
	12	0 ·029	—	—	—
A=14 B=14	14	10 ·049	9 ·020	8 ·008	7 ·003
	13	8 ·038	7 ·016	6 ·006	5 ·002
	12	6 ·023	6 ·023	5 ·009	4 ·003
	11	5 ·027	4 ·011	3 ·004	3 ·004
	10	4 ·028	3 ·011	2 ·003	2 ·003
	9	3 ·027	2 ·009	2 ·009	1 ·002
	8	2 ·023	2 ·023	1 ·006	0 ·001
	7	1 ·016	1 ·016	0 ·003	0 ·003
	6	1 ·038	0 ·008	0 ·008	—
	5	0 ·020	0 ·020	—	—
	4	0 ·049	—	—	—
13	14	9 ·041	8 ·016	7 ·006	6 ·002
	13	7 ·029	6 ·011	5 ·004	5 ·004
	12	6 ·037	5 ·015+	4 ·005+	3 ·002
	11	5 ·041	4 ·017	3 ·006	2 ·001
	10	4 ·041	3 ·016	2 ·005−	2 ·005−
	9	3 ·038	2 ·013	1 ·003	1 ·003
	8	2 ·031	1 ·009	1 ·009	0 ·001

Table 38. *Significance tests in a* 2 × 2 *contingency table (continued)*

	a	Probability			
		0·05	0·025	0·01	0·005
A=14 B=13	7	**1** ·021	**1** ·021	**0** ·004	**0** ·004
	6	**1** ·048	**0** ·010+	—	—
	5	**0** ·025−	**0** ·025−	—	—
12	14	**8** ·033	**7** ·012	**6** ·004	**6** ·004
	13	**6** ·021	**6** ·021	**5** ·007	**4** ·002
	12	**5** ·025+	**4** ·009	**4** ·009	**3** ·003
	11	**4** ·026	**3** ·009	**3** ·009	**2** ·002
	10	**3** ·024	**3** ·024	**2** ·007	**1** ·002
	9	**2** ·019	**2** ·019	**1** ·005−	**1** ·005−
	8	**2** ·042	**1** ·012	**0** ·002	**0** ·002
	7	**1** ·028	**0** ·005+	**0** ·005+	—
	6	**0** ·013	**0** ·013	—	—
	5	**0** ·030	—	—	—
11	14	**7** ·026	**6** ·009	**6** ·009	**5** ·003
	13	**6** ·039	**5** ·014	**4** ·004	**4** ·004
	12	**5** ·043	**4** ·016	**3** ·005−	**3** ·005−
	11	**4** ·042	**3** ·015−	**2** ·004	**2** ·004
	10	**3** ·036	**2** ·011	**1** ·003	**1** ·003
	9	**2** ·027	**1** ·007	**1** ·007	**0** ·001
	8	**1** ·017	**1** ·017	**0** ·003	**0** ·003
	7	**1** ·038	**0** ·007	**0** ·007	—
	6	**0** ·017	**0** ·017	—	—
	5	**0** ·038	—	—	—
10	14	**6** ·020	**6** ·020	**5** ·006	**4** ·002
	13	**5** ·028	**4** ·009	**4** ·009	**3** ·002
	12	**4** ·028	**3** ·009	**3** ·009	**2** ·002
	11	**3** ·024	**3** ·024	**2** ·007	**1** ·001
	10	**2** ·018	**2** ·018	**1** ·004	**1** ·004
	9	**2** ·040	**1** ·011	**0** ·002	**0** ·002
	8	**1** ·024	**1** ·024	**0** ·004	**0** ·004
	7	**0** ·010−	**0** ·010−	**0** ·010−	—
	6	**0** ·022	**0** ·022	—	—
	5	**0** ·047	—	—	—
9	14	**6** ·047	**5** ·014	**4** ·004	**4** ·004
	13	**4** ·018	**4** ·018	**3** ·005−	**3** ·005−
	12	**3** ·017	**3** ·017	**2** ·004	**2** ·004
	11	**3** ·042	**2** ·012	**1** ·002	**1** ·002
	10	**2** ·029	**1** ·007	**1** ·007	**0** ·001
	9	**1** ·017	**1** ·017	**0** ·002	**0** ·002
	8	**1** ·036	**0** ·006	**0** ·006	—
	7	**0** ·014	**0** ·014	—	—
	6	**0** ·030	—	—	—
8	14	**5** ·036	**4** ·010−	**4** ·010−	**3** ·002
	13	**4** ·039	**3** ·011	**2** ·002	**2** ·002
	12	**3** ·032	**2** ·008	**2** ·008	**1** ·001
	11	**2** ·022	**2** ·022	**1** ·005−	**1** ·005−
	10	**2** ·048	**1** ·012	**0** ·002	**0** ·002
	9	**1** ·026	**0** ·004	**0** ·004	**0** ·004
	8	**0** ·009	**0** ·009	**0** ·009	—
	7	**0** ·020	**0** ·020	—	—
	6	**0** ·040	—	—	—

	a	Probability			
		0·05	0·025	0·01	0·005
A=14 B=7	14	**4** ·026	**3** ·006	**3** ·006	**2** ·001
	13	**3** ·025	**2** ·006	**2** ·006	**1** ·001
	12	**2** ·017	**2** ·017	**1** ·003	**1** ·003
	11	**2** ·041	**1** ·009	**1** ·009	**0** ·001
	10	**1** ·021	**1** ·021	**0** ·003	**0** ·003
	9	**1** ·043	**0** ·007	**0** ·007	—
	8	**0** ·015−	**0** ·015−	—	—
	7	**0** ·030	—	—	—
6	14	**3** ·018	**3** ·018	**2** ·003	**2** ·003
	13	**2** ·014	**2** ·014	**1** ·002	**1** ·002
	12	**2** ·037	**1** ·007	**1** ·007	**0** ·001
	11	**1** ·018	**1** ·018	**0** ·002	**0** ·002
	10	**1** ·038	**0** ·005+	**0** ·005+	—
	9	**0** ·012	**0** ·012	—	—
	8	**0** ·024	**0** ·024	—	—
	7	**0** ·044	—	—	—
5	14	**2** ·010+	**2** ·010+	**1** ·001	**1** ·001
	13	**2** ·037	**1** ·006	**1** ·006	**0** ·001
	12	**1** ·017	**1** ·017	**0** ·002	**0** ·002
	11	**1** ·038	**0** ·005−	**0** ·005−	**0** ·005−
	10	**0** ·011	**0** ·011	—	—
	9	**0** ·022	**0** ·022	—	—
	8	**0** ·040	—	—	—
4	14	**2** ·039	**1** ·005−	**1** ·005−	**1** ·005−
	13	**1** ·019	**1** ·019	**0** ·002	**0** ·002
	12	**1** ·044	**0** ·005−	**0** ·005−	**0** ·005−
	11	**0** ·011	**0** ·011	—	—
	10	**0** ·023	**0** ·023	—	—
	9	**0** ·041	—	—	—
3	14	**1** ·022	**1** ·022	**0** ·001	**0** ·001
	13	**0** ·006	**0** ·006	**0** ·006	—
	12	**0** ·015−	**0** ·015−	—	—
	11	**0** ·029	—	—	—
2	14	**0** ·008	**0** ·008	**0** ·008	—
	13	**0** ·025	**0** ·025	—	—
	12	**0** ·050	—	—	—
A=15 B=15	15	**11** ·050−	**10** ·021	**9** ·008	**8** ·003
	14	**9** ·040	**8** ·018	**7** ·007	**6** ·003
	13	**7** ·025+	**6** ·010+	**5** ·004	**5** ·004
	12	**6** ·030	**5** ·013	**4** ·005−	**4** ·005−
	11	**5** ·033	**4** ·013	**3** ·005−	**3** ·005−
	10	**4** ·033	**3** ·013	**2** ·004	**2** ·004
	9	**3** ·030	**2** ·010+	**1** ·003	**1** ·003
	8	**2** ·025+	**1** ·007	**1** ·007	**0** ·001
	7	**1** ·018	**1** ·018	**0** ·003	**0** ·003
	6	**1** ·040	**0** ·008	**0** ·008	—
	5	**0** ·021	**0** ·021	—	—
	4	**0** ·050−	—	—	—

The table shows: (1) In bold type, for given a, A and B, the value of b ($<a$) which is just significant at the probability level quoted (single-tail test).

(2) In small type, for given A, B and $r=a+b$, the exact probability (if there is independence) that b is equal to or less than the integer shown in bold type.

Table 38 (*continued*)

	a	Probability 0·05	0·025	0·01	0·005
A=15 B=14	15	10 ·042	9 ·017	8 ·006	7 ·002
	14	8 ·031	7 ·013	6 ·005−	6 ·005−
	13	7 ·041	6 ·017	5 ·007	4 ·002
	12	6 ·046	5 ·020	4 ·007	3 ·002
	11	5 ·048	4 ·020	3 ·007	2 ·002
	10	4 ·046	3 ·018	2 ·006	1 ·001
	9	3 ·041	2 ·014	1 ·004	1 ·004
	8	2 ·033	1 ·009	1 ·009	0 ·001
	7	1 ·022	1 ·022	0 ·004	0 ·004
	6	1 ·049	0 ·011	—	—
	5	0 ·025+	—	—	—
13	15	9 ·035−	8 ·013	7 ·005−	7 ·005−
	14	7 ·023	7 ·023	6 ·009	5 ·003
	13	6 ·029	5 ·011	4 ·004	4 ·004
	12	5 ·031	4 ·012	3 ·004	3 ·004
	11	4 ·030	3 ·011	2 ·003	2 ·003
	10	3 ·026	2 ·008	2 ·008	1 ·002
	9	2 ·020	2 ·020	1 ·005+	0 ·001
	8	2 ·043	1 ·013	0 ·002	0 ·002
	7	1 ·029	0 ·005+	0 ·005+	—
	6	0 ·013	0 ·013	—	—
	5	0 ·031	—	—	—
12	15	8 ·028	7 ·010−	7 ·010−	6 ·003
	14	7 ·043	6 ·016	5 ·006	4 ·002
	13	6 ·049	5 ·019	4 ·007	3 ·002
	12	5 ·049	4 ·019	3 ·006	2 ·002
	11	4 ·045+	3 ·017	2 ·005−	2 ·005−
	10	3 ·038	2 ·012	1 ·003	1 ·003
	9	2 ·028	1 ·007	1 ·007	0 ·001
	8	1 ·018	1 ·018	0 ·003	0 ·003
	7	1 ·038	0 ·007	0 ·007	—
	6	0 ·017	0 ·017	—	—
	5	0 ·037	—	—	—
11	15	7 ·022	7 ·022	6 ·007	5 ·002
	14	6 ·032	5 ·011	4 ·003	4 ·003
	13	5 ·034	4 ·012	3 ·003	3 ·003
	12	4 ·032	3 ·010+	2 ·003	2 ·003
	11	3 ·026	2 ·008	2 ·008	1 ·002
	10	2 ·019	2 ·019	1 ·004	1 ·004
	9	2 ·040	1 ·011	0 ·002	0 ·002
	8	1 ·024	1 ·024	0 ·004	0 ·004
	7	1 ·049	0 ·010−	0 ·010−	—
	6	0 ·022	0 ·022	—	—
	5	0 ·046	—	—	—
10	15	6 ·017	6 ·017	5 ·005−	5 ·005−
	14	5 ·023	5 ·023	4 ·007	3 ·002
	13	4 ·022	4 ·022	3 ·007	2 ·001
	12	3 ·018	3 ·018	2 ·005−	2 ·005−
	11	3 ·042	2 ·013	1 ·003	1 ·003
	10	2 ·029	1 ·007	1 ·007	0 ·001
	9	1 ·016	1 ·016	0 ·002	0 ·002
	8	1 ·034	0 ·006	0 ·006	—
	7	0 ·013	0 ·013	—	—
	6	0 ·028	—	—	—
9	15	6 ·042	5 ·012	4 ·003	4 ·003
	14	5 ·047	4 ·015−	3 ·004	3 ·004

	a	Probability 0·05	0·025	0·01	0·005
A=15 B=9	13	4 ·042	3 ·013	2 ·003	2 ·003
	12	3 ·032	2 ·009	2 ·009	1 ·002
	11	2 ·021	2 ·021	1 ·005−	1 ·005−
	10	2 ·045−	1 ·011	0 ·002	0 ·002
	9	1 ·024	1 ·024	0 ·004	0 ·004
	8	1 ·048	0 ·009	0 ·009	—
	7	0 ·019	0 ·019	—	—
	6	0 ·037	—	—	—
8	15	5 ·032	4 ·008	4 ·008	3 ·002
	14	4 ·033	3 ·009	3 ·009	2 ·002
	13	3 ·026	2 ·006	2 ·006	1 ·001
	12	2 ·017	2 ·017	1 ·003	1 ·003
	11	2 ·037	1 ·008	1 ·008	0 ·001
	10	1 ·019	1 ·019	0 ·003	0 ·003
	9	1 ·038	0 ·006	0 ·006	—
	8	0 ·013	0 ·013	—	—
	7	0 ·026	—	—	—
	6	0 ·050−	—	—	—
7	15	4 ·023	4 ·023	3 ·005−	3 ·005−
	14	3 ·021	3 ·021	2 ·004	2 ·004
	13	2 ·014	2 ·014	1 ·002	1 ·002
	12	2 ·032	1 ·007	1 ·007	0 ·001
	11	1 ·015+	1 ·015+	0 ·002	0 ·002
	10	1 ·032	0 ·005−	0 ·005−	0 ·005−
	9	0 ·010+	0 ·010+	—	—
	8	0 ·020	0 ·020	—	—
	7	0 ·038	—	—	—
6	15	3 ·015+	3 ·015+	2 ·003	2 ·003
	14	2 ·011	2 ·011	1 ·002	1 ·002
	13	2 ·031	1 ·006	1 ·006	0 ·001
	12	1 ·014	1 ·014	0 ·002	0 ·002
	11	1 ·029	0 ·004	0 ·004	0 ·004
	10	0 ·009	0 ·009	0 ·009	—
	9	0 ·017	0 ·017	—	—
	8	0 ·032	—	—	—
5	15	2 ·009	2 ·009	2 ·009	1 ·001
	14	2 ·032	1 ·005−	1 ·005−	1 ·005−
	13	1 ·014	1 ·014	0 ·001	0 ·001
	12	1 ·031	0 ·004	0 ·004	0 ·004
	11	0 ·008	0 ·008	0 ·008	—
	10	0 ·016	0 ·016	—	—
	9	0 ·030	—	—	—
4	15	2 ·035+	1 ·004	1 ·004	1 ·004
	14	1 ·016	1 ·016	0 ·001	0 ·001
	13	1 ·037	0 ·004	0 ·004	0 ·004
	12	0 ·009	0 ·009	0 ·009	—
	11	0 ·018	0 ·018	—	—
	10	0 ·033	—	—	—
3	15	1 ·020	1 ·020	0 ·001	0 ·001
	14	0 ·005−	0 ·005−	0 ·005−	0 ·005−
	13	0 ·012	0 ·012	—	—
	12	0 ·025−	0 ·025−	—	—
	11	0 ·043	—	—	—
2	15	0 ·007	0 ·007	0 ·007	—
	14	0 ·022	0 ·022	—	—
	13	0 ·044	—	—	—

Table 39. *Individual terms, $e^{-m} m^i/i!$, of the Poisson distribution*

i	*m* 0·1	0·2	0·3	0·4	0·5	0·6	0·7	0·8	0·9	1·0	*i*
0	·904837	·818731	·740818	·670320	·606531	·548812	·496585	·449329	·406570	·367879	0
1	·090484	·163746	·222245	·268128	·303265	·329287	·347610	·359463	·365913	·367879	1
2	·004524	·016375	·033337	·053626	·075816	·098786	·121663	·143785	·164661	·183940	2
3	·000151	·001092	·003334	·007150	·012636	·019757	·028388	·038343	·049398	·061313	3
4	·000004	·000055	·000250	·000715	·001580	·002964	·004968	·007669	·011115	·015328	4
5	—	·000002	·000015	·000057	·000158	·000356	·000696	·001227	·002001	·003066	5
6	—	—	·000001	·000004	·000013	·000036	·000081	·000164	·000300	·000511	6
7	—	—	—	—	·000001	·000003	·000008	·000019	·000039	·000073	7
8	—	—	—	—	—	—	·000001	·000002	·000004	·000009	8
9	—	—	—	—	—	—	—	—	—	·000001	9
	1·1	1·2	1·3	1·4	1·5	1·6	1·7	1·8	1·9	2·0	
0	·332871	·301194	·272532	·246597	·223130	.201897	·182684	·165299	·149569	·135335	0
1	·366158	·361433	·354291	·345236	·334695	.323034	·310562	·297538	·284180	·270671	1
2	·201387	·216860	·230289	·241665	·251021	.258428	·263978	·267784	·269971	·270671	2
3	·073842	·086744	·099792	·112777	·125510	·137828	·149587	·160671	·170982	·180447	3
4	·020307	·026023	·032432	·039472	·047067	·055131	·063575	·072302	·081216	·090224	4
5	·004467	·006246	·008432	·011052	·014120	·017642	·021615	·026029	·030862	·036089	5
6	·000819	·001249	·001827	·002579	·003530	·004705	·006124	·007809	·009773	·012030	6
7	·000129	·000214	·000339	·000516	·000756	·001075	·001487	·002008	·002653	·003437	7
8	·000018	·000032	·000055	·000090	·000142	·000215	·000316	·000452	·000630	·000859	8
9	·000002	·000004	·000008	·000014	·000024	·000038	·000060	·000090	·000133	·000191	9
10	—	·000001	·000001	·000002	·000004	·000006	·000010	·000016	·000025	·000038	10
11	—	—	—	—	—	·000001	·000002	·000003	·000004	·000007	11
12	—	—	—	—	—	—	—	—	·000001	·000001	12
	2·1	2·2	2·3	2·4	2·5	2·6	2·7	2·8	2·9	3·0	
0	·122456	·110803	·100259	·090718	·082085	·074274	·067206	·060810	·055023	·049787	0
1	·257159	·243767	·230595	·217723	·205212	·193111	·181455	·170268	·159567	·149361	1
2	·270016	·268144	·265185	·261268	·256516	·251045	·244964	·238375	·231373	·224042	2
3	·189012	·196639	·203308	·209014	·213763	·217572	·220468	·222484	·223660	·224042	3
4	·099231	·108151	·116902	·125409	·133602	·141422	·148816	·155739	·162154	·168031	4
5	·041677	·047587	·053775	·060196	·066801	·073539	·080360	·087214	·094049	·100819	5
6	·014587	·017448	·020614	·024078	·027834	·031867	·036162	·040700	·045457	·050409	6
7	·004376	·005484	·006773	·008255	·009941	·011836	·013948	·016280	·018832	·021604	7
8	·001149	·001508	·001947	·002477	·003106	·003847	·004708	·005698	·006827	·008102	8
9	·000268	·000369	·000498	·000660	·000863	·001111	·001412	·001773	·002200	·002701	9
10	·000056	·000081	·000114	·000158	·000216	·000289	·000381	·000496	·000638	·000810	10
11	·000011	·000016	·000024	·000035	·000049	·000068	·000094	·000126	·000168	·000221	11
12	·000002	·000003	·000005	·000007	·000010	·000015	·000021	·000029	·000041	·000055	12
13	—	·000001	·000001	·000001	·000002	·000003	·000004	·000006	·000009	·000013	13
14	—	—	—	—	—	·000001	·000001	·000001	·000002	·000003	14
15	—	—	—	—	—	—	—	—	—	·000001	15

Table 39 (*continued*)

i	m										*i*
	3·1	*3·2*	*3·3*	*3·4*	*3·5*	*3·6*	*3·7*	*3·8*	*3·9*	*4·0*	
0	·045049	·040762	·036883	·033373	·030197	·027324	·024724	·022371	·020242	·018316	*0*
1	·139653	·130439	·121714	·113469	·105691	·098365	·091477	·085009	·078943	·073263	*1*
2	·216461	·208702	·200829	·192898	·184959	·177058	·169233	·161517	·153940	·146525	*2*
3	·223677	·222616	·220912	·218617	·215785	·212469	·208720	·204588	·200122	·195367	*3*
4	·173350	·178093	·182252	·185825	·188812	·191222	·193066	·194359	·195119	·195367	*4*
5	·107477	·113979	·120286	·126361	·132169	·137680	·142869	·147713	·152193	·156293	*5*
6	·055530	·060789	·066158	·071604	·077098	·082608	·088102	·093551	·098925	·104196	*6*
7	·024592	·027789	·031189	·034779	·038549	·042484	·046568	·050785	·055115	·059540	*7*
8	·009529	·011116	·012865	·014781	·016865	·019118	·021538	·024123	·026869	·029770	*8*
9	·003282	·003952	·004717	·005584	·006559	·007647	·008854	·010185	·011643	·013231	*9*
10	·001018	·001265	·001557	·001899	·002296	·002753	·003276	·003870	·004541	·005292	*10*
11	·000287	·000368	·000467	·000587	·000730	·000901	·001102	·001337	·001610	·001925	*11*
12	·000074	·000098	·000128	·000166	·000213	·000270	·000340	·000423	·000523	·000642	*12*
13	·000018	·000024	·000033	·000043	·000057	·000075	·000097	·000124	·000157	·000197	*13*
14	·000004	·000006	·000008	·000011	·000014	·000019	·000026	·000034	·000044	·000056	*14*
15	·000001	·000001	·000002	·000002	·000003	·000005	·000006	·000009	·000011	·000015	*15*
16	—	—	—	·000001	·000001	·000001	·000001	·000002	·000003	·000004	*16*
17	—	—	—	—	—	—	—	—	·000001	·000001	*17*
	4·1	*4·2*	*4·3*	*4·4*	*4·5*	*4·6*	*4·7*	*4·8*	*4·9*	*5·0*	
0	·016573	·014996	·013569	·012277	·011109	·010052	·009095	·008230	·007447	·006738	*0*
1	·067948	·062981	·058345	·054020	·049990	·046238	·042748	·039503	·036488	·033690	*1*
2	·139293	·132261	·125441	·118845	·112479	·106348	·100457	·094807	·089396	·084224	*2*
3	·190368	·185165	·179799	·174305	·168718	·163068	·157383	·151691	·146014	·140374	*3*
4	·195127	·194424	·193284	·191736	·189808	·187528	·184925	·182029	·178867	·175467	*4*
5	·160004	·163316	·166224	·168728	·170827	·172525	·173830	·174748	·175290	·175467	*5*
6	·109336	·114321	·119127	·123734	·128120	·132270	·136167	·139798	·143153	·146223	*6*
7	·064040	·068593	·073178	·077775	·082363	·086920	·091426	·095862	·100207	·104445	*7*
8	·032820	·036011	·039333	·042776	·046329	·049979	·053713	·057517	·061377	·065278	*8*
9	·014951	·016805	·018793	·020913	·023165	·025545	·028050	·030676	·033416	·036266	*9*
10	·006130	·007058	·008081	·009202	·010424	·011751	·013184	·014724	·016374	·018133	*10*
11	·002285	·002695	·003159	·003681	·004264	·004914	·005633	·006425	·007294	·008242	*11*
12	·000781	·000943	·001132	·001350	·001599	·001884	·002206	·002570	·002978	·003434	*12*
13	·000246	·000305	·000374	·000457	·000554	·000667	·000798	·000949	·001123	·001321	*13*
14	·000072	·000091	·000115	·000144	·000178	·000219	·000268	·000325	·000393	·000472	*14*
15	·000020	·000026	·000033	·000042	·000053	·000067	·000084	·000104	·000128	·000157	*15*
16	·000005	·000007	·000009	·000012	·000015	·000019	·000025	·000031	·000039	·000049	*16*
17	·000001	·000902	·000002	·000003	·000004	·000005	·000007	·000009	·000011	·000014	*17*
18	—	—	·000001	·000001	·000001	·000001	·000002	·000002	·000003	·000004	*18*
19	—	—	—	—	—	—	—	·000001	·000001	·000001	*19*
	5·1	*5·2*	*5·3*	*5·4*	*5·5*	*5·6*	*5·7*	*5·8*	*5·9*	*6·0*	
0	·006097	·005517	·004992	·004517	·004087	·003698	·003346	·003028	·002739	·002479	*0*
1	·031093	·028686	·026455	·024390	·022477	·020708	·019072	·017560	·016163	·014873	*1*
2	·079288	·074584	·070107	·065852	·061812	·057982	·054355	·050923	·047680	·044618	*2*
3	·134790	·129279	·123856	·118533	·113323	·108234	·103275	·098452	·093771	·089235	*3*

Table 39. *Individual terms of the Poisson distribution* (*continued*)

i	*m* 5·1	5·2	5·3	5·4	5·5	5·6	5·7	5·8	5·9	6·0	*i*
4	·171857	·168063	·164109	·160020	·155819	·151528	·147167	·142755	·138312	·133853	4
5	·175294	·174785	·173955	·172821	·171401	·169711	·167770	·165596	·163208	·160623	5
6	·149000	·151480	·153660	·155539	·157117	·158397	·159382	·160076	·160488	·160623	6
7	·108557	·112528	·116343	·119987	·123449	·126717	·129782	·132635	·135268	·137677	7
8	·069205	·073143	·077077	·080991	·084871	·088702	·092470	·096160	·099760	·103258	8
9	·039216	·042261	·045390	·048595	·051866	·055192	·058564	·061970	·065398	·068838	9
10	·020000	·021976	·024057	·026241	·028526	·030908	·033382	·035943	·038585	·041303	10
11	·009273	·010388	·011591	·012882	·014263	·015735	·017298	·018952	·020696	·022529	11
12	·003941	·004502	·005119	·005797	·006537	·007343	·008216	·009160	·010175	·011264	12
13	·001546	·001801	·002087	·002408	·002766	·003163	·003603	·004087	·004618	·005199	13
14	·000563	·000669	·000790	·000929	·001087	·001265	·001467	·001693	·001946	·002228	14
15	·000191	·000232	·000279	·000334	·000398	·000472	·000557	·000655	·000766	·000891	15
16	·000061	·000075	·000092	·000113	·000137	·000165	·000199	·000237	·000282	·000334	16
17	·000018	·000023	·000029	·000036	·000044	·000054	·000067	·000081	·000098	·000118	17
18	·000005	·000007	·000008	·000011	·000014	·000017	·000021	·000026	·000032	·000039	18
19	·000001	·000002	·000002	·000003	·000004	·000005	·000006	·000008	·000010	·000012	19
20	—	—	·000001	·000001	·000001	·000001	·000002	·000002	·000003	·000004	20
21	—	—	—	—	—	—	—	·000001	·000001	·000001	21

i	6·1	6·2	6·3	6·4	6·5	6·6	6·7	6·8	6·9	7·0	*i*
0	·002243	·002029	·001836	·001662	·001503	·001360	·001231	·001114	·001008	·000912	0
1	·013682	·012582	·011569	·010634	·009772	·008978	·008247	·007574	·006954	·006383	1
2	·041729	·039006	·036441	·034029	·031760	·029629	·027628	·025751	·023990	·022341	2
3	·084848	·080612	·076527	·072595	·068814	·065183	·061702	·058368	·055178	·052129	3
4	·129393	·124948	·120530	·116151	·111822	·107553	·103351	·099225	·095182	·091226	4
5	·157860	·154936	·151868	·148674	·145369	·141969	·138490	·134946	·131351	·127717	5
6	·160491	·160100	·159461	·158585	·157483	·156166	·154648	·152939	·151053	·149003	6
7	·139856	·141803	·143515	·144992	·146234	·147243	·148020	·148569	·148895	·149003	7
8	·106640	·109897	·113018	·115994	·118815	·121475	·123967	·126284	·128422	·130377	8
9	·072278	·075707	·079113	·082484	·085811	·089082	·092286	·095415	·098457	·101405	9
10	·044090	·046938	·049841	·052790	·055777	·058794	·061832	·064882	·067935	·070983	10
11	·024450	·026456	·028545	·030714	·032959	·035276	·037661	·040109	·042614	·045171	11
12	·012429	·013669	·014986	·016381	·017853	·019402	·021028	·022728	·024503	·026350	12
13	·005832	·006519	·007263	·008064	·008926	·009850	·010837	·011889	·013005	·014188	13
14	·002541	·002887	·003268	·003687	·004144	·004644	·005186	·005774	·006410	·007094	14
15	·001033	·001193	·001373	·001573	·001796	·002043	·002317	·002618	·002949	·003311	15
16	·000394	·000462	·000540	·000629	·000730	·000843	·000970	·001113	·001272	·001448	16
17	·000141	·000169	·000200	·000237	·000279	·000327	·000382	·000445	·000516	·000596	17
18	·000048	·000058	·000070	·000084	·000101	·000120	·000142	·000168	·000198	·000232	18
19	·000015	·000019	·000023	·000028	·000034	·000042	·000050	·000060	·000072	·000085	19
20	·000005	·000006	·000007	·000009	·000011	·000014	·000017	·000020	·000025	·000030	20
21	·000001	·000002	·000002	·000003	·000003	·000004	·000005	·000007	·000008	·000010	21
22	—	—	·000001	·000001	·000001	·000001	·000002	·000002	·000003	·000003	22
23	—	—	—	—	—	—	—	·000001	·000001	·000001	23

Table 39 (*continued*)

i	*m*										*i*
	7·1	*7·2*	*7·3*	*7·4*	*7·5*	*7·6*	*7·7*	*7·8*	*7·9*	*8·0*	
0	·000825	·000747	·000676	·000611	·000553	·000500	·000453	·000410	·000371	·000335	*0*
1	·005858	·005375	·004931	·004523	·004148	·003803	·003487	·003196	·002929	·002684	*1*
2	·020797	·019352	·018000	·016736	·015555	·014453	·013424	·012464	·011569	·010735	*2*
3	·049219	·046444	·043799	·041282	·038889	·036614	·034455	·032407	·030465	·028626	*3*
4	·087364	·083598	·079934	·076372	·072916	·069567	·066326	·063193	·060169	·057252	*4*
5	·124057	·120382	·116703	·113031	·109375	·105742	·102142	·098581	·095067	·091604	*5*
6	·146800	·144458	·141989	·139405	·136718	·133940	·131082	·128156	·125171	·122138	*6*
7	·148897	·148586	·148074	·147371	·146484	·145421	·144191	·142802	·141264	·139587	*7*
8	·132146	·133727	·135118	·136318	·137329	·138150	·138783	·139232	·139499	·139587	*8*
9	·104249	·106982	·109596	·112084	·114440	·116660	·118737	·120668	·122449	·124077	*9*
10	·074017	·077027	·080005	·082942	·085830	·088661	·091427	·094121	·096735	·099262	*10*
11	·047774	·050418	·053094	·055797	·058521	·061257	·063999	·066740	·069473	·072190	*11*
12	·028267	·030251	·032299	·034408	·036575	·038796	·041066	·043381	·045736	·048127	*12*
13	·015438	·016754	·018137	·019586	·021101	·022681	·024324	·026029	·027794	·029616	*13*
14	·007829	·008616	·009457	·010353	·011304	·012312	·013378	·014502	·015684	·016924	*14*
15	·003706	·004136	·004603	·005107	·005652	·006238	·006867	·007541	·008260	·009026	*15*
16	·001644	·001861	·002100	·002362	·002649	·002963	·003305	·003676	·004078	·004513	*16*
17	·000687	·000788	·000902	·001028	·001169	·001325	·001497	·001687	·001895	·002124	*17*
18	·000271	·000315	·000366	·000423	·000487	·000559	·000640	·000731	·000832	·000944	*18*
19	·000101	·000119	·000141	·000165	·000192	·000224	·000259	·000300	·000346	·000397	*19*
20	·000036	·000043	·000051	·000061	·000072	·000085	·000100	·000117	·000137	·000159	*20*
21	·000012	·000015	·000018	·000021	·000026	·000031	·000037	·000043	·000051	·000061	*21*
22	·000004	·000005	·000006	·000007	·000009	·000011	·000013	·000015	·000018	·000022	*22*
23	·000001	·000002	·000002	·000002	·000003	·000004	·000004	·000005	·000006	·000008	*23*
24	—	—	·000001	·000001	·000001	·000001	·000001	·000002	·000002	·000003	*24*
25	—	—	—	—	—	—	—	·000001	·000001	·000001	*25*
	8·1	*8·2*	*8·3*	*8·4*	*8·5*	*8·6*	*8·7*	*8·8*	*8·9*	*9·0*	
0	·000304	·000275	·000249	·000225	·000203	·000184	·000167	·000151	·000136	·000123	*0*
1	·002459	·002252	·002063	·001889	·001729	·001583	·001449	·001326	·001214	·001111	*1*
2	·009958	·009234	·008560	·007933	·007350	·006808	·006304	·005836	·005402	·004998	*2*
3	·026885	·025239	·023683	·022213	·020826	·019517	·018283	·017120	·016025	·014994	*3*
4	·054443	·051740	·049142	·046648	·044255	·041961	·039765	·037664	·035656	·033737	*4*
5	·088198	·084854	·081576	·078368	·075233	·072174	·069192	·066289	·063467	·060727	*5*
6	·119067	·115967	·112847	·109716	·106581	·103449	·100328	·097224	·094143	·091090	*6*
7	·137778	·135848	·133805	·131659	·129419	·127094	·124693	·122224	·119696	·117116	*7*
8	·139500	·139244	·138823	·138242	·137508	·136626	·135604	·134446	·133161	·131756	*8*
9	·125550	·126866	·128025	·129026	·129869	·130554	·131084	·131459	·131682	·131756	*9*
10	·101696	·104031	·106261	·108382	·110388	·112277	·114043	·115684	·117197	·118580	*10*
11	·074885	·077550	·080179	·082764	·085300	·087780	·090197	·092547	·094823	·097020	*11*
12	·050547	·052993	·055457	·057935	·060421	·062909	·065393	·067868	·070327	·072765	*12*
13	·031495	·033426	·035407	·037435	·039506	·041617	·043763	·045941	·048147	·050376	*13*
14	·018222	·019578	·020991	·022461	·023986	·025565	·027196	·028877	·030608	·032384	*14*
15	·009840	·010703	·011615	·012578	·013592	·014657	·015773	·016941	·018161	·019431	*15*
16	·004981	·005485	·006025	·006604	·007221	·007878	·008577	·009318	·010102	·010930	*16*
17	·002373	·002646	·002942	·003263	·003610	·003985	·004389	·004823	·005289	·005786	*17*
18	·001068	·001205	·001356	·001523	·001705	·001904	·002121	·002358	·002615	·002893	*18*
19	·000455	·000520	·000593	·000673	·000763	·000862	·000971	·001092	·001225	·001370	*19*
20	·000184	·000213	·000246	·000283	·000324	·000371	·000423	·000481	·000545	·000617	*20*

Table 39. *Individual terms of the Poisson distribution* (*continued*)

i	*m* 8·1	8·2	8·3	8·4	8·5	8·6	8·7	8·8	8·9	9·0	*i*
21	·000071	·000083	·000097	·000113	·000131	·000152	·000175	·000201	·000231	·000264	21
22	·000026	·000031	·000037	·000043	·000051	·000059	·000069	·000081	·000093	·000108	22
23	·000009	·000011	·000013	·000016	·000019	·000022	·000026	·000031	·000036	·000042	23
24	·000003	·000004	·000005	·000006	·000007	·000008	·000009	·000011	·000013	·000016	24
25	·000001	·000001	·000002	·000002	·000002	·000003	·000003	·000004	·000005	·000006	25
26	—	—	—	·000001	·000001	·000001	·000001	·000001	·000002	·000002	26
27	—	—	—	—	—	—	—	—	·000001	·000001	27
	9·1	9·2	9·3	9·4	9·5	9·6	9·7	9·8	9·9	10·0	
0	·000112	·000101	·000091	·000083	·000075	·000068	·000061	·000055	·000050	·000045	0
1	·001016	·000930	·000850	·000778	·000711	·000650	·000594	·000543	·000497	·000454	1
2	·004624	·004276	·003954	·003655	·003378	·003121	·002883	·002663	·002459	·002270	2
3	·014025	·013113	·012256	·011452	·010696	·009987	·009322	·008698	·008114	·007567	3
4	·031906	·030160	·028496	·026911	·025403	·023969	·022606	·021311	·020082	·018917	4
5	·058069	·055494	·053002	·050593	·048266	·046020	·043855	·041770	·039763	·037833	5
6	·088072	·085091	·082154	·079262	·076421	·073632	·070899	·068224	·065609	·063055	6
7	·114493	·111834	·109147	·106438	·103714	·100981	·098246	·095514	·092790	·090079	7
8	·130236	·128609	·126883	·125065	·123160	·121178	·119123	·117004	·114827	·112599	8
9	·131683	·131467	·131113	·130623	·130003	·129256	·128388	·127405	·126310	·125110	9
10	·119832	·120950	·121935	·122786	·123502	·124086	·124537	·124857	·125047	·125110	10
11	·099133	·101158	·103090	·104926	·106661	·108293	·109819	·111236	·112542	·113736	11
12	·075176	·077555	·079895	·082192	·084440	·086634	·088770	·090843	·092847	·094780	12
13	·052623	·054885	·057156	·059431	·061706	·063976	·066236	·068481	·070707	·072908	13
14	·034205	·036067	·037968	·039904	·041872	·043869	·045892	·047937	·050000	·052077	14
15	·020751	·022121	·023540	·025006	·026519	·028076	·029677	·031319	·033000	·034718	15
16	·011802	·012720	·013683	·014691	·015746	·016846	·017992	·019183	·020419	·021699	16
17	·006318	·006884	·007485	·008123	·008799	·009513	·010266	·011058	·011891	·012764	17
18	·003194	·003518	·003867	·004242	·004644	·005074	·005532	·006021	·006540	·007091	18
19	·001530	·001704	·001893	·002099	·002322	·002563	·002824	·003105	·003408	·003732	19
20	·000696	·000784	·000880	·000986	·001103	·001230	·001370	·001522	·001687	·001866	20
21	·000302	·000343	·000390	·000442	·000499	·000563	·000633	·000710	·000795	·000889	21
22	·000125	·000144	·000165	·000189	·000215	·000245	·000279	·000316	·000358	·000404	22
23	·000049	·000057	·000067	·000077	·000089	·000102	·000118	·000135	·000154	·000176	23
24	·000019	·000022	·000026	·000030	·000035	·000041	·000048	·000055	·000064	·000073	24
25	·000007	·000008	·000010	·000011	·000013	·000016	·000018	·000022	·000025	·000029	25
26	·000002	·000003	·000003	·000004	·000005	·000006	·000007	·000008	·000010	·000011	26
27	·000001	·000001	·000001	·000001	·000002	·000002	·000002	·000003	·000004	·000004	27
28	—	—	—	—	·000001	·000001	·000001	·000001	·000001	·000001	28
29	—	—	—	—	—	—	—	—	—	·000001	29
	10·1	10·2	10·3	10·4	10·5	10·6	10·7	10·8	10·9	11·0	
0	·000041	·000037	·000034	·000030	·000028	·000025	·000023	·000020	·000018	·000017	0
1	·000415	·000379	·000346	·000317	·000289	·000264	·000241	·000220	·000201	·000184	1
2	·002095	·001934	·001784	·001646	·001518	·001400	·001291	·001190	·001097	·001010	2
3	·007054	·006574	·006125	·005705	·005313	·004946	·004603	·004283	·003984	·003705	3

Table 39 (*continued*)

i	m										i
	10·1	*10·2*	*10·3*	*10·4*	*10·5*	*10·6*	*10·7*	*10·8*	*10·9*	*11·0*	
4	·017811	·016764	·015773	·014834	·013946	·013107	·012313	·011564	·010856	·010189	*4*
5	·035979	·034199	·032492	·030855	·029287	·027786	·026350	·024978	·023667	·022415	*5*
6	·060565	·058139	·055777	·053482	·051252	·049089	·046991	·044960	·042995	·041095	*6*
7	·087387	·084716	·082072	·079458	·076878	·074334	·071830	·069367	·066949	·064577	*7*
8	·110326	·108013	·105668	·103296	·100902	·098493	·096072	·093646	·091218	·088794	*8*
9	·123810	·122415	·120931	·119364	·117720	·116003	·114219	·112375	·110475	·108526	*9*
10	·125048	·124863	·124559	·124139	·123606	·122963	·122215	·121365	·120418	·119378	*10*
11	·114817	·115782	·116633	·117368	·117987	·118492	·118882	·119159	·119323	·119378	*11*
12	·096637	·098415	·100110	·101719	·103239	·104667	·106003	·107243	·108386	·109430	*12*
13	·075080	·077218	·079318	·081375	·083385	·085344	·087248	·089094	·090877	·092595	*13*
14	·054165	·056259	·058355	·060450	·062539	·064618	·066683	·068730	·070754	·072753	*14*
15	·036471	·038256	·040071	·041912	·043777	·045663	·047567	·049485	·051415	·053352	*15*
16	·023022	·024388	·025795	·027243	·028729	·030252	·031810	·033403	·035026	·036680	*16*
17	·013678	·014633	·015629	·016666	·017744	·018863	·020022	·021220	·022458	·023734	*17*
18	·007675	·008292	·008943	·009629	·010351	·011108	·011902	·012732	·013600	·014504	*18*
19	·004080	·004451	·004848	·005271	·005720	·006197	·006703	·007237	·007802	·008397	*19*
20	·002060	·002270	·002497	·002741	·003003	·003285	·003586	·003908	·004252	·004618	*20*
21	·000991	·001103	·001225	·001357	·001502	·001658	·001827	·002010	·002207	·002419	*21*
22	·000455	·000511	·000573	·000642	·000717	·000799	·000889	·000987	·001093	·001210	*22*
23	·000200	·000227	·000257	·000290	·000327	·000368	·000413	·000463	·000518	·000578	*23*
24	·000084	·000096	·000110	·000126	·000143	·000163	·000184	·000208	·000235	·000265	*24*
25	·000034	·000039	·000045	·000052	·000060	·000069	·000079	·000090	·000103	·000117	*25*
26	·000013	·000015	·000018	·000021	·000024	·000028	·000032	·000037	·000043	·000049	*26*
27	·000005	·000006	·000007	·000008	·000009	·000011	·000013	·000015	·000017	·000020	*27*
28	·000002	·000002	·000003	·000003	·000004	·000004	·000005	·000006	·000007	·000008	*28*
29	·000001	·000001	·000001	·000001	·000001	·000002	·000002	·000002	·000003	·000003	*29*
30	—	—	—	—	—	·000001	·000001	·000001	·000001	·000001	*30*
	11·1	*11·2*	*11·3*	*11·4*	*11·5*	*11·6*	*11·7*	*11·8*	*11·9*	*12·0*	
0	·000015	·000014	·000012	·000011	·000010	·000009	·000008	·000008	·000007	·000006	*0*
1	·000168	·000153	·000140	·000128	·000116	·000106	·000097	·000089	·000081	·000074	*1*
2	·000931	·000858	·000790	·000727	·000670	·000617	·000568	·000522	·000481	·000442	*2*
3	·003445	·003202	·002976	·002764	·002568	·002385	·002214	·002055	·001907	·001770	*3*
4	·009559	·008965	·008406	·007879	·007382	·006915	·006476	·006062	·005674	·005309	*4*
5	·021221	·020082	·018997	·017963	·016979	·016043	·015153	·014307	·013504	·012741	*5*
6	·039259	·037487	·035778	·034130	·032544	·031017	·029549	·028137	·026782	·025481	*6*
7	·062253	·059979	·057755	·055584	·053465	·051400	·049388	·047432	·045530	·043682	*7*
8	·086376	·083970	·081579	·079206	·076856	·074529	·072231	·069962	·067725	·065523	*8*
9	·106531	·104496	·102427	·100328	·098204	·096060	·093900	·091728	·089548	·087364	*9*
10	·118249	·117036	·115743	·114374	·112935	·111430	·109863	·108239	·106562	·104837	*10*
11	·119324	·119164	·118899	·118533	·118068	·117508	·116854	·116110	·115281	·114368	*11*
12	·110375	·111220	·111964	·112607	·113149	·113591	·113933	·114175	·114320	·114363	*12*
13	·094243	·095820	·097322	·098747	·100093	·101358	·102539	·103636	·104647	·105570	*13*
14	·074721	·076656	·078553	·080409	·082219	·083982	·085694	·087350	·088950	·090489	*14*
15	·055294	·057236	·059177	·061110	·063035	·064946	·066841	·068716	·070567	·072391	*15*
16	·038360	·040065	·041793	·043541	·045306	·047086	·048877	·050678	·052484	·054293	*16*
17	·025047	·026396	·027780	·029198	·030648	·032129	·033639	·035176	·036739	·038325	*17*
18	·015446	·016424	·017440	·018492	·019581	·020706	·021865	·023060	·024288	·025550	*18*
19	·009023	·009682	·010372	·011095	·011852	·012641	·013465	·014322	·015212	·016137	*19*

Table 39. *Individual terms of the Poisson distribution (continued)*

i	*m* = 11·1	11·2	11·3	11·4	11·5	11·6	11·7	11·8	11·9	12·0	*i*
20	·005008	·005422	·005860	·006324	·006815	·007332	·007877	·008450	·009051	·009682	20
21	·002647	·002892	·003153	·003433	·003732	·004050	·004388	·004748	·005129	·005533	21
22	·001336	·001472	·001620	·001779	·001951	·002136	·002334	·002547	·002774	·003018	22
23	·000645	·000717	·000796	·000882	·000975	·001077	·001187	·001307	·001435	·001575	23
24	·000298	·000335	·000375	·000419	·000467	·000521	·000579	·000642	·000712	·000787	24
25	·000132	·000150	·000169	·000191	·000215	·000242	·000271	·000303	·000339	·000378	25
26	·000057	·000065	·000074	·000084	·000095	·000108	·000122	·000138	·000155	·000174	26
27	·000023	·000027	·000031	·000035	·000041	·000046	·000053	·000060	·000068	·000078	27
28	·000009	·000011	·000012	·000014	·000017	·000019	·000022	·000025	·000029	·000033	28
29	·000004	·000004	·000005	·000006	·000007	·000008	·000009	·000010	·000012	·000014	29
30	·000001	·000002	·000002	·000002	·000003	·000003	·000003	·000004	·000005	·000005	30
31	—	·000001	·000001	·000001	·000001	·000001	·000001	·000002	·000002	·000002	31
32	—	—	—	—	—	—	—	·000001	·000001	·000001	32

i	12·1	12·2	12·3	12·4	12·5	12·6	12·7	12·8	12·9	13·0	*i*
0	·000006	·000005	·000005	·000004	·000004	·000003	·000003	·000003	·000002	·000002	0
1	·000067	·000061	·000056	·000051	·000047	·000042	·000039	·000035	·000032	·000029	1
2	·000407	·000374	·000344	·000317	·000291	·000268	·000246	·000226	·000208	·000191	2
3	·001641	·001522	·001412	·001309	·001213	·001124	·001042	·000965	·000894	·000828	3
4	·004966	·004643	·004341	·004057	·003791	·003541	·003307	·003088	·002882	·002690	4
5	·012017	·011330	·010679	·010062	·009477	·008924	·008400	·007905	·007436	·006994	5
6	·024233	·023037	·021892	·020794	·019744	·018740	·017781	·016864	·015988	·015153	6
7	·041889	·040151	·038467	·036836	·035258	·033733	·032259	·030837	·029464	·028141	7
8	·063358	·061230	·059142	·057095	·055091	·053129	·051212	·049339	·047511	·045730	8
9	·085181	·083000	·080828	·078665	·076515	·074381	·072266	·070171	·068100	·066054	9
10	·103069	·101261	·099418	·097544	·095644	·093720	·091777	·089819	·087849	·085870	10
11	·113376	·112308	·111168	·109959	·108686	·107352	·105961	·104516	·103023	·101483	11
12	·114321	·114180	·113947	·113624	·113215	·112720	·112142	·111484	·110749	·109940	12
13	·106406	·107153	·107811	·108380	·108860	·109251	·109554	·109769	·109897	·109940	13
14	·091965	·093376	·094720	·095994	·097197	·098326	·099381	·100360	·101263	·102087	14
15	·074185	·075946	·077670	·079355	·080997	·082594	·084143	·085641	·087086	·088475	15
16	·056103	·057909	·059709	·061500	·063279	·065043	·066788	·068513	·070213	·071886	16
17	·039932	·041558	·043201	·044859	·046529	·048208	·049895	·051586	·053279	·054972	17
18	·026843	·028167	·029521	·030903	·032312	·033746	·035204	·036683	·038183	·039702	18
19	·017095	·018086	·019111	·020168	·021258	·022379	·023531	·024713	·025925	·027164	19
20	·010342	·011033	·011753	·012504	·013286	·014099	·014942	·015816	·016721	·017657	20
21	·005959	·006409	·006884	·007383	·007908	·008459	·009036	·009640	·010272	·010930	21
22	·003278	·003554	·003849	·004162	·004493	·004845	·005216	·005609	·006023	·006459	22
23	·001724	·001885	·002058	·002244	·002442	·002654	·002880	·003122	·003378	·003651	23
24	·000869	·000958	·001055	·001159	·001272	·001393	·001524	·001665	·001816	·001977	24
25	·000421	·000468	·000519	·000575	·000636	·000702	·000774	·000852	·000937	·001028	25
26	·000196	·000219	·000246	·000274	·000306	·000340	·000378	·000420	·000465	·000514	26
27	·000088	·000099	·000112	·000126	·000142	·000159	·000178	·000199	·000222	·000248	27
28	·000038	·000043	·000049	·000056	·000063	·000071	·000081	·000091	·000102	·000115	28
29	·000016	·000018	·000021	·000024	·000027	·000031	·000035	·000040	·000046	·000052	29
30	·000006	·000007	·000009	·000010	·000011	·000013	·000015	·000017	·000020	·000022	30
31	·000002	·000003	·000003	·000004	·000005	·000005	·000006	·000007	·000008	·000009	31
32	·000001	·000001	·000001	·000002	·000002	·000002	·000002	·000003	·000003	·000004	32
33	—	—	—	·000001	·000001	·000001	·000001	·000001	·000001	·000002	33
34	—	—	—	—	—	—	—	—	—	·000001	34

Table 39 (*continued*)

i	m										i
	13·1	*13·2*	*13·3*	*13·4*	*13·5*	*13·6*	*13·7*	*13·8*	*13·9*	*14·0*	
0	·000002	·000002	·000002	·000002	·000001	·000001	·000001	·000001	·000001	·000001	0
1	·000027	·000024	·000022	·000020	·000019	·000017	·000015	·000014	·000013	·000012	1
2	·000175	·000161	·000148	·000136	·000125	·000115	·000105	·000097	·000089	·000081	2
3	·000766	·000709	·000657	·000608	·000562	·000520	·000481	·000445	·000411	·000380	3
4	·002510	·002341	·002183	·002035	·001897	·001768	·001648	·001535	·001429	·001331	4
5	·006575	·006180	·005807	·005455	·005123	·004810	·004514	·004236	·003974	·003727	5
6	·014356	·013596	·012872	·012183	·011526	·010902	·010308	·009743	·009206	·008696	6
7	·026867	·025639	·024458	·023322	·022230	·021181	·020173	·019207	·018280	·017392	7
8	·043994	·042304	·040661	·039064	·037512	·036007	·034547	·033132	·031762	·030435	8
9	·064036	·062046	·060088	·058161	·056269	·054410	·052588	·050802	·049054	·047344	9
10	·083887	·081901	·079916	·077936	·075963	·073998	·072046	·070107	·068185	·066282	10
11	·099901	·098281	·096626	·094940	·093227	·091489	·089730	·087953	·086162	·084359	11
12	·109059	·108109	·107094	·106017	·104880	·103687	·102441	·101146	·099804	·098418	12
13	·109898	·109773	·109566	·109279	·108914	·108473	·107957	·107370	·106713	·105989	13
14	·102833	·103500	·104087	·104595	·105024	·105373	·105644	·105836	·105951	·105989	14
15	·089807	·091080	·092291	·093439	·094522	·095539	·096488	·097369	·098181	·098923	15
16	·073530	·075141	·076717	·078255	·079753	·081208	·082618	·083981	·085295	·086558	16
17	·056661	·058345	·060019	·061683	·063333	·064966	·066580	·068173	·069741	·071283	17
18	·041237	·042786	·044348	·045920	·047500	·049086	·050675	·052266	·053856	·055442	18
19	·028432	·029725	·031043	·032385	·033750	·035135	·036539	·037962	·039400	·040852	19
20	·018623	·019619	·020644	·021698	·022781	·023892	·025030	·026193	·027383	·028597	20
21	·011617	·012332	·013074	·013846	·014645	·015473	·016329	·017213	·018125	·019064	21
22	·006917	·007399	·007904	·008433	·008987	·009565	·010168	·010797	·011452	·012132	22
23	·003940	·004246	·004571	·004913	·005275	·005656	·006057	·006478	·006921	·007385	23
24	·002151	·002336	·002533	·002743	·002967	·003205	·003457	·003725	·004008	·004308	24
25	·001127	·001233	·001348	·001470	·001602	·001744	·001895	·002056	·002229	·002412	25
26	·000568	·000626	·000689	·000758	·000832	·000912	·000998	·001091	·001191	·001299	26
27	·000275	·000306	·000340	·000376	·000416	·000459	·000507	·000558	·000613	·000674	27
28	·000129	·000144	·000161	·000180	·000201	·000223	·000248	·000275	·000305	·000337	28
29	·000058	·000066	·000074	·000083	·000093	·000105	·000117	·000131	·000146	·000163	29
30	·000025	·000029	·000033	·000037	·000042	·000047	·000053	·000060	·000068	·000076	30
31	·000011	·000012	·000014	·000016	·000018	·000021	·000024	·000027	·000030	·000034	31
32	·000004	·000005	·000006	·000007	·000008	·000009	·000010	·000012	·000013	·000015	32
33	·000002	·000002	·000002	·000003	·000003	·000004	·000004	·000005	·000006	·000006	33
34	·000001	·000001	·000001	·000001	·000001	·000001	·000002	·000002	·000002	·000003	34
35	—	—	—	—	—	·000001	·000001	·000001	·000001	·000001	35
	14·1	*14·2*	*14·3*	*14·4*	*14·5*	*14·6*	*14·7*	*14·8*	*14·9*	*15·0*	
0	·000001	·000001	·000001	·000001	·000001	—	—	—	—	—	0
1	·000011	·000010	·000009	·000008	·000007	·000007	·000006	·000006	·000005	·000005	1
2	·000075	·000069	·000063	·000058	·000053	·000049	·000045	·000041	·000038	·000034	2
3	·000352	·000325	·000300	·000277	·000256	·000237	·000219	·000202	·000186	·000172	3
4	·001239	·001153	·001073	·000999	·000929	·000864	·000803	·000747	·000694	·000645	4
5	·003494	·003275	·003070	·002876	·002694	·002523	·002362	·002211	·002069	·001936	5
6	·008212	·007752	·007316	·006902	·006510	·006139	·005787	·005454	·005138	·004839	6
7	·016541	·015726	·014946	·014199	·013486	·012804	·012152	·011530	·010937	·010370	7
8	·029153	·027913	·026715	·025559	·024443	·023367	·022330	·021331	·020370	·019444	8
9	·045673	·044040	·042447	·040894	·039380	·037907	·036472	·035078	·033723	·032407	9
10	·064399	·062537	·060700	·058887	·057101	·055343	·053614	·051915	·050247	·048611	10
11	·082547	·080730	·078910	·077089	·075270	·073456	·071648	·069850	·068062	·066287	11

Table 39. *Individual terms of the Poisson distribution* (*continued*)

i	*m*										*i*
	14·1	*14·2*	*14·3*	*14·4*	*14·5*	*14·6*	*14·7*	*14·8*	*14·9*	*15·0*	
12	·096993	·095530	·094034	·092507	·090951	·089371	·087769	·086148	·084510	·082859	*12*
13	·105200	·104349	·103437	·102469	·101446	·100371	·099247	·098076	·096862	·095607	*13*
14	·105951	·105839	·105654	·105396	·105069	·104672	·104209	·103681	·103089	·102436	*14*
15	·099594	·100195	·100723	·101181	·101567	·101881	·102125	·102298	·102402	·102436	*15*
16	·087768	·088923	·090021	·091063	·092045	·092967	·093827	·094626	·095361	·096034	*16*
17	·072795	·074277	·075724	·077135	·078509	·079842	·081133	·082380	·083581	·084736	*17*
18	·057023	·058596	·060158	·061708	·063243	·064761	·066259	·067735	·069187	·070613	*18*
19	·042317	·043793	·045277	·046768	·048264	·049763	·051263	·052762	·054257	·055747	*19*
20	·029834	·031093	·032373	·033673	·034992	·036327	·037678	·039044	·040422	·041810	*20*
21	·020031	·021025	·022045	·023090	·024161	·025256	·026375	·027517	·028680	·029865	*21*
22	·012838	·013570	·014329	·015114	·015924	·016761	·017623	·018511	·019424	·020362	*22*
23	·007870	·008378	·008909	·009462	·010039	·010640	·011264	·011911	·012584	·013280	*23*
24	·004624	·004957	·005308	·005677	·006065	·006472	·006899	·007345	·007812	·008300	*24*
25	·002608	·002816	·003036	·003270	·003518	·003780	·004057	·004348	·004656	·004980	*25*
26	·001414	·001538	·001670	·001811	·001962	·002123	·002294	·002475	·002668	·002873	*26*
27	·000739	·000809	·000884	·000966	·001054	·001148	·001249	·001357	·001473	·001596	*27*
28	·000372	·000410	·000452	·000497	·000546	·000598	·000656	·000717	·000784	·000855	*28*
29	·000181	·000201	·000223	·000247	·000273	·000301	·000332	·000366	·000403	·000442	*29*
30	·000085	·000095	·000106	·000118	·000132	·000147	·000163	·000181	·000200	·000221	*30*
31	·000039	·000044	·000049	·000055	·000062	·000069	·000077	·000086	·000096	·000107	*31*
32	·000017	·000019	·000022	·000025	·000028	·000032	·000035	·000040	·000045	·000050	*32*
33	·000007	·000008	·000009	·000011	·000012	·000014	·000016	·000018	·000020	·000023	*33*
34	·000003	·000003	·000004	·000005	·000005	·000006	·000007	·000008	·000009	·000010	*34*
35	·000001	·000001	·000002	·000002	·000002	·000002	·000003	·000003	·000004	·000004	*35*
36	—	·000001	·000001	·000001	·000001	·000001	·000001	·000001	·000002	·000002	*36*
37	—	—	—	—	—	—	—	·000001	·000001	·000001	*37*

Table 40. *Confidence limits for the expectation of a Poisson variable*

$1-2\alpha$	0·998		0·99		0·98		0·95		0·90		$1-2\alpha$
α	0·001		0·005		0·01		0·025		0·05		α
c	Lower	Upper	Lower	Upper	Lower	Upper	Lower	Upper	Lower	Upper	c
0	0·00000	6·91	0·00000	5·30	0·0000	4·61	0·0000	3·69	0·0000	3·00	**0**
1	·00100	9·23	·00501	7·43	·0101	6·64	·0253	5·57	·0513	4·74	**1**
2	·0454	11·23	·103	9·27	·149	8·41	·242	7·22	·355	6·30	**2**
3	·191	13·06	·338	10·98	·436	10·05	·619	8·77	·818	7·75	**3**
4	·429	14·79	·672	12·59	·823	11·60	1·09	10·24	1·37	9·15	**4**
5	0·739	16·45	1·08	14·15	1·28	13·11	1·62	11·67	1·97	10·51	**5**
6	1·11	18·06	1·54	15·66	1·79	14·57	2·20	13·06	2·61	11·84	**6**
7	1·52	19·63	2·04	17·13	2·33	16·00	2·81	14·42	3·29	13·15	**7**
8	1·97	21·16	2·57	18·58	2·91	17·40	3·45	15·76	3·98	14·43	**8**
9	2·45	22·66	3·13	20·00	3·51	18·78	4·12	17·08	4·70	15·71	**9**
10	2·96	24·13	3·72	21·40	4·13	20·14	4·80	18·39	5·43	16·96	**10**
11	3·49	25·59	4·32	22·78	4·77	21·49	5·49	19·68	6·17	18·21	**11**
12	4·04	27·03	4·94	24·14	5·43	22·82	6·20	20·96	6·92	19·44	**12**
13	4·61	28·45	5·58	25·50	6·10	24·14	6·92	22·23	7·69	20·67	**13**
14	5·20	29·85	6·23	26·84	6·78	25·45	7·65	23·49	8·46	21·89	**14**
15	5·79	31·24	6·89	28·16	7·48	26·74	8·40	24·74	9·25	23·10	**15**
16	6·41	32·62	7·57	29·48	8·18	28·03	9·15	25·98	10·04	24·30	**16**
17	7·03	33·99	8·25	30·79	8·89	29·31	9·90	27·22	10·83	25·50	**17**
18	7·66	35·35	8·94	32·09	9·62	30·58	10·67	28·45	11·63	26·69	**18**
19	8·31	36·70	9·64	33·38	10·35	31·85	11·44	29·67	12·44	27·88	**19**
20	8·96	38·04	10·35	34·67	11·08	33·10	12·22	30·89	13·25	29·06	**20**
21	9·62	39·38	11·07	35·95	11·82	34·36	13·00	32·10	14·07	30·24	**21**
22	10·29	40·70	11·79	37·22	12·57	35·60	13·79	33·31	14·89	31·42	**22**
23	10·96	42·02	12·52	38·48	13·33	36·84	14·58	34·51	15·72	32·59	**23**
24	11·65	43·33	13·25	39·74	14·09	38·08	15·38	35·71	16·55	33·75	**24**
25	12·34	44·64	14·00	41·00	14·85	39·31	16·18	36·90	17·38	34·92	**25**
26	13·03	45·94	14·74	42·25	15·62	40·53	16·98	38·10	18·22	36·08	**26**
27	13·73	47·23	15·49	43·50	16·40	41·76	17·79	39·28	19·06	37·23	**27**
28	14·44	48·52	16·24	44·74	17·17	42·98	18·61	40·47	19·90	38·39	**28**
29	15·15	49·80	17·00	45·98	17·96	44·19	19·42	41·65	20·75	39·54	**29**
30	15·87	51·08	17·77	47·21	18·74	45·40	20·24	42·83	21·59	40·69	**30**
35	19·52	57·42	21·64	53·32	22·72	51·41	24·38	48·68	25·87	46·40	**35**
40	23·26	63·66	25·59	59·36	26·77	57·35	28·58	54·47	30·20	52·07	**40**
45	27·08	69·83	29·60	65·34	30·88	63·23	32·82	60·21	34·56	57·69	**45**
50	30·96	75·94	33·66	71·27	35·03	69·07	37·11	65·92	38·96	63·29	**50**

If c is the observed frequency or count and m_A, m_B are the lower and upper confidence limits for its expectation, m, then

$$\Pr\{m_A \leqslant m \leqslant m_B\} \leqslant 1-2\alpha.$$

Table 41. *Chart providing confidence limits for p in binomial sampling, given a sample fraction c/n. Confidence coefficient,* $1-2\alpha = 0{\cdot}95$.

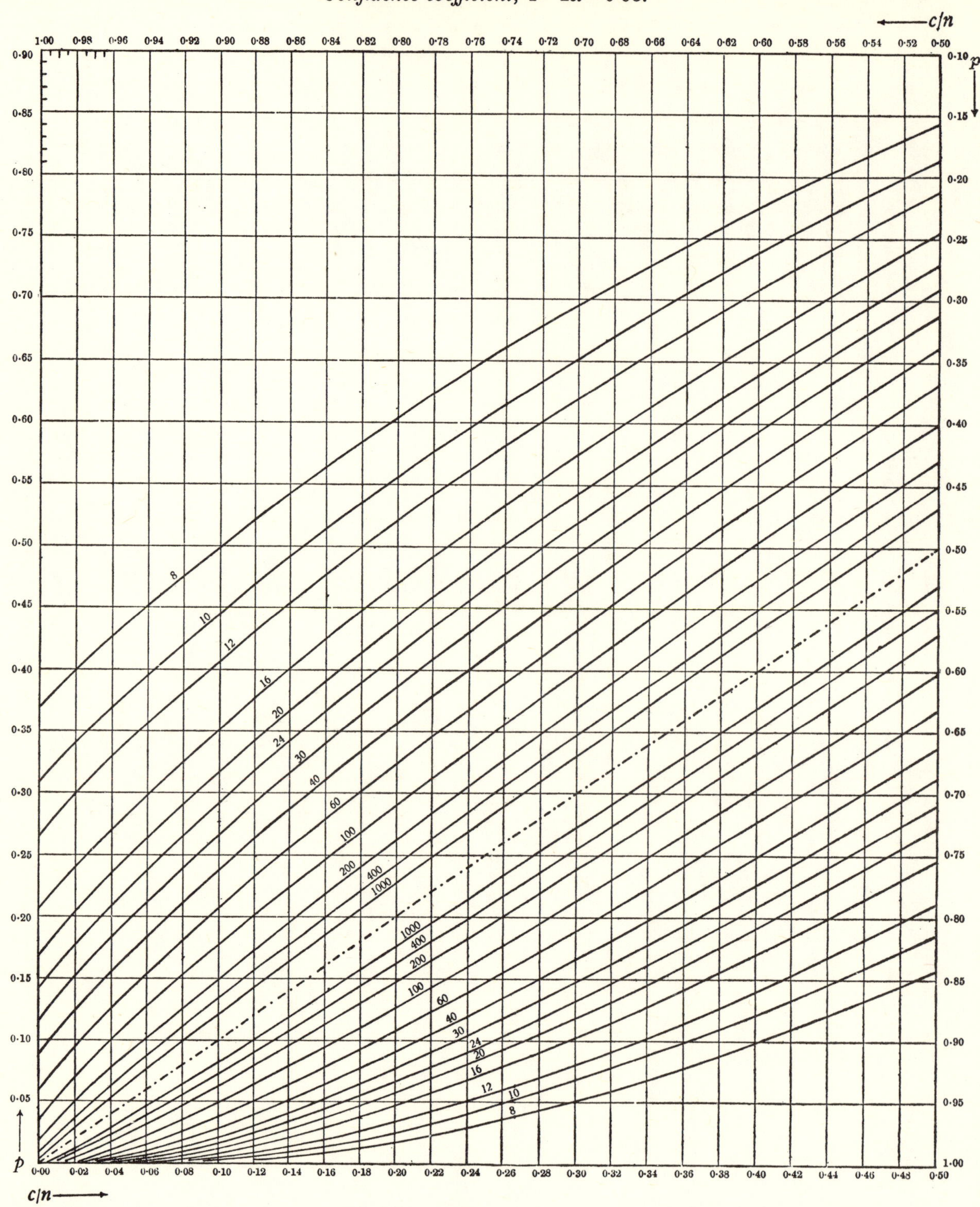

The numbers printed along the curves indicate the sample size n. If for a given value of the abscissa c/n, p_A and p_B are the ordinates read from (or interpolated between) the appropriate lower and upper curves, then

$$\Pr\{p_A \leqslant p \leqslant p_B\} \geqslant 1-2\alpha.$$

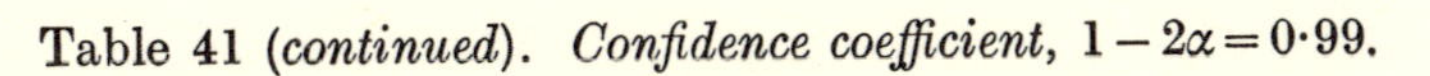

Table 41 (*continued*). *Confidence coefficient*, $1-2\alpha=0{\cdot}99$.

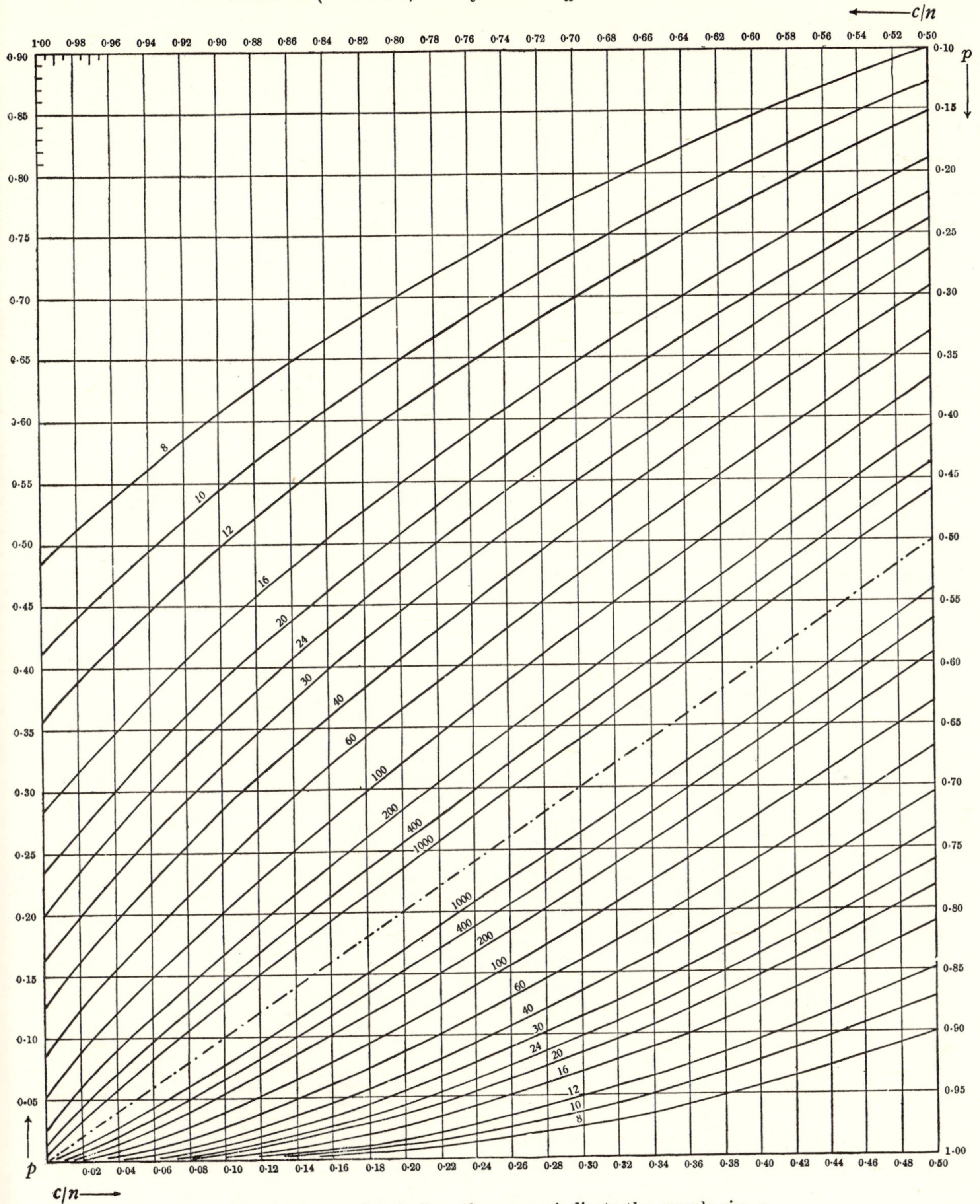

The numbers printed along the curves indicate the sample size n.

Note: the process of reading from the curves can be simplified with the help of the right-angled corner of a loose sheet of paper or thin card, along the edges of which are marked off the scales shown in the top left-hand corner of each Chart.

Table 42. *Percentage points of Pearson curves for given* β_1, β_2, *expressed in standardized measure*

Lower 5 % *points of the standardized deviate* $(x_P - \mu)/\sigma$, $(P = 0{\cdot}05)$

(Note that for positive skewness, i.e. $\mu_3 > 0$, the deviates in this table are negative.)

β_2 \ β_1	0·00	0·01	0·03	0·05	0·10	0·15	0·20	0·30	0·40	0·50	0·60	0·70	0·80	0·90	1·00
1·8	1·56	—	—	—	—	—	—	—	—	—	—	—	—	—	—
2·0	1·61	1·56	1·51	1·47	—	—	—	—	—	—	—	—	—	—	—
2·2	1·64	1·59	1·55	1·52	1·46	1·40	1·35	—	—	—	—	—	—	—	—
2·4	1·65	1·61	1·58	1·55	1·50	1·45	1·41	1·33	—	—	—	—	—	—	—
2·6	1·65	1·61	1·59	1·57	1·53	1·49	1·45	1·38	1·30	—	—	—	—	—	—
2·8	1·65	1·62	1·59	1·57	1·54	1·51	1·48	1·42	1·35	1·29	—	—	—	—	—
3·0	1·64	1·62	1·59	1·58	1·55	1·52	1·49	1·44	1·39	1·33	1·27	—	—	—	—
3·2	1·64	1·61	1·59	1·58	1·55	1·53	1·50	1·46	1·42	1·37	1·31	1·25	1·19	—	—
3·4	1·64	1·61	1·59	1·58	1·55	1·53	1·51	1·47	1·43	1·39	1·35	1·30	1·24	1·19	—
3·6	1·63	1·61	1·59	1·58	1·55	1·53	1·52	1·48	1·45	1·41	1·37	1·33	1·28	1·23	1·18
3·8	1·63	1·60	1·59	1·58	1·55	1·54	1·52	1·49	1·46	1·42	1·39	1·35	1·31	1·27	1·22
4·0	1·62	1·60	1·59	1·57	1·55	1·54	1·52	1·49	1·46	1·43	1·40	1·37	1·34	1·30	1·26
4·2	1·62	1·60	1·58	1·57	1·55	1·54	1·52	1·49	1·47	1·44	1·42	1·39	1·36	1·32	1·28
4·4	1·61	1·60	1·58	1·57	1·55	1·54	1·52	1·50	1·47	1·45	1·42	1·40	1·37	1·34	1·31
4·6	1·61	1·59	1·58	1·57	1·55	1·54	1·52	1·50	1·48	1·45	1·43	1·41	1·38	1·35	1·33
4·8	1·60	1·59	1·58	1·57	1·55	1·53	1·52	1·50	1·48	1·46	1·44	1·41	1·39	1·37	1·34
5·0	1·60	1·58	1·57	1·56	1·55	1·53	1·52	1·50	1·48	1·46	1·44	1·42	1·40	1·38	1·35

Upper 5 % *points of the standardized deviate* $(x_P - \mu)/\sigma$, $(P = 0{\cdot}95)$

β_2 \ β_1	0·00	0·01	0·03	0·05	0·10	0·15	0·20	0·30	0·40	0·50	0·60	0·70	0·80	0·90	1·00
1·8	1·56	—	—	—	—	—	—	—	—	—	—	—	—	—	—
2·0	1·61	1·66	1·70	1·72	—	—	—	—	—	—	—	—	—	—	—
2·2	1·64	1·68	1·71	1·74	1·77	1·80	1·83	—	—	—	—	—	—	—	—
2·4	1·65	1·69	1·71	1·74	1·77	1·80	1·83	1·87	—	—	—	—	—	—	—
2·6	1·65	1·68	1·71	1·73	1·76	1·79	1·81	1·86	1·90	—	—	—	—	—	—
2·8	1·65	1·68	1·70	1·72	1·75	1·77	1·80	1·84	1·88	1·92	—	—	—	—	—
3·0	1·64	1·67	1·69	1·71	1·74	1·76	1·78	1·82	1·86	1·90	1·93	—	—	—	—
3·2	1·64	1·67	1·69	1·70	1·73	1·75	1·77	1·80	1·84	1·87	1·91	1·94	1·98	—	—
3·4	1·64	1·66	1·68	1·69	1·72	1·74	1·76	1·79	1·82	1·85	1·88	1·92	1·95	1·98	—
3·6	1·63	1·65	1·67	1·68	1·71	1·73	1·74	1·77	1·80	1·83	1·86	1·89	1·92	1·95	1·98
3·8	1·63	1·65	1·66	1·68	1·70	1·72	1·73	1·76	1·79	1·82	1·84	1·87	1·90	1·93	1·96
4·0	1·62	1·64	1·66	1·67	1·69	1·71	1·72	1·75	1·78	1·80	1·83	1·85	1·88	1·90	1·93
4·2	1·62	1·64	1·65	1·66	1·68	1·70	1·71	1·74	1·76	1·79	1·81	1·83	1·86	1·88	1·91
4·4	1·61	1·63	1·65	1·66	1·68	1·69	1·70	1·73	1·75	1·78	1·80	1·82	1·84	1·86	1·89
4·6	1·61	1·63	1·64	1·65	1·67	1·68	1·70	1·72	1·74	1·76	1·78	1·80	1·82	1·84	1·87
4·8	1·60	1·62	1·64	1·65	1·66	1·68	1·69	1·71	1·73	1·75	1·77	1·79	1·81	1·83	1·85
5·0	1·60	1·62	1·63	1·64	1·66	1·67	1·68	1·71	1·73	1·74	1·76	1·78	1·80	1·82	1·84

Table 42 (*continued*)

Lower 2·5% *points of the standardized deviate* $(x_P - \mu)/\sigma$, $(P = 0{\cdot}025)$

(Note that for positive skewness, i.e. $\mu_3 > 0$, the deviates in this table are negative.)

β_2 \ β_1	0·00	0·01	0·03	0·05	0·10	0·15	0·20	0·30	0·40	0·50	0·60	0·70	0·80	0·90	1·00
1·8	1·65	—	—	—	—	—	—	—	—	—	—	—	—	—	—
2·0	1·76	1·68	1·62	1·56	—	—	—	—	—	—	—	—	—	—	—
2·2	1·83	1·76	1·71	1·66	1·57	1·49	1·41	—	—	—	—	—	—	—	—
2·4	1·88	1·82	1·77	1·73	1·65	1·58	1·51	1·39	—	—	—	—	—	—	—
2·6	1·92	1·86	1·82	1·78	1·71	1·64	1·58	1·47	1·37	—	—	—	—	—	—
2·8	1·94	1·89	1·85	1·82	1·76	1·70	1·65	1·55	1·45	1·35	—	—	—	—	—
3·0	1·96	1·91	1·87	1·84	1·79	1·74	1·69	1·60	1·52	1·42	1·33	—	—	—	—
3·2	1·97	1·93	1·89	1·86	1·81	1·77	1·72	1·65	1·57	1·49	1·40	1·32	1·24	—	—
3·4	1·98	1·94	1·90	1·88	1·83	1·79	1·75	1·68	1·61	1·54	1·46	1·39	1·31	1·23	—
3·6	1·99	1·95	1·91	1·89	1·85	1·81	1·77	1·71	1·65	1·58	1·51	1·44	1·38	1·30	1·23
3·8	1·99	1·95	1·92	1·90	1·86	1·82	1·79	1·73	1·67	1·62	1·56	1·49	1·43	1·36	1·29
4·0	1·99	1·96	1·93	1·91	1·87	1·84	1·81	1·75	1·70	1·64	1·59	1·53	1·47	1·41	1·35
4·2	2·00	1·96	1·93	1·91	1·88	1·84	1·82	1·76	1·72	1·67	1·62	1·56	1·51	1·45	1·40
4·4	2·00	1·96	1·94	1·92	1·88	1·85	1·83	1·78	1·73	1·69	1·64	1·59	1·54	1·49	1·44
4·6	2·00	1·96	1·94	1·92	1·89	1·86	1·83	1·79	1·75	1·70	1·66	1·62	1·57	1·52	1·47
4·8	2·00	1·97	1·94	1·93	1·89	1·87	1·84	1·80	1·76	1·72	1·68	1·64	1·59	1·55	1·50
5·0	2·00	1·97	1·94	1·93	1·90	1·87	1·85	1·81	1·77	1·73	1·69	1·65	1·61	1·57	1·53

Upper 2·5% *points of the standardized deviate* $(x_P - \mu)/\sigma$, $(P = 0{\cdot}975)$

β_2 \ β_1	0·00	0·01	0·03	0·05	0·10	0·15	0·20	0·30	0·40	0·50	0·60	0·70	0·80	0·90	1·00
1·8	1·65	—	—	—	—	—	—	—	—	—	—	—	—	—	—
2·0	1·76	1·82	1·86	1·89	—	—	—	—	—	—	—	—	—	—	—
2·2	1·83	1·89	1·93	1·96	2·00	2·04	2·06	—	—	—	—	—	—	—	—
2·4	1·88	1·94	1·98	2·01	2·05	2·08	2·11	2·15	—	—	—	—	—	—	—
2·6	1·92	1·97	2·01	2·03	2·08	2·11	2·14	2·18	2·22	—	—	—	—	—	—
2·8	1·94	1·99	2·03	2·05	2·09	2·13	2·15	2·20	2·24	2·27	—	—	—	—	—
3·0	1·96	2·01	2·04	2·06	2·10	2·13	2·16	2·21	2·25	2·28	2·32	—	—	—	—
3·2	1·97	2·02	2·05	2·07	2·11	2·14	2·16	2·21	2·25	2·29	2·32	2·35	2·38	—	—
3·4	1·98	2·02	2·05	2·07	2·11	2·14	2·16	2·21	2·25	2·28	2·32	2·35	2·38	2·41	—
3·6	1·99	2·02	2·05	2·07	2·11	2·14	2·16	2·20	2·24	2·28	2·31	2·34	2·37	2·41	2·44
3·8	1·99	2·03	2·05	2·07	2·11	2·13	2·16	2·20	2·24	2·27	2·30	2·33	2·36	2·40	2·43
4·0	1·99	2·03	2·05	2·07	2·11	2·13	2·15	2·19	2·23	2·26	2·29	2·32	2·35	2·38	2·42
4·2	2·00	2·03	2·05	2·07	2·10	2·13	2·15	2·19	2·22	2·25	2·28	2·31	2·34	2·37	2·40
4·4	2·00	2·03	2·05	2·07	2·10	2·13	2·15	2·18	2·22	2·25	2·28	2·31	2·33	2·36	2·39
4·6	2·00	2·03	2·05	2·07	2·10	2·12	2·14	2·18	2·21	2·24	2·27	2·30	2·32	2·35	2·38
4·8	2·00	2·03	2·05	2·07	2·10	2·12	2·14	2·17	2·21	2·23	2·26	2·29	2·31	2·34	2·37
5·0	2·00	2·03	2·05	2·07	2·09	2·12	2·14	2·17	2·20	2·23	2·25	2·28	2·30	2·33	2·35

Table 42. *Percentage points of Pearson curves (continued)*

Lower 1% *points of the standardized deviate* $(x_P - \mu)/\sigma$, $(P = 0{\cdot}01)$

(Note that for positive skewness, i.e. $\mu_3 > 0$, the deviates in this table are negative.)

β_2 \ β_1	0·00	0·01	0·03	0·05	0·10	0·15	0·20	0·30	0·40	0·50	0·60	0·70	0·80	0·90	1·00
1·8	1·70	—	—	—	—	—	—	—	—	—	—	—	—	—	—
2·0	1·87	1·77	1·69	1·62	—	—	—	—	—	—	—	—	—	—	—
2·2	2·01	1·91	1·83	1·76	1·64	1·55	1·45	—	—	—	—	—	—	—	—
2·4	2·12	2·03	1·95	1·89	1·77	1·68	1·59	1·43	—	—	—	—	—	—	—
2·6	2·21	2·12	2·05	1·99	1·88	1·79	1·70	1·55	1·41	—	—	—	—	—	—
2·8	2·27	2·19	2·13	2·08	1·98	1·89	1·81	1·66	1·52	1·39	—	—	—	—	—
3·0	2·33	2·25	2·19	2·14	2·05	1·97	1·90	1·76	1·62	1·50	1·38	—	—	—	—
3·2	2·37	2·29	2·24	2·19	2·11	2·03	1·96	1·84	1·71	1·59	1·48	1·37	1·26	—	—
3·4	2·40	2·33	2·28	2·24	2·16	2·09	2·02	1·90	1·79	1·68	1·57	1·46	1·36	1·26	—
3·6	2·43	2·36	2·31	2·27	2·20	2·13	2·07	1·96	1·86	1·76	1·65	1·55	1·45	1·35	1·26
3·8	2·45	2·39	2·34	2·30	2·23	2·17	2·11	2·01	1·91	1·82	1·72	1·62	1·53	1·43	1·34
4·0	2·47	2·41	2·36	2·33	2·26	2·20	2·15	2·05	1·96	1·87	1·78	1·69	1·60	1·51	1·42
4·2	2·49	2·43	2·38	2·35	2·28	2·23	2·18	2·09	2·00	1·92	1·83	1·75	1·66	1·58	1·49
4·4	2·50	2·44	2·40	2·37	2·31	2·25	2·21	2·12	2·04	1·96	1·88	1·80	1·72	1·64	1·56
4·6	2·51	2·46	2·42	2·38	2·32	2·27	2·23	2·15	2·07	2·00	1·92	1·84	1·77	1·70	1·62
4·8	2·52	2·47	2·43	2·40	2·34	2·29	2·25	2·17	2·10	2·03	1·96	1·88	1·81	1·74	1·67
5·0	2·53	2·48	2·44	2·41	2·36	2·31	2·27	2·19	2·12	2·06	1·99	1·92	1·85	1·79	1·72

Upper 1% *points of the standardized deviate* $(x_P - \mu)/\sigma$, $(P = 0{\cdot}99)$

β_2 \ β_1	0·00	0·01	0·03	0·05	0·10	0·15	0·20	0·30	0·40	0·50	0·60	0·70	0·80	0·90	1·00
1·8	1·70	—	—	—	—	—	—	—	—	—	—	—	—	—	—
2·0	1·87	1·95	2·00	2·03	—	—	—	—	—	—	—	—	—	—	—
2·2	2·01	2·10	2·15	2·18	2·22	2·24	2·25	—	—	—	—	—	—	—	—
2·4	2·12	2·20	2·25	2·28	2·33	2·36	2·38	2·40	—	—	—	—	—	—	—
2·6	2·21	2·28	2·33	2·36	2·42	2·45	2·48	2·51	2·52	—	—	—	—	—	—
2·8	2·27	2·34	2·39	2·43	2·48	2·52	2·55	2·59	2·61	2·63	—	—	—	—	—
3·0	2·33	2·40	2·44	2·48	2·53	2·56	2·59	2·64	2·68	2·70	2·71	—	—	—	—
3·2	2·37	2·44	2·48	2·51	2·56	2·60	2·63	2·68	2·72	2·75	2·77	2·79	2·80	—	—
3·4	2·40	2·47	2·51	2·54	2·59	2·63	2·66	2·71	2·75	2·79	2·82	2·84	2·86	2·87	—
3·6	2·43	2·49	2·53	2·56	2·61	2·65	2·68	2·74	2·78	2·81	2·85	2·87	2·90	2·91	2·93
3·8	2·45	2·51	2·55	2·58	2·63	2·67	2·70	2·75	2·80	2·83	2·87	2·90	2·92	2·95	2·97
4·0	2·47	2·53	2·57	2·60	2·65	2·68	2·71	2·77	2·81	2·85	2·88	2·91	2·94	2·97	2·99
4·2	2·49	2·54	2·58	2·61	2·66	2·69	2·73	2·78	2·82	2·86	2·89	2·92	2·95	2·98	3·01
4·4	2·50	2·56	2·59	2·62	2·67	2·70	2·73	2·78	2·83	2·86	2·90	2·93	2·96	2·99	3·02
4·6	2·51	2·57	2·60	2·63	2·68	2·71	2·74	2·79	2·83	2·87	2·90	2·94	2·97	3·00	3·03
4·8	2·52	2·58	2·61	2·64	2·68	2·72	2·75	2·80	2·84	2·87	2·91	2·94	2·97	3·00	3·03
5·0	2·53	2·58	2·62	2·64	2·69	2·72	2·75	2·80	2·84	2·88	2·91	2·95	2·97	3·00	3·03

Table 42 (*continued*)

Lower 0·5% *points of the standardized deviate* $(x_P - \mu)/\sigma$, $(P = 0{\cdot}005)$

(Note that for positive skewness, i.e. $\mu_3 > 0$, the deviates in this table are negative.)

β_2 \ β_1	0·00	0·01	0·03	0·05	0·10	0·15	0·20	0·30	0·40	0·50	0·60	0·70	0·80	0·90	1·00
1·8	1·71	—	—	—	—	—	—	—	—	—	—	—	—	—	—
2·0	1·92	1·80	1·71	1·64	—	—	—	—	—	—	—	—	—	—	—
2·2	2·10	1·99	1·89	1·82	1·68	1·56	1·46	—	—	—	—	—	—	—	—
2·4	2·26	2·14	2·04	1·97	1·83	1·71	1·62	1·44	—	—	—	—	—	—	—
2·6	2·38	2·27	2·18	2·12	1·98	1·87	1·77	1·58	1·42	—	—	—	—	—	—
2·8	2·49	2·38	2·30	2·23	2·10	1·99	1·89	1·71	1·55	1·41	—	—	—	—	—
3·0	2·58	2·48	2·39	2·33	2·21	2·11	2·01	1·84	1·68	1·53	1·40	—	—	—	—
3·2	2·65	2·55	2·48	2·42	2·30	2·20	2·11	1·95	1·79	1·65	1·51	1·39	1·27	—	—
3·4	2·71	2·61	2·54	2·48	2·38	2·28	2·20	2·04	1·90	1·76	1·62	1·50	1·38	1·27	—
3·6	2·76	2·67	2·60	2·54	2·44	2·35	2·27	2·13	1·99	1·85	1·72	1·60	1·48	1·37	1·27
3·8	2·80	2·71	2·65	2·60	2·50	2·41	2·34	2·20	2·07	1·94	1·82	1·70	1·58	1·47	1·36
4·0	2·83	2·75	2·69	2·64	2·54	2·47	2·39	2·26	2·14	2·02	1·90	1·78	1·67	1·56	1·45
4·2	2·87	2·79	2·72	2·68	2·59	2·51	2·44	2·32	2·20	2·09	1·97	1·86	1·75	1·65	1·54
4·4	2·89	2·82	2·76	2·71	2·62	2·55	2·49	2·37	2·25	2·15	2·04	1·93	1·83	1·73	1·62
4·6	2·92	2·85	2·79	2·74	2·66	2·59	2·52	2·41	2·30	2·20	2·10	2·00	1·90	1·80	1·70
4·8	2·94	2·87	2·81	2·77	2·69	2·62	2·56	2·45	2·35	2·25	2·15	2·05	1·96	1·87	1·77
5·0	2·96	2·89	2·83	2·79	2·71	2·65	2·59	2·48	2·39	2·29	2·20	2·11	2·01	1·92	1·84

Upper 0·5% *points of the standardized deviate* $(x_P - \mu)/\sigma$, $(P = 0{\cdot}995)$

β_2 \ β_1	0·00	0·01	0·03	0·05	0·10	0·15	0·20	0·30	0·40	0·50	0·60	0·70	0·80	0·90	1·00
1·8	1·71	—	—	—	—	—	—	—	—	—	—	—	—	—	—
2·0	1·92	2·01	2·06	2·09	—	—	—	—	—	—	—	—	—	—	—
2·2	2·10	2·19	2·24	2·27	2·31	2·33	2·35	—	—	—	—	—	—	—	—
2·4	2·26	2·35	2·41	2·44	2·49	2·52	2·53	2·53	—	—	—	—	—	—	—
2·6	2·38	2·48	2·54	2·57	2·63	2·66	2·68	2·70	2·69	—	—	—	—	—	—
2·8	2·49	2·58	2·64	2·68	2·73	2·77	2·80	2·83	2·84	2·83	—	—	—	—	—
3·0	2·58	2·66	2·72	2·76	2·82	2·86	2·89	2·93	2·95	2·96	2·95	—	—	—	—
3·2	2·65	2·73	2·79	2·83	2·89	2·93	2·96	3·01	3·04	3·06	3·07	3·06	3·04	—	—
3·4	2·71	2·79	2·85	2·88	2·95	2·99	3·02	3·07	3·11	3·13	3·15	3·16	3·15	3·14	—
3·6	2·76	2·84	2·89	2·93	2·99	3·03	3·07	3·12	3·16	3·19	3·22	3·23	3·24	3·24	3·24
3·8	2·80	2·88	2·93	2·97	3·03	3·07	3·11	3·16	3·20	3·24	3·27	3·29	3·30	3·31	3·32
4·0	2·83	2·91	2·96	3·00	3·06	3·10	3·14	3·20	3·24	3·28	3·31	3·34	3·36	3·37	3·38
4·2	2·87	2·94	2·99	3·03	3·09	3·13	3·17	3·22	3·27	3·31	3·34	3·37	3·40	3·42	3·43
4·4	2·89	2·97	3·02	3·05	3·11	3·15	3·19	3·25	3·29	3·33	3·37	3·40	3·42	3·45	3·47
4·6	2·92	2·99	3·04	3·07	3·13	3·17	3·21	3·27	3·31	3·36	3·39	3·42	3·44	3·47	3·50
4·8	2·94	3·01	3·06	3·09	3·15	3·19	3·23	3·28	3·33	3·37	3·41	3·44	3·47	3·49	3·53
5·0	2·96	3·03	3·07	3·11	3·16	3·21	3·24	3·30	3·35	3·39	3·43	3·46	3·49	3·52	3·55

Table 43. *Chart relating the type of Pearson frequency curve to the values of* β_1, β_2.

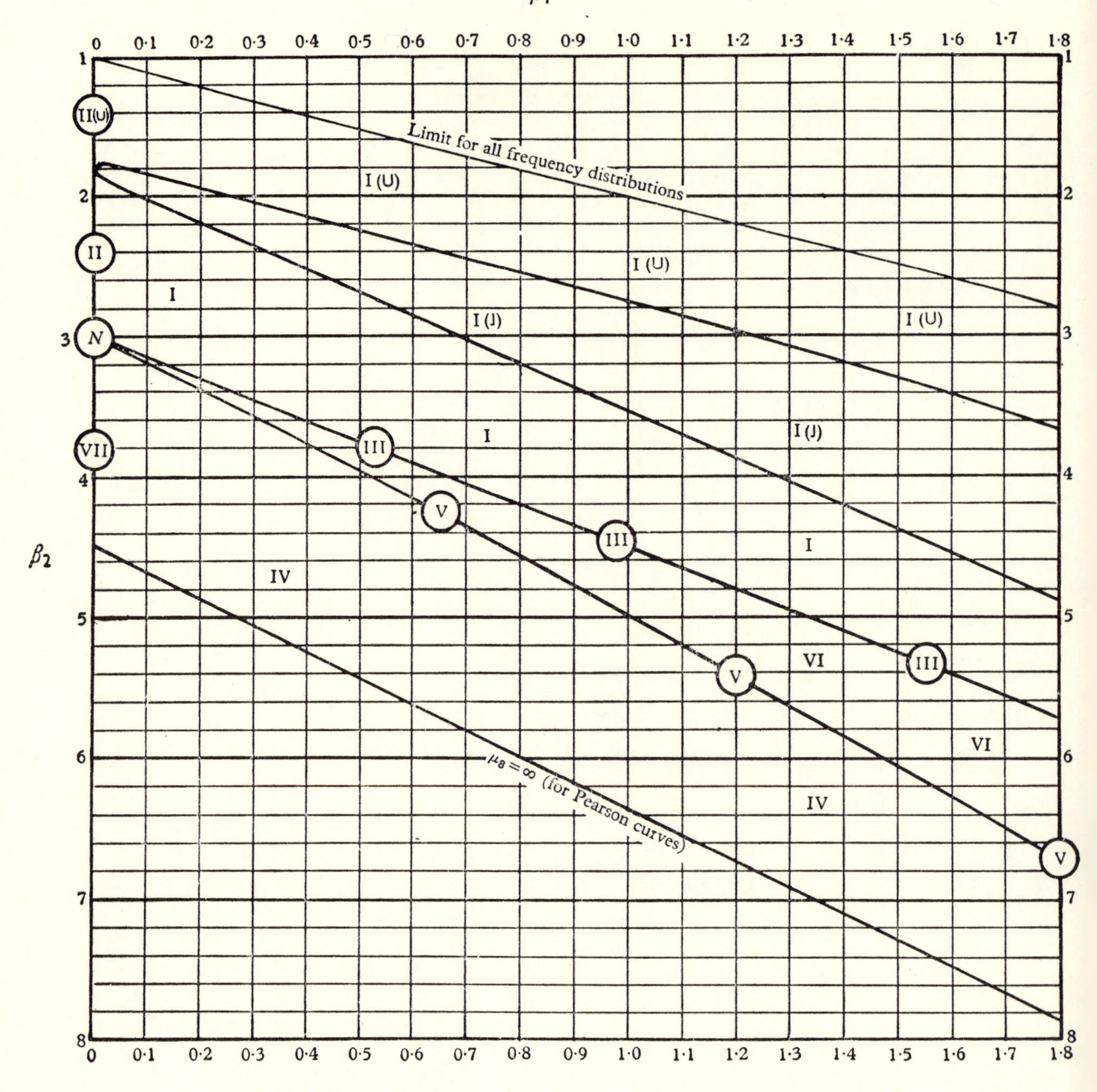

Equations of bounding curves

Upper limit for all frequency distributions: $\beta_2 - \beta_1 - 1 = 0$.

Boundary of $I(J)$ area: $4(4\beta_2 - 3\beta_1)(5\beta_2 - 6\beta_1 - 9)^2 = \beta_1(\beta_2 + 3)^2(8\beta_2 - 9\beta_1 - 12)$.

Type III line: $2\beta_2 - 3\beta_1 - 6 = 0$.

Type V line: $\beta_1(\beta_2 + 3)^2 = 4(4\beta_2 - 3\beta_1)(2\beta_2 - 3\beta_1 - 6)$.

Line on and below which $\mu_8 = \infty$ (for Pearson curves): $8\beta_2 - 15\beta_1 - 36 = 0$.

Table 44. *Distribution of Spearman's rank correlation coefficient, r_s, in random rankings. Probability $Q = 1 - P(S_r \mid n)$ that a given value of S_r will be attained or exceeded.* $r_s = 1 - 6S_r/(n^3 - n)$

$n = 4$		$n = 5$		$n = 6$		$n = 7$		$n = 8$		$n = 9$		$n = 10$	
S_r	Q	S_r	Q	S_r	Q	S_r	Q	S_r	Q	S_r	Q	S_r	Q
12	0·458	22	0·475	50	0·210	74	0·249	108	0·250	156	0·218	208	0·235
14	·375	24	·392	52	·178	78	·198	114	·195	164	·168	218	·184
16	·208	26	·342	54	·149	82	·151	120	·150	172	·125	228	·139
18	·167	28	·258	56	·121	86	·118	126	·108	180	·089	238	·102
20	·042	30	·225	58	·088	90	·083	132	·076	188	·060	248	·072
		32	0·175	60	0·068	94	0·055	138	0·048	196	0·038	258	0·048
		34	·117	62	·051	98	·033	144	·029	204	·022	268	·030
		36	·067	64	·029	102	·017	150	·014	212	·011	278	·017
		38	·042	66	·017	106	·0062	156	·0054	220	·0041	288	·0087
		40	·0083	68	·0083	110	·0014	162	·0011	228	·0010	298	·0036
				70	0·0014							308	0·0011
20		40		70		112		168		240		330	

Tail area at the lower end of the distribution. If S_r and Q are two corresponding tabular entries for a given sample size n, and if $S_r' = \frac{1}{3}(n^3 - n) - S_r$, then the probability is Q that the sum of squares of the rank differences will be $\leqslant S_r'$. Values of $\frac{1}{3}(n^3 - n)$ are shown in the bottom row of the table. E.g. if $n = 8$, the probability is 0·048 that the sum of squares of the rank differences will be less than or equal to $168 - 138 = 30$.

Table 45. *Distribution of Kendall's rank correlation coefficient, t_k, in random rankings. Probability, $Q = 1 - P(S_t \mid n)$ that a given value of S_t will be attained or exceeded.* $t_k = 2S_t/(n^2 - n)$

S_t	Values of n				S_t	Values of n		
	4	5	8	9		6	7	10
0	0·625	0·592	0·548	0·540	1	0·500	0·500	0·500
2	·375	·408	·452	·460	3	·360	·386	·431
4	·167	·242	·360	·381	5	·235	·281	·364
6	·042	·117	·274	·306	7	·136	·191	·300
8		·042	·199	·238	9	·068	·119	·242
10		0·0083	0·138	0·179	11	0·028	0·068	0·190
12			·089	·130	13	·0083	·035	·146
14			·054	·090	15	·0014	·015	·108
16			·031	·060	17		·0054	·078
18			·016	·038	19		·0014	·054
20			0·0071	0·022	21		0·0002	0·036
22			·0028	·012	23			·023
24			·0009	·0063	25			·014
26			·0002	·0029	27			·0083
28				·0012	29			·0046
30				0·0004	31			0·0023
					33			·0011
					35			·0005

The distribution of S_t is symmetrical so that values of the probability for negative S_t can be obtained by appropriate subtraction from unity; e.g. for $n = 9$,

$$\Pr\{S_t \geqslant -14\} = 1 - \Pr\{S_t \geqslant 16\} = 1 - 0{\cdot}060 = 0{\cdot}940.$$

Table 46. *Distribution of the concordance coefficient, W, in random rankings. Probability $Q = 1 - P(S_W \mid n, m)$ that a given value of S_W will be attained or exceeded.* $W = 12S_W/\{m^2(n^3 - n)\}$

(*a*) Case $n = 3$

$m = 3$		$m = 5$		$m = 6$		$m = 7$		$m = 8$		$m = 9$		$m = 10$	
S_W	Q	S_W	Q	S_W	Q	S_W	Q	S_W	Q	S_W	Q	S_W	Q
6	0·528	14	0·367	18	0·252	24	0·237	26	0·236	32	0·187	32	0·222
8	·361	18	·182	24	·184	26	·192	32	·149	38	·154	42	·135
14	·194	24	·124	26	·142	32	·112	38	·120	42	·107	50	·092
18	·028	26	·093	32	·072	38	·085	42	·079	50	·069	56	·066
		32	·039	38	·052	42	·051	50	·047	56	·048	62	·046
$m = 4$		38	0·024	42	0·029	50	0·027	56	0·030	62	0·031	74	0·026
		42	·0085	50	·012	56	·016	72	·0099	78	·010	86	·012
		50	·0008	54	·0081	62	·0084	78	·0048	86	·0060	96	·0075
S_W	Q			56	·0055	72	·0036	86	·0024	93	·0029	104	·0034
				62	·0017	78	·0012	98	·0009	104	·0013	122	·0013
				72	0·0001	96	0·0003			114	0·0007	126	0·0008
8	0·431												
14	·273												
18	·125												
24	·069												
26	·042												
32	·0046												

(*b*) Case $n = 4$								(*c*) Case $n = 5$	
$m = 3$		$m = 4$		$m = 5$		$m = 6$		$m = 3$	
S_W	Q	S_W	Q	S_W	Q	S_W	Q	S_W	Q
19	0·342	32	0·200	41	0·210	46	0·218	46	0·213
21	·300	36	·158	43	·162	52	·163	50	·163
25	·207	40	·105	51	·107	62	·108	56	·096
27	·175	46	·068	57	·075	68	·073	60	·063
29	·148	50	·052	61	·055	74	·056	62	·056
33	0·075	54	0·033	67	0·034	80	0·037	66	0·038
35	·054	62	·012	81	·012	100	·010	74	·015
37	·033	66	·0062	85	·0067	108	·0061	78	·0053
41	·017	70	·0027	93	·0023	118	·0028	82	·0028
45	0·0017	74	·0009	101	·0014	128	·0009	86	·0009
				105	0·0006				

Table 47. *Orthogonal polynomials*

Definitions:

$$\phi_0(x)=1,\quad \phi_1(x)=\lambda_1 x,\quad \phi_2(x)=\lambda_2\{x^2-\tfrac{1}{12}(n^2-1)\},$$
$$\phi_3(x)=\lambda_3\{x^3-\tfrac{1}{20}(3n^2-7)\,x\},\quad \phi_4(x)=\lambda_4\{x^4-\tfrac{1}{14}(3n^2-13)\,x^2+\tfrac{3}{560}(n^2-1)\,(n^2-9)\},$$
$$\phi_5(x)=\lambda_5\{x^5-\tfrac{5}{18}(n^2-7)\,x^3+\tfrac{1}{1008}(15n^4-230n^2+407)\,x\},$$
$$\phi_6(x)=\lambda_6\{x^6-\tfrac{5}{44}(3n^2-31)\,x^4+\tfrac{1}{176}(5n^4-110n^2+329)\,x^2-\tfrac{5}{14784}(n^2-1)\,(n^2-9)\,(n^2-25)\}.$$

The table gives the values of these polynomials, $\phi_i(x)$ for $i=1\ (1)\ 6$* and is arranged in fifty sections corresponding to sample sizes $n=3\ (1)\ 52$. The arguments $x=x_t\equiv t-\tfrac{1}{2}(n+1)$ cover the full observational range $t=1\ (1)\ n$ for $n=3\ (1)\ 12$ and the half-range $t=1\ (1)\left[\frac{n+1}{2}\right]$ for $n>12$, when use is made of the symmetry (antisymmetry) relations:

$$\phi_i(x)=\phi_i(-x)\ \text{for } i \text{ even},\quad \phi_i(x)=-\phi_i(-x)\ \text{for } i \text{ odd}.$$

The argument x is not shown as such in the table, but the first column of each section gives $\phi_1=\lambda_1 x$, where λ_1 is 1 for n odd and 2 for n even. The leading coefficients λ_i are chosen so that the $\phi_i(x_t)$ are positive or negative integers throughout, and are given in the bottom line of each section. Also shown, in the line above the λ_i, are the sums of squares $\sum_{t=1}^{n}\{\phi_i(x_t)\}^2$ for the full range of x_t values. Note the arrangements of the sections which progress from left to right through consecutive pages up to $n=28$. From this point onwards the sections progress from left to right through the top halves of the pages up to $n=40$ and then return from right to left through the bottom halves of the pages.

Some formulae:

Estimate of coefficient of $\phi_i(x)$: $A_i=\sum_t y_t\,\phi_i(x_t)/\sum_t\{\phi_i(x_t)\}^2$.

Variance of $A_i=\sigma_y^2/\sum_t\{\phi_i(x_t)\}^2$.

Estimate of σ_y^2: $s^2=\left(\sum_t y_t^2-\sum_{i=0}^{k}\{A_i^2\sum_t[\phi_i(x_t)]^2\}\right)\Big/(n-k-1)$.

* Except that for $n\leqslant 6$, i is not taken beyond $n-1$.

n = 3

ϕ_1	ϕ_2
−1	1
0	−2
1	1
2	6
1	3

n = 4

ϕ_1	ϕ_2	ϕ_3
−3	1	−1
−1	−1	3
1	−1	−3
3	1	1
20	4	20
2	1	$\frac{10}{3}$

n = 5

ϕ_1	ϕ_2	ϕ_3	ϕ_4
−2	2	−1	1
−1	−1	2	−4
0	−2	0	6
1	−1	−2	−4
2	2	1	1
10	14	10	70
1	1	$\frac{5}{6}$	$\frac{35}{12}$

n = 6

ϕ_1	ϕ_2	ϕ_3	ϕ_4	ϕ_5
−5	5	−5	1	−1
−3	−1	7	−3	5
−1	−4	4	2	−10
1	−4	−4	2	10
3	−1	−7	−3	−5
5	5	5	1	1
70	84	180	28	252
2	$\frac{3}{2}$	$\frac{5}{3}$	$\frac{7}{12}$	$\frac{21}{10}$

n = 7

ϕ_1	ϕ_2	ϕ_3	ϕ_4	ϕ_5	ϕ_6
−3	5	−1	3	−1	1
−2	0	1	−7	4	−6
−1	−3	1	1	−5	15
0	−4	0	6	0	−20
1	−3	−1	1	5	15
2	0	−1	−7	−4	−6
3	5	1	3	1	1
28	84	6	154	84	924
1	1	$\frac{1}{6}$	$\frac{7}{12}$	$\frac{7}{20}$	$\frac{77}{60}$

n = 8

ϕ_1	ϕ_2	ϕ_3	ϕ_4	ϕ_5	ϕ_6
−7	7	−7	7	−7	1
−5	1	5	−13	23	−5
−3	−3	7	−3	−17	9
−1	−5	3	9	−15	−5
1	−5	−3	9	15	−5
3	−3	−7	−3	17	9
5	1	−5	−13	−23	−5
7	7	7	7	7	1
168	168	264	616	2,184	264
2	1	$\frac{2}{3}$	$\frac{7}{12}$	$\frac{7}{10}$	$\frac{11}{60}$

n = 9

ϕ_1	ϕ_2	ϕ_3	ϕ_4	ϕ_5	ϕ_6
−4	28	−14	14	−4	4
−3	7	7	−21	11	−17
−2	−8	13	−11	−4	22
−1	−17	9	9	−9	1
0	−20	0	18	0	−20
1	−17	−9	9	9	1
2	−8	−13	−11	4	22
3	7	−7	−21	−11	−17
4	28	14	14	4	4
60	2,772	990	2,002	468	1,980
1	3	$\frac{5}{6}$	$\frac{7}{12}$	$\frac{3}{20}$	$\frac{11}{60}$

n = 10

ϕ_1	ϕ_2	ϕ_3	ϕ_4	ϕ_5	ϕ_6
−9	6	−42	18	−6	3
−7	2	14	−22	14	−11
−5	−1	35	−17	−1	10
−3	−3	31	3	−11	6
−1	−4	12	18	−6	−8
1	−4	−12	18	6	−8
3	−3	−31	3	11	6
5	−1	−35	−17	1	10
7	2	−14	−22	−14	−11
9	6	42	18	6	3
330	132	8,580	2,860	780	660
2	$\frac{1}{2}$	$\frac{5}{3}$	$\frac{5}{12}$	$\frac{1}{10}$	$\frac{11}{240}$

Table 47 (*continued*)

$n=11$					
ϕ_1	ϕ_2	ϕ_3	ϕ_4	ϕ_5	ϕ_6
−5	15	−30	6	−3	15
−4	6	6	−6	6	−48
−3	−1	22	−6	1	29
−2	−6	23	−1	−4	36
−1	−9	14	4	−4	−12
0	−10	0	6	0	−40
1	−9	−14	4	4	−12
2	−6	−23	−1	4	36
3	−1	−22	−6	−1	29
4	6	−6	−6	−6	−48
5	15	30	6	3	15
110	858	4,290	286	156	11,220
1	1	$\frac{5}{6}$	$\frac{1}{12}$	$\frac{1}{40}$	$\frac{11}{120}$

$n=12$					
ϕ_1	ϕ_2	ϕ_3	ϕ_4	ϕ_5	ϕ_6
−11	55	−33	33	−33	11
−9	25	3	−27	57	−31
−7	1	21	−33	21	11
−5	−17	25	−13	−29	25
−3	−29	19	12	−44	4
−1	−35	7	28	−20	−20
1	−35	−7	28	20	−20
3	−29	−19	12	44	4
5	−17	−25	−13	29	25
7	1	−21	−33	−21	11
9	25	−3	−27	−57	−31
11	55	33	33	33	11
572	12,012	5,148	8,008	15,912	4,488
2	3	$\frac{2}{3}$	$\frac{7}{24}$	$\frac{3}{20}$	$\frac{11}{360}$

$n=13$					
ϕ_1	ϕ_2	ϕ_3	ϕ_4	ϕ_5	ϕ_6
−6	22	−11	99	−22	22
−5	11	0	−66	33	−55
−4	2	6	−96	18	8
−3	−5	8	−54	−11	43
−2	−10	7	11	−26	22
−1	−13	4	64	−20	−20
0	−14	0	84	0	−40
182	2,002	572	68,068	6,188	14,212
1	1	$\frac{1}{6}$	$\frac{7}{12}$	$\frac{7}{120}$	$\frac{11}{360}$

$n=14$					
ϕ_1	ϕ_2	ϕ_3	ϕ_4	ϕ_5	ϕ_6
−13	13	−143	143	−143	143
−11	7	−11	−77	187	−319
−9	2	66	−132	132	−11
−7	−2	98	−92	−28	227
−5	−5	95	−13	−139	185
−3	−7	67	63	−145	−25
−1	−8	24	108	−60	−200
910	728	97,240	136,136	235,144	497,420
2	$\frac{1}{2}$	$\frac{5}{3}$	$\frac{7}{12}$	$\frac{7}{30}$	$\frac{77}{720}$

$n=15$					
ϕ_1	ϕ_2	ϕ_3	ϕ_4	ϕ_5	ϕ_6
−7	91	−91	1,001	−1,001	143
−6	52	−13	−429	1,144	−286
−5	19	35	−869	979	−55
−4	−8	58	−704	44	176
−3	−29	61	−249	−751	197
−2	−44	49	251	−1,000	50
−1	−53	27	621	−675	−125
0	−56	0	756	0	−200
280	37,128	39,780	6,466,460	10,581,480	426,360
1	3	$\frac{5}{6}$	$\frac{35}{12}$	$\frac{21}{20}$	$\frac{11}{180}$

$n=16$					
ϕ_1	ϕ_2	ϕ_3	ϕ_4	ϕ_5	ϕ_6
−15	35	−455	273	−143	65
−13	21	−91	−91	143	−117
−11	9	143	−221	143	−39
−9	−1	267	−201	33	59
−7	−9	301	−101	−77	87
−5	−15	265	23	−131	45
−3	−19	179	129	−115	−25
−1	−21	63	189	−45	−75
1,360	5,712	1,007,760	470,288	201,552	77,520
2	1	$\frac{10}{3}$	$\frac{7}{12}$	$\frac{1}{10}$	$\frac{1}{60}$

Table 47. *Orthogonal polynomials (continued)*

$n=17$

ϕ_1	ϕ_2	ϕ_3	ϕ_4	ϕ_5	ϕ_6
−8	40	−28	52	−104	104
−7	25	−7	−13	91	−169
−6	12	7	−39	104	−78
−5	1	15	−39	39	65
−4	−8	18	−24	−36	128
−3	−15	17	−3	−83	93
−2	−20	13	17	−88	2
−1	−23	7	31	−55	−85
0	−24	0	36	0	−120
408	7,752	3,876	16,796	100,776	178,296
1	1	$\frac{1}{6}$	$\frac{1}{12}$	$\frac{1}{20}$	$\frac{1}{60}$

$n=18$

ϕ_1	ϕ_2	ϕ_3	ϕ_4	ϕ_5	ϕ_6
−17	68	−68	68	−884	442
−15	44	−20	−12	676	−650
−13	23	13	−47	871	−377
−11	5	33	−51	429	169
−9	−10	42	−36	−156	481
−7	−22	42	−12	−588	439
−5	−31	35	13	−733	145
−3	−37	23	33	−583	−209
−1	−40	8	44	−220	−440
1,938	23,256	23,256	28,424	6,953,544	2,941,884
2	$\frac{3}{2}$	$\frac{1}{3}$	$\frac{1}{12}$	$\frac{3}{10}$	$\frac{11}{240}$

$n=19$

ϕ_1	ϕ_2	ϕ_3	ϕ_4	ϕ_5	ϕ_6
−9	51	−204	612	−102	1,326
−8	34	−68	−68	68	−1,768
−7	19	28	−388	98	−1,222
−6	6	89	−453	58	234
−5	−5	120	−354	−3	1,235
−4	−14	126	−168	−54	1,352
−3	−21	112	42	−79	729
−2	−26	83	227	−74	−214
−1	−29	44	352	−44	−1,012
0	−30	0	396	0	−1,320
570	13,566	213,180	2,288,132	89,148	24,515,700
1	1	$\frac{5}{6}$	$\frac{7}{12}$	$\frac{1}{40}$	$\frac{11}{120}$

$n=20$

ϕ_1	ϕ_2	ϕ_3	ϕ_4	ϕ_5	ϕ_6
−19	57	−969	1,938	−1,938	1,938
−17	39	−357	−102	1,122	−2,346
−15	23	85	−1,122	1,802	−1,870
−13	9	377	−1,402	1,222	6
−11	−3	539	−1,187	187	1,497
−9	−13	591	−687	−771	1,931
−7	−21	553	−77	−1,351	1,353
−5	−27	445	503	−1,441	195
−3	−31	287	948	−1,076	−988
−1	−33	99	1,188	−396	−1,716
2,660	17,556	4,903,140	22,881,320	31,201,800	49,031,400
2	1	$\frac{10}{3}$	$\frac{35}{24}$	$\frac{7}{20}$	$\frac{11}{120}$

$n=21$

ϕ_1	ϕ_2	ϕ_3	ϕ_4	ϕ_5	ϕ_6
−10	190	−285	969	−3,876	6,460
−9	133	−114	0	1,938	−7,106
−8	82	12	−510	3,468	−6,392
−7	37	98	−680	2,618	−918
−6	−2	149	−615	788	3,996
−5	−35	170	−406	−1,063	6,075
−4	−62	166	−130	−2,354	5,088
−3	−83	142	150	−2,819	2,001
−2	−98	103	385	−2,444	−1,716
−1	−107	54	540	−1,404	−4,628
0	−110	0	594	0	−5,720
770	201,894	432,630	5,720,330	121,687,020	514,829,700
1	3	$\frac{5}{6}$	$\frac{7}{12}$	$\frac{21}{40}$	$\frac{77}{360}$

$n=22$

ϕ_1	ϕ_2	ϕ_3	ϕ_4	ϕ_5	ϕ_6
−21	35	−133	1,197	−2,261	646
−19	25	−57	57	969	−646
−17	16	0	−570	1,938	−646
−15	8	40	−810	1,598	−170
−13	1	65	−775	663	306
−11	−5	77	−563	−363	558
−9	−10	78	−258	−1,158	537
−7	−14	70	70	−1,554	303
−5	−17	55	365	−1,509	−30
−3	−19	35	585	−1,079	−338
−1	−20	12	702	−390	−520
3,542	7,084	96,140	8,748,740	40,562,340	4,903,140
2	$\frac{1}{2}$	$\frac{1}{3}$	$\frac{7}{12}$	$\frac{7}{30}$	$\frac{11}{720}$

Table 47 (*continued*)

$n=23$

ϕ_1	ϕ_2	ϕ_3	ϕ_4	ϕ_5	ϕ_6
-11	77	-77	1,463	-209	3,553
-10	56	-35	133	76	-3,230
-9	37	-3	-627	171	-3,553
-8	20	20	-950	152	-1,292
-7	5	35	-955	77	1,207
-6	-8	43	-747	-12	2,754
-5	-19	45	-417	-87	2,985
-4	-28	42	-42	-132	2,076
-3	-35	35	315	-141	501
-2	-40	25	605	-116	-1,166
-1	-43	13	793	-65	-2,405
0	-44	0	858	0	-2,860
1,012	35,420	32,890	13,123,110	340,860	142,191,060
1	1	$\frac{1}{6}$	$\frac{7}{12}$	$\frac{1}{60}$	$\frac{11}{180}$

$n=24$

ϕ_1	ϕ_2	ϕ_3	ϕ_4	ϕ_5	ϕ_6
-23	253	-1,771	253	-4,807	4,807
-21	187	-847	33	1,463	-3,971
-19	127	-133	-97	3,743	-4,769
-17	73	391	-157	3,553	-2,147
-15	25	745	-165	2,071	1,045
-13	-17	949	-137	169	3,271
-11	-53	1,023	-87	-1,551	3,957
-9	-83	987	-27	-2,721	3,183
-7	-107	861	33	-3,171	1,419
-5	-125	665	85	-2,893	-695
-3	-137	419	123	-2,005	-2,525
-1	-143	143	143	-715	-3,575
4,600	394,680	17,760,600	394,680	177,928,920	250,925,400
2	3	$\frac{10}{3}$	$\frac{1}{12}$	$\frac{3}{10}$	$\frac{11}{180}$

$n=25$

ϕ_1	ϕ_2	ϕ_3	ϕ_4	ϕ_5	ϕ_6
-12	92	-506	1,518	-1,012	19,228
-11	69	-253	253	253	-14,421
-10	48	-55	-517	748	-18,810
-9	29	93	-897	753	-9,899
-8	12	196	-982	488	2,052
-7	-3	259	-857	119	11,229
-6	-16	287	-597	-236	15,142
-5	-27	285	-267	-501	13,635
-4	-36	258	78	-636	8,028
-3	-43	211	393	-631	391
-2	-48	149	643	-500	-7,050
-1	-51	77	803	-275	-12,375
0	-52	0	858	0	-14,300
1,300	53,820	1,480,050	14,307,150	7,803,900	3,889,343,700
1	1	$\frac{5}{6}$	$\frac{5}{12}$	$\frac{1}{20}$	$\frac{11}{60}$

$n=26$

ϕ_1	ϕ_2	ϕ_3	ϕ_4	ϕ_5	ϕ_6
-25	50	-1,150	2,530	-2,530	6,325
-23	38	-598	506	506	-4,301
-21	27	-161	-759	1,771	-6,072
-19	17	171	-1,419	1,881	-3,608
-17	8	408	-1,614	1,326	46
-15	0	560	-1,470	482	3,090
-13	-7	637	-1,099	-377	4,672
-11	-13	649	-599	-1,067	4,624
-9	-18	606	-54	-1,482	3,231
-7	-22	518	466	-1,582	1,033
-5	-25	395	905	-1,381	-1,340
-3	-27	247	1,221	-935	-3,300
-1	-28	84	1,386	-330	-4,400
5,850	16,380	7,803,900	40,060,020	48,384,180	409,404,600
2	$\frac{1}{2}$	$\frac{5}{3}$	$\frac{7}{12}$	$\frac{1}{10}$	$\frac{11}{240}$

$n=27$

ϕ_1	ϕ_2	ϕ_3	ϕ_4	ϕ_5	ϕ_6
-13	325	-130	2,990	-16,445	1,495
-12	250	-70	690	2,530	-920
-11	181	-22	-782	10,879	-1,403
-10	118	15	-1,587	12,144	-920
-9	61	42	-1,872	9,174	-122
-8	10	60	-1,770	4,188	592
-7	-35	70	-1,400	-1,162	1,018
-6	-74	73	-867	-5,728	1,096
-5	-107	70	-262	-8,803	865
-4	-134	62	338	-10,058	424
-3	-155	50	870	-9,479	-101
-2	-170	35	1,285	-7,304	-584
-1	-179	18	1,548	-3,960	-920
0	-182	0	1,638	0	-1,040
1,638	712,530	101,790	56,448,210	2,032,135,560	22,331,160
1	3	$\frac{1}{6}$	$\frac{7}{12}$	$\frac{21}{40}$	$\frac{1}{120}$

$n=28$

ϕ_1	ϕ_2	ϕ_3	ϕ_4	ϕ_5	ϕ_6
-27	117	-585	1,755	-13,455	13,455
-25	91	-325	455	1,495	-7,475
-23	67	-115	-395	8,395	-12,305
-21	45	49	-879	9,821	-8,763
-19	25	171	-1,074	7,866	-2,162
-17	7	255	-1,050	4,182	4,138
-15	-9	305	-870	22	8,310
-13	-23	325	-590	-3,718	9,682
-11	-35	319	-259	-6,457	8,401
-9	-45	291	81	-7,887	5,139
-7	-53	245	395	-7,931	841
-5	-59	185	655	-6,701	-3,485
-3	-63	115	840	-4,456	-6,936
-1	-65	39	936	-1,560	-8,840
7,308	95,004	2,103,660	19,634,160	1,354,757,040	1,771,605,360
2	1	$\frac{2}{3}$	$\frac{7}{24}$	$\frac{7}{20}$	$\frac{7}{120}$

Table 47. *Orthogonal polynomials (continued)*

n=29

ϕ_1	ϕ_2	ϕ_3	ϕ_4	ϕ_5	ϕ_6
-14	126	-819	4,095	-8,190	26,910
-13	99	-468	1,170	585	-13,455
-12	74	-182	-780	4,810	-23,920
-11	51	44	-1,930	5,885	-18,285
-10	30	215	-2,441	4,958	-6,210
-9	11	336	-2,460	2,946	6,026
-8	-6	412	-2,120	556	14,832
-7	-21	448	-1,540	-1,694	18,678
-6	-34	449	-825	-3,454	17,534
-5	-45	420	-66	-4,521	12,375
-4	-54	366	660	-4,818	4,752
-3	-61	292	1,290	-4,373	-3,571
-2	-66	203	1,775	-3,298	-10,914
-1	-69	104	2,080	-1,768	-15,912
0	-70	0	2,184	0	-17,680
2,030	113,274	4,207,320	107,987,880	500,671,080	6,959,878,200
1	1	$\frac{5}{6}$	$\frac{7}{12}$	$\frac{7}{40}$	$\frac{11}{120}$

n=30

ϕ_1	ϕ_2	ϕ_3	ϕ_4	ϕ_5	ϕ_6
-29	203	-1,827	23,751	-16,965	5,655
-27	161	-1,071	7,371	585	-2,535
-25	122	-450	-3,744	9,360	-4,875
-23	86	46	-10,504	11,960	-3,965
-21	53	427	-13,749	10,535	-1,655
-19	23	703	-14,249	6,821	823
-17	-4	884	-12,704	2,176	2,734
-15	-28	980	-9,744	-2,384	3,730
-13	-49	1,001	-5,929	-6,149	3,751
-11	-67	957	-1,749	-8,679	2,937
-9	-82	858	2,376	-9,768	1,551
-7	-94	714	6,096	-9,408	-87
-5	-103	535	9,131	-7,753	-1,655
-3	-109	331	11,271	-5,083	-2,873
-1	-112	112	12,376	-1,768	-3,536
8,990	302,064	21,360,240	3,671,587,920	2,145,733,200	302,603,400
2	$\frac{3}{2}$	$\frac{5}{3}$	$\frac{35}{12}$	$\frac{3}{10}$	$\frac{11}{720}$

n=52

ϕ_1	ϕ_2	ϕ_3	ϕ_4	ϕ_5	ϕ_6
-51	425	-4,165	3,570	-55,930	1,286,390
-49	375	-3,185	2,170	-23,030	227,010
-47	327	-2,303	1,022	658	-408,618
-45	281	-1,515	102	16,638	-724,270
-43	237	-817	-613	26,273	-807,507
-41	195	-205	-1,145	30,791	-731,193
-39	155	325	-1,515	31,291	-554,947
-37	117	777	-1,743	28,749	-326,529
-35	81	1,155	-1,848	24,024	-83,160
-33	47	1,463	-1,848	17,864	147,224
-31	15	1,705	-1,760	10,912	344,784
-29	-15	1,885	-1,600	3,712	496,656
-27	-43	2,007	-1,383	-3,285	595,895
-25	-69	2,075	-1,123	-9,715	640,485
-23	-93	2,093	-833	-15,295	632,415
-21	-115	2,065	-525	-19,817	576,821
-19	-135	1,995	-210	-23,142	481,194
-17	-153	1,887	102	-25,194	354,654
-15	-169	1,745	402	-25,954	207,290
-13	-183	1,573	682	-25,454	49,566
-11	-195	1,375	935	-23,771	-108,207
-9	-205	1,155	1,155	-21,021	-256,333
-7	-213	917	1,337	-17,353	-386,127
-5	-219	665	1,477	-12,943	-490,245
-3	-223	403	1,572	-7,988	-562,948
-1	-225	135	1,620	-2,700	-600,300
46,852	2,108,340	162,342,180	108,228,120	26,358,466,680	14,876,313,079,320
2	1	$\frac{2}{3}$	$\frac{1}{24}$	$\frac{1}{20}$	$\frac{11}{120}$

n=51

ϕ_1	ϕ_2	ϕ_3	ϕ_4	ϕ_5	ϕ_6
-25	1,225	-4,900	46,060	-75,670	378,350
-24	1,078	-3,724	27,636	-30,268	60,536
-23	937	-2,668	12,596	2,162	-127,558
-22	802	-1,727	611	23,782	-218,362
-21	673	-896	-8,634	36,547	-239,131
-20	550	-170	-15,440	42,214	-212,440
-19	433	456	-20,094	42,351	-156,657
-18	322	987	-22,869	38,346	-86,394
-17	217	1,428	-24,024	31,416	-12,936
-16	118	1,784	-23,804	22,616	55,352
-15	25	2,060	-22,440	12,848	112,640
-14	-62	2,261	-20,149	2,870	155,270
-13	-143	2,392	-17,134	-6,695	181,415
-12	-218	2,458	-13,584	-15,350	190,760
-11	-287	2,464	-9,674	-22,715	184,205
-10	-350	2,415	-5,565	-28,518	163,590
-9	-407	2,316	-1,404	-32,586	131,442
-8	-458	2,172	2,676	-34,836	90,744
-7	-503	1,988	6,556	-35,266	44,726
-6	-542	1,769	10,131	-33,946	-3,322
-5	-575	1,520	13,310	-31,009	-50,215
-4	-302	1,246	16,016	-26,642	-93,016
-3	-623	952	18,186	-21,077	-129,157
-2	-638	643	19,771	-14,582	-156,538
-1	-647	324	20,736	-7,452	-173,604
0	-650	0	21,060	0	-179,400
11,050	17,218,110	221,375,700	17,803,525,740	47,861,426,340	1,282,440,782,700
1	3	$\frac{5}{6}$	$\frac{7}{12}$	$\frac{3}{40}$	$\frac{11}{360}$

Table 47 (*continued*)

$n=31$					
ϕ_1	ϕ_2	ϕ_3	ϕ_4	ϕ_5	ϕ_6
−15	145	−1,015	783	−1,131	28,275
−14	116	−609	261	0	−11,310
−13	89	−273	−99	585	−23,595
−12	64	−2	−324	780	−20,280
−11	41	209	−439	715	−9,815
−10	20	365	−467	496	2,050
−9	1	471	−429	207	11,759
−8	−16	532	−344	−88	17,488
−7	−31	553	−229	−343	18,727
−6	−44	539	−99	−528	15,906
−5	−55	495	33	−627	10,065
−4	−64	426	156	−636	2,568
−3	−71	337	261	−561	−5,139
−2	−76	233	341	−416	−11,726
−1	−79	119	391	−221	−16,133
0	−80	0	408	0	−17,680
2,480	158,224	6,724,520	4,034,712	9,536,592	7,464,217,200
1	1	$\frac{5}{6}$	$\frac{1}{12}$	$\frac{1}{60}$	$\frac{11}{180}$

$n=32$					
ϕ_1	ϕ_2	ϕ_3	ϕ_4	ϕ_5	ϕ_6
−31	155	−899	899	−2,697	35,061
−29	125	−551	319	−87	−12,441
−27	97	−261	−87	1,305	−28,275
−25	71	−25	−347	1,815	−25,545
−23	47	161	−487	1,725	−13,845
−21	25	301	−531	1,267	169
−19	5	399	−501	627	12,251
−17	−13	459	−417	−51	20,081
−15	−29	485	−297	−661	22,825
−13	−43	481	−157	−1,131	20,739
−11	−55	451	−11	−1,419	14,817
−9	−65	399	129	−1,509	6,483
−7	−73	329	253	−1,407	−2,673
−5	−79	245	353	−1,137	−11,115
−3	−83	151	423	−737	−17,537
−1	−85	51	459	−255	−20,995
10,912	185,504	5,379,616	5,379,616	54,285,216	11,345,610,144
2	1	$\frac{2}{3}$	$\frac{1}{12}$	$\frac{1}{30}$	$\frac{11}{180}$

$n=50$					
ϕ_1	ϕ_2	ϕ_3	ϕ_4	ϕ_5	ϕ_6
−49	196	−9,212	211,876	−211,876	15,134
−47	172	−6,956	125,396	−82,156	2,162
−45	149	−4,935	55,131	9,729	−5,405
−43	127	−3,139	−529	70,219	−8,947
−41	106	−1,558	−43,124	105,124	−9,619
−39	86	−182	−74,124	119,652	−8,373
−37	67	999	−94,929	118,437	−5,979
−35	49	1,995	−106,869	105,567	−3,045
−33	32	2,816	−111,204	84,612	−36
−31	16	3,472	−109,124	58,652	2,708
−29	1	3,973	−101,749	30,305	4,955
−27	−13	4,329	−90,129	1,755	6,565
−25	−26	4,550	−75,244	−25,220	7,475
−23	−38	4,646	−58,004	−49,220	7,685
−21	−49	4,627	−39,249	−69,195	7,245
−19	−59	4,503	−19,749	−84,417	6,243
−17	−68	4,284	−204	−94,452	4,794
−15	−76	3,980	18,756	−99,132	3,030
−13	−83	3,601	36,571	−98,527	1,091
−11	−89	3,157	52,751	−92,917	−883
−9	−94	2,658	66,876	−82,764	−2,759
−7	−98	2,114	78,596	−68,684	−4,417
−5	−101	1,535	87,631	−51,419	−5,755
−3	−103	931	93,771	−31,809	−6,693
−1	−104	312	96,876	−10,764	−7,176
41,650	433,160	770,715,400	372,255,538,200	372,255,538,200	2,045,360,100
2	$\frac{1}{2}$	$\frac{5}{3}$	$\frac{35}{12}$	$\frac{7}{30}$	$\frac{1}{720}$

$n=49$					
ϕ_1	ϕ_2	ϕ_3	ϕ_4	ϕ_5	ϕ_6
−24	376	−4,324	38,916	−95,128	371,864
−23	329	−3,243	22,701	−35,673	46,483
−22	284	−2,277	9,591	6,072	−140,438
−21	241	−1,421	−729	33,187	−225,019
−20	200	−670	−8,560	48,444	−237,360
−19	161	−19	−14,189	54,321	−202,143
−18	124	537	−17,889	53,016	−139,206
−17	89	1,003	−19,919	46,461	−64,089
−16	56	1,384	−20,524	36,336	11,448
−15	25	1,685	−19,935	24,083	78,935
−14	−4	1,911	−18,369	10,920	132,730
−13	−31	2,067	−16,029	−2,145	169,585
−12	−56	2,158	−13,104	−14,300	188,240
−11	−79	2,189	−9,769	−24,915	189,045
−10	−100	2,165	−6,185	−33,528	173,610
−9	−119	2,091	−2,499	−39,831	144,483
−8	−136	1,972	1,156	−43,656	104,856
−7	−151	1,813	4,661	−44,961	58,299
−6	−164	1,619	7,911	−43,816	8,522
−5	−175	1,395	10,815	−40,389	−40,835
−4	−184	1,146	13,296	−34,932	−86,384
−3	−191	877	15,291	−27,767	−125,143
−2	−196	593	16,751	−19,272	−154,662
−1	−199	299	17,641	−9,867	−173,121
0	−200	0	17,940	0	−179,400
9,800	1,566,040	167,230,700	12,408,517,940	74,451,107,640	1,231,306,780,200
1	1	$\frac{5}{6}$	$\frac{7}{12}$	$\frac{7}{60}$	$\frac{7}{180}$

Table 47. *Orthogonal polynomials* (*continued*)

n=33

ϕ_1	ϕ_2	ϕ_3	ϕ_4	ϕ_5	ϕ_6
−16	496	−248	7,192	−14,384	43,152
−15	403	−155	2,697	− 899	−13,485
−14	316	− 77	− 493	6,496	−33,582
−13	235	− 13	−2,581	9,425	−31,755
−12	160	38	−3,756	9,260	−18,840
−11	91	77	−4,193	7,139	− 2,487
−10	28	105	−4,053	3,984	12,290
− 9	− 29	123	−3,483	519	22,607
− 8	− 80	132	−2,616	− 2,712	27,248
− 7	−125	133	−1,571	− 5,327	26,247
− 6	−164	127	− 453	− 7,088	20,514
− 5	−197	115	647	− 7,883	11,505
− 4	−224	98	1,652	− 7,708	936
− 3	−245	77	2,499	− 6,649	− 9,459
− 2	−260	53	3,139	− 4,864	−18,126
− 1	−269	27	3,537	− 2,565	−23,845
0	−272	0	3,672	0	−25,840
2,992	1,947,792	417,384	348,330,136	1,547,128,656	17,018,415,216
1	3	$\frac{1}{6}$	$\frac{7}{12}$	$\frac{3}{20}$	$\frac{11}{180}$

n=34

ϕ_1	ϕ_2	ϕ_3	ϕ_4	ϕ_5	ϕ_6
−33	88	−2,728	8,184	−79,112	39,556
−31	72	−1,736	3,224	− 7,192	−10,788
−29	57	− 899	− 341	33,263	−29,667
−27	43	− 207	−2,721	50,373	−29,261
−25	30	350	−4,112	51,040	−18,705
−23	18	782	−4,696	41,032	− 4,551
−21	7	1,099	−4,641	25,067	8,803
−19	− 3	1,311	−4,101	6,897	18,717
−17	−12	1,428	−3,216	−10,608	23,946
−15	−20	1,460	−2,112	−25,376	24,310
−13	−27	1,417	− 901	−36,049	20,397
−11	−33	1,309	319	−41,899	13,299
− 9	−38	1,146	1,464	−42,744	4,381
− 7	−42	938	2,464	−38,864	− 4,917
− 5	−45	695	3,263	−30,917	−13,245
− 3	−47	427	3,819	−19,855	−19,475
− 1	−48	144	4,104	− 6,840	−22,800
13,090	62,832	51,477,360	456,432,592	46,929,569,232	14,182,012,680
2	$\frac{1}{2}$	$\frac{5}{8}$	$\frac{7}{12}$	$\frac{7}{10}$	$\frac{11}{240}$

n=48

ϕ_1	ϕ_2	ϕ_3	ϕ_4	ϕ_5	ϕ_6
−47	1,081	−3,243	35,673	−1,533,939	511,313
−45	943	−2,415	20,493	− 554,829	54,395
−43	811	−1,677	8,283	126,291	−203,863
−41	685	−1,025	− 1,265	562,397	−316,437
−39	565	− 455	− 8,445	801,047	−327,273
−37	451	37	−13,537	884,633	−272,211
−35	343	455	−16,807	850,633	−179,865
−33	241	803	−18,507	731,863	− 72,459
−31	145	1,085	−18,875	556,729	33,381
−29	55	1,305	−18,135	349,479	125,879
−27	− 29	1,467	−16,497	130,455	197,405
−25	−107	1,575	−14,157	− 83,655	243,815
−23	−179	1,633	−11,297	− 279,565	263,835
−21	−245	1,645	− 8,085	− 447,139	258,489
−19	−305	1,615	− 4,675	− 579,139	230,571
−17	−359	1,547	− 1,207	− 670,973	184,161
−15	−407	1,445	2,193	− 720,443	124,185
−13	−449	1,313	5,413	− 727,493	56,019
−11	−485	1,155	8,355	− 693,957	− 14,863
− 9	−515	975	10,935	− 623,307	− 83,197
− 7	−539	777	13,083	− 520,401	−144,193
− 5	−557	565	14,743	− 391,231	−193,755
− 3	−569	343	15,873	− 242,671	−228,657
− 1	−575	115	16,445	− 82,225	−246,675
36,848	12,712,560	92,620,080	10,301,411,120	19,208,385,771,120	2,321,892,785,520
2	3	$\frac{2}{3}$	$\frac{7}{12}$	$\frac{21}{10}$	$\frac{11}{180}$

n=47

ϕ_1	ϕ_2	ϕ_3	ϕ_4	ϕ_5	ϕ_6
−23	345	−759	32,637	−32,637	1,338,117
−22	300	−561	18,447	−11,352	116,358
−21	257	−385	7,095	3,311	−562,397
−20	216	−230	− 1,720	12,556	−846,240
−19	177	− 95	− 8,285	17,461	−857,433
−18	140	21	−12,873	18,984	−695,114
−17	105	119	−15,743	17,969	−437,871
−16	72	200	−17,140	15,152	−146,184
−15	41	265	−17,295	11,167	135,265
−14	12	315	−16,425	6,552	375,414
−13	− 15	351	−14,733	1,755	554,775
−12	− 40	374	−12,408	− 2,860	663,520
−11	− 63	385	− 9,625	− 7,007	699,699
−10	− 84	385	− 6,545	−10,472	667,590
− 9	−103	375	− 3,315	−13,107	576,181
− 8	−120	356	− 68	−14,824	437,784
− 7	−135	329	3,077	−15,589	266,781
− 6	−148	295	6,015	−15,416	78,502
− 5	−159	255	8,655	−14,361	−111,765
− 4	−168	210	10,920	−12,516	−289,632
− 3	−175	161	12,747	−10,003	−442,337
− 2	−180	109	14,087	− 6,968	−559,338
− 1	−183	55	14,905	− 3,575	−632,775
0	−184	0	15,180	0	−657,800
8,648	1,271,256	4,994,220	8,518,474,580	8,629,104,120	15,866,267,367,720
1	1	$\frac{1}{6}$	$\frac{7}{12}$	$\frac{1}{20}$	$\frac{11}{60}$

Table 47 (*continued*)

$n=35$					
ϕ_1	ϕ_2	ϕ_3	ϕ_4	ϕ_5	ϕ_6
−17	187	−1,496	46,376	−23,188	672,452
−16	154	− 968	19,096	− 2,728	−158,224
−15	123	− 520	− 744	9,052	−485,460
−14	94	− 147	−14,229	14,322	−498,046
−13	67	156	−22,374	14,937	−339,097
−12	42	394	−26,124	12,458	−112,752
−11	19	572	−26,354	8,173	109,589
−10	− 2	695	−23,869	3,118	283,490
− 9	− 21	768	−19,404	− 1,902	386,166
− 8	− 38	796	−13,624	− 6,292	411,632
− 7	− 53	784	− 7,124	− 9,646	366,314
− 6	− 66	737	− 429	−11,726	265,122
− 5	− 77	660	6,006	−12,441	127,985
− 4	− 86	558	11,796	−11,826	− 23,152
− 3	− 93	436	16,626	−10,021	−166,869
− 2	− 98	299	20,251	− 7,250	−284,350
− 1	−101	152	22,496	− 3,800	−361,000
0	−102	0	23,256	0	−387,600
3,570	290,598	15,775,320	14,834,059,240	4,045,652,520	4,070,237,639,160
1	1	$\frac{5}{6}$	$\frac{35}{12}$	$\frac{7}{40}$	$\frac{77}{120}$

$n=36$					
ϕ_1	ϕ_2	ϕ_3	ϕ_4	ϕ_5	ϕ_6
−35	595	−6,545	5,236	−162,316	115,940
−33	493	−4,301	2,244	− 23,188	−23,188
−31	397	−2,387	44	58,652	−80,476
−29	307	− 783	−1,476	97,092	−85,684
−27	223	531	−2,421	104,067	−61,597
−25	145	1,575	−2,889	89,685	−25,015
−23	73	2,369	−2,971	62,353	12,323
−21	7	2,933	−2,751	28,903	42,881
−19	− 53	3,287	−2,306	− 5,282	62,534
−17	−107	3,451	−1,706	− 36,142	69,842
−15	−155	3,445	−1,014	− 60,814	65,390
−13	−197	3,289	− 286	− 77,506	51,194
−11	−233	3,003	429	− 85,371	30,173
− 9	−263	2,607	1,089	− 84,381	5,687
− 7	−287	2,121	1,659	− 75,201	−18,859
− 5	−305	1,565	2,111	− 59,063	−40,345
− 3	−317	959	2,424	− 37,640	−56,200
− 1	−323	323	2,584	− 12,920	−64,600
15,540	3,011,652	307,618,740	191,407,216	199,046,103,984	120,302,590,320
2	3	$\frac{10}{3}$	$\frac{7}{24}$	$\frac{21}{20}$	$\frac{11}{120}$

$n=46$					
ϕ_1	ϕ_2	ϕ_3	ϕ_4	ϕ_5	ϕ_6
−45	165	−7,095	4,257	−58,179	290,895
−43	143	−5,203	2,365	−19,393	19,393
−41	122	−3,526	860	7,052	−128,699
−39	102	−2,054	− 300	23,452	−187,821
−37	83	− 777	−1,155	31,857	−186,263
−35	65	315	−1,743	34,083	−146,825
−33	48	1,232	−2,100	31,724	− 87,444
−31	32	1,984	−2,260	26,164	− 21,788
−29	17	2,581	−2,255	18,589	40,183
−27	3	3,033	−2,115	9,999	91,689
−25	− 10	3,350	−1,868	1,220	128,635
−23	− 22	3,542	−1,540	− 7,084	149,149
−21	− 33	3,619	−1,155	−14,399	153,153
−19	− 43	3,591	− 735	−20,349	141,967
−17	− 52	3,468	− 300	−24,684	117,946
−15	− 60	3,260	132	−27,268	84,150
−13	− 67	2,977	545	−28,067	44,047
−11	− 73	2,629	925	−27,137	1,249
− 9	− 78	2,226	1,260	−24,612	− 40,719
− 7	− 82	1,778	1,540	−20,692	− 78,617
− 5	− 85	1,295	1,757	−15,631	−109,655
− 3	− 87	787	1,905	− 9,725	−131,625
− 1	− 88	264	1,980	− 3,300	−143,000
32,430	285,384	429,502,920	143,167,640	27,214,866,840	748,408,838,100
2	$\frac{1}{2}$	$\frac{5}{3}$	[illegible]/12	$\frac{1}{10}$	$\frac{11}{240}$

$n=45$					
ϕ_1	ϕ_2	ϕ_3	ϕ_4	ϕ_5	ϕ_6
−22	946	−3,311	19,393	−38,786	504,218
−21	817	−2,408	10,578	−12,341	22,919
−20	694	−1,610	3,608	5,494	−234,520
−19	577	− 912	− 1,722	16,359	−332,059
−18	466	− 309	− 5,607	21,714	−321,958
−17	361	204	− 8,232	22,848	−246,064
−16	262	632	− 9,772	20,888	−137,032
−15	169	980	−10,392	16,808	− 19,480
−14	82	1,253	−10,247	11,438	88,922
−13	1	1,456	− 9,482	5,473	176,429
−12	− 74	1,594	− 8,232	− 518	236,264
−11	−143	1,672	− 6,622	− 6,083	265,727
−10	−206	1,695	− 4,767	−10,878	265,370
− 9	−263	1,668	− 2,772	−14,658	238,238
− 8	−314	1,596	− 732	−17,268	189,176
− 7	−359	1,484	1,268	−18,634	124,202
− 6	−398	1,337	3,153	−18,754	49,946
− 5	−431	1,160	4,858	−17,689	− 26,845
− 4	−458	958	6,328	−15,554	− 99,736
− 3	−479	736	7,518	−12,509	−162,967
− 2	−494	499	8,393	− 8,750	−211,750
− 1	−503	252	8,928	− 4,500	−242,500
0	−506	0	9,108	0	−253,000
7,590	9,203,634	92,036,340	2,934,936,620	12,006,558,900	2,245,226,514,300
1	3	$\frac{5}{6}$	$\frac{5}{12}$	$\frac{3}{40}$	$\frac{11}{120}$

31-2

Table 47. *Orthogonal polynomials (continued)*

$n=37$					
ϕ_1	ϕ_2	ϕ_3	ϕ_4	ϕ_5	ϕ_6
-18	210	-357	11,781	-4,488	139,128
-17	175	-238	5,236	-748	-23,188
-16	142	-136	374	1,496	-92,752
-15	111	-50	-3,036	2,596	-102,300
-14	82	21	-5,211	2,856	-77,128
-13	55	78	-6,354	2,535	-36,115
-12	30	122	-6,654	1,850	7,320
-11	7	154	-6,286	979	44,327
-10	-14	175	-5,411	64	69,800
-9	-33	186	-4,176	-786	81,618
-8	-50	188	-2,714	-1,492	79,952
-7	-65	182	-1,144	-2,002	66,638
-6	-78	169	429	-2,288	44,616
-5	-89	150	1,914	-2,343	17,435
-4	-98	126	3,234	-2,178	-11,176
-3	-105	98	4,326	-1,819	-37,671
-2	-110	67	5,141	-1,304	-58,984
-1	-113	34	5,644	-680	-72,760
0	-114	0	5,814	0	-77,520
4,218	383,838	932,178	980,961,982	152,877,192	172,433,712,792
1	1	$\frac{1}{6}$	$\frac{7}{12}$	$\frac{1}{40}$	$\frac{11}{120}$

$n=38$					
ϕ_1	ϕ_2	ϕ_3	ϕ_4	ϕ_5	ϕ_6
-37	111	-777	1,887	-20,757	7,548
-35	93	-525	867	-3,927	-1,020
-33	76	-308	102	6,358	-4,828
-31	60	-124	-442	11,594	-5,508
-29	45	29	-797	13,079	-4,332
-27	31	153	-993	11,925	-2,260
-25	18	250	-1,058	9,070	15
-23	6	322	-1,018	5,290	2,025
-21	-5	371	-897	1,211	3,488
-19	-15	399	-717	-2,679	4,272
-17	-24	408	-498	-6,018	4,362
-15	-32	400	-258	-8,558	3,830
-13	-39	377	-13	-10,153	2,808
-11	-45	341	223	-10,747	1,464
-9	-50	294	438	-10,362	-19
-7	-54	238	622	-9,086	-1,461
-5	-57	175	767	-7,061	-2,700
-3	-59	107	867	-4,471	-3,604
-1	-60	36	918	-1,530	-4,080
18,278	109,668	4,496,388	25,479,532	3,286,859,628	505,670,712
2	$\frac{1}{2}$	$\frac{1}{3}$	$\frac{1}{12}$	$\frac{1}{10}$	$\frac{1}{240}$

$n=44$					
ϕ_1	ϕ_2	ϕ_3	ϕ_4	ϕ_5	ϕ_6
-43	301	-12,341	12,341	-22,919	435,461
-41	259	-8,897	6,601	-6,929	10,127
-39	219	-5,863	2,091	3,731	-212,667
-37	181	-3,219	-1,329	10,101	-292,201
-35	145	-945	-3,792	13,104	-276,640
-33	111	979	-5,424	13,552	-204,288
-31	79	2,573	-6,344	12,152	-104,776
-29	49	3,857	-6,664	9,512	-184
-27	21	4,851	-6,489	6,147	93,903
-25	-5	5,575	-5,917	2,485	167,405
-23	-29	6,049	-5,039	-1,127	214,823
-21	-51	6,293	-3,939	-4,417	234,381
-19	-71	6,327	-2,694	-7,182	227,234
-17	-89	6,171	-1,374	-9,282	196,742
-15	-105	5,845	-42	-10,634	147,810
-13	-119	5,369	1,246	-11,206	86,294
-11	-131	4,763	2,441	-11,011	18,473
-9	-141	4,047	3,501	-10,101	-49,413
-7	-149	3,241	4,391	-8,561	-111,559
-5	-155	2,365	5,083	-6,503	-162,925
-3	-159	1,439	5,556	-4,060	-199,500
-1	-161	483	5,796	-1,380	-218,500
28,380	913,836	1,257,829,980	1,173,974,648	4,162,273,752	1,672,913,873,400
2	1	$\frac{10}{3}$	$\frac{7}{24}$	$\frac{1}{20}$	$\frac{11}{120}$

$n=43$					
ϕ_1	ϕ_2	ϕ_3	ϕ_4	ϕ_5	ϕ_6
-21	287	-574	22,386	-70,889	374,699
-20	246	-410	11,726	-20,254	0
-19	207	-266	3,406	13,091	-191,919
-18	170	-141	-2,847	32,604	-255,892
-17	135	-34	-7,292	41,344	-236,208
-16	102	56	-10,174	41,992	-167,832
-15	71	130	-11,724	36,872	-77,560
-14	42	189	-12,159	27,972	14,892
-13	15	234	-11,682	16,965	95,865
-12	-10	266	-10,482	5,230	156,800
-11	-33	286	-8,734	-6,127	193,347
-10	-54	295	-6,599	-16,252	204,540
-9	-73	294	-4,224	-24,522	192,038
-8	-90	284	-1,742	-30,524	159,432
-7	-105	266	728	-34,034	111,618
-6	-118	241	3,081	-34,996	54,236
-5	-129	210	5,226	-33,501	-6,825
-4	-138	174	7,086	-29,766	-65,856
-3	-145	134	8,598	-24,113	-117,691
-2	-150	91	9,713	-16,948	-158,004
-1	-153	46	10,396	-8,740	-183,540
0	-154	0	10,626	0	-192,280
6,622	814,506	2,676,234	3,815,417,606	39,541,600,644	1,237,956,266,316
1	1	$\frac{1}{6}$	$\frac{7}{12}$	$\frac{7}{40}$	$\frac{11}{120}$

Table 47 (*continued*)

$n=39$

ϕ_1	ϕ_2	ϕ_3	ϕ_4	ϕ_5	ϕ_6
−19	703	−2,109	2,109	−35,853	11,951
−18	592	−1,443	999	−7,548	−1,258
−17	487	−867	159	10,047	−7,327
−16	388	−376	−446	19,312	−8,636
−15	295	35	−849	22,321	−7,055
−14	208	371	−1,081	20,860	−4,010
−13	127	637	−1,171	16,445	−545
−12	52	838	−1,146	10,340	2,620
−11	−17	979	−1,031	3,575	5,035
−10	−80	1,065	−849	−3,036	6,470
−9	−137	1,101	−621	−8,877	6,869
−8	−188	1,092	−366	−13,512	6,308
−7	−233	1,043	−101	−16,667	4,957
−6	−272	959	159	−18,212	3,046
−5	−305	845	401	−18,143	835
−4	−332	706	614	−16,564	−1,412
−3	−353	547	789	−13,669	−3,449
−2	−368	373	919	−9,724	−5,066
−1	−377	189	999	−5,049	−6,103
0	−380	0	1,026	0	−6,460
4,940	4,496,388	33,722,910	32,224,114	9,860,578,884	1,264,176,780
1	3	$\frac{5}{6}$	$\frac{1}{12}$	$\frac{3}{20}$	$\frac{1}{180}$

$n=40$

ϕ_1	ϕ_2	ϕ_3	ϕ_4	ϕ_5	ϕ_6
−39	247	−9,139	82,251	−9,139	155,363
−37	209	−6,327	40,071	−2,109	−11,951
−35	173	−3,885	7,881	2,331	−91,205
−33	139	−1,793	−15,579	4,741	−110,959
−31	107	−31	−31,499	5,611	−93,823
−29	77	1,421	−40,999	5,365	−57,205
−27	49	2,583	−45,129	4,365	−14,015
−25	23	3,475	−44,869	2,915	26,675
−23	−1	4,117	−41,129	1,265	59,015
−21	−23	4,529	−34,749	−385	79,805
−19	−43	4,731	−26,499	−1,881	87,967
−17	−61	4,743	−17,079	−3,111	84,061
−15	−77	4,585	−7,119	−4,001	69,845
−13	−91	4,277	2,821	−4,511	47,879
−11	−103	3,839	12,251	−4,631	21,173
−9	−113	3,291	20,751	−4,377	−7,121
−7	−121	2,653	27,971	−3,787	−33,973
−5	−127	1,945	33,631	−2,917	−56,695
−3	−131	1,187	37,521	−1,837	−73,117
−1	−133	399	39,501	−627	−81,719
21,320	567,112	644,482,280	49,625,135,560	644,482,280	213,224,483,560
2	1	$\frac{10}{3}$	$\frac{35}{12}$	$\frac{1}{30}$	$\frac{11}{180}$

$n=42$

ϕ_1	ϕ_2	ϕ_3	ϕ_4	ϕ_5	ϕ_6
−41	410	−1,066	20,254	−749,398	374,699
−39	350	−754	10,374	−201,058	−9,139
−37	293	−481	2,717	155,363	−201,058
−35	239	−245	−2,983	359,233	−260,110
−33	188	−44	−6,978	445,258	−233,692
−31	140	124	−9,506	443,734	−158,932
−29	95	261	−10,791	380,799	−63,998
−27	53	369	−11,043	278,685	30,670
−25	14	450	−10,458	155,970	111,205
−23	−22	506	−9,218	27,830	169,235
−21	−55	539	−7,491	−93,709	200,882
−19	−85	551	−5,431	−199,519	205,838
−17	−112	544	−3,178	−283,118	186,518
−15	−136	520	−858	−340,418	147,290
−13	−157	481	1,417	−369,473	93,782
−11	−175	429	3,549	−370,227	32,266
−9	−190	366	5,454	−344,262	−30,881
−7	−202	294	7,062	−294,546	−89,639
−5	−211	215	8,317	−225,181	−138,730
−3	−217	131	9,177	−141,151	−173,926
−1	−220	44	9,614	−48,070	−192,280
24,682	1,629,012	9,075,924	3,084,805,724	4,389,117,671,484	1,237,956,266,316
2	$\frac{3}{2}$	$\frac{1}{3}$	$\frac{7}{12}$	$\frac{21}{10}$	$\frac{77}{720}$

$n=41$

ϕ_1	ϕ_2	ϕ_3	ϕ_4	ϕ_5	ϕ_6
−20	260	−2,470	18,278	−36,556	182,780
−19	221	−1,729	9,139	−9,139	−9,139
−18	184	−1,083	2,109	8,436	−102,638
−17	149	−527	−3,071	18,241	−128,797
−16	116	−56	−6,646	22,096	−112,396
−15	85	335	−8,847	21,583	−72,685
−14	56	651	−9,891	18,060	−24,110
−13	29	897	−9,981	12,675	23,005
−12	4	1,078	−9,306	6,380	61,820
−11	−19	1,199	−8,041	−55	88,385
−10	−40	1,265	−6,347	−6,028	101,090
−9	−59	1,281	−4,371	−11,091	100,159
−8	−76	1,252	−2,246	−14,936	87,188
−7	−91	1,183	−91	−17,381	64,727
−6	−104	1,079	1,989	−18,356	35,906
−5	−115	945	3,903	−17,889	4,105
−4	−124	786	5,574	−16,092	−27,332
−3	−131	607	6,939	−13,147	−55,339
−2	−136	413	7,949	−9,292	−77,326
−1	−139	209	8,569	−4,807	−91,333
0	−140	0	8,778	0	−96,140
5,740	641,732	47,900,710	2,481,256,778	10,376,164,708	294,751,491,980
1	1	$\frac{5}{6}$	$\frac{7}{12}$	$\frac{7}{60}$	$\frac{11}{180}$

Table 48. *Powers of integers*

n	n^2	n^3	n^4	n^5	n^6	n^7	n
1	1	1	1	1	1	1	1
2	4	8	16	32	64	128	2
3	9	27	81	243	729	2187	3
4	16	64	256	1024	4096	16384	4
5	25	125	625	3125	15625	78125	5
6	36	216	1296	7776	46656	279936	6
7	49	343	2401	16807	117649	823543	7
8	64	512	4096	32768	262144	2097152	8
9	81	729	6561	59049	531441	4782969	9
10	100	1000	10000	100000	1000000	10000000	10
11	121	1331	14641	161051	1771561	19487171	11
12	144	1728	20736	248832	2985984	35831808	12
13	169	2197	28561	371293	4826809	62748517	13
14	196	2744	38416	537824	7529536	105413504	14
15	225	3375	50625	759375	11390625	170859375	15
16	256	4096	65536	1048576	16777216	268435456	16
17	289	4913	83521	1419857	24137569	410338673	17
18	324	5832	104976	1889568	34012224	612220032	18
19	361	6859	130321	2476099	47045881	893871739	19
20	400	8000	160000	3200000	64000000	1280000000	20
21	441	9261	194481	4084101	85766121	1801088541	21
22	484	10648	234256	5153632	113379904	2494357888	22
23	529	12167	279841	6436343	148035889	3404825447	23
24	576	13824	331776	7962624	191102976	4586471424	24
25	625	15625	390625	9765625	244140625	6103515625	25
26	676	17576	456976	11881376	308915776	8031810176	26
27	729	19683	531441	14348907	387420489	10460353203	27
28	784	21952	614656	17210368	481890304	13492928512	28
29	841	24389	707281	20511149	594823321	17249876309	29
30	900	27000	810000	24300000	729000000	21870000000	30
31	961	29791	923521	28629151	887503681	27512614111	31
32	1024	32768	1048576	33554432	1073741824	34359738368	32
33	1089	35937	1185921	39135393	1291467969	42618442977	33
34	1156	39304	1336336	45435424	1544804416	52523350144	34
35	1225	42875	1500625	52521875	1838265625	64339296875	35
36	1296	46656	1679616	60466176	2176782336	78364164096	36
37	1369	50653	1874161	69343957	2565726409	94931877133	37
38	1444	54872	2085136	79235168	3010936384	114415582592	38
39	1521	59319	2313441	90224199	3518743761	137231006679	39
40	1600	64000	2560000	102400000	4096000000	163840000000	40
41	1681	68921	2825761	115856201	4750104241	194754273881	41
42	1764	74088	3111696	130691232	5489031744	230539333248	42
43	1849	79507	3418801	147008443	6321363049	271818611107	43
44	1936	85184	3748096	164916224	7256313856	319277809664	44
45	2025	91125	4100625	184528125	8303765625	373669453125	45
46	2116	97336	4477456	205962976	9474296896	435817657216	46
47	2209	103823	4879681	229345007	10779215329	506623120463	47
48	2304	110592	5308416	254803968	12230590464	587068342272	48
49	2401	117649	5764801	282475249	13841287201	678223072849	49
50	2500	125000	6250000	312500000	15625000000	781250000000	50

Table 48 (*continued*)

n	n^2	n^3	n^4	n^5	n^6	n^7	n
51	2601	132651	6765201	345025251	17596287801	897410677851	51
52	2704	140608	7311616	380204032	19770609664	1028071702528	52
53	2809	148877	7890481	418195493	22164361129	1174711139837	53
54	2916	157464	8503056	459165024	24794911296	1338925209984	54
55	3025	166375	9150625	503284375	27680640625	1522435234375	55
56	3136	175616	9834496	550731776	30840979456	1727094849536	56
57	3249	185193	10556001	601692057	34296447249	1954897493193	57
58	3364	195112	11316496	656356768	38068692544	2207984167552	58
59	3481	205379	12117361	714924299	42180533641	2488651484819	59
60	3600	216000	12960000	777600000	46656000000	2799360000000	60
61	3721	226981	13845841	844596301	51520374361	3142742836021	61
62	3844	238328	14776336	916132832	56800235584	3521614606208	62
63	3969	250047	15752961	992436543	62523502209	3938980639167	63
64	4096	262144	16777216	1073741824	68719476736	4398046511104	64
65	4225	274625	17850625	1160290625	75418890625	4902227890625	65
66	4356	287496	18974736	1252332576	82653950016	5455160701056	66
67	4489	300763	20151121	1350125107	90458382169	6060711605323	67
68	4624	314432	21381376	1453933568	98867482624	6722988818432	68
69	4761	328509	22667121	1564031349	107918163081	7446353252589	69
70	4900	343000	24010000	1680700000	117649000000	8235430000000	70
71	5041	357911	25411681	1804229351	128100283921	9095120158391	71
72	5184	373248	26873856	1934917632	139314069504	10030613004288	72
73	5329	389017	28398241	2073071593	151334226289	11047398519097	73
74	5476	405224	29986576	2219006624	164206490176	12151280273024	74
75	5625	421875	31640625	2373046875	177978515625	13348388671875	75
76	5776	438976	33362176	2535525376	192699928576	14645194571776	76
77	5929	456533	35153041	2706784157	208422380089	16048523266853	77
78	6084	474552	37015056	2887174368	225199600704	17565568854912	78
79	6241	493039	38950081	3077056399	243087455521	19203908986159	79
80	6400	512000	40960000	3276800000	262144000000	20971520000000	80
81	6561	531441	43046721	3486784401	282429536481	22876792454961	81
82	6724	551368	45212176	3707398432	304006671424	24928547056768	82
83	6889	571787	47458321	3939040643	326940373369	27136050989627	83
84	7056	592704	49787136	4182119424	351298031616	29509034655744	84
85	7225	614125	52200625	4437053125	377149515625	32057708828125	85
86	7396	636056	54700816	4704270176	404567235136	34792782221696	86
87	7569	658503	57289761	4984209207	433626201009	37725479487783	87
88	7744	681472	59969536	5277319168	464404086784	40867559636992	88
89	7921	704969	62742241	5584059449	496981290961	44231334895529	89
90	8100	729000	65610000	5904900000	531441000000	47829690000000	90
91	8281	753571	68574961	6240321451	567869252041	51676101935731	91
92	8464	778688	71639296	6590815232	606355001344	55784660123648	92
93	8649	804357	74805201	6956883693	646990183449	60170087060757	93
94	8836	830584	78074896	7339040224	689869781056	64847759419264	94
95	9025	857375	81450625	7737809375	735091890625	69833729609375	95
96	9216	884736	84934656	8153726976	782757789696	75144747810816	96
97	9409	912673	88529281	8587340257	832972004929	80798284478113	97
98	9604	941192	92236816	9039207968	885842380864	86812553324672	98
99	9801	970299	96059601	9509900499	941480149401	93206534790699	99
100	10000	1000000	100000000	10000000000	1000000000000	100000000000000	100

Table 49. *Sums of powers of integers*

n	$S(n)$	$S(n^2)$	$S(n^3)$	$S(n^4)$	$S(n^5)$	$S(n^6)$	$S(n^7)$	n
1	1	1	1	1	1	1	1	*1*
2	3	5	9	17	33	65	129	*2*
3	6	14	36	98	276	794	2316	*3*
4	10	30	100	354	1300	4890	18700	*4*
5	15	55	225	979	4425	20515	96825	*5*
6	21	91	441	2275	12201	67171	376761	*6*
7	28	140	784	4676	29008	184820	1200304	*7*
8	36	204	1296	8772	61776	446964	3297456	*8*
9	45	285	2025	15333	120825	978405	8080425	*9*
10	55	385	3025	25333	220825	1978405	18080425	*10*
11	66	506	4356	39974	381876	3749966	37567596	*11*
12	78	650	6084	60710	630708	6735950	73399404	*12*
13	91	819	8281	89271	1002001	11562759	136147921	*13*
14	105	1015	11025	127687	1539825	19092295	241561425	*14*
15	120	1240	14400	178312	2299200	30482920	412420800	*15*
16	136	1496	18496	243848	3347776	47260136	680856256	*16*
17	153	1785	23409	327369	4767633	71397705	1091194929	*17*
18	171	2109	29241	432345	6657201	105409929	1703414961	*18*
19	190	2470	36100	562666	9133300	152455810	2597286700	*19*
20	210	2870	44100	722666	12333300	216455810	3877286700	*20*
21	231	3311	53361	917147	16417401	302221931	5678375241	*21*
22	253	3795	64009	1151403	21571033	415601835	8172733129	*22*
23	276	4324	76176	1431244	28007376	563637724	11577558576	*23*
24	300	4900	90000	1763020	35970000	754740700	16164030000	*24*
25	325	5525	105625	2153645	45735625	998881325	22267545625	*25*
26	351	6201	123201	2610621	57617001	1307797101	30299355801	*26*
27	378	6930	142884	3142062	71965908	1695217590	40759709004	*27*
28	406	7714	164836	3756718	89176276	2177107894	54252637516	*28*
29	435	8555	189225	4463999	109687425	2771931215	71502513825	*29*
30	465	9455	216225	5273999	133987425	3500931215	93372513825	*30*
31	496	10416	246016	6197520	162616576	4388434896	120885127936	*31*
32	528	11440	278784	7246096	196171008	5462176720	155244866304	*32*
33	561	12529	314721	8432017	135306401	6753644689	197863309281	*33*
34	595	13685	354025	9768353	180741825	8298449105	250386659425	*34*
35	630	14910	396900	11268978	333263700	10136714730	314725956300	*35*
36	666	16206	443556	12948594	393729876	12313497066	393090120396	*36*
37	703	17575	494209	14822755	463073833	14879223475	488021997529	*37*
38	741	19019	549081	16907891	542309001	17890159859	602437580121	*38*
39	780	20540	608400	19221332	632533200	21408903620	739668586800	*39*
40	820	22140	672400	21781332	734933200	25504903620	903508586800	*40*
41	861	23821	741321	24607093	850789401	30255007861	1098262860681	*41*
42	903	25585	815409	27718789	981480633	35744039605	1328802193929	*42*
43	946	27434	894916	31137590	1128489076	42065402654	1600620805036	*43*
44	990	29370	980100	34885686	1293405300	49321716510	1919898614700	*44*
45	1035	31395	1071225	38986311	1477933425	57625482135	2293568067825	*45*
46	1081	33511	1168561	43463767	1683896401	67099779031	2729385725041	*46*
47	1128	35720	1272384	48343448	1913241408	77878994360	3236008845504	*47*
48	1176	38024	1382976	53651864	2168045376	90109584824	3823077187776	*48*
49	1225	40425	1500625	59416665	2450520625	103950872025	4501300260625	*49*
50	1275	42925	1625625	65666665	2763020625	119575872025	5282550260625	*50*

Table 49 (*continued*)

n	$S(n)$	$S(n^2)$	$S(n^3)$	$S(n^4)$	$S(n^5)$	$S(n^6)$	$S(n^7)$	n
51	1326	45526	1758276	72431866	3108045876	137172159826	6179960938476	*51*
52	1378	48230	1898884	79743482	3488249908	156942769490	7208032641004	*52*
53	1431	51039	2047761	87633963	3906445401	179107130619	8382743780841	*53*
54	1485	53955	2205225	96137019	4365610425	203902041915	9721668990825	*54*
55	1540	56980	2371600	105287644	4868894800	231582682540	11244104225200	*55*
56	1596	60116	2547216	115122140	5419626576	262423661996	12971199074736	*56*
57	1653	63365	2732409	125678141	6021318633	296720109245	14926096567929	*57*
58	1711	66729	2927521	136994637	6677675401	334788801789	17134080735481	*58*
59	1770	70210	3132900	149111998	7392599700	376969335430	19622732220300	*59*
60	1830	73810	3348900	162071998	8170199700	423625335430	22422092220300	*60*
61	1891	77531	3575881	175917839	9014796001	475145709791	25564835056321	*61*
62	1953	81375	3814209	190694175	9930928833	531945945375	29086449662529	*62*
63	2016	85344	4064256	206447136	10923365376	594469447584	33025430301696	*63*
64	2080	89440	4326400	223224352	11997107200	663188924320	37423476812800	*64*
65	2145	93665	4601025	241074977	13157397825	738607814945	42325704703425	*65*
66	2211	98021	4888521	260049713	14409730401	821261764961	47780865404481	*66*
67	2278	102510	5189284	280200834	15759855508	911720147130	53841577009804	*67*
68	2346	107134	5503716	301582210	17213789076	1010587629754	60564565828236	*68*
69	2415	111895	5832225	324249331	18777820425	1118505792835	68010919080825	*69*
70	2485	116795	6175225	348259331	20458520425	1236154792835	76246349080825	*70*
71	2556	121836	6533136	373671012	22262749776	1364255076756	85341469239216	*71*
72	2628	127020	6906384	400544868	24197667408	1503569146260	95372082243504	*72*
73	2701	132349	7295401	428943109	26270739001	1654903372549	106419480762601	*73*
74	2775	137825	7700625	458929685	28489745625	1819109862725	118570761035625	*74*
75	2850	143450	8122500	490570310	30862792500	1997088378350	131919149707500	*75*
76	2926	149226	8561476	523932486	33398317876	2189788306926	146564344279276	*76*
77	3003	155155	9018009	559085527	36105102033	2398210687015	162612867546129	*77*
78	3081	161239	9492561	596100583	38992276401	2623410287719	180178436401041	*78*
79	3160	167480	9985600	635050664	42069332800	2866497743240	199382345387200	*79*
80	3240	173880	10497600	676010664	45346132800	3128641743240	220353865387200	*80*
81	3321	180441	11029041	719057385	48832917201	3411071279721	243230657842161	*81*
82	3403	187165	11580409	764269561	52540315633	3715077951145	268159204898929	*82*
83	3486	194054	12152196	811727882	56479356276	4042018324514	295295255888556	*83*
84	3570	201110	12744900	861515018	60661475700	4393316356130	324804290544300	*84*
85	3655	208335	13359025	913715643	65098528825	4770465871755	356861999372425	*85*
86	3741	215731	13995081	968416459	69802799001	5175033106891	391654781594121	*86*
87	3828	223300	14653584	1025706220	74787008208	5608659307900	429380261081904	*87*
88	3916	231044	15335056	1085675756	80064327376	6073063394684	470247820718896	*88*
89	4005	238965	16040025	1148417997	85648386825	6570044685645	514479155614425	*89*
90	4095	247065	16769025	1214027997	91553286825	7101485685645	562308845614425	*90*
91	4186	255346	17522596	1282602958	97793608276	7669354937686	613984947550156	*91*
92	4278	263810	18301284	1354242254	104384423508	8275709939030	669769607673804	*92*
93	4371	272459	19105641	1429047455	111341307201	8922700122479	729939694734561	*93*
94	4465	281295	19936225	1507122351	118680347425	9612569903535	794787454153825	*94*
95	4560	290320	20793600	1588572976	126418156800	10347661794160	864621183763200	*95*
96	4656	299536	21678336	1673507632	134571883776	11130419583856	939765931574016	*96*
97	4753	308945	22591009	1762036913	143159224033	11963391588785	1020564216052129	*97*
98	4851	318549	23532201	1854273729	152198432001	12849233969649	1107376769376801	*98*
99	4950	328350	24502500	1950333330	161708332500	13790714119050	1200583304167500	*99*
100	5050	338350	25502500	2050333330	171708332500	14790714119050	1300583304167500	*100*

Table 50. *Squares of integers*

n	0	1	2	3	4	5	6	7	8	9
0	0	1	4	9	16	25	36	49	64	81
1	100	121	144	169	196	225	256	289	324	361
2	400	441	484	529	576	625	676	729	784	841
3	900	961	1024	1089	1156	1225	1296	1369	1444	1521
4	1600	1681	1764	1849	1936	2025	2116	2209	2304	2401
5	2500	2601	2704	2809	2916	3025	3136	3249	3364	3481
6	3600	3721	3844	3969	4096	4225	4356	4489	4624	4761
7	4900	5041	5184	5329	5476	5625	5776	5929	6084	6241
8	6400	6561	6724	6889	7056	7225	7396	7569	7744	7921
9	8100	8281	8464	8649	8836	9025	9216	9409	9604	9801
10	10000	10201	10404	10609	10816	11025	11236	11449	11664	11881
11	12100	12321	12544	12769	12996	13225	13456	13689	13924	14161
12	14400	14641	14884	15129	15376	15625	15876	16129	16384	16641
13	16900	17161	17424	17689	17956	18225	18496	18769	19044	19321
14	19600	19881	20164	20449	20736	21025	21316	21609	21904	22201
15	22500	22801	23104	23409	23716	24025	24336	24649	24964	25281
16	25600	25921	26244	26569	26896	27225	27556	27889	28224	28561
17	28900	29241	29584	29929	30276	30625	30976	31329	31684	32041
18	32400	32761	33124	33489	33856	34225	34596	34969	35344	35721
19	36100	36481	36864	37249	37636	38025	38416	38809	39204	39601
20	40000	40401	40804	41209	41616	42025	42436	42849	43264	43681
21	44100	44521	44944	45369	45796	46225	46656	47089	47524	47961
22	48400	48841	49284	49729	50176	50625	51076	51529	51984	52441
23	52900	53361	53824	54289	54756	55225	55696	56169	56644	57121
24	57600	58081	58564	59049	59536	60025	60516	61009	61504	62001
25	62500	63001	63504	64009	64516	65025	65536	66049	66564	67081
26	67600	68121	68644	69169	69696	70225	70756	71289	71824	72361
27	72900	73441	73984	74529	75076	75625	76176	76729	77284	77841
28	78400	78961	79524	80089	80656	81225	81796	82369	82944	83521
29	84100	84681	85264	85849	86436	87025	87616	88209	88804	89401
30	90000	90601	91204	91809	92416	93025	93636	94249	94864	95481
31	96100	96721	97344	97969	98596	99225	99856	100489	101124	101761
32	102400	103041	103684	104329	104976	105625	106276	106929	107584	108241
33	108900	109561	110224	110889	111556	112225	112896	113569	114244	114921
34	115600	116281	116964	117649	118336	119025	119716	120409	121104	121801
35	122500	123201	123904	124609	125316	126025	126736	127449	128164	128881
36	129600	130321	131044	131769	132496	133225	133956	134689	135424	136161
37	136900	137641	138384	139129	139876	140625	141376	142129	142884	143641
38	144400	145161	145924	146689	147456	148225	148996	149769	150544	151321
39	152100	152881	153664	154449	155236	156025	156816	157609	158404	159201
40	160000	160801	161604	162409	163216	164025	164836	165649	166464	167281
41	168100	168921	169744	170569	171396	172225	173056	173889	174724	175561
42	176400	177241	178084	178929	179776	180625	181476	182329	183184	184041
43	184900	185761	186624	187489	188356	189225	190096	190969	191844	192721
44	193600	194481	195364	196249	197136	198025	198916	199809	200704	201601
45	202500	203401	204304	205209	206116	207025	207936	208849	209764	210681
46	211600	212521	213444	214369	215296	216225	217156	218089	219024	219961
47	220900	221841	222784	223729	224676	225625	226576	227529	228484	229441
48	230400	231361	232324	233289	234256	235225	236196	237169	238144	239121
49	240100	241081	242064	243049	244036	245025	246016	247009	248004	249001

Table 50 (*continued*)

n	0	1	2	3	4	5	6	7	8	9
50	250000	251001	252004	253009	254016	255025	256036	257049	258064	259081
51	260100	261121	262144	263169	264196	265225	266256	267289	268324	269361
52	270400	271441	272484	273529	274576	275625	276676	277729	278784	279841
53	280900	281961	283024	284089	285156	286225	287296	288369	289444	290521
54	291600	292681	293764	294849	295936	297025	298116	299209	300304	301401
55	302500	303601	304704	305809	306916	308025	309136	310249	311364	312481
56	313600	314721	315844	316969	318096	319225	320356	321489	322624	323761
57	324900	326041	327184	328329	329476	330625	331776	332929	334084	335241
58	336400	337561	338724	339889	341056	342225	343396	344569	345744	346921
59	348100	349281	350464	351649	352836	354025	355216	356409	357604	358801
60	360000	361201	362404	363609	364816	366025	367236	368449	369664	370881
61	372100	373321	374544	375769	376996	378225	379456	380689	381924	383161
62	384400	385641	386884	388129	389376	390625	391876	393129	394384	395641
63	396900	398161	399424	400689	401956	403225	404496	405769	407044	408321
64	409600	410881	412164	413449	414736	416025	417316	418609	419904	421201
65	422500	423801	425104	426409	427716	429025	430336	431649	432964	434281
66	435600	436921	438244	439569	440896	442225	443556	444889	446224	447561
67	448900	450241	451584	452929	454276	455625	456976	458329	459684	461041
68	462400	463761	465124	466489	467856	469225	470596	471969	473344	474721
69	476100	477481	478864	480249	481636	483025	484416	485809	487204	488601
70	490000	491401	492804	494209	495616	497025	498436	499849	501264	502681
71	504100	505521	506944	508369	509796	511225	512656	514089	515524	516961
72	518400	519841	521284	522729	524176	525625	527076	528529	529984	531441
73	532900	534361	535824	537289	538756	540225	541696	543169	544644	546121
74	547600	549081	550564	552049	553536	555025	556516	558009	559504	561001
75	562500	564001	565504	567009	568516	570025	571536	573049	574564	576081
76	577600	579121	580644	582169	583696	585225	586756	588289	589824	591361
77	592900	594441	595984	597529	599076	600625	602176	603729	605284	606841
78	608400	609961	611524	613089	614656	616225	617796	619369	620944	622521
79	624100	625681	627264	628849	630436	632025	633616	635209	636804	638401
80	640000	641601	643204	644809	646416	648025	649636	651249	652864	654481
81	656100	657721	659344	660969	662596	664225	665856	667489	669124	670761
82	672400	674041	675684	677329	678976	680625	682276	683929	685584	687241
83	688900	690561	692224	693889	695556	697225	698896	700569	702244	703921
84	705600	707281	708964	710649	712336	714025	715716	717409	719104	720801
85	722500	724201	725904	727609	729316	731025	732736	734449	736164	737881
86	739600	741321	743044	744769	746496	748225	749956	751689	753424	755161
87	756900	758641	760384	762129	763876	765625	767376	769129	770884	772641
88	774400	776161	777924	779689	781456	783225	784996	786769	788544	790321
89	792100	793881	795664	797449	799236	801025	802816	804609	806404	808201
90	810000	811801	813604	815409	817216	819025	820836	822649	824464	826281
91	828100	829921	831744	833569	835396	837225	839056	840889	842724	844561
92	846400	848241	850084	851929	853776	855625	857476	859329	861184	863041
93	864900	866761	868624	870489	872356	874225	876096	877969	879844	881721
94	883600	885481	887364	889249	891136	893025	894916	896809	898704	900601
95	902500	904401	906304	908209	910116	912025	913936	915849	917764	919681
96	921600	923521	925444	927369	929296	931225	933156	935089	937024	938961
97	940900	942841	944784	946729	948676	950625	952576	954529	956484	958441
98	960400	962361	964324	966289	968256	970225	972196	974169	976144	978121
99	980100	982081	984064	986049	988036	990025	992016	994009	996004	998001

Table 51. *Factorials of integers, their logarithms; square roots; and their reciprocals*

n	$n!$*	$\log_{10} n!$	$\frac{1}{n!}$†	$\sqrt{n}$	$\frac{1}{\sqrt{n}}$	$\frac{1}{n}$
1	1	0·000 0000	1·000 000	1·000 0000	1·000 0000	1·000 0000
2	2	0·301 0300	0·500 000	1·414 2136	0·707 1068	0·500 0000
3	6	0·778 1513	·166 667	1·732 0508	·577 3503	·333 3333
4	24	1·380 2112	·416 667	2·000 0000	·500 0000	·250 0000
5	120	2·079 1812	·833 333	2·236 0680	·447 2136	·200 0000
6	720	2·857 3325	0·138 889	2·449 4897	0·408 2483	0·166 6667
7	5040	3·702 4305	·198 413	2·645 7513	·377 9645	·142 8571
8	40320	4·605 5205	·248 016	2·828 4271	·353 5534	·125 0000
9	362880	5·559 7630	·275 573	3·000 0000	·333 3333	·111 1111
10	3·62880	6·559 7630	·275 573	3·162 2777	·316 2278	·100 0000
11	3·99168	7·601 1557	0·250 521	3·316 6248	0·301 5113	0·090 9091
12	4·79002	8·680 3370	·208 768	3·464 1016	·288 6751	·083 3333
13	6·22702	9·794 2803	·160 590	3·605 5513	·277 3501	·076 9231
14	8·71783	10·940 4084	·114 707	3·741 6574	·267 2612	·071 4286
15	1·30767	12·116 4996	·764 716	3·872 9833	·258 1989	·066 6667
16	2·09228	13·320 6196	0·477 948	4·000 0000	0·250 0000	0·062 5000
17	3·55687	14·551 0685	·281 146	4·123 1056	·242 5356	·058 8235
18	6·40237	15·806 3410	·156 192	4·242 6407	·235 7023	·055 5556
19	1·21645	17·085 0946	·822 064	4·358 8989	·229 4157	·052 6316
20	2·43290	18·386 1246	·411 032	4·472 1360	·223 6068	·050 0000
21	5·10909	19·708 3439	0·195 729	4·582 5757	0·218 2179	0·047 6190
22	1·12400	21·050 7666	·889 679	4·690 4158	·213 2007	·045 4545
23	2·58520	22·412 4944	·386 817	4·795 8315	·208 5144	·043 4783
24	6·20448	23·792 7057	·161 174	4·898 9795	·204 1241	·041 6667
25	1·55112	25·190 6457	·644 695	5·000 0000	·200 0000	·040 0000
26	4·03291	26·605 6190	0·247 960	5·099 0195	0·196 1161	0·038 4615
27	1·08889	28·036 9828	·918 369	5·196 1524	·192 4501	·037 0370
28	3·04888	29·484 1408	·327 989	5·291 5026	·188 9822	·035 7143
29	8·84176	30·946 5388	·113 100	5·385 1648	·185 6953	·034 4828
30	2·65253	32·423 6601	·376 999	5·477 2256	·182 5742	·033 3333
31	8·22284	33·915 0218	0·121 613	5·567 7644	0·179 6053	0·032 2581
32	2·63131	35·420 1717	·380 039	5·656 8542	·176 7767	·031 2500
33	8·68332	36·938 6857	·115 163	5·744 5626	·174 0777	·030 3030
34	2·95233	38·470 1646	·338 716	5·830 9519	·171 4986	·029 4118
35	1·03331	40·014 2326	·967 759	5·916 0798	·169 0309	·028 5714
36	3·71993	41·570 5351	0·268 822	6·000 0000	0·166 6667	0·027 7778
37	1·37638	43·138 7369	·726 546	6·082 7625	·164 3990	·027 0270
38	5·23023	44·718 5205	·191 196	6·164 4140	·162 2214	·026 3158
39	2·03979	46·309 5851	·490 247	6·244 9980	·160 1282	·025 6410
40	8·15915	47·911 6451	·122 562	6·324 5553	·158 1139	·025 0000
41	3·34525	49·524 4289	0·298 931	6·403 1242	0·156 1738	0·024 3902
42	1·40501	51·147 6782	·711 741	6·480 7407	·154 3034	·023 8095
43	6·04153	52·781 1467	·165 521	6·557 4385	·152 4986	·023 2558
44	2·65827	54·424 5993	·376 184	6·633 2496	·150 7557	·022 7273
45	1·19622	56·077 8119	·835 965	6·708 2039	·149 0712	·022 2222
46	5·50262	57·740 5697	0·181 732	6·782 3300	0·147 442[illegible]	0·021 7391
47	2·58623	59·412 6676	·386 663	6·855 6546	·145 8650	·021 2766
48	1·24139	61·093 9088	·805 548	6·928 2032	·144 3376	·020 8333
49	6·08282	62·784 1049	·164 397	7·000 0000	·142 8571	·020 4082
50	3·04141	64·483 0749	·328 795	7·071 0678	·141 4214	·020 0000

* For $n > 9$, multiply by 10^c, where c is the characteristic of $\log n!$ shown alongside in the next column.
† Multiply by 10^{-c}, where c is the characteristic of $\log n!$ shown in the preceding column.

Table 51 (*continued*)

n	$n!$*	$\log_{10} n!$	$\frac{1}{n!}$†	$\sqrt{n}$	$\frac{1}{\sqrt{n}}$	$\frac{1}{n}$
51	1·55112	66·190 6450	0·644 696	7·141 4284	0·140 0280	0·019 6078
52	8·06582	67·906 6484	·123 980	7·211 1026	·138 6750	·019 2308
53	4·27488	69·630 9243	·233 925	7·280 1099	·137 3606	·018 8679
54	2·30844	71·363 3180	·433 194	7·348 4692	·136 0828	·018 5185
55	1·26964	73·103 6807	·787 625	7·416 1985	·134 8400	·018 1818
56	7·10999	74·851 8687	0·140 647	7·483 3148	0·133 6306	0·017 8571
57	4·05269	76·607 7436	·246 750	7·549 8344	·132 4532	·017 5439
58	2·35056	78·371 1716	·425 430	7·615 7731	·131 3064	·017 2414
59	1·38683	80·142 0236	·721 068	7·681 1457	·130 1889	·016 9492
60	8·32099	81·920 1748	·120 178	7·745 9667	·129 0994	·016 6667
61	5·07580	83·705 5047	0·197 013	7·810 2497	0·128 0369	0·016 3934
62	3·14700	85·497 8964	·317 763	7·874 0079	·127 0001	·016 1290
63	1·98261	87·297 2369	·504 386	7·937 2539	·125 9882	·015 8730
64	1·26887	89·103 4169	·788 103	8·000 0000	·125 0000	·015 6250
65	8·24765	90·916 3303	·121 247	8·062 2577	·124 0347	·015 3846
66	5·44345	92·735 8742	0·183 707	8·124 0384	0·123 0915	0·015 1515
67	3·64711	94·561 9490	·274 190	8·185 3528	·122 1694	·014 9254
68	2·48004	96·394 4579	·403 220	8·246 2113	·121 2678	·014 7059
69	1·71122	98·233 3070	·584 377	8·306 6239	·120 3859	·014 4928
70	1·19786	100·078 4050	·834 824	8·366 6003	·119 5229	·014 2857
71	8·50479	101·929 6634	0·117 581	8·426 1498	0·118 6782	0·014 0845
72	6·12345	103·786 9959	·163 307	8·485 2814	·117 8511	·013 8889
73	4·47012	105·650 3187	·223 708	8·544 0037	·117 0411	·013 6986
74	3·30789	107·519 5505	·302 308	8·602 3253	·116 2476	·013 5135
75	2·48091	109·394 6117	·403 077	8·660 2540	·115 4701	·013 3333
76	1·88549	111·275 4253	0·530 365	8·717 7979	0·114 7079	0·013 1579
77	1·45183	113·161 9160	·688 785	8·774 9644	·113 9606	·012 9870
78	1·13243	115·054 0106	·883 058	8·831 7609	·113 2277	·012 8205
79	8·94618	116·951 6377	·111 780	8·888 1944	·112 5088	·012 6582
80	7·15695	118·854 7277	·139 724	8·944 2719	·111 8034	·012 5000
81	5·79713	120·763 2127	0·172 499	9·000 0000	0·111 1111	0·012 3457
82	4·75364	122·677 0266	·210 365	9·055 3851	·110 4315	·012 1951
83	3·94552	124·596 1047	·253 452	9·110 4336	·109 7643	·012 0482
84	3·31424	126·520 3840	·301 728	9·165 1514	·109 1089	·011 9048
85	2·81710	128·449 8029	·354 974	9·219 5445	·108 4652	·011 7647
86	2·42271	130·384 3013	0·412 761	9·273 6185	0·107 8328	0·011 6279
87	2·10776	132·323 8206	·474 438	9·327 3791	·107 2113	·011 4943
88	1·85483	134·268 3033	·539 134	9·380 8315	·106 6004	·011 3636
89	1·65080	136·217 6933	·605 769	9·433 9811	·105 9998	·011 2360
90	1·48572	138·171 9358	·673 076	9·486 8330	·105 4093	·011 1111
91	1·35200	140·130 9772	0·739 644	9·539 3920	0·104 8285	0·010 9890
92	1·24384	142·094 7650	·803 961	9·591 6630	·104 2572	·010 8696
93	1·15677	144·063 2480	·864 474	9·643 6508	·103 6952	·010 7527
94	1·08737	146·036 3758	·919 653	9·695 3597	·103 1421	·010 6383
95	1·03300	148·014 0994	·968 056	9·746 7943	·102 5978	·010 5263
96	9·91678	149·996 3707	0·100 839	9·797 9590	0·102 0621	0·010 4167
97	9·61928	151·983 1424	·103 958	9·848 8578	·101 5346	·010 3093
98	9·42689	153·974 3685	·106 080	9·899 4949	·101 0153	·010 2041
99	9·33262	155·970 0037	·107 151	9·949 8744	·100 5038	·010 1010
100	9·33262	157·970 0037	·107 151	10·000 0000	·100 0000	·010 0000

* Multiply by 10^c, where c is the characteristic of log $n!$ shown alongside in the next column.
† Multiply by 10^{-c}, where c is the characteristic of log $n!$ shown in the preceding column.

Table 51. *Factorials of integers and their logarithms* (*continued*)

n	$n!$*	$\log_{10} n!$	n	$n!$*	$\log_{10} n!$	n	$n!$*	$\log_{10} n!$
101	9·42595	159·974 3250	*151*	8·62721	264·935 8704	*201*	1·58520	377·200 0847
102	9·61447	161·982 9252	*152*	1·31134	267·117 7139	*202*	3·20211	379·505 4361
103	9·90290	163·995 7624	*153*	2·00634	269·302 4054	*203*	6·50028	381·812 9321
104	1·02990	166·012 7958	*154*	3·08977	271·489 9261	*204*	1·32606	384·122 5623
105	1·08140	168·033 9851	*155*	4·78914	273·680 2578	*205*	2·71842	386·434 3161
106	1·14628	170·059 2909	*156*	7·47106	275·873 3824	*206*	5·59994	388·748 1834
107	1·22652	172·088 6747	*157*	1·17296	278·069 2820	*207*	1·15919	391·064 1537
108	1·32464	174·122 0985	*158*	1·85327	280·267 9391	*208*	2·41111	393·382 2170
109	1·44386	176·159 5250	*159*	2·94670	282·469 3363	*209*	5·03922	395·702 3633
110	1·58825	178·200 9176	*160*	4·71472	284·673 4562	*210*	1·05824	398·024 5826
111	1·76295	180·246 2406	*161*	7·59071	286·880 2821	*211*	2·23288	400·348 8651
112	1·97451	182·295 4586	*162*	1·22969	289·089 7971	*212*	4·73370	402·675 2009
113	2·23119	184·348 5371	*163*	2·00440	291·301 9847	*213*	1·00828	405·003 5805
114	2·54356	186·405 4419	*164*	3·28722	293·516 8286	*214*	2·15772	407·333 9943
115	2·92509	188·466 1398	*165*	5·42391	295·734 3125	*215*	4·63909	409·666 4328
116	3·39311	190·530 5978	*166*	9·00369	297·954 4206	*216*	1·00204	412·000 8865
117	3·96994	192·598 7836	*167*	1·50362	300·177 1371	*217*	2·17443	414·337 3463
118	4·68453	194·670 6656	*168*	2·52608	302·402 4464	*218*	4·74027	416·675 8027
119	5·57459	196·746 2126	*169*	4·26907	304·630 3331	*219*	1·03812	419·016 2469
120	6·68950	198·825 3938	*170*	7·25742	306·860 7820	*220*	2·28386	421·358 6695
121	8·09430	200·908 1792	*171*	1·24102	309·093 7781	*221*	5·04733	423·703 0618
122	9·87504	202·994 5390	*172*	2·13455	311·329 3066	*222*	1·12051	426·049 4148
123	1·21463	205·084 4442	*173*	3·69277	313·567 3527	*223*	2·49873	428·397 7197
124	1·50614	207·177 8658	*174*	6·42543	315·807 9019	*224*	5·59716	430·747 9677
125	1·88268	209·274 7759	*175*	1·12445	318·050 9400	*225*	1·25936	433·100 1502
126	2·37217	211·375 1464	*176*	1·97903	320·296 4526	*226*	2·84616	435·454 2586
127	3·01266	213·478 9501	*177*	3·50289	322·544 4259	*227*	6·46077	437·810 2845
128	3·85620	215·586 1601	*178*	6·23514	324·794 8459	*228*	1·47306	440·168 2193
129	4·97450	217·696 7498	*179*	1·11609	327·047 6989	*229*	3·37330	442·528 0548
130	6·46686	219·810 6932	*180*	2·00896	329·302 9714	*230*	7·75859	444·889 7827
131	8·47158	221·927 9645	*181*	3·63622	331·560 6500	*231*	1·79223	447·253 3946
132	1·11825	224·048 5384	*182*	6·61792	333·820 7214	*232*	4·15798	449·618 8826
133	1·48727	226·172 3900	*183*	1·21108	336·083 1725	*233*	9·68810	451·986 2385
134	1·99294	228·299 4948	*184*	2·22839	338·347 9903	*234*	2·26702	454·355 4544
135	2·69047	230·429 8286	*185*	4·12251	340·615 1620	*235*	5·32749	456·726 5223
136	3·65904	232·563 3675	*186*	7·66787	342·884 6750	*236*	1·25729	459·099 4343
137	5·01289	234·700 0881	*187*	1·43389	345·156 5166	*237*	2·97977	461·474 1826
138	6·91779	236·839 9672	*188*	2·69572	347·430 6744	*238*	7·09185	463·850 7596
139	9·61572	238·982 9820	*189*	5·09491	349·707 1362	*239*	1·69495	466·229 1575
140	1·34620	241·129 1100	*190*	9·68032	351·985 8898	*240*	4·06789	468·609 3687
141	1·89814	243·278 3291	*191*	1·84894	354·266 9232	*241*	9·80360	470·991 3857
142	2·69536	245·430 6174	*192*	3·54997	356·550 2244	*242*	2·37247	473·375 2011
143	3·85437	247·585 9535	*193*	6·85144	358·835 7817	*243*	5·76511	475·760 8074
144	5·55029	249·744 3160	*194*	1·32918	361·123 5835	*244*	1·40669	478·148 1972
145	8·04793	251·905 6840	*195*	2·59190	363·413 6181	*245*	3·44638	480·537 3633
146	1·17500	254·070 0368	*196*	5·08012	365·705 8742	*246*	8·47810	482·928 2984
147	1·72725	256·237 3542	*197*	1·00078	368·000 3404	*247*	2·09409	485·320 9954
148	2·55632	258·407 6159	*198*	1·98155	370·297 0056	*248*	5·19334	487·715 4470
149	3·80892	260·580 8022	*199*	3·94329	372·595 8586	*249*	1·29314	490·111 6464
150	5·71338	262·756 8934	*200*	7·88658	374·896 8886	*250*	3·23286	492·509 5864

* Multiply by 10^c, where c is the characteristic of $\log n!$ shown alongside in the next column.

Table 51 (*continued*)

n	$\log_{10} n!$	n	$\log_{10} n!$	n	$\log_{10} n!$	n	$\log_{10} n!$	n	$\log_{10} n!$
251	494·909 2601	301	616·964 3695	351	742·637 2813	401	871·409 5586	451	1002·893 0675
252	497·310 6607	302	619·444 3765	352	745·183 8240	402	874·013 7846	452	1005·548 2059
253	499·713 7812	303	621·925 8191	353	747·731 5987	403	876·619 0896	453	1008·204 3041
254	502·118 6149	304	624·408 6927	354	750·280 6020	404	879·225 4710	454	1010·861 3600
255	504·525 1551	305	626·892 9925	355	752·830 8303	405	881·832 9260	455	1013·519 3714
256	506·933 3950	306	629·378 7140	356	755·382 2803	406	884·441 4521	456	1016·178 3362
257	509·343 3282	307	631·865 8523	357	757·934 9485	407	887·051 0465	457	1018·838 2524
258	511·754 9479	308	634·354 4031	358	760·488 8316	408	889·661 7066	458	1021·499 1179
259	514·168 2476	309	636·844 3615	359	763·043 9260	409	892·273 4300	459	1024·160 9306
260	516·583 2210	310	639·335 7232	360	765·600 2285	410	894·886 2138	460	1026·823 6884
261	518·999 8615	311	641·828 4836	361	768·157 7357	411	897·500 0556	461	1029·487 3893
262	521·418 1628	312	644·322 6382	362	770·716 4443	412	900·114 9528	462	1032·152 0313
263	523·838 1185	313	646·818 1825	363	773·276 3509	413	902·730 9029	463	1034·817 6123
264	526·259 7225	314	649·315 1122	364	775·837 4523	414	905·347 9032	464	1037·484 1303
265	528·682 9683	315	651·813 4227	365	778·399 7452	415	907·965 9513	465	1040·151 5832
266	531·107 8500	316	654·313 1098	366	780·963 2262	416	910·585 0447	466	1042·819 9692
267	533·534 3612	317	656·814 1691	367	783·527 8923	417	913·205 1807	467	1045·489 2860
268	535·962 4960	318	659·316 5962	368	786·093 7401	418	915·826 3570	468	1048·159 5319
269	538·392 2483	319	661·820 3869	369	788·660 7665	419	918·448 5710	469	1050·830 7047
270	540·823 6121	320	664·325 5369	370	791·228 9682	420	921·071 8203	470	1053·502 8026
271	543·256 5814	321	666·832 0419	371	793·798 3421	421	923·696 1024	471	1056·175 8235
272	545·691 1503	322	669·339 8978	372	796·368 8851	422	926·321 4149	472	1058·849 7655
273	548·127 3129	323	671·849 1003	373	798·940 5939	423	928·947 7552	473	1061·524 6266
274	550·565 0635	324	674·359 6453	374	801·513 4655	424	931·575 1211	474	1064·200 4050
275	553·004 3962	325	676·871 5287	375	804·087 4968	425	934·203 5100	475	1066·877 0986
276	555·445 3052	326	679·384 7463	376	806·662 6846	426	936·832 9196	476	1069·554 7056
277	557·887 7850	327	681·899 2940	377	809·239 0260	427	939·463 3475	477	1072·233 2239
278	560·331 8298	328	684·415 1679	378	811·816 5178	428	942·094 7913	478	1074·912 6518
279	562·777 4340	329	686·932 3638	379	814·395 1570	429	944·727 2486	479	1077·592 9873
280	565·224 5920	330	689·450 8777	380	816·974 9406	430	947·360 7170	480	1080·274 2286
281	567·673 2984	331	691·970 7057	381	819·555 8655	431	949·995 1943	481	1082·956 3737
282	570·123 5475	332	694·491 8438	382	822·137 9289	432	952·630 6780	482	1085·639 4207
283	572·575 3339	333	697·014 2880	383	824·721 1277	433	955·267 1659	483	1088·323 3678
284	575·028 6523	334	699·538 0345	384	827·305 4589	434	957·904 6557	484	1091·008 2132
285	577·483 4971	335	702·063 0793	385	829·890 9196	435	960·543 1449	485	1093·693 9549
286	579·939 8631	336	704·589 4186	386	832·477 5069	436	963·182 6314	486	1096·380 5912
287	582·397 7450	337	707·117 0485	387	835·065 2179	437	965·823 1128	487	1099·068 1202
288	584·857 1375	338	709·645 9652	388	837·654 0496	438	968·464 5869	488	1101·756 5400
289	587·318 0354	339	712·176 1649	389	840·243 9992	439	971·107 0515	489	1104·445 8488
290	589·780 4334	340	714·707 6438	390	842·835 0638	440	973·750 5041	490	1107·136 0449
291	592·244 3264	341	717·240 3982	391	845·427 2406	441	976·394 9427	491	1109·827 1264
292	594·709 7092	342	719·774 4243	392	848·020 5267	442	979·040 3650	492	1112·519 0915
293	597·176 5768	343	722·309 7184	393	850·614 9192	443	981·686 7687	493	1115·211 9384
294	599·644 9242	344	724·846 2768	394	853·210 4154	444	984·334 1517	494	1117·905 6654
295	602·114 7462	345	727·384 0959	395	855·807 0125	445	986·982 5117	495	1120·600 2706
296	604·586 0379	346	729·923 1720	396	858·404 7077	446	989·631 8466	496	1123·295 7523
297	607·058 7943	347	732·463 5015	397	861·003 4982	447	992·282 1541	497	1125·992 1086
298	609·533 0106	348	735·005 0807	398	863·603 3813	448	994·933 4321	498	1128·689 3380
299	612·008 6818	349	737·547 9062	399	866·204 3542	449	997·585 6784	499	1131·387 4385
300	614·485 8030	350	740·091 9742	400	868·806 4142	450	1000·238 8910	500	1134·086 4085

Table 51. *Logarithms of factorials of integers* (*continued*)

n	$\log_{10} n!$	n	$\log_{10} n!$	n	$\log_{10} n!$	n	$\log_{10} n!$	n	$\log_{10} n!$
501	1136·786 2463	551	1272·848 0029	601	1410·881 1614	651	1550·721 4519	701	1692·229 8994
502	1139·486 9500	552	1275·589 9419	602	1413·660 7579	652	1553·535 6995	702	1695·076 2365
503	1142·188 5180	553	1278·332 6671	603	1416·441 0752	653	1556·350 6126	703	1697·923 1918
504	1144·890 9485	554	1281·076 1768	604	1419·222 1122	654	1559·166 1904	704	1700·770 7644
505	1147·594 2399	555	1283·820 4698	605	1422·003 8676	655	1561·982 4317	705	1703·618 9536
506	1150·298 3904	556	1286·565 5446	606	1424·786 3402	656	1564·799 3355	706	1706·467 7583
507	1153·003 3984	557	1289·311 3998	607	1427·569 5289	657	1567·616 9009	707	1709·317 1777
508	1155·709 2621	558	1292·058 0340	608	1430·353 4324	658	1570·435 1268	708	1712·167 2109
509	1158·415 9798	559	1294·805 4458	609	1433·138 0497	659	1573·254 0122	709	1715·017 8572
510	1161·123 5500	560	1297·553 6338	610	1435·923 3796	660	1576·073 5561	710	1717·869 1155
511	1163·831 9709	561	1300·302 5967	611	1438·709 4208	661	1578·893 7576	711	1720·720 9851
512	1166·541 2409	562	1303·052 3330	612	1441·496 1722	662	1581·714 6156	712	1723·573 4651
513	1169·251 3583	563	1305·802 8414	613	1444·283 6327	663	1584·536 1291	713	1726·426 5546
514	1171·962 3214	564	1308·554 1205	614	1447·071 8011	664	1587·358 2972	714	1729·280 2529
515	1174·674 1286	565	1311·306 1690	615	1449·860 6762	665	1590·181 1188	715	1732·134 5589
516	1177·386 7783	566	1314·058 9854	616	1452·650 2569	666	1593·004 5931	716	1734·989 4719
517	1180·100 2688	567	1316·812 5684	617	1455·440 5420	667	1595·828 7189	717	1737·844 9911
518	1182·814 5986	568	1319·566 9168	618	1458·231 5305	668	1598·653 4954	718	1740·701 1155
519	1185·529 7660	569	1322·322 0290	619	1461·023 2212	669	1601·478 9215	719	1743·557 8444
520	1188·245 7693	570	1325·077 9039	620	1463·815 6129	670	1604·304 9963	720	1746·415 1769
521	1190·962 6070	571	1327·834 5400	621	1466·608 7045	671	1607·131 7188	721	1749·273 1122
522	1193·680 2775	572	1330·591 9360	622	1469·402 4948	672	1609·959 0881	722	1752·131 6494
523	1196·398 7792	573	1333·350 0907	623	1472·196 9829	673	1612·787 1031	723	1754·990 7877
524	1199·118 1105	574	1336·109 0026	624	1474·992 1675	674	1615·615 7630	724	1757·850 5262
525	1201·838 2698	575	1338·868 6704	625	1477·788 0475	675	1618·445 0668	725	1760·710 8642
526	1204·559 2556	576	1341·629 0929	626	1480·584 6218	676	1621·275 0135	726	1763·571 8009
527	1207·281 0662	577	1344·390 2687	627	1483·381 8894	677	1624·105 6022	727	1766·433 3353
528	1210·003 7001	578	1347·152 1965	628	1486·179 8490	678	1626·936 8319	728	1769·295 4667
529	1212·727 1558	579	1349·914 8751	629	1488·978 4997	679	1629·768 7016	729	1772·158 1942
530	1215·451 4316	580	1352·678 3031	630	1491·777 8402	680	1632·601 2106	730	1775·021 5170
531	1218·176 5262	581	1355·442 4792	631	1494·577 8696	681	1635·434 3577	731	1777·885 4344
532	1220·902 4378	582	1358·207 4022	632	1497·378 5866	682	1638·268 1420	732	1780·749 9455
533	1223·629 1650	583	1360·973 0708	633	1500·179 9904	683	1641·102 5627	733	1783·615 0495
534	1226·356 7063	584	1363·739 4836	634	1502·982 0796	684	1643·937 6189	734	1786·480 7455
535	1229·085 0600	585	1366·506 6395	635	1505·784 8533	685	1646·773 3094	735	1789·347 0329
536	1231·814 2248	586	1369·274 5371	636	1508·588 3105	686	1649·609 6335	736	1792·213 9107
537	1234·544 1991	587	1372·043 1752	637	1511·392 4499	687	1652·446 5903	737	1795·081 3782
538	1237·274 9814	588	1374·812 5525	638	1514·197 2706	688	1655·284 1787	738	1797·949 4345
539	1240·006 5702	589	1377·582 6678	639	1517·002 7714	689	1658·122 3979	739	1800·818 0790
540	1242·738 9639	590	1380·353 5198	640	1519·808 9514	690	1660·961 2470	740	1803·687 3107
541	1245·472 1612	591	1383·125 1073	641	1522·615 8094	691	1663·800 7251	741	1806·557 1289
542	1248·206 1605	592	1385·897 4290	642	1525·423 3445	692	1666·640 8312	742	1809·427 5328
543	1250·940 9603	593	1388·670 4837	643	1528·231 5554	693	1669·481 5644	743	1812·298 5216
544	1253·676 5592	594	1391·444 2702	644	1531·040 4413	694	1672·322 9239	744	1815·170 0946
545	1256·412 9557	595	1394·218 7871	645	1533·850 0010	695	1675·164 9087	745	1818·042 2508
546	1259·150 1483	596	1396·994 0334	646	1536·660 2335	696	1678·007 5179	746	1820·914 9897
547	1261·888 1357	597	1399·770 0077	647	1539·471 1378	697	1680·850 7507	747	1823·788 3103
548	1264·626 9162	598	1402·546 7089	648	1542·282 7128	698	1683·694 6061	748	1826·662 2119
549	1267·366 4886	599	1405·324 1357	649	1545·094 9575	699	1686·539 0833	749	1829·536 6937
550	1270·106 8513	600	1408·102 2870	650	1547·907 8709	700	1689·384 1813	750	1832·411 7549

Table 51 (*continued*)

n	$\log_{10} n!$	n	$\log_{10} n!$	n	$\log_{10} n!$	n	$\log_{10} n!$	n	$\log_{10} n!$
751	1835·287 3949	*801*	1979·790 7168	*851*	2125·649 5488	*901*	2272·784 2010	*951*	2421·123 8376
752	1838·163 6127	*802*	1982·694 8911	*852*	2128·579 9884	*902*	2275·739 4075	*952*	2424·102 4745
753	1841·040 4077	*803*	1985·599 6067	*853*	2131·510 9374	*903*	2278·695 0953	*953*	2427·081 5674
754	1843·917 7790	*804*	1988·504 8627	*854*	2134·442 3953	*904*	2281·651 2637	*954*	2430·061 1158
755	1846·795 7260	*805*	1991·410 6586	*855*	2137·374 3614	*905*	2284·607 9123	*955*	2433·041 1192
756	1849·674 2478	*806*	1994·316 9936	*856*	2140·306 8352	*906*	2287·565 0405	*956*	2436·021 5771
757	1852·553 3437	*807*	1997·223 8672	*857*	2143·239 8160	*907*	2290·522 6478	*957*	2439·002 4890
758	1855·433 0129	*808*	2000·131 2785	*858*	2146·173 3033	*908*	2293·480 7336	*958*	2441·983 8545
759	1858·313 2546	*809*	2003·039 2271	*859*	2149·107 2964	*909*	2296·439 2975	*959*	2444·965 6731
760	1861·194 0682	*810*	2005·947 7121	*860*	2152·041 7949	*910*	2299·398 3389	*960*	2447·947 9443
761	1864·075 4529	*811*	2008·856 7329	*861*	2154·976 7980	*911*	2302·357 8573	*961*	2450·930 6677
762	1866·957 4079	*812*	2011·766 2890	*862*	2157·912 3053	*912*	2305·317 8521	*962*	2453·913 8428
763	1869·839 9324	*813*	2014·676 3795	*863*	2160·848 3161	*913*	2308·278 3229	*963*	2456·897 4691
764	1872·723 0258	*814*	2017·587 0039	*864*	2163·784 8298	*914*	2311·239 2691	*964*	2459·881 5461
765	1875·606 6872	*815*	2020·498 1615	*865*	2166·721 8459	*915*	2314·200 6902	*965*	2462·866 0734
766	1878·490 9160	*816*	2023·409 8517	*866*	2169·659 3638	*916*	2317·162 5856	*966*	2465·851 0506
767	1881·375 7113	*817*	2026·322 0737	*867*	2172·597 3829	*917*	2320·124 9550	*967*	2468·836 4770
768	1884·261 0726	*818*	2029·234 8270	*868*	2175·535 9027	*918*	2323·087 7977	*968*	2471·822 3524
769	1887·146 9989	*819*	2032·148 1109	*869*	2178·474 9224	*919*	2326·051 1132	*969*	2474·808 6762
770	1890·033 4896	*820*	2035·061 9248	*870*	2181·414 4417	*920*	2329·014 9010	*970*	2477·795 4479
771	1892·920 5440	*821*	2037·976 2679	*871*	2184·354 4598	*921*	2331·979 1606	*971*	2480·782 6671
772	1895·808 1613	*822*	2040·891 1398	*872*	2187·294 9763	*922*	2334·943 8915	*972*	2483·770 3334
773	1898·696 3408	*823*	2043·806 5396	*873*	2190·235 9906	*923*	2337·909 0932	*973*	2486·758 4462
774	1901·585 0817	*824*	2046·722 4668	*874*	2193·177 5020	*924*	2340·874 7652	*974*	2489·747 0052
775	1904·474 3835	*825*	2049·638 9208	*875*	2196·119 5101	*925*	2343·840 9069	*975*	2492·736 0098
776	1907·364 2452	*826*	2052·555 9008	*876*	2199·062 0142	*926*	2346·807 5179	*976*	2495·725 4596
777	1910·254 6662	*827*	2055·473 4063	*877*	2202·005 0138	*927*	2349·774 5977	*977*	2498·715 3542
778	1913·145 6458	*828*	2058·391 4367	*878*	2204·948 5083	*928*	2352·742 1456	*978*	2501·705 6930
779	1916·037 1832	*829*	2061·309 9912	*879*	2207·892 4971	*929*	2355·710 1614	*979*	2504·696 4757
780	1918·929 2778	*830*	2064·229 0693	*880*	2210·836 9798	*930*	2358·678 6443	*980*	2507·687 7018
781	1921·821 9289	*831*	2067·148 6703	*881*	2213·781 9557	*931*	2361·647 5940	*981*	2510·679 3708
782	1924·715 1356	*832*	2070·068 7936	*882*	2216·727 4243	*932*	2364·617 0099	*982*	2513·671 4823
783	1927·608 8974	*833*	2072·989 4386	*883*	2219·673 3850	*933*	2367·586 8915	*983*	2516·664 0358
784	1930·503 2135	*834*	2075·910 6047	*884*	2222·619 8373	*934*	2370·557 2384	*984*	2519·657 0309
785	1933·398 0831	*835*	2078·832 2912	*885*	2225·566 7805	*935*	2373·528 0500	*985*	2522·650 4672
786	1936·293 5057	*836*	2081·754 4974	*886*	2228·514 2143	*936*	2376·499 3259	*986*	2525·644 3441
787	1939·189 4804	*837*	2084·677 2229	*887*	2231·462 1379	*937*	2379·471 0655	*987*	2528·638 6612
788	1942·086 0066	*838*	2087·600 4669	*888*	2234·410 5509	*938*	2382·443 2683	*988*	2531·633 4182
789	1944·983 0836	*839*	2090·524 2289	*889*	2237·359 4526	*939*	2385·415 9339	*989*	2534·628 6145
790	1947·880 7107	*840*	2093·448 5082	*890*	2240·308 8426	*940*	2388·389 0618	*990*	2537·624 2497
791	1950·778 8872	*841*	2096·373 3042	*891*	2243·258 7203	*941*	2391·362 6514	*991*	2540·620 3233
792	1953·677 6124	*842*	2099·298 6162	*892*	2246·209 0852	*942*	2394·336 7023	*992*	2543·616 8350
793	1956·576 8856	*843*	2102·224 4438	*893*	2249·159 9366	*943*	2397·311 2140	*993*	2546·613 7842
794	1959·476 7061	*844*	2105·150 7863	*894*	2252·111 2742	*944*	2400·286 1860	*994*	2549·611 1706
795	1962·377 0732	*845*	2108·077 6430	*895*	2255·063 0972	*945*	2403·261 6178	*995*	2552·608 9937
796	1965·277 9863	*846*	2111·005 0133	*896*	2258·015 4052	*946*	2406·237 5089	*996*	2555·607 2530
797	1968·179 4446	*847*	2113·932 8967	*897*	2260·968 1976	*947*	2409·213 8589	*997*	2558·605 9482
798	1971·081 4475	*848*	2116·861 2926	*898*	2263·921 4740	*948*	2412·190 6672	*998*	2561·605 0787
799	1973·983 9943	*849*	2119·790 2003	*899*	2266·875 2337	*949*	2415·167 9334	*999*	2564·604 6442
800	1976·887 0842	*850*	2122·719 6192	*900*	2269·829 4762	*950*	2418·145 6570	*1000*	2567·604 6442

Table 52. *Miscellaneous functions of p and $q=1-p$ over the unit range*

p	$\Gamma(1+p)$	$\log_{10}\Gamma(1+p)$	$1-p^2$	$\sqrt{(1-p^2)}$	$\frac{1}{\sqrt{(1-p^2)}}$	pq	$\sqrt{(pq)}$	p^2+q^2	$q=1-p$
0·00	1·0000 000	0·0000 000	1·0000	1·00000	1·00000	0·0000	0·00000	1·0000	**1·00**
0·01	0·9943 259	$\bar{1}$·9975 287	0·9999	0·99995	1·00005	0·0099	0·09950	0·9802	**0·99**
0·02	0·9888 442	$\bar{1}$·9951 279	0·9996	0·99980	1·00020	0·0196	0·14000	0·9608	**0·98**
0·03	0·9835 500	$\bar{1}$·9927 964	0·9991	0·99955	1·00045	0·0291	0·17059	0·9418	**0·97**
0·04	0·9784 382	$\bar{1}$·9905 334	0·9984	0·99920	1·00080	0·0384	0·19596	0·9232	**0·96**
0·05	0·9735 043	$\bar{1}$·9883 379	0·9975	0·99875	1·00125	0·0475	0·21794	0·9050	**0·95**
0·06	0·9687 436	$\bar{1}$·9862 089	0·9964	0·99820	1·00180	0·0564	0·23749	0·8872	**0·94**
0·07	0·9641 520	$\bar{1}$·9841 455	0·9951	0·99755	1·00246	0·0651	0·25515	0·8698	**0·93**
0·08	0·9597 253	$\bar{1}$·9821 469	0·9936	0·99679	1·00322	0·0736	0·27129	0·8528	**0·92**
0·09	0·9554 595	$\bar{1}$·9802 123	0·9919	0·99594	1·00407	0·0819	0·28618	0·8362	**0·91**
0·10	0·9513 508	$\bar{1}$·9783 407	0·9900	0·99499	1·00504	0·0900	0·30000	0·8200	**0·90**
0·11	0·9473 955	$\bar{1}$·9765 313	0·9879	0·99393	1·00611	0·0979	0·31289	0·8042	**0·89**
0·12	0·9435 902	$\bar{1}$·9747 834	0·9856	0·99277	1·00728	0·1056	0·32496	0·7888	**0·88**
0·13	0·9399 314	$\bar{1}$·9730 962	0·9831	0·99151	1·00856	0·1131	0·33630	0·7738	**0·87**
0·14	0·9364 161	$\bar{1}$·9714 689	0·9804	0·99015	1·00995	0·1204	0·34699	0·7592	**0·86**
0·15	0·9330 409	$\bar{1}$·9699 007	0·9775	0·98869	1·01144	0·1275	0·35707	0·7450	**0·85**
0·16	0·9298 031	$\bar{1}$·9683 910	0·9744	0·98712	1·01305	0·1344	0·36661	0·7312	**0·84**
0·17	0·9266 996	$\bar{1}$·9669 390	0·9711	0·98544	1·01477	0·1411	0·37563	0·7178	**0·83**
0·18	0·9237 278	$\bar{1}$·9655 440	0·9676	0·98367	1·01660	0·1476	0·38419	0·7048	**0·82**
0·19	0·9208 850	$\bar{1}$·9642 054	0·9639	0·98178	1·01855	0·1539	0·39230	0·6922	**0·81**
0·20	0·9181 687	$\bar{1}$·9629 225	0·9600	0·97980	1·02062	0·1600	0·40000	0·6800	**0·80**
0·21	0·9155 765	$\bar{1}$·9616 946	0·9559	0·97770	1·02281	0·1659	0·40731	0·6682	**0·79**
0·22	0·9131 059	$\bar{1}$·9605 212	0·9516	0·97550	1·02512	0·1716	0·41425	0·6568	**0·78**
0·23	0·9107 549	$\bar{1}$·9594 015	0·9471	0·97319	1·02755	0·1771	0·42083	0·6458	**0·77**
0·24	0·9085 211	$\bar{1}$·9583 350	0·9424	0·97077	1·03011	0·1824	0·42708	0·6352	**0·76**
0·25	0·9064 025	$\bar{1}$·9573 211	0·9375	0·96825	1·03280	0·1875	0·43301	0·6250	**0·75**
0·26	0·9043 971	$\bar{1}$·9563 592	0·9324	0·96561	1·03562	0·1924	0·43863	0·6152	**0·74**
0·27	0·9025 031	$\bar{1}$·9554 487	0·9271	0·96286	1·03857	0·1971	0·44396	0·6058	**0·73**
0·28	0·9007 185	$\bar{1}$·9545 891	0·9216	0·96000	1·04167	0·2016	0·44900	0·5968	**0·72**
0·29	0·8990 416	$\bar{1}$·9537 798	0·9159	0·95703	1·04490	0·2059	0·45376	0·5882	**0·71**
0·30	0·8974 707	$\bar{1}$·9530 203	0·9100	0·95394	1·04828	0·2100	0·45826	0·5800	**0·70**
0·31	0·8960 042	$\bar{1}$·9523 100	0·9039	0·95074	1·05182	0·2139	0·46249	0·5722	**0·69**
0·32	0·8946 405	$\bar{1}$·9516 485	0·8976	0·94742	1·05550	0·2176	0·46648	0·5648	**0·68**
0·33	0·8933 781	$\bar{1}$·9510 353	0·8911	0·94398	1·05934	0·2211	0·47021	0·5578	**0·67**
0·34	0·8922 155	$\bar{1}$·9504 698	0·8844	0·94043	1·06335	0·2244	0·47371	0·5512	**0·66**
0·35	0·8911 514	$\bar{1}$·9499 515	0·8775	0·93675	1·06752	0·2275	0·47697	0·5450	**0·65**
0·36	0·8901 845	$\bar{1}$·9494 800	0·8704	0·93295	1·07187	0·2304	0·48000	0·5392	**0·64**
0·37	0·8893 135	$\bar{1}$·9490 549	0·8631	0·92903	1·07639	0·2331	0·48280	0·5338	**0·63**
0·38	0·8885 371	$\bar{1}$·9486 756	0·8556	0·92499	1·08110	0·2356	0·48539	0·5288	**0·62**
0·39	0·8878 543	$\bar{1}$·9483 417	0·8479	0·92081	1·08599	0·2379	0·48775	0·5242	**0·61**
0·40	0·8872 638	$\bar{1}$·9480 528	0·8400	0·91652	1·09109	0·2400	0·48990	0·5200	**0·60**
0·41	0·8867 647	$\bar{1}$·9478 084	0·8319	0·91209	1·09639	0·2419	0·49183	0·5162	**0·59**
0·42	0·8863 558	$\bar{1}$·9476 081	0·8236	0·90752	1·10190	0·2436	0·49356	0·5128	**0·58**
0·43	0·8860 362	$\bar{1}$·9474 515	0·8151	0·90283	1·10763	0·2451	0·49508	0·5098	**0·57**
0·44	0·8858 051	$\bar{1}$·9473 382	0·8064	0·89800	1·11359	0·2464	0·49639	0·5072	**0·56**
0·45	0·8856 614	$\bar{1}$·9472 677	0·7975	0·89303	1·11979	0·2475	0·49749	0·5050	**0·55**
0·46	0·8856 043	$\bar{1}$·9472 397	0·7884	0·88792	1·12623	0·2484	0·49840	0·5032	**0·54**
0·47	0·8856 331	$\bar{1}$·9472 539	0·7791	0·88267	1·13293	0·2491	0·49910	0·5018	**0·53**
0·48	0·8857 470	$\bar{1}$·9473 097	0·7696	0·87727	1·13990	0·2496	0·49960	0·5008	**0·52**
0·49	0·8859 451	$\bar{1}$·9474 068	0·7599	0·87172	1·14715	0·2499	0·49990	0·5002	**0·51**
0·50	0·8862 269	$\bar{1}$·9475 449	0·7500	0·86603	1·15470	0·2500	0·50000	0·5000	**0·50**

Table 52 (*continued*)

p	$\Gamma(1+p)$	$\log_{10}\Gamma(1+p)$	$1-p^2$	$\sqrt{(1-p^2)}$	$\frac{1}{\sqrt{(1-p^2)}}$	p	$\sqrt{(1-p^2)}$	$\frac{1}{\sqrt{(1-p^2)}}$	$\sqrt{(pq)}$
0·50	0·8862 269	$\bar{1}$·9475 449	0·7500	0·86603	1·15470	**0·910**	0·41461	2·41192	0·28618
0·51	0·8865 917	$\bar{1}$·9477 237	0·7399	0·86017	1·16255	**0·912**	0·41019	2·43789	0·28329
0·52	0·8870 388	$\bar{1}$·9479 426	0·7296	0·85417	1·17073	**0·914**	0·40571	2·46479	0·28036
0·53	0·8875 676	$\bar{1}$·9482 015	0·7191	0·84800	1·17925	**0·916**	0·40118	2·49266	0·27739
0·54	0·8881 777	$\bar{1}$·9484 998	0·7084	0·84167	1·18812	**0·918**	0·39658	2·52156	0·27436
0·55	0·8888 683	$\bar{1}$·9488 374	0·6975	0·83516	1·19737	**0·920**	0·39192	2·55155	0·27129
0·56	0·8896 392	$\bar{1}$·9492 139	0·6864	0·82849	1·20701	**0·922**	0·38719	2·58271	0·26817
0·57	0·8904 897	$\bar{1}$·9496 289	0·6751	0·82164	1·21707	**0·924**	0·38239	2·61511	0·26500
0·58	0·8914 196	$\bar{1}$·9500 822	0·6636	0·81462	1·22757	**0·926**	0·37752	2·64884	0·26177
0·59	0·8924 282	$\bar{1}$·9505 733	0·6519	0·80740	1·23854	**0·928**	0·37258	2·68399	0·25849
0·60	0·8935 153	$\bar{1}$·9511 020	0·6400	0·80000	1·25000	**0·930**	0·36756	2·72065	0·25515
0·61	0·8946 806	$\bar{1}$·9516 680	0·6279	0·79240	1·26199	**0·932**	0·36246	2·75894	0·25175
0·62	0·8959 237	$\bar{1}$·9522 710	0·6156	0·78460	1·27453	**0·934**	0·35727	2·79898	0·24828
0·63	0·8972 442	$\bar{1}$·9529 107	0·6031	0·77660	1·28767	**0·936**	0·35200	2·84091	0·24475
0·64	0·8986 420	$\bar{1}$·9535 867	0·5904	0·76837	1·30145	**0·938**	0·34664	2·88488	0·24116
0·65	0·9001 168	$\bar{1}$·9542 989	0·5775	0·75993	1·31590	**0·940**	0·34117	2·93105	0·23749
0·66	0·9016 684	$\bar{1}$·9550 468	0·5644	0·75127	1·33109	**0·942**	0·33561	2·97962	0·23374
0·67	0·9032 965	$\bar{1}$·9558 303	0·5511	0·74236	1·34705	**0·944**	0·32995	3·03080	0·22992
0·68	0·9050 010	$\bar{1}$·9566 491	0·5376	0·73321	1·36386	**0·946**	0·32417	3·08483	0·22602
0·69	0·9067 818	$\bar{1}$·9575 028	0·5239	0·72381	1·38158	**0·948**	0·31827	3·14198	0·22203
0·70	0·9086 387	$\bar{1}$·9583 912	0·5100	0·71414	1·40028	**0·950**	0·31225	3·20256	0·21794
0·71	0·9105 717	$\bar{1}$·9593 141	0·4959	0·70420	1·42005	**0·952**	0·30610	3·26693	0·21377
0·72	0·9125 806	$\bar{1}$·9602 712	0·4816	0·69397	1·44098	**0·954**	0·29981	3·33548	0·20949
0·73	0·9146 654	$\bar{1}$·9612 622	0·4671	0·68345	1·46317	**0·956**	0·29337	3·40870	0·20510
0·74	0·9168 260	$\bar{1}$·9622 869	0·4524	0·67261	1·48675	**0·958**	0·28677	3·48714	0·20059
0·75	0·9190 625	$\bar{1}$·9633 451	0·4375	0·66144	1·51186	**0·960**	0·28000	3·57143	0·19596
0·76	0·9213 749	$\bar{1}$·9644 364	0·4224	0·64992	1·53864	**0·962**	0·27305	3·66234	0·19120
0·77	0·9237 631	$\bar{1}$·9655 606	0·4071	0·63804	1·56729	**0·964**	0·26590	3·76078	0·18629
0·78	0·9262 273	$\bar{1}$·9667 176	0·3916	0·62578	1·59801	**0·966**	0·25854	3·86784	0·18123
0·79	0·9287 675	$\bar{1}$·9679 070	0·3759	0·61311	1·63104	**0·968**	0·25095	3·98485	0·17600
0·80	0·9313 838	$\bar{1}$·9691 287	0·3600	0·60000	1·66667	**0·970**	0·24310	4·11345	0·17059
0·81	0·9340 763	$\bar{1}$·9703 823	0·3439	0·58643	1·70523	**0·972**	0·23498	4·25567	0·16497
0·82	0·9368 451	$\bar{1}$·9716 678	0·3276	0·57236	1·74714	**0·974**	0·22655	4·41408	0·15914
0·83	0·9396 904	$\bar{1}$·9729 848	0·3111	0·55776	1·79287	**0·976**	0·21777	4·59199	0·15305
0·84	0·9426 124	$\bar{1}$·9743 331	0·2944	0·54259	1·84302	**0·978**	0·20860	4·79375	0·14668
0·85	0·9456 112	$\bar{1}$·9757 126	0·2775	0·52678	1·89832	**0·980**	0·19900	5·02519	0·14000
0·86	0·9486 870	$\bar{1}$·9771 230	0·2604	0·51029	1·95965	**0·982**	0·18888	5·29434	0·13295
0·87	0·9518 402	$\bar{1}$·9785 640	0·2431	0·49305	2·02818	**0·984**	0·17817	5·61267	0·12548
0·88	0·9550 709	$\bar{1}$·9800 356	0·2256	0·47497	2·10538	**0·986**	0·16675	5·99717	0·11749
0·89	0·9583 793	$\bar{1}$·9815 374	0·2079	0·45596	2·19317	**0·988**	0·15445	6·47442	0·10889
0·90	0·9617 658	$\bar{1}$·9830 693	0·1900	0·43589	2·29416	**0·990**	0·14107	7·08881	0·09950
0·91	0·9652 307	$\bar{1}$·9846 311	0·1719	0·41461	2·41192	**0·991**	0·13386	7·47039	0·09444
0·92	0·9687 743	$\bar{1}$·9862 226	0·1536	0·39192	2·55155	**0·992**	0·12624	7·92155	0·08908
0·93	0·9723 969	$\bar{1}$·9878 436	0·1351	0·36756	2·72065	**0·993**	0·11811	8·46637	0·08337
0·94	0·9760 989	$\bar{1}$·9894 938	0·1164	0·34117	2·93105	**0·994**	0·10938	9·14243	0·07723
0·95	0·9798 807	$\bar{1}$·9911 732	0·0975	0·31225	3·20256	**0·995**	0·09987	10·01252	0·07053
0·96	0·9837 425	$\bar{1}$·9928 815	0·0784	0·28000	3·57143	**0·996**	0·08935	11·19154	0·06312
0·97	0·9876 850	$\bar{1}$·9946 185	0·0591	0·24310	4·11345	**0·997**	0·07740	12·91964	0·05469
0·98	0·9917 084	$\bar{1}$·9963 840	0·0396	0·19900	5·02519	**0·998**	0·06321	15·81930	0·04468
0·99	0·9958 133	$\bar{1}$·9981 779	0·0199	0·14107	7·08881	**0·999**	0·04471	22·36627	0·03161
1·00	1·0000 000	0·0000 000	0·0000	0·00000	∞	**1·000**	0·00000	∞	0·00000

Table 53. *Natural logarithms*, $\log_e x$

1	2	3	4	5		00	05	10	15	20	25	30	35	40	45
98	196	293	391	489	**1·0**	0·00 000	00 499	00 995	01 489	01 980	02 469	02 956	03 440	03 922	04 402
89	178	267	356	445	**1·1**	0·09 531	09 985	10 436	10 885	11 333	11 778	12 222	12 663	13 103	13 540
82	164	245	327	409	**1·2**	0·18 232	18 648	19 062	19 474	19 885	20 294	20 701	21 107	21 511	21 914
76	151	227	303	378	**1·3**	0·26 236	26 620	27 003	27 384	27 763	28 141	28 518	28 893	29 267	29 639
70	141	211	281	352	**1·4**	0·33 647	34 004	34 359	34 713	35 066	35 417	35 767	36 116	36 464	36 811
66	131	197	263	329	**1·5**	0·40 547	40 879	41 211	41 542	41 871	42 199	42 527	42 853	43 178	43 502
62	123	185	247	308	**1·6**	0·47 000	47 312	47 623	47 933	48 243	48 551	48 858	49 164	49 470	49 774
58	116	174	232	290	**1·7**	0·53 063	53 357	53 649	53 941	54 232	54 523	54 812	55 101	55 389	55 675
55	110	165	220	274	**1·8**	0·58 779	59 056	59 333	59 609	59 884	60 158	60 432	60 704	60 977	61 248
52	104	156	208	260	**1·9**	0·64 185	64 448	64 710	64 972	65 233	65 493	65 752	66 011	66 269	66 526

1	2	3	4	5		0	1	2	3	4	5	6	7	8	9
49	98	147	196	244	**2·0**	0·69 315	69 813	70 310	70 804	71 295	71 784	72 271	72 755	73 237	73 716
47	93	140	186	233	**2·1**	0·74 194	74 669	75 142	75 612	76 081	76 547	77 011	77 473	77 932	78 390
45	89	134	178	223	**2·2**	0·78 846	79 299	79 751	80 200	80 648	81 093	81 536	81 978	82 418	82 855
43	85	128	171	213	**2·3**	0·83 291	83 725	84 157	84 587	85 015	85 442	85 866	86 289	86 710	87 129
41	82	123	164	204	**2·4**	0·87 547	87 963	88 377	88 789	89 200	89 609	90 016	90 422	90 826	91 228
39	79	118	157	196	**2·5**	0·91 629	92 028	92 426	92 822	93 216	93 609	94 001	94 391	94 779	95 166
38	76	113	151	189	**2·6**	0·95 551	95 935	96 317	96 698	97 078	97 456	97 833	98 208	98 582	98 954
36	73	109	146	182	**2·7**	0·99 325	99 695	**00 063**	**00 430**	**00 796**	**01 160**	**01 523**	**01 885**	**02 245**	**02 604**
35	70	105	141	176	**2·8**	1·02 962	03 318	03 674	04 028	04 380	04 732	05 082	05 431	05 779	06 126
34	68	102	136	170	**2·9**	1·06 471	06 815	07 158	07 500	07 841	08 181	08 519	08 856	09 192	09 527
33	66	99	131	164	**3·0**	1·09 861	10 194	10 526	10 856	11 186	11 514	11 841	12 168	12 493	12 817
32	64	95	127	159	**3·1**	1·13 140	13 462	13 783	14 103	14 422	14 740	15 057	15 373	15 688	16 002
31	62	92	123	154	**3·2**	1·16 315	16 627	16 938	17 248	17 557	17 865	18 173	18 479	18 784	19 089
30	60	90	120	150	**3·3**	1·19 392	19 695	19 996	20 297	20 597	20 896	21 194	21 491	21 788	22 083
29	58	87	116	145	**3·4**	1·22 378	22 671	22 964	23 256	23 547	23 837	24 127	24 415	24 703	24 990
28	56	85	113	141	**3·5**	1·25 276	25 562	25 846	26 130	26 413	26 695	26 976	27 257	27 536	27 815
27	55	82	110	137	**3·6**	1·28 093	28 371	28 647	28 923	29 198	29 473	29 746	30 019	30 291	30 563
27	53	80	107	134	**3·7**	1·30 833	31 103	31 372	31 641	31 909	32 176	32 442	32 708	32 972	33 237
26	52	78	104	130	**3·8**	1·33 500	33 763	34 025	34 286	34 547	34 807	35 067	35 325	35 584	35 841
25	51	76	101	127	**3·9**	1·36 098	36 354	36 609	36 864	37 118	37 372	37 624	37 877	38 128	38 379
25	49	74	99	124	**4·0**	1·38 629	38 879	39 128	39 377	39 624	39 872	40 118	40 364	40 610	40 854
24	48	72	96	121	**4·1**	1·41 099	41 342	41 585	41 828	42 070	42 311	42 552	42 792	43 031	43 270
24	47	71	94	118	**4·2**	1·43 508	43 746	43 984	44 220	44 456	44 692	44 927	45 161	45 395	45 629
23	46	69	92	115	**4·3**	1·45 862	46 094	46 326	46 557	46 787	47 018	47 247	47 476	47 705	47 933
22	45	68	90	112	**4·4**	1·48 160	48 387	48 614	48 840	49 065	49 290	49 515	49 739	49 962	50 185
22	44	66	88	110	**4·5**	1·50 408	50 630	50 851	51 072	51 293	51 513	51 732	51 951	52 170	52 388
22	43	65	86	108	**4·6**	1·52 606	52 823	53 039	53 256	53 471	53 687	53 902	54 116	54 330	54 543
21	42	63	84	105	**4·7**	1·54 756	54 969	55 181	55 393	55 604	55 814	56 025	56 235	56 444	56 653
21	41	62	83	103	**4·8**	1·56 862	57 070	57 277	57 485	57 691	57 898	58 104	58 309	58 515	58 719
20	40	61	81	101	**4·9**	1·58 924	59 127	59 331	59 534	59 737	59 939	60 141	60 342	60 543	60 744
20	40	59	79	99	**5·0**	1·60 944	61 144	61 343	61 542	61 741	61 939	62 137	62 334	62 531	62 728
19	39	58	78	97	**5·1**	1·62 924	63 120	63 315	63 511	63 705	63 900	64 094	64 287	64 481	64 673
19	38	57	76	95	**5·2**	1·64 866	65 058	65 250	65 441	65 632	65 823	66 013	66 203	66 393	66 582
19	37	56	75	94	**5·3**	1·66 771	66 959	67 147	67 335	67 523	67 710	67 896	68 083	68 269	68 455
18	37	55	73	92	**5·4**	1·68 640	68 825	69 010	69 194	69 378	69 562	69 745	69 928	70 111	70 293
18	36	54	72	90	**5·5**	1·70 475	70 656	70 838	71 019	71 199	71 380	71 560	71 740	71 919	72 098
18	35	53	71	89	**5·6**	1·72 277	72 455	72 633	72 811	72 988	73 166	73 342	73 519	73 695	73 871
17	35	52	70	87	**5·7**	1·74 047	74 222	74 397	74 572	74 746	74 920	75 094	75 267	75 440	75 613
17	34	51	68	86	**5·8**	1·75 786	75 958	76 130	76 302	76 473	76 644	76 815	76 985	77 156	77 326
17	34	50	67	84	**5·9**	1·77 495	77 665	77 834	78 002	78 171	78 339	78 507	78 675	78 842	79 009

$\log_e 10 = 2{\cdot}30259$, $\log_e 10^2 = 4{\cdot}60517$, $\log_e 10^3 = 6{\cdot}90776$, $\log_e 10^4 = 9{\cdot}21034$, $\log_e 10^5 = 11{\cdot}51293$, $\log_e 10^6 = 13{\cdot}81551$.

N.B. Proportional parts should be applied to the *nearest* tabular entry.

Examples: $\log 1{\cdot}099 = 0{\cdot}09531 - 0{\cdot}00089 = 0{\cdot}09442$ (correct value, 0·09440),

$\log 11{\cdot}435 = 2{\cdot}30259 + 0{\cdot}13540 - 0{\cdot}00089 - \frac{1}{10}(0{\cdot}00445) = 2{\cdot}43666$ (correct value, 2·43668).

Table 53 (*continued*)

	50	55	60	65	70	75	80	85	90	95	1	2	3	4	5
1·0	0·04 879	05 354	05 827	06 297	06 766	07 232	07 696	08 158	08 618	09 075	93	187	280	373	466
1·1	0·13 976	14 410	14 842	15 272	15 700	16 127	16 551	16 974	17 395	17 815	85	171	256	341	426
1·2	0·22 314	22 714	23 111	23 507	23 902	24 295	24 686	25 076	25 464	25 851	79	157	236	314	393
1·3	0·30 010	30 380	30 748	31 115	31 481	31 845	32 208	32 570	32 930	33 289	73	146	219	291	364
1·4	0·37 156	37 501	37 844	38 186	38 526	38 866	39 204	39 541	39 878	40 213	68	136	204	272	340
1·5	0·43 825	44 148	44 469	44 789	45 108	45 426	45 742	46 058	46 373	46 687	64	127	191	254	318
1·6	0·50 078	50 380	50 682	50 983	51 282	51 581	51 879	52 177	52 473	52 768	60	120	179	239	299
1·7	0·55 962	56 247	56 531	56 815	57 098	57 380	57 661	57 942	58 222	58 501	56	113	169	226	282
1·8	0·61 519	61 788	62 058	62 326	62 594	62 861	63 127	63 393	63 658	63 922	53	107	160	214	267
1·9	0·66 783	67 039	67 294	67 549	67 803	68 057	68 310	68 562	68 813	69 064	51	101	152	203	253

	0	1	2	3	4	5	6	7	8	9	1	2	3	4	5
6·0	1·79 176	79 342	79 509	79 675	79 840	80 006	80 171	80 336	80 500	80 665	17	33	50	66	83
6·1	1·80 829	80 993	81 156	81 319	81 482	81 645	81 808	81 970	82 132	82 294	16	33	49	65	81
6·2	1·82 455	82 616	82 777	82 938	83 098	83 258	83 418	83 578	83 737	83 896	16	32	48	64	80
6·3	1·84 055	84 214	84 372	84 530	84 688	84 845	85 003	85 160	85 317	85 473	16	32	47	63	79
6·4	1·85 630	85 786	85 942	86 097	86 253	86 408	86 563	86 718	86 872	87 026	16	31	47	62	78
6·5	1·87 180	87 334	87 487	87 641	87 794	87 947	88 099	88 251	88 403	88 555	15	31	46	61	76
6·6	1·88 707	88 858	89 010	89 160	89 311	89 462	89 612	89 762	89 912	90 061	15	30	45	60	75
6·7	1·90 211	90 360	90 509	90 658	90 806	90 954	91 102	91 250	91 398	91 545	15	30	44	59	74
6·8	1·91 692	91 839	91 986	92 132	92 279	92 425	92 571	92 716	92 862	93 007	15	29	44	58	73
6·9	1·93 152	93 297	93 442	93 586	93 730	93 874	94 018	94 162	94 305	94 448	14	29	43	58	72
7·0	1·94 591	94 734	94 876	95 019	95 161	95 303	95 445	95 586	95 727	95 869	14	28	43	57	71
7·1	1·96 009	96 150	96 291	96 431	96 571	96 711	96 851	96 991	97 130	97 269	14	28	42	56	70
7·2	1·97 408	97 547	97 685	97 824	97 962	98 100	98 238	98 376	98 513	98 650	14	28	41	55	69
7·3	1·98 787	98 924	99 061	99 198	99 334	99 470	99 606	99 742	99 877	**00 013**	14	27	41	54	68
7·4	2·00 148	00 283	00 418	00 553	00 687	00 821	00 956	01 089	01 223	01 357	13	27	40	54	67
7·5	2·01 490	01 624	01 757	01 890	02 022	02 155	02 287	02 419	02 551	02 683	13	27	40	53	66
7·6	2·02 815	02 946	03 078	03 209	03 340	03 471	03 601	03 732	03 862	03 992	13	26	39	52	65
7·7	2·04 122	04 252	04 381	04 511	04 640	04 769	04 898	05 027	05 156	05 284	13	26	39	52	65
7·8	2·05 412	05 540	05 668	05 796	05 924	06 051	06 179	06 306	06 433	06 560	13	25	38	51	64
7·9	2·06 686	06 813	06 939	07 065	07 191	07 317	07 443	07 568	07 694	07 819	13	25	38	50	63
8·0	2·07 944	08 069	08 194	08 318	08 443	08 567	08 691	08 815	08 939	09 063	12	25	37	50	62
8·1	2·09 186	09 310	09 433	09 556	09 679	09 802	09 924	10 047	10 169	10 291	12	25	37	49	61
8·2	2·10 413	10 535	10 657	10 779	10 900	11 021	11 142	11 263	11 384	11 505	12	24	36	49	61
8·3	2·11 626	11 746	11 866	11 986	12 106	12 226	12 346	12 465	12 585	12 704	12	24	36	48	60
8·4	2·12 823	12 942	13 061	13 180	13 298	13 417	13 535	13 653	13 771	13 889	12	24	36	47	59
8·5	2·14 007	14 124	14 242	14 359	14 476	14 593	14 710	14 827	14 943	15 060	12	23	35	47	58
8·6	2·15 176	15 292	15 409	15 524	15 640	15 756	15 871	15 987	16 102	16 217	12	23	35	46	58
8·7	2·16 332	16 447	16 562	16 677	16 791	16 905	17 020	17 134	17 248	17 361	11	23	34	46	57
8·8	2·17 475	17 589	17 702	17 816	17 929	18 042	18 155	18 267	18 380	18 493	11	23	34	45	56
8·9	2·18 605	18 717	18 830	18 942	19 054	19 165	19 277	19 389	19 500	19 611	11	22	34	45	56
9·0	2·19 722	19 834	19 944	20 055	20 166	20 276	20 387	20 497	20 607	20 717	11	22	33	44	55
9·1	2·20 827	20 937	21 047	21 157	21 266	21 375	21 485	21 594	21 703	21 812	11	22	33	44	55
9·2	2·21 920	22 029	22 138	22 246	22 354	22 462	22 570	22 678	22 786	22 894	11	22	32	43	54
9·3	2·23 001	23 109	23 216	23 324	23 431	23 538	23 645	23 751	23 858	23 965	11	21	32	43	54
9·4	2·24 071	24 177	24 284	24 390	24 496	24 601	24 707	24 813	24 918	25 024	11	21	32	42	53
9·5	2·25 129	25 234	25 339	25 444	25 549	25 654	25 759	25 863	25 968	26 072	10	21	31	42	52
9·6	2·26 176	26 280	26 384	26 488	26 592	26 696	26 799	26 903	27 006	27 109	10	21	31	41	52
9·7	2·27 213	27 316	27 419	27 521	27 624	27 727	27 829	27 932	28 034	28 136	10	21	31	41	51
9·8	2·28 238	28 340	28 442	28 544	28 646	28 747	28 849	28 950	29 051	29 152	10	20	30	41	51
9·9	2·29 253	29 354	29 455	29 556	29 657	29 757	29 858	29 958	30 058	30 158	10	20	30	40	50

$\log_e 10^{-1} = \bar{3}{\cdot}69741$, $\log_e 10^{-2} = \bar{5}{\cdot}39483$, $\log_e 10^{-3} = \bar{7}{\cdot}09224$, $\log_e 10^{-4} = \overline{10}{\cdot}78966$, $\log_e 10^{-5} = \overline{12}{\cdot}48707$, $\log_e 10^{-6} = \overline{14}{\cdot}18449$.

Table 54. *Useful constants*

Mathematical constants

	Number	Reciprocal	Logarithm
π	3·14159 26536	0·31830 98862	0·49714 98727
2π	6·28318 53072	0·15915 49431	0·79817 98684
$\frac{1}{2}\pi$	1·57079 63268	0·63661 97724	0·19611 98770
$\frac{1}{4}\pi$	0·78539 81634	1·27323 95447	$\bar{1}$·89508 98814
$\sqrt{\pi}$	1·77245 38509	0·56418 95835	0·24857 49363
$\sqrt{(2\pi)}$	2·50662 82746	0·39894 22804	0·39908 99342
$\sqrt{(\frac{1}{2}\pi)}$	1·25331 41373	0·79788 45608	0·09805 99385
$\frac{1}{2}\sqrt{\pi}$	0·88622 69255	1·12837 91671	$\bar{1}$·94754 49407
e	2·71828 18285	0·36787 94412	0·43429 44819
e^2	7·38905 60989	0·13533 52832	0·86858 89638
$\sqrt{2}$	1·41421 35624	0·70710 67812	0·15051 49978
$\sqrt{3}$	1·73205 08076	0·57735 02692	0·23856 06274
$\sqrt{10}$	3·16227 76602	0·31622 77660	0·50000 00000
$\log_{10} e$	0·43429 44819	2·30258 50930	$\bar{1}$·63778 43113
1 radian	57°·29577 95131	0·01745 32925	1·75812 26324

Binomial coefficients $\binom{n}{i}$

Value of n: 4	5	6	7	8	9	10	11	12	i
1	1	1	1	1	1	1	1	1	0
4	5	6	7	8	9	10	11	12	1
6	10	15	21	28	36	45	55	66	2
4	10	20	35	56	84	120	165	220	3
1	5	15	35	70	126	210	330	495	4
	1	6	21	56	126	252	462	792	5
		1	7	28	84	210	462	924	6
			1	8	36	120	330	792	7
				1	9	45	165	495	8
					1	10	55	220	9
						1	11	66	10
							1	12	11
								1	12

Conversion Tables

Length

Metric to British		British to Metric	
1 mm.	= 0·039 370 in. (inch)	1 in.	= 25·399 978 mm.
1 cm.	= 0·393 701 in.		= 2·539 998 cm.
1 m.	= 39·370 113 in.	1 ft.	= 30·479 973 cm.
	= 3·280 843 ft. (foot)		= 0·304 800 m. (metre)
	= 1·093 614 yd. (yard)	1 yd.	= 0·914 399 m.
1 km.	= 0·621 372 miles	1 mile	= 1·609 343 km.

Area

Metric to British		British to Metric	
1 sq.cm.	= 0·155 001 sq.in.	1 sq.in.	= 6·451 589 sq.cm.
1 sq.m.	= 10·763 929 sq.ft.	1 sq.ft.	= 0·092 903 sq.m.
	= 1·195 992 sq.yd.	1 sq.yd.	= 0·836 126 sq.m.
1 sq.km.	= 0·386 103 sq.miles	1 sq.mile	= 2·589 984 sq.km.
1 hectare	= 2·471 058 acres	1 acre	= 0·404 685 hectares

Volume

Metric to British		British to Metric	
1 cu.cm.	= 0·061 024 cu.in.	1 cu.in.	= 16·387 021 cu.cm.
1 cu.m.	= 35·314 759 cu.ft.	1 cu.ft.	= 0·028 317 cu.m.
	= 1·307 954 cu.yd.	1 cu.yd.	= 0·764 553 cu.m.

Weight

Metric to British		British to Metric	
1 g.	= 15·432 356 gr. (grain)	1 gr.	= 0·064 799 g. (gram)
	= 0·035 274 oz. (ounce)	1 oz.	= 28·349 527 g.
1 kg.	= 2·204 622 lb. (pound)	1 lb.	= 0·453 592 kg.
1000 kg.	= 0·984 206 tons	1 ton	= 1016·047 kg.

Capacity

Metric to British		British to Metric	
1 litre	= 1·759 803 pints	1 pint	= 0·568 245 litres
	= 0·879 902 quarts	1 quart	= 1·136 491 litres
	= 0·219 975 gallons	1 gallon	= 4·545 963 litres

ADDENDUM

Table 31*c*. *Percentage points of the ratio* $w_{\max}/w_{\min}$

$n \backslash k$	2	3	4	5	6	7	8	9	10	11	12
					Upper 5% points						
3	6·267	9·392	11·99	14·30	16·40	18·35	20·19	21·93	23·59	25·18	26·71
4	3·971	5·335	6·371	7·237	7·992	8·669	9·285	9·854	10·38	10·88	11·35
5	3·157	4·018	4·643	5·149	5·580	5·958	6·298	6·607	6·892	7·156	7·404
6	2·744	3·381	3·831	4·187	4·487	4·747	4·978	5·187	5·377	5·553	5·717
7	2·494	3·007	3·362	3·640	3·871	4·070	4·245	4·403	4·546	4·678	4·800
8	2·325	2·760	3·056	3·286	3·476	3·638	3·781	3·909	4·024	4·130	4·227
9	2·203	2·584	2·841	3·039	3·201	3·340	3·461	3·568	3·666	3·755	3·837
10	2·110	2·452	2·681	2·855	2·998	3·120	3·226	3·320	3·404	3·482	3·553
11	2·037	2·349	2·556	2·714	2·842	2·951	3·046	3·129	3·205	3·274	3·337
13	1·928	2·198	2·375	2·508	2·617	2·708	2·787	2·857	2·920	2·977	3·029
16	1·820	2·049	2·198	2·309	2·399	2·475	2·540	2·597	2·648	2·695	2·737
21	1·709	1·900	2·022	2·113	2·186	2·246	2·299	2·344	2·385	2·422	2·456
31	1·592	1·745	1·841	1·912	1·969	2·016	2·056	2·091	2·123	2·151	2·176
61	1·459	1·571	1·641	1·692	1·732	1·765	1·794	1·818	1·840	1·860	1·867
∞	1·000	1·000	1·000	1·000	1·000	1·000	1·000	1·000	1·000	1·000	1·000
					Upper 2·5% points						
3	8·920	13·38	17·08	20·37	23·36	26·15	28·76	31·24	33·60	35·87	38·05
4	5·078	6·805	8·116	9·212	10·17	11·02	11·80	12·52	13·19	13·82	14·42
5	3·837	4·863	5·607	6·209	6·722	7·173	7·577	7·945	8·284	8·599	8·893
6	3·238	3·970	4·485	4·894	5·238	5·537	5·802	6·041	6·260	6·461	6·649
7	2·886	3·461	3·858	4·169	4·428	4·650	4·847	5·023	5·184	5·331	5·467
8	2·654	3·133	3·439	3·711	3·920	4·098	4·255	4·395	4·522	4·639	4·746
9	2·489	2·903	3·182	3·396	3·572	3·722	3·853	3·970	4·076	4·173	4·261
10	2·365	2·733	2·978	3·166	3·319	3·449	3·563	3·664	3·755	3·838	3·914
11	2·269	2·601	2·822	2·989	3·126	3·242	3·342	3·432	3·512	3·585	3·652
13	2·127	2·411	2·596	2·737	2·851	2·946	3·029	3·103	3·169	3·229	3·283
16	1·988	2·226	2·380	2·495	2·589	2·667	2·734	2·794	2·847	2·[illegible]95	2·939
21	1·848	2·043	2·168	2·261	2·335	2·398	2·451	2·498	2·540	2·577	2·612
31	1·703	1·857	1·954	2·026	2·083	2·130	2·171	2·206	2·238	2·266	2·292
61	1·540	1·652	1·721	1·772	1·812	1·845	1·874	1·898	1·920	1·940	1·958
∞	1·000	1·000	1·000	1·000	1·000	1·000	1·000	1·000	1·000	1·000	1·000

Table 31c (*continued*)

n \ k	2	3	4	5	6	7	8	9	10	11	12
					Upper 1% points						
3	14·16	21·24	27·12	32·33	37·09	41·53	45·73	49·74	53·55	57·12	60·41
4	6·970	9·316	11·10	12·59	13·89	15·05	16·11	17·08	17·99	18·85	19·66
5	4·914	6·202	7·137	7·893	8·537	9·104	9·612	10·07	10·50	10·90	11·27
6	3·985	4·860	5·477	5·967	6·378	6·736	7·053	7·340	7·602	7·843	8·068
7	3·461	4·127	4·587	4·947	5·247	5·505	5·733	5·938	6·124	6·295	6·452
8	3·126	3·667	4·035	4·321	4·558	4·760	4·938	5·096	5·240	5·372	5·492
9	2·982	3·352	3·661	3·900	4·096	4·263	4·409	4·539	4·657	4·764	4·863
10	2·720	3·122	3·391	3·597	3·765	3·908	4·032	4·143	4·243	4·334	4·418
11	2·588	2·948	3·186	3·368	3·516	3·642	3·751	3·848	3·935	4·014	4·087
13	2·396	2·698	2·896	3·045	3·166	3·269	3·357	3·436	3·506	3·570	3·628
16	2·211	2·460	2·621	2·742	2·840	2·922	2·993	3·055	3·111	3·161	3·206
21	2·029	2·230	2·358	2·454	2·531	2·595	2·650	2·698	2·741	2·780	2·815
31	1·845	2·000	2·099	2·171	2·229	2·277	2·318	2·354	2·386	2·414	2·440
61	1·642	1·752	1·821	1·874	1·914	1·947	1·976	2·001	2·023	2·043	2·054
∞	1·000	1·000	1·000	1·000	1·000	1·000	1·000	1·000	1·000	1·000	1·000
					Upper 0·5% points						
3	20·05	30·09	38·43	45·82	52·56	58·82	64·70	70·28	75·60	80·71	85·62
4	8·825	11·78	14·03	15·91	17·55	19·01	20·35	21·58	22·73	23·81	24·83
5	5·897	7·427	8·537	9·436	10·20	10·88	11·48	12·03	12·54	13·01	13·45
6	4·636	5·638	6·345	6·907	7·379	7·789	8·153	8·482	8·783	9·060	9·319
7	3·947	4·691	5·205	5·608	5·943	6·232	6·488	6·717	6·925	7·117	7·294
8	3·516	4·109	4·514	4·828	5·088	5·310	5·506	5·681	5·839	5·984	6·118
9	3·220	3·717	4·053	4·311	4·523	4·705	4·863	5·005	5·133	5·249	5·357
10	3·005	3·435	3·723	3·944	4·124	4·278	4·412	4·531	4·638	4·736	4·826
11	2·840	3·223	3·476	3·669	3·827	3·961	4·077	4·180	4·273	4·358	4·436
13	2·606	2·922	3·129	3·286	3·414	3·521	3·614	3·697	3·771	3·838	3·900
16	2·383	2·640	2·807	2·932	3·033	3·118	3·192	3·257	3·315	3·367	3·415
21	2·167	2·371	2·502	2·600	2·678	2·744	2·800	2·850	2·894	2·934	2·971
31	1·951	2·107	2·206	2·279	2·337	2·386	2·427	2·463	2·496	2·525	2·551
61	1·719	1·830	1·899	1·950	1·990	2·024	2·053	2·078	2·101	2·121	2·140
∞	1·000	1·000	1·000	1·000	1·000	1·000	1·000	1·000	1·000	1·000	1·000

Table 32*a*. *Test for heterogeneity of variance for equal degrees of freedom: percentage points of M.*

ν \ k	α	3		4		5		6		7	
1	0·001	0·003	16·600	0·036	19·630	0·131	22·351	0·299	24·886	0·538	27·292
	·01	0·030	11·460	0·168	14·087	0·426	16·471	0·785	18·709	1·222	20·847
	·05	0·153	7·739	0·510	9·992	1·010	12·066	1·605	14·031	2·267	15·923
	·10	0·313	6·080	0·842	8·129	1·501	10·033	2·240	11·851	3·034	13·610
2	·001	0·002	15·810	0·030	18·590	0·110	21·090	0·254	23·421	0·458	25·635
	·01	0·025	10·740	0·140	13·157	0·360	15·351	0·668	17·410	1·047	19·376
	·05	0·127	7·108	0·430	9·185	0·860	11·092	1·378	12·899	1·958	14·637
	·10	0·261	5·512	0·713	7·398	1·285	9·147	1·933	10·815	2·634	12·429
3	·001	0·002	15·360	0·028	18·033	0·104	20·439	0·239	22·683	0·432	24·814
	·01	0·023	10·353	0·132	12·684	0·339	14·798	0·629	16·782	0·987	18·676
	·05	0·119	6·796	0·403	8·799	0·809	10·637	1·299	12·377	1·850	14·051
	·10	0·244	5·246	0·669	7·063	1·211	8·747	1·825	10·352	2·491	11·905
4	·001	0·002	15·069	0·027	17·686	0·100	20·042	0·231	22·238	0·419	24·324
	·01	0·022	10·116	0·127	12·401	0·328	14·472	0·610	16·415	0·958	18·270
	·05	0·115	6·616	0·390	8·578	0·784	10·378	1·260	12·082	1·795	13·720
	·10	0·235	5·098	0·647	6·875	1·173	8·524	1·771	10·095	2·419	11·614
5	·001	0·002	14·867	0·026	17·450	0·098	19·774	0·227	21·942	0·411	24·000
	·01	0·022	9·957	0·125	12·214	0·322	14·259	0·599	16·176	0·940	18·006
	·05	0·112	6·500	0·382	8·436	0·769	10·213	1·237	11·894	1·763	13·510
	·10	0·230	5·004	0·635	6·757	1·151	8·383	1·738	9·932	2·375	11·430
6	·001	0·002	14·719	0·026	17·279	0·097	19·583	0·224	21·731	0·406	23·770
	·01	0·022	9·845	0·123	12·082	0·317	14·109	0·591	16·009	0·929	17·822
	·05	0·110	6·420	0·377	8·338	0·759	10·098	1·221	11·764	1·741	13·364
	·10	0·227	4·940	0·626	6·676	1·136	8·286	1·717	9·821	2·347	11·304
7	·001	0·002	14·606	0·026	17·151	0·096	19·440	0·222	21·573	0·402	23·598
	·01	0·021	9·761	0·122	11·984	0·315	13·998	0·586	15·885	0·921	17·686
	·05	0·109	6·361	0·373	8·267	0·752	10·015	1·210	11·669	1·726	13·258
	·10	0·224	4·893	0·620	6·617	1·126	8·216	1·701	9·740	2·326	11·213
8	·001	0·002	14·518	0·026	17·050	0·095	19·328	0·221	21·451	0·400	23·466
	·01	0·021	9·697	0·121	11·909	0·312	13·913	0·582	15·791	0·914	17·583
	·05	0·108	6·317	0·371	8·212	0·747	9·951	1·202	11·596	1·715	13·178
	·10	0·223	4·858	0·615	6·572	1·118	8·163	1·690	9·678	2·311	11·144
9	·001	0·002	14·447	0·026	16·970	0·095	19·240	0·219	21·354	0·397	23·361
	·01	0·021	9·646	0·120	11·850	0·311	13·845	0·579	15·716	0·910	17·501
	·05	0·108	6·282	0·368	8·169	0·743	9·901	1·196	11·539	1·706	13·114
	·10	0·221	4·830	0·612	6·537	1·112	8·121	1·681	9·630	2·299	11·089
10	·001	0·002	14·389	0·026	16·905	0·095	19·167	0·219	21·275	0·396	23·276
	·01	0·021	9·604	0·120	11·801	0·309	13·791	0·576	15·655	0·906	17·434
	·05	0·107	6·253	0·367	8·135	0·740	9·861	1·191	11·494	1·699	13·063
	·10	0·220	4·808	0·609	6·509	1·107	8·087	1·674	9·592	2·289	11·046

If $s_1^2, \ldots, s_k^2$ are independent estimates of variance each with ν degrees of freedom and $s^2 = \Sigma s_i^2/k$, then $M = k\nu \log s^2 - \nu \Sigma \log s_i^2$. For each value of k the left-hand column contains the lower percentage points and the right-hand column the upper percentage points.

Table ~~31a~~ 32a (*continued*)

ν \ k	α	8		9		10		11		12	
1	0·001	0·838	29·603	1·193	31·839	1·596	34·016	2·039	36·143	2·519	38·227
	·01	1·722	22·911	2·273	24·917	2·867	26·877	3·496	28·799	4·156	30·687
	·05	2·980	17·762	3·732	19·558	4·517	21·321	5·328	23·055	6·162	24·766
	·10	3·867	15·326	4·733	17·009	5·623	18·665	6·535	20·299	7·464	21·914
2	·001	0·717	27·761	1·024	29·819	1·373	31·822	1·760	33·778	2·179	35·695
	·01	1·482	21·274	1·964	23·119	2·484	24·920	3·038	26·686	3·620	28·420
	·05	2·585	16·325	3·250	17·974	3·946	19·591	4·667	21·181	5·409	22·750
	·10	3·373	14·002	4·143	15·544	4·938	17·061	5·752	18·557	6·584	20·036
3	·001	0·677	26·861	0·967	28·842	1·298	30·769	1·665	32·652	2·062	34·496
	·01	1·400	20·504	1·857	22·280	2·351	24·013	2·877	25·712	3·431	27·381
	·05	2·446	15·676	3·078	17·262	3·740	18·817	4·427	20·347	5·135	21·856
	·10	3·195	13·418	3·928	14·901	4·685	16·359	5·462	17·798	6·256	19·219
4	·001	0·657	26·328	0·939	28·266	1·261	30·153	1·617	31·995	2·004	33·800
	·01	1·359	20·059	1·803	21·797	2·284	23·494	2·797	25·157	3·336	26·790
	·05	2·375	15·310	2·991	16·862	3·636	18·384	4·305	19·881	4·995	21·357
	·10	3·104	13·094	3·818	14·545	4·556	15·971	5·313	17·378	6·087	18·768
5	·001	0·645	25·976	0·922	27·889	1·239	29·749	1·589	31·566	1·970	33·346
	·01	1·334	19·772	1·771	21·486	2·245	23·160	2·748	24·800	3·279	26·410
	·05	2·333	15·078	2·939	16·609	3·573	18·110	4·232	19·586	4·910	21·041
	·10	3·049	12·890	3·752	14·320	4·478	15·727	5·224	17·114	5·985	18·485
6	·001	0·637	25·728	0·911	27·622	1·224	29·465	1·570	31·265	1·946	33·028
	·01	1·318	19·571	1·750	21·270	2·218	22·927	2·716	24·552	3·241	26·147
	·05	2·305	14·918	2·904	16·434	3·532	17·921	4·183	19·383	4·854	20·824
	·10	3·013	12·750	3·708	14·167	4·426	15·560	5·164	16·934	5·917	18·291
7	·001	0·631	25·543	0·903	27·425	1·214	29·255	1·557	31·042	1·930	32·793
	·01	1·307	19·423	1·735	21·110	2·199	22·757	2·693	24·370	3·214	25·954
	·05	2·285	14·801	2·879	16·307	3·502	17·783	4·148	19·235	4·814	20·666
	·10	2·987	12·649	3·677	14·055	4·389	15·438	5·121	16·802	5·868	18·150
8	·001	0·627	25·401	0·898	27·273	1·206	29·094	1·547	30·872	1·917	32·613
	·01	1·298	19·311	1·724	20·989	2·185	22·627	2·676	24·231	3·193	25·807
	·05	2·270	14·712	2·861	16·210	3·479	17·678	4·122	19·122	4·784	20·545
	·10	2·968	12·572	3·653	13·971	4·361	15·346	5·089	16·703	5·832	18·043
9	·001	0·624	25·288	0·893	27·152	1·200	28·966	—	—	—	—
	·01	1·291	19·222	1·715	20·893	2·174	22·524	—	—	—	—
	·05	2·259	14·642	2·847	16·134	3·462	17·596	—	—	—	—
	·10	2·953	12·511	3·635	13·904	4·340	15·274	—	—	—	—
10	·001	0·621	25·197	0·889	27·055	1·195	28·862	—	—	—	—
	·01	1·286	19·149	1·708	20·815	2·165	22·441	—	—	—	—
	·05	2·250	14·586	2·835	16·073	3·448	17·530	—	—	—	—
	·10	2·941	12·463	3·620	13·851	4·322	15·216	—	—	—	—

If $s_1^2, \ldots, s_k^2$ are independent estimates of variance each with ν degrees of freedom, and $s^2 = \Sigma s_i^2/k$, then $M = k\nu \log s^2 - \nu \Sigma \log s_i^2$. For each value of k the left-hand column contains the lower percentage points and the right-hand column the upper percentage points.

INDEX TO TABLES

Date Due

Oversize
QA
276
.P33
55175

AUTHOR
PEARSON, E.S.
TITLE
Biometrika tables for statisticians

DATE DUE | BORROWER'S NAME

Oversize
QA
276
.P33
55175

Pearson, E. S.
Biometrika tables for statisticians